Symmetry and Complexity 2019

Symmetry and Complexity 2019

Editor

Carlo Cattani

MDPI • Basel • Beijing • Wuhan • Barcelona • Belgrade • Manchester • Tokyo • Cluj • Tianjin

Editor
Carlo Cattani
Largo dell'Università
Italy

Editorial Office
MDPI
St. Alban-Anlage 66
4052 Basel, Switzerland

This is a reprint of articles from the Special Issue published online in the open access journal *Symmetry* (ISSN 2073-8994) (available at: https://www.mdpi.com/journal/symmetry/special_issues/Symmetry_Complexity_2019).

For citation purposes, cite each article independently as indicated on the article page online and as indicated below:

LastName, A.A.; LastName, B.B.; LastName, C.C. Article Title. *Journal Name* **Year**, *Article Number*, Page Range.

ISBN 978-3-03936-844-0 (Hbk)
ISBN 978-3-03936-845-7 (PDF)

Contents

About the Editor

Carlo Cattani has been a Professor of Mathematical Physics and Applied Mathematics at the Engineering School (DEIM) of Tuscia University (VT)-Italy, since 2015. He was previously a professor at the Dept. of Mathematics, University of Rome "La Sapienza" (1980–2004) and the Dept. of Mathematics, University of Salerno (2004–2015). His scientific research interests focus on wavelets, dynamical systems, fractals, fractional and stochastic equations, computational and numerical methods, nonlinear analysis, complexity of living systems, pattern analysis, data mining, and artificial intelligence. He has been recognized as an honorary professor at the Azerbaijan University Baku (2019) and (in 2009) at the BSP University, Ufa (Russia). Since 2018, he is also an adjunct Professor at the Ton Duc Thang University – HCMC Vietnam. He has been visiting professor at the Dep. De Matematica Aplicada, EUTII, Politecnico di Valencia (2002), East China University (Shanghai, 2007, 2009), BSP University (Ufa, 2008,2010), and Research Fellow at the Physics Institute of the Stockholm University (1987–1988).

Preface to "Symmetry and Complexity 2019"

Symmetry and complexity are two fundamental features of almost all phenomena in nature and science. Any complex physical model is characterized by the existence of some symmetry groups at different scales. On the other hand, breaking the symmetry of a scientific model has always been considered the most challenging direction for discoveries. Modeling complexity has recently become an increasingly popular subject, with impressive growth in applications. The main goal of modeling complexity is to search for hidden or broken symmetries. Usually, complexity is modeled by dealing with big data or dynamical systems, depending on a large number of parameters. Nonlinear dynamical systems and chaotic dynamical systems are also used for modeling complexity. Complex models are often represented by un-smooth objects, non-differentiable objects, fractals, pseudo-random phenomena, and stochastic processes. The discovery of complexity and symmetry in mathematics, physics, engineering, economics, biology, and medicine have opened new challenging fields of research. Therefore, new mathematical tools have been developed to obtain quantitative information from models, newly reformulated in terms of nonlinear differential equations. This Special Issue focuses on the most recent advances in calculus, applied to dynamical problems, linear and nonlinear (fractional, stochastic) ordinary and partial differential equations, integral differential equations, and stochastic integral problems, arising in all fields of science, engineering applications, and other applied fields dealing with complexity.

Carlo Cattani
Editor

symmetry

MDPI

Article

Numerical Inverse Laplace Transform for Solving a Class of Fractional Differential Equations

Dimple Rani [1], **Vinod Mishra** [1] **and Carlo Cattani** [2,3,*]

[1] Department of Mathematics, Sant Longowal Institute of Engineering and Technology, Longowal-148106, Punjab, India; chawla23dimple@gmail.com (D.R.); vinodmishra.2011@rediffmail.com (V.M.)

[2] Engineering School (DEIM), University of Tuscia, 01100 Viterbo, Italy

[3] Ton Duc Thang University, HCMC 700000, Vietnam

* Correspondence: cattani@unitus.it

Received: 3 March 2019; Accepted: 10 April 2019; Published: 12 April 2019

Abstract: This paper discusses the applications of numerical inversion of the Laplace transform method based on the Bernstein operational matrix to find the solution to a class of fractional differential equations. By the use of Laplace transform, fractional differential equations are firstly converted to system of algebraic equations then the numerical inverse of a Laplace transform is adopted to find the unknown function in the equation by expanding it in a Bernstein series. The advantages and computational implications of the proposed technique are discussed and verified in some numerical examples by comparing the results with some existing methods. We have also combined our technique to the standard Laplace Adomian decomposition method for solving nonlinear fractional order differential equations. The method is given with error estimation and convergence criterion that exclude the validity of our method.

Keywords: numerical inverse Laplace transform; orthonormalized Bernstein polynomials; operational matrices; fractional differential equations

1. Introduction

Over the years, researchers have been attracted to study the scientific problems modelled in fractional differential equations due to their constant appearance in the different disciplines of mathematical sciences and engineering such as fluid mechanics, viscoelasticity, mathematical physics, mathematical biology, system identification, control theory, electrochemistry and signal processing [1–6]. Several analytical and numerical techniques have been developed to solve such kind of equations in the literature. Among these methods, Li and Sun [7] derived the generalized block pulse operational matrix to find the solution of fractional differential equations in terms of block pulse function. A truncated Legendre series together with generalized Legendre operational matrix is used to solve fractional differential equations by Saadatmandi and Dehgan in [8] and they also have presented shifted Legendre-tau method for finding the solution of fractional diffusion equations with variable coefficients in [9]. Doha et al. [10] used the shifted Jacobi operational matrix of fractional derivatives applied together with spectral tau-method for solving fractional differential equations. Kazem et al. [11] constructed an orthogonal fractional order Legendre function based on Legendre polynomials to solve fractional differential equations. Mokhtary et al. [12] provided the operational tau method based on Muntz–Legendre polynomials for solving fractional differential equations. Bernoulli wavlet operational matrix of fractional order integration has been derived to approximate the numerical solution of fractional differential equations in [13]. Fractional-order Lagrange polynomials have been proposed to solve the fractional differential equations in [14]. Albadarneh et al. [15] adopted the fractional finite difference method for solving linear and nonlinear fractional differential equations.

Garrappa [16] provided a detailed survey on the two methods, i.e., product integration rules (PI) and Lubich's fractional linear multi step methods (FLMMs) for solving fractional differential equations. Dehghan et al. [17] adopted homotopy analysis method to solve linear fractional partial differential equations, a meshless approximation strategy for solving fractional partial differential equations based on radial basis function is used in [18]. Haar wavlets have been employed to obtain the solutions of boundary value problems for linear fractional partial differential equations by Rehman and Khan [19]. Li et al. [20] solved the linear fractional partial differential equations based on operational matrix of fractional Bernstein polynomials. Yang et al. [21] discussed the solution of fractional order diffusion equations within the negative Prabhakar kernel comprise with the Laplace transform and the series solutions in terms of general Mittag–Leffler functions. In [22], a new factorization technique has been adopted for nonlinear differential equations involving local fractional derivatives and found the exact solutions for nonlinear local fractional FitzHugh–Nagumo and Newell–Whitehead equations. Cesarano [23] proposed the Hermite polynomials based operational method to solve fractional diffusive equations. From the last few decades, the Laplace transform method has become popular and adopted by many researchers to solve differential and integral equations. Since then it is necessary to find the inverse Laplace transform for finding the solution in its original domain. There exist a number of analytical and numerical methods for inverting a Laplace transform. For details one can refer [24–41]. The Laplace transform method reduces the differential or integral equation into a system of algebraic equations due to the Heaviside's operational method. Further, Hasio and Chen [42] extended the idea of Heaviside's operational method to operational matrix of integration. In the literature, few papers reported the use of operational matrices for inversion of Laplace transform: Chen et al. [39] obtained the walsh operational matrices and applied to distributed system, Wu et al. [40] adopted Haar wavlet operational matrices for numerical inverse Laplace transform of certain functions, Aznam and Hussin [38] modified Haar wavlet operational matrices by using generalized block pulse functions, Babolian and Shamloo [24] used operational matrices based on piecewise-constant block pulse function to invert the Laplace transform and solved Volterra integral equations, Maleknejad and Nouri [25] improved the operational matrices based on block pulse functions and Shamloo et al. [41] adopted this method to solve first kind of integral equations. The main purpose of using an operational matrix is that it converts the differential or integral equations into a system of algebraic equations that is simple and easy to solve.

Bernstein polynomials and its operational matrix is used to solve many differential, integral and integro-differential equations [43–49]. But these are not adopted with Laplace transform for inverting the Laplace transform. A Bernstein operational matrix of integration has been developed to find the numerical inverse Laplace transform of certain functions in our paper [50]. The proposed method expresses the solution of equations in terms of truncated Bernstein expansion and then using its operational matrix of integration, numerical inverse Laplace transform is obtained. The operational matrix of integration of Bernstein polynomials is easily calculated using a single formula of integration rather than Haar or block pulse function where the order of matrix is taken too large, i.e., 8, 16, 128. Here in our method, we achieve the accuracy using a matrix of order 6 or 7. In the present research, our aim is to find the numerical solutions of linear fractional ordinary and partial differential equations, nonlinear fractional differential equations and system of fractional differential equations using numerical inverse Laplace transform method based on Bernstein operational matrix of integration developed in [50]. At first, for linear problems, we transform the fractional differential equations to system of linear algebraic equations using Laplace transform then the numerical approach for calculating the inverse Laplace transform is used to retrieve the time-domain. Here, we extend our numerical approach to solve some nonlinear fractional differential equations together with a well-known iterative method, i.e., a Laplace Adomian decomposition method [51] (briefly explained in the Section 4). One more advantage of using our proposed method is that there is no need of any fractional order matrix of integration for numerical inversion to solve fractional order differential equations.

The paper is organized as follows: In Section 2, some necessary definitions and preliminaries of the fractional calculus theory and Laplace transform are given. Section 3 reveals the basics of Bernstein polynomials, derivation of operational matrix of integration and function approximation in Bernstein polynomials. In Section 4, the proposed method is explained and presented the application to a class of fractional differential equations. Section 5 represents the error estimation and convergence analysis. In Section 6 the illustrative examples are given to show the applicability of the method. Section 7 is refers to conclusions.

2. Preliminaries

In this section, we give some basic definitions and properties of a Laplace transform [36] and fractional calculus as follows [1,2,4]:

Definition 1. *The Laplace transform of a continuous or piecewise continuous function $f(t)$ in $[0,\infty)$ is defined as*

$$L(f(t)) = F(s) = \int_0^\infty e^{-st} f(t)dt, \tag{1}$$

where s is known as Laplace variable.

Definition 2. *A real function $f(t)$, $t > 0$ is said to be in the space C_μ if $\mu \in \mathbb{R}$, there exists a real number $p > \mu$ and the function $f_1(t) \in C[0,\infty)$ such that $f(t) = t^p f_1(t)$. Moreover, if $f^{(n)} \in C_\mu$ then $f(t)$ is said to be in the space C_μ^n, $n \in N$.*

Definition 3. *The Riemann–Liouville fractional integral of order $\alpha \geq 0$ for a function $f(t)$ is defined as*

$$J^\alpha f(t) = \begin{cases} \dfrac{1}{\Gamma(\alpha)} \displaystyle\int_0^t (t-\tau)^{\alpha-1} f(\tau)d\tau & , \alpha > 0 \\ f(t) & , \alpha = 0, \end{cases}$$

where $\Gamma(.)$ denotes the Gamma function.

Definition 4. *The Riemann–Liouville fractional derivative of order $\alpha > 0$ for a function $f(t)$ is defined as*

$$D^\alpha f(t) = \frac{d^n}{dt^n} J^{n-\alpha} f(t), n \in N, n-1 < \alpha \leq n. \tag{2}$$

Definition 5. *The Caputo fractional derivative of order $\alpha > 0$ is defined as*

$$D^\alpha f(t) = \begin{cases} \dfrac{d^n f(t)}{dt^n} & , \alpha = n, n \in N \\ \dfrac{1}{\Gamma(n-\alpha)} \displaystyle\int_0^t \frac{f^{(n)}(\tau)}{(t-\tau)^{\alpha-n+1}} d\tau & , 0 \leq n-1 < \alpha < n, \end{cases}$$

where n is an integer, $t > 0$, and $f(t) \in C_1^n$.

Definition 6. *The Laplace transform of a function $f(x,t)$, denoted as $F(x,s)$, $t \geqslant 0$ is defined by*

$$L(f(x,t)) = F(x,s) = \int_0^\infty e^{-st} f(x,t)dt,$$

where s is the transformed parameter.

Definition 7. *The Caputo fractional partial derivative operator of order $\alpha > 0$ is defined as*

$$\frac{\partial^\alpha f(x,t)}{\partial t^\alpha} = \begin{cases} \dfrac{\partial^n f(x,t)}{\partial t^n} & , \alpha = n, n \in N \\ \dfrac{1}{\Gamma(n-\alpha)} \displaystyle\int_0^t \frac{\frac{\partial^n f(x,\tau)}{\partial t^n}}{(t-\tau)^{\alpha-n+1}} d\tau & , 0 \le n-1 < \alpha < n, \end{cases}$$

where n is an integer, $t > 0$.

Property 1. *The Laplace transform of the Caputo fractional derivative $\mathcal{D}^\alpha f(t)$ can be found as*

$$L(\mathcal{D}^\alpha f(t)) = \frac{1}{s^{m-\alpha}} \left(s^m L(f(t)) - s^{m-1} f(0) - s^{m-2} f'(0) - \ldots - f^{(m-1)}(0) \right).$$

3. Bernstein Polynomials and Function Approximation

To explain the operational matrices of Bernstein polynomials, we need to give some basic definitions and properties following [52].

Definition 8. *The Bernstein basis polynomials of degree n are defined over the interval $[0,1]$:*

$$b_{i,n}(t) = \binom{n}{i} t^i (1-t)^{n-i} \quad , \quad i = 0, 1, \ldots, n. \tag{3}$$

Some useful results of Bernstein polynomials are as follows:

- $b_{i,n}(t) = 0$, if $i < 0$ or $i > n$.
- $b_{i,n}(t) \ge 0$ for $t \in [0,1]$.
- The Bernstein polynomials form a partition of unity i.e., $\sum_{i=0}^n b_{i,n}(t) = 1$.

Bernstein polynomials have one more important property that these polynomials are not orthogonal therefore to conquer this difficulty Bernstein polynomials of degree n are orthonormailized using Gram–Schmidt orthonormalization procedure and denoted as $B_{i,n}(t), i = 0., 1 \ldots n$.

We can find the function approximation based on Bernstein polynomials. A function $f(t)$, $f(t) \in L^2[0,1]$ can be expressed in terms of orthonormal Bernstein polynomials [47]:

$$f(t) = \lim_{n \to \infty} \sum_{i=0}^{n} c_i B_{i,n}(t), \tag{4}$$

where $c_i = \langle f, B_{i,n} \rangle$ and $\langle .,. \rangle$ denotes the standard inner product.

If the infinite series is truncated at $n = k$, then the approximate solution can be expressed as

$$f_k(t) = \sum_{i=0}^{k} c_i B_{i,k}(t) = C^T B(t), \tag{5}$$

where $C = \begin{bmatrix} c_0 & c_1 & c_2 & \ldots & c_k \end{bmatrix}^T$ and $B(t) = \begin{bmatrix} B_{0,k}(t) & B_{1,k}(t) & B_{2,k}(t) & \ldots & B_{k,k}(t) \end{bmatrix}^T$.

Here, the operational matrix of integration of orthonormal Bernstein polynomials is introduced which depends on the integral property of basis vector, i.e., suppose we have a column vector $\phi(t) = [\phi_0(t), \phi_1(t), \ldots \phi_k(t)]$ where $\phi_0(t), \phi_1(t), \ldots \phi_k(t)$ are the basis functions orthogonal on some interval $[a,b]$, then the property states that

$$\int_a^t \cdots \int_a^t \phi(x)(dx)^m = A_{k+1}^m \phi(t), \tag{6}$$

where A_{k+1}^m is the operational matrix of integration of $\phi(t)$ which is a constant matrix of order $(k+1) \times (k+1)$. Now adopting this property on vector $B(t)$, we get

$$\int_0^t B(x)dx = I_{k+1}B(t), \tag{7}$$

where I_{k+1} is the operational matrix of integration of Bernstein polynomials defined as

$$\int_0^t B_{i,k}(x)dx = \alpha_i = \sum_{j=0}^{k} a_{jk}^i B_{j,k} \quad , \quad i = 0, 1, \ldots, k, 0 \leq t < 1. \tag{8}$$

Therefore

$$I_{k+1} = (a_{jk}^i) = \langle \alpha_i, B_{j,k} \rangle \quad , \quad i, j = 0, 1, \ldots, k. \tag{9}$$

4. The Method for Numerical Inverse Laplace Transform

In this section, we describe the algorithm proposed in [50] by considering the linear time varying system

$$f'(t) + \alpha f(t) = u(t) \quad , \quad f(0) = 0 \tag{10}$$

where $u(t)$ is the unit step function.
Here, we convert this differential equation to integral equation

$$f(t) + \int_0^t \alpha f(x)dx = \int_0^t u(x)dx, f(0) = 0. \tag{11}$$

Performing Laplace transform on both sides of (11), we get

$$F(s) = \frac{1}{s(s+\alpha)}. \tag{12}$$

We can rewrite (12) as

$$F(s) = \frac{\frac{1}{s^2}}{(1 + \frac{\alpha}{s})} = \bar{F}\left(\frac{1}{s}\right). \tag{13}$$

Here, we use result from Laplace theory [36]: if $L(f(t)) = F(s)$, then

$$L\left(\int_0^t f(x)dx\right) = \frac{1}{s}F(s). \tag{14}$$

This result can be explained as the integration in time-domain is corresponding to multiplication of $1/s$ in $s - domain$.

We can say that here three domains are introduced: one is time-domain, by performing Laplace transform we move from time-domain, $f(t)$ to $s - domain$ $F(s)$, or $\bar{F}\left(\frac{1}{s}\right)$, and then to matrix domain where we define the functional $\tilde{F}(I_{k+1})$ (say) defined on the space of matrices.

The integration in time domain of the inverse Laplace function is corresponding to the definition of the functional $\tilde{F}$ defined onto the space of Bernstein operational matrices. Therefore (13) becomes

$$\tilde{F}\left(I_{k+1}\right) = I_{k+1}^2(I + \alpha I_{k+1})^{-1}. \tag{15}$$

To solve the integral Equation (11), we approximate

$$\int_0^t f_k(x)dx = C^T I_{k+1}B(t) \tag{16}$$

Also

$$\int_0^t u(x)dx = d^T I_{k+1} B(t), \tag{17}$$

where $d = \begin{bmatrix} d_0 & d_1 & d_2 & \dots & d_k \end{bmatrix}^T$ defined by

$$d_i = \int_0^t B_{i,k}(t)dt \quad , \quad i = 0,1,2\dots,k, 0 \leq t < 1. \tag{18}$$

Therefore, the integral Equation (11) becomes

$$C^T B(t) + \alpha C^T I_{k+1} B(t) = d^T I_{k+1} B(t)$$

$$C^T = d^T I_{k+1}(I + \alpha I_{k+1}))^{-1}$$

$$C^T = d^T I_{k+1}^{-1} \tilde{F}(I_{k+1}). \tag{19}$$

Thus, the unknown vector C^T is calculated, where $\tilde{F}(I_{k+1})$ can be taken from Equation (15). Consequently, the approximate solution $f_k(t)$ can be obtained in terms of Bernstein polynomials by substituting the unknown vector C^T into (5).

In all these computations for finding the solution, we used simple algebraic operations of matrices, which have been computed using MATLAB R2014a on Intel® Core^{TM} i3 processor (2328M)(2nd Gen.).

4.1. *Application to a Class of Fractional Differential Equations*

Here, we present the fundamental importance of proposed method by applying it to linear fractional differential equations, linear partial fractional differential equations and nonlinear fractional differential equations.

4.1.1. Application to Linear Fractional Differential Equations

Consider the linear ordinary fractional differential equation

$$\mathcal{D}^\alpha f(t) = M(f(t)) + g(t), \tag{20}$$

with initial conditions $f^{(i)}(0) = f_i, i = 0,1,\dots,m-1$.

Here $\mathcal{D}^\alpha$ denotes the Caputo fractional order derivative and $M(f(t))$ represents the linear operator, and may also contain the other derivatives than order α.

To solve the initial problem, Laplace transform is applied to Equation (20) and we attain

$$\frac{1}{s^{m-\alpha}}\left[s^m L(f(t)) - s^{m-1}f(0) - s^{m-2}f'(0)\dots - f^{(m-1)}(0)\right] = L[M(f(t)) + g(t)] = G(s)$$

$$L(f(t)) = \frac{1}{s^m}\left[s^{m-\alpha}G(s) + s^{m-1}f(0) + s^{m-2}f'(0)\dots + f^{(m-1)}(0)\right]. \tag{21}$$

Therefore, $f(t) = L^{-1}(H(s))$, where $H(s) = \frac{1}{s^m}\left[s^{m-\alpha}G(s) + s^{m-1}f(0) + s^{m-2}f'(0)\dots + f^{(m-1)}(0)\right]$. Hence, $f(t)$, i.e., the solution can be obtained by finding the inverse Laplace transform of $H(s)$ using the above described procedure.

4.1.2. Application to Linear Partial Fractional Differential Equations

Consider the linear partial fractional differential equation of the form

$$\frac{\partial^\alpha f}{\partial t^\alpha} + A(x)\frac{\partial f}{\partial x} + B(x)\frac{\partial^2 f}{\partial x^2} + C(x)f = g(x,t), \tag{22}$$

with initial condition $\frac{\partial^k f}{\partial t^k}(x,0) = h_k(x), k = 0,1,\ldots m-1$.

Taking Laplace transform to both sides of Equation (22), we have

$$\frac{1}{s^{m-\alpha}}\left[s^m F(x,s) - s^{m-1}h_0(x) - s^{m-2}h_1(x)\ldots - h_{m-1}(x)\right] + A(x)\frac{dF(x,s)}{dx} + B(x)\frac{d^2F(x,s)}{dx^2} + C(x)F(x,s) = G(x,s),$$

where $F(x,s)$ and $G(x,s)$ denote the Laplace transform of $f(x,t)$ and $g(x,t)$ respectively. The above equation can be written as

$$B(x)\frac{d^2F(x,s)}{dx^2} + A(x)\frac{dF(x,s)}{dx} + (s^\alpha + C(x))F(x,s) = G(x,s) + \frac{1}{s^{m-\alpha}}\left[s^m F(x,s) + s^{m-1}h_0(x) + s^{m-2}h_1(x)\ldots + h_{m-1}(x)\right]. \quad (23)$$

Now Equation (23) has become a second order ordinary differential equation in $F(x,s)$, that can be easily solved by any classical method for variable x, while keeping s as Laplace variable. The obtained solution $F(x,s)$ can be inverted to $f(x,t)$ using our developed technique.

4.1.3. Application to Nonlinear Fractional Differential Equations

In view to solve, nonlinear fractional order differential equations using standard Laplace adomian decomposition method, we briefly recall the Laplace adomian decomposition method here.

Consider the nonlinear fractional order differential equation

$$\mathcal{D}^\alpha f(t) + M(f(t)) + N(f(t)) = g(t), m-1 \leq \alpha < m, \quad (24)$$

with initial condition $f^{(i)}(0) = f_i, i = 0,1,\ldots,m-1$.

Here $M(f(t))$ represents the linear operator which may include other derivatives than order α and $N(f(t))$ be the nonlinear operator of $f(t)$.

The standard Laplace Adomian Decomposition Method (LADM) procedure begins with taking Laplace transform to this nonlinear equation and using the properties of Laplace transform, we have

$$\frac{1}{s^{m-\alpha}}\left[s^m L(f(t)) - s^{m-1}f(0) - s^{m-2}f'(0)\ldots - f^{(m-1)}(0)\right] + L(M(f(t))) + L(N(f(t))) = L(g(t))$$

$$L(f(t)) = \frac{a}{s^\alpha} - \frac{1}{s^\alpha}L(M(f(t))) - \frac{1}{s^\alpha}L(N(f(t))) + \frac{1}{s^\alpha}L(g(t)), \quad (25)$$

where $a = \sum_{i=0}^{m-1} s^{\alpha-i-1} f^{(i)}(0)$.

The method describes the series solution as

$$f(t) = \sum_{n=0}^{\infty} f_n(t). \quad (26)$$

Therefore, truncated series solution takes the form

$$f_m(t) = \sum_{n=0}^{m} f_n(t), \quad (27)$$

and the nonlinear term is decomposed as follows:

$$N(f(t)) = \sum_{n=0}^{\infty} A_n, \quad (28)$$

where A_n's are the Adomian polynomials, given by

$$A_n = \frac{1}{n!}\frac{d^n}{d\lambda^n}\left[N\left\{\sum_{i=0}^{n}\lambda^i(f_i)\right\}\right]_{\lambda=0}. \tag{29}$$

Consequently, Equation (25) leads to following recurrence relation

$$f_0(t) = L^{-1}\left[\frac{a}{s^\alpha} - \frac{1}{s^\alpha}L\{M(f(t))\} + \frac{1}{s^\alpha}L\{g(t)\}\right] \tag{30}$$

$$f_1(t) = L^{-1}\left[-\frac{1}{s^\alpha}L\left\{A_0(f_0(t))\right\}\right]. \tag{31}$$

In general

$$f_m(t) = L^{-1}\left[-\frac{1}{s^\alpha}L\left\{A_{m-1}(f_0(t), f_1(t)\dots f_{m-1}(t))\right\}\right]. \tag{32}$$

The new development is that we find the values of $f_0(t), f_1(t)\dots$ by finding inverse Laplace transform using our proposed technique, i.e., Bernstein operational matrix, as described above.

5. Error Estimation and Convergence Analysis

5.1. Error Estimation via RK45 Method

We have investigated error estimation of the proposed method [24]. The error function $e_k(t)$ of the truncated Bernstein expansion $f_k(t)$ is defined as

$$e_k(t) = f(t) - f_k(t). \tag{33}$$

From the following theorem the absolute error bound can be estimated:

Theorem 1. *Let $f(t)$ be the function defined on $[0,1]$, then the upper bound for the errors in truncated Bernstein expansion can be estimated.*

Proof. Suppose that $f_k(t)$ and $f_{k+1}(t)$ are two consecutive approximate solutions of given differential equation. If the error sequence is increasing or decreasing, then we use the triangle inequality

$$|\|f(t) - f_{k_1}(t)\|_\infty - \|f(t) - f_{k_2}(t)\|_\infty| \le \|f_{k_1}(t) - f_{k_2}(t)\|_\infty, \tag{34}$$

to the approximate solutions at k and $k+1$ and we achieve

$$|\|f(t) - f_{k+1}(t)\|_\infty - \|f(t) - f_k(t)\|_\infty| = \beta\|f(t) - f_m(t)\|_\infty \le \|f_{k+1}(t) - f_k(t)\|_\infty, 0 \le \beta < 1, \tag{35}$$

where

$$\|f(t) - f_m(t)\|_\infty = \max\{\|f(t) - f_k(t)\|_\infty, \|f(t) - f_{k+1}(t)\|_\infty\}$$

and $\beta = \frac{\|f_{k+1}(t) - f_k(t)\|_\infty}{\|f(t) - f_m(t)\|}$.

Hence, it can be said that the error can be bounded from above. One of the absolute errors $e_k(t)$ or $e_{k+1}(t)$ are bounded by $\|f_{k+1}(t) - f_k(t)\|_\infty$, if the error sequence is monotone. The computable error bound is represented for $0 \le \beta < 1$. But, the solution diverges when the Bernstein series diverges for $\beta > 1$. □

5.2. Convergence Analysis

In order to show the effectiveness of our method, a residual function of a linear time varying system in the Banach space for the values of k is adopted to interpret the convergence of the Bernstein polynomials solution as described in [47,53].

Suppose $f_k(t)$ are the approximate solution of (10), we write the residual function as

$$R_k(t) \simeq f_k'(t) + \alpha f_k(t) - u(t) \quad , \quad t \in [0,1]. \tag{36}$$

The Bernstein polynomial-based numerical solution or the residual function can be expressed in terms of Taylor series expansion as:

$$R_k(t) = r_0 + r_1 t + r_2 t^2 \ldots + r_k t^k = \sum_{j=0}^{k} r_j t^j. \tag{37}$$

Now, we desire to prove that the residual function sequence is convergent in Banach space and satisfying the condition: $\|R_{k+1}\| \leq \zeta_k \|R_k\|$, for $0 \leq \zeta_k < 1$.

For the convergence, we prove here that $\{R_{k+1}(t)\}, k = 0, 1, \ldots \infty$ is a Cauchy sequence.

Let us take that

$$\|R_{k+1}(t)\| = \| \sum_{j=0}^{k+1} r_j t^j \| \leq \sup \left\{ \sum_{j=0}^{k+1} \|r_j t^j\| : t \in [a,b] \right\} = |R_{k+1}(b)|.$$

Therefore, the given condition can be written as

$$|R_{k+1}(b)| \leq \zeta_k |R_k(b)|,$$

and this can be easily shown as follows. Let us write explicitly as functions of Bernstein polynomials

$$f_k'(t) = \sum_{i=0}^{k} f\left(\frac{i}{k}\right) B_{i,k}'(t) \tag{38}$$

$$f_k(t) = \sum_{i=0}^{k} f\left(\frac{i}{k}\right) B_{i,k}(t) \,. \tag{39}$$

If we define

$$m_k = \max_{i=0,1,\ldots k} \left[f\left(\frac{i}{k}\right) \right], \tag{40}$$

we have

$$f_k'(t) = m_k \sum_{i=0}^{k} B_{i,k}'(t).$$

Here we used the following property of Bernstein polynomials

$$B_{i,k}'(t) = k \left[B_{i-1,k-1}(t) - B_{i,k-1}(t) \right], \tag{41}$$

so that

$$f_k'(t) < m_k \sum_{i=1}^{k-1} k \left[B_{i-1,k-1}(t) - B_{i,k-1}(t) \right].$$

From where, we get

$$f_k'(t) < m_k k \left[B_{0,k-1}(t) - B_{k-1,k-1}(t) \right]$$

$$f_k'(t) < m_k k \left[(1-t)^{k-1} - t^{k-1} \right].$$

Since the function in brackets is a decreasing function in $[-1,1]$, with $\max_{t \in [0,1]} \left[(1-t)^{k-1} - t^{k-1} \right] = 1$ there follows

$$f_k'(t) < k m_k \tag{42}$$

Analogously, by using the properties of Bernstein polynomials, we can also estimate

$$f_k(t) < m_k \sum_{i=0}^{k} B_{i,k}(t),$$

and since Bernstein polynomials are partition of unit there follows

$$f_k(t) < m_k \quad . \tag{43}$$

Let us combine Equation (36) with the condition in Equations (42) and (43), then we get

$$R_k < m_k k + \alpha m_k - u,$$

and analogously

$$R_{k+1} < m_{k+1} k + \alpha m_{k+1} - u.$$

This can be written as

$$R_{k+1} < \frac{m_{k+1}}{m_k}\,[m_k k + \alpha m_k - u] + \frac{m_{k+1}}{m_k} u - u$$

$$R_{k+1} < \frac{m_{k+1}}{m_k} R_k + \left[\frac{m_{k+1}}{m_k} u - u\right].$$

Here

$$\lim_{k\to\infty} \frac{m_{k+1}}{m_k} = 1,$$

which implies

$$\lim_{k\to\infty} \left[\frac{m_{k+1}}{m_k} u - u\right] = 0,$$

so that $R_{k+1} < \zeta_k R_k$, with $\zeta_k = \frac{m_{k+1}}{m_k}$, $0 < \zeta_k < 1$, $\lim_{k\to\infty} \zeta_k = 1$.

Now, we begin from the above inequality

$$|R_{k+1}(b)| \leq |R_{k+1}(b) - R_k(b)| \leq (\zeta_k - 1)|R_k(b)|.$$

Thus, we generalize the inequality and attain

$$\begin{aligned} |R_{k+1}(b) - R_k(b)| &\leq (\zeta_k - 1)|R_k(b)| \leq (\zeta_k - 1)(\zeta_{k-1} - 1)|R_{k-1}(b)| \dots \\ &\leq (\zeta_k - 1)(\zeta_{k-1} - 1) \dots (\zeta_0 - 1)|R_0(b)|. \end{aligned} \tag{44}$$

For $k, s \in N$ and $k \geq s$, to prove that the sequence is Cauchy sequence, we take

$$\begin{aligned} |R_k(b) - R_s(b)| &\leq |R_k(b) - R_{k-1}(b) + R_{k-1}(b) - R_{k-2}(b) + \dots R_{s+1}(b) - R_s(b)| \\ &\leq |R_k(b) - R_{k-1}(b)| + |R_{k-1}(b) - R_{k-2}(b)| + \dots |R_{s+1}(b) - R_s(b)| \\ &\leq \epsilon_{k-1}|R_0(b)| + \epsilon_{k-2}|R_0(b)| + \dots \epsilon_{s+1}|R_0(b)| \\ &\leq (\epsilon_{k-1} + \epsilon_{k-2} \dots \epsilon_{s+1})|R_0(b)| \\ &\leq \eta |R_0(b)|, \end{aligned} \tag{45}$$

where $\epsilon_k = \prod_{i=0}^{k}(\zeta_i - 1)$ and $\eta = (\epsilon_{k-1} + \epsilon_{k-2} \dots \epsilon_{s+1})$. Hence, for $0 \leq \eta < 1$, $|R_k(b) - R_s(b)| \to 0$ as $k, s \to \infty$, that proves the residual function sequence is Cauchy sequence and convergent. Therefore the approximate solution is convergent.

6. Illustrative Examples

Here, we present some examples to demonstrate the applicability of the presented technique. The relative errors and L_∞-errors are plotted at the different values of k. Also the error estimation is discussed in each example.

Example 1. *Consider the following fractional differential equation [12]*

$$\mathcal{D}f(t) + \mathcal{D}^{0.25}f(t) + f(t) = t^{5/2} + \frac{5}{2}t^{3/2} + \frac{15}{8}\frac{\sqrt{\pi}}{\Gamma(13/4)}t^{9/4} \quad , \quad f(0) = 0, \tag{46}$$

with the exact solution $f(t) = t^2\sqrt{t}$.

The relative error for different values of k are plotted in Figure 1 which shows that the method is giving good results as compared to analytic solution. Using the error estimation method, we estimated the upper bound of errors and calculate

$$\begin{aligned} \|e_6\|_\infty &= 1.07 \times 10^{-3} \\ \|e_7\|_\infty &= 6.00 \times 10^{-5} \\ \|f_6 - f_7\|_\infty &= 1.07 \times 10^{-3}. \end{aligned}$$

Therefore, it is clear from the data that the error is estimated and bounded above by $\|f_6 - f_7\|_\infty$.

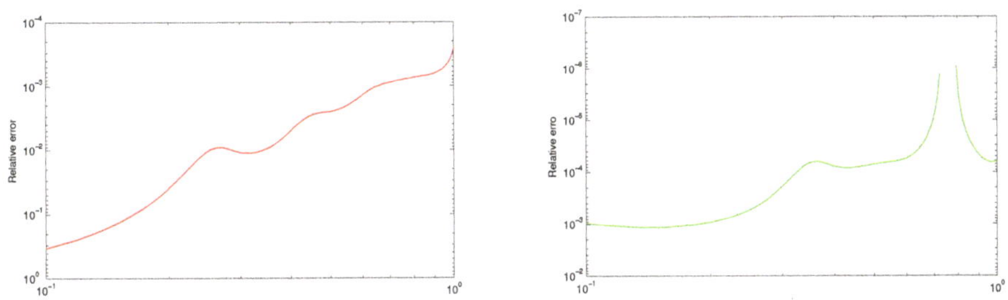

Figure 1. The relative errors for $k = 5$ (**left**) and $k = 6$ (**right**) in Example 1.

Example 2. *Consider the following fractional differential equation*

$$\mathcal{D}^{1.5}f(t) + 3f(t) = 3t^3 + \frac{4}{\Gamma(1.5)}t^{1.5} \quad , \quad f(0) = f'(0) = 0, \tag{47}$$

with exact solution $f(t) = t^3$.

The relative errors for different values of k are plotted in Figure 2. Using the error estimation method, we estimated the upper bound of errors and calculated

$$\begin{aligned} \|e_6\|_\infty &= 1.71 \times 10^{-3} \\ \|e_7\|_\infty &= 3.21 \times 10^{-5} \\ \|f_6 - f_7\|_\infty &= 1.72 \times 10^{-3}. \end{aligned}$$

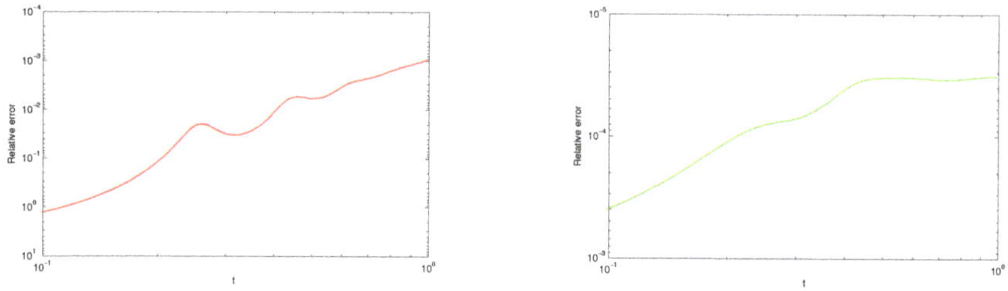

Figure 2. The relative errors for $k = 5$ (**left**) and $k = 6$ (**right**) in Example 2.

Hence, in this example $max\{\|e_6\|_\infty, \|e_7\|_\infty\} \le \|f_6 - f_7\|_\infty$ that justify the result of error analysis.

Example 3. *Consider the following fractional differential equation [15]*

$$\mathcal{D}^{0.5} f(t) + f(t) = t^2 + \frac{2}{\Gamma(2.5)} t^{1.5} \quad , \quad f(0) = 0, \tag{48}$$

with the exact solution $f(t) = t^2$.

In Figure 3, we plotted the relative errors for $k = 5$ and $k = 6$. To show the efficiency, absolute errors compared to fractional finite difference method (FFDM) [15] are given in Table 1 which clearly states that our method gives more accurate results. Using the error estimation method, we also estimate the upper bound of errors and calculate $\|e_6\|_\infty = 8.17 \times 10^{-4}$, $\|e_7\|_\infty = 2.97 \times 10^{-5}$ and $\|f_6 - f_7\|_\infty = 8.17 \times 10^{-4}$. Therefore $max\{\|e_6\|_\infty, \|e_7\|_\infty\} \le \|f_6 - f_7\|_\infty$ that follows the estimated upper bound of error.

Table 1. Comparison of absolute error in Example 3.

t	Present Method at k = 5	Present Method at k = 6	FFDM [15]
0.1	8.17×10^{-4}	2.02×10^{-7}	1.16×10^{-4}
0.2	5.36×10^{-4}	3.09×10^{-6}	1.56×10^{-4}
0.3	3.64×10^{-4}	4.59×10^{-6}	1.81×10^{-4}
0.4	2.65×10^{-4}	4.87×10^{-6}	2.00×10^{-4}
0.5	2.13×10^{-4}	6.01×10^{-6}	2.15×10^{-4}
0.6	1.88×10^{-4}	9.61×10^{-6}	2.27×10^{-4}
0.7	1.77×10^{-4}	1.56×10^{-5}	2.37×10^{-4}
0.8	1.72×10^{-4}	2.23×10^{-5}	2.46×10^{-4}
0.9	1.68×10^{-4}	2.73×10^{-5}	2.54×10^{-4}
1	1.64×10^{-4}	2.97×10^{-5}	2.61×10^{-4}

Example 4. *Consider the following mathematical model, which is developed for a micro-electro mechanical system instrument, that has been designed to measure the viscosity of the fluids that are encountered during oil exploration [54]:*

$$\mathcal{D}^2 f(t) + \beta\sqrt{\pi}\mathcal{D}^{1.5} f(t) + f(t) = 0 \quad , \quad f(0) = 1 \quad , \quad f'(0) = 0 \tag{49}$$

In Figure 4, we plotted the relative errors for $k = 5$ and $k = 6$, taking $\beta = 1$. To show the efficiency, absolute errors compared to cubic spline method [54] are given in Table 2 which clearly states that our method gives more accurate results than the method in [54]. Using the error estimation method, we also estimate the upper bound of errors and calculate $||e_6||_\infty = 2.61 \times 10^{-4}$, $||e_7||_\infty = 7.43 \times 10^{-5}$ and $||f_6 - f_7||_\infty = 1.90 \times 10^{-4}$. Therefore $||e_7||_\infty \leq ||f_6 - f_7||_\infty$ that follows the estimated upper bound of error.

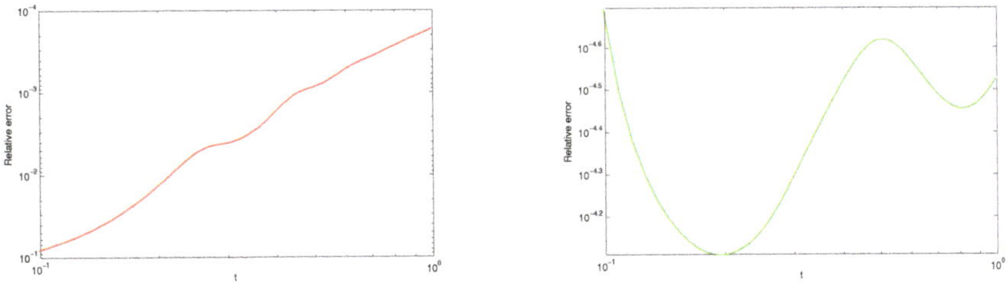

Figure 3. The relative errors for $k = 5$ (**left**) and $k = 6$ (**right**) in Example 3.

Table 2. Comparison of Absolute error in Example 4.

t	Present Method at k = 6	Cubic Spline Method [54]
0.125	4.49×10^{-5}	1.24×10^{-3}
0.250	1.18×10^{-6}	5.12×10^{-3}
0.375	1.80×10^{-5}	1.39×10^{-2}
0.500	1.53×10^{-5}	2.61×10^{-2}
0.625	9.29×10^{-6}	4.04×10^{-2}
0.75	7.63×10^{-6}	5.58×10^{-2}
0.875	2.47×10^{-5}	7.15×10^{-2}
1	7.43×10^{-5}	8.72×10^{-2}

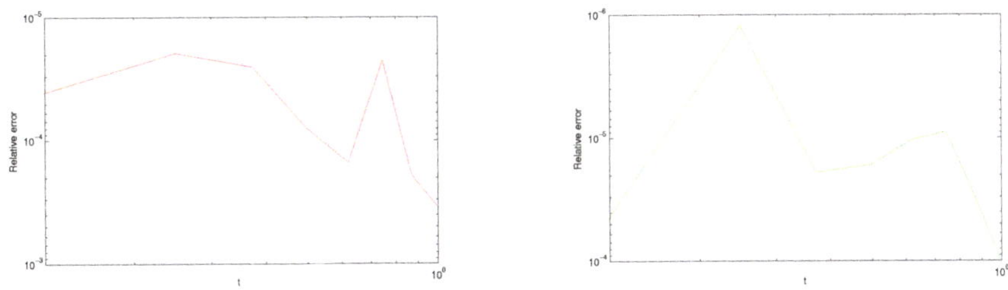

Figure 4. The relative errors for $k = 5$ (**left**) and $k = 6$ (**right**) in Example 4.

Example 5. *Consider the following system of fractional differential equations [14,55]*

$$\mathcal{D}^\alpha f_1(t) = f_1(t) + f_2(t), \tag{50}$$

$$\mathcal{D}^\beta f_2(t) = -f_1(t) + f_2(t), \tag{51}$$

subject to the conditions

$$f_1(0) = 0 \quad , \quad f_2(0) = 1, \tag{52}$$

where the exact solution of system at $\alpha = \beta = 1$ *is* $f_1(t) = e^t \sin t$ *and* $f_2(t) = e^t \cos t$.

The relative error for different values of k are plotted in Figures 5 and 6. Using the error estimation method, we estimate the upper bound of errors for $f_1(t)$ and calculate

$$\begin{aligned} \|e_6\|_\infty &= 1.59 \times 10^{-3} \\ \|e_7\|_\infty &= 6.18 \times 10^{-5} \\ \|f_6 - f_7\|_\infty &= 1.59 \times 10^{-3} \end{aligned}$$

We observe that the result $max\{\|e_6\|_\infty, \|e_7\|_\infty\} \leq \|f_6 - f_7\|_\infty$ is satisfied here. Similarly for the function $f_2(t)$ the results are reported as

$$\begin{aligned} \|e_6\|_\infty &= 9.96 \times 10^{-4} \\ \|e_7\|_\infty &= 2.98 \times 10^{-4} \\ \|f_6 - f_7\|_\infty &= 1.08 \times 10^{-3}. \end{aligned}$$

The numbers clearly show that the error is bounded from above, i.e., $max\{\|e_6\|_\infty, \|e_7\|_\infty\} \leq \|f_6 - f_7\|_\infty$. To show the efficiency, absolute errors compared to variation iteration method (VIM) [55] are given in Table 3 and 4 which excludes that our method gives more accurate results than the method in [55].

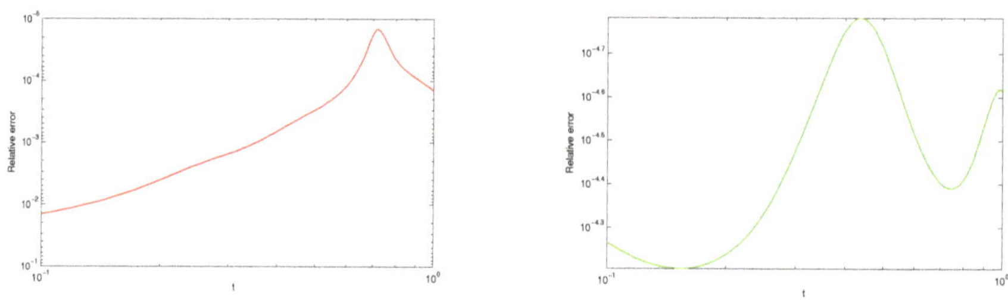

Figure 5. The relative errors for $k = 5$ (**left**) and $k = 6$ (**right**) for $f_1(t)$ in Example 5.

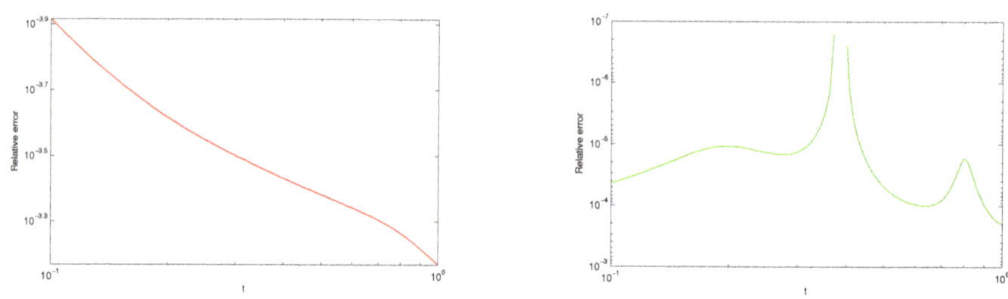

Figure 6. The relative errors for $k = 5$ (**left**) and $k = 6$ (**right**) for $f_2(t)$ in Example 5.

Example 6. *Consider the following linear time fractional wave equation [18]*

$$\frac{\partial^\alpha f}{\partial t^\alpha} = \frac{1}{2}x^2\frac{\partial^2 f}{\partial x^2}, \tag{53}$$

having $f(x,0) = x$, $\frac{\partial f(x,0)}{\partial t} = x^2$.

The exact solution is not known.

Here, we first convert the fractional partial differential equation to ordinary differential equation as described in Section 4, then the solution is obtained by our proposed method for $\alpha = 1.5$, $k = 6$ presented in Table 5 and have compared with the method VIM (Table 6) and Inverse Multiquadric-Radial Basis Functions (IMQ-RBF) (Table 7) which shows that the solution is in good agreement with VIM and IMQ-RBF [18].

Table 3. Comparison of Absolute errors in $f_1(t)$ in Example 5.

t	Present Method at k = 5	Present Method at k = 6	VIM [55]
0.1	1.59×10^{-3}	5.99×10^{-6}	1.66×10^{-4}
0.2	9.84×10^{-4}	1.40×10^{-5}	1.32×10^{-3}
0.3	6.08×10^{-4}	1.32×10^{-5}	4.39×10^{-3}
0.4	3.80×10^{-4}	1.02×10^{-5}	1.02×10^{-2}
0.5	2.34×10^{-4}	1.53×10^{-5}	1.93×10^{-2}
0.6	1.22×10^{-4}	3.16×10^{-5}	3.21×10^{-2}
0.7	2.24×10^{-5}	5.15×10^{-5}	4.86×10^{-2}
0.8	7.21×10^{-5}	6.18×10^{-5}	6.81×10^{-2}
0.9	1.73×10^{-4}	5.60×10^{-5}	8.95×10^{-2}
1	3.29×10^{-4}	5.56×10^{-5}	1.11×10^{-1}

Table 4. Comparison of Absolute errors in $f_2(t)$ in Example 5.

t	Present Method at k = 5	Present Method at k = 6	VIM [55]
0.1	1.33×10^{-4}	4.81×10^{-5}	1.79×10^{-4}
0.2	2.89×10^{-4}	1.31×10^{-5}	1.54×10^{-4}
0.3	4.07×10^{-4}	1.74×10^{-5}	5.57×10^{-3}
0.4	5.09×10^{-4}	3.42×10^{-7}	1.41×10^{-2}
0.5	6.02×10^{-4}	7.26×10^{-5}	2.94×10^{-2}
0.6	6.85×10^{-4}	1.45×10^{-4}	5.40×10^{-2}
0.7	7.64×10^{-4}	1.25×10^{-4}	9.11×10^{-2}
0.8	8.44×10^{-4}	2.68×10^{-5}	1.44×10^{-1}
0.9	9.28×10^{-4}	1.53×10^{-4}	2.17×10^{-1}
1	9.96×10^{-4}	2.98×10^{-4}	3.13×10^{-1}

Table 5. Solution by proposed method for different values of x and t in Example 6.

x	t = 0	t = 0.06	t = 0.13	t = 0.29	t = 0.50
0.00	0.00	0.00	0.00	0.00	0.00
0.11	0.11	0.110729	0.111595	0.113677	0.116726
0.31	0.31	0.315791	0.322671	0.339207	0.363420
0.88	0.88	0.926665	0.982110	1.115365	1.310479
1.00	1.00	1.060260	1.131857	1.303932	1.555887

Table 6. Solution by variation iteration method (VIM) for different values of x and t in Example 6.

x	t = 0	t = 0.06	t = 0.13	t = 0.29	t = 0.50
0.00	0.00	0.00	0.00	0.00	0.00
0.11	0.11	0.110729	0.111595	0.113677	0.116726
0.31	0.31	0.315791	0.322670	0.339207	0.363419
0.88	0.88	0.926669	0.982101	1.115360	1.310469
1.00	1.00	1.060265	1.131845	1.303926	1.555874

Table 7. Solution by IMQ-RBF for different values of x and t in Example 6.

x	t = 0	t = 0.06	t = 0.13	t = 0.29	t = 0.50
0.00	0.00	0.00	0.00	0.00	0.00
0.11	0.11	0.110729	0.111595	0.113679	0.116710
0.31	0.31	0.315794	0.322675	0.339223	0.363298
0.88	0.88	0.926696	0.982140	1.115490	1.309495
1.00	1.00	1.060300	1.131896	1.304094	1.554617

Example 7. *Consider the nonlinear fractional differential equation*

$$\mathcal{D}^4 f(t) + \mathcal{D}^{7/2} f(t) + f^3(t) = t^9 \quad , \quad f(0) = f'(0) = f''(0) = 0 \quad , \quad f'''(0) = 6, \tag{54}$$

with exact solution $f(t) = t^3$.

The relative error for different values of k are plotted in Figure 7 which show that our method gives a very close solution to the analytic solution. Using the error estimation method, we estimated the upper bound of errors and calculated

$$\begin{aligned} \|e_6\|_\infty &= 1.71 \times 10^{-3} \\ \|e_7\|_\infty &= 7.72 \times 10^{-5} \\ \|f_6 - f_7\|_\infty &= 1.72 \times 10^{-3}. \end{aligned}$$

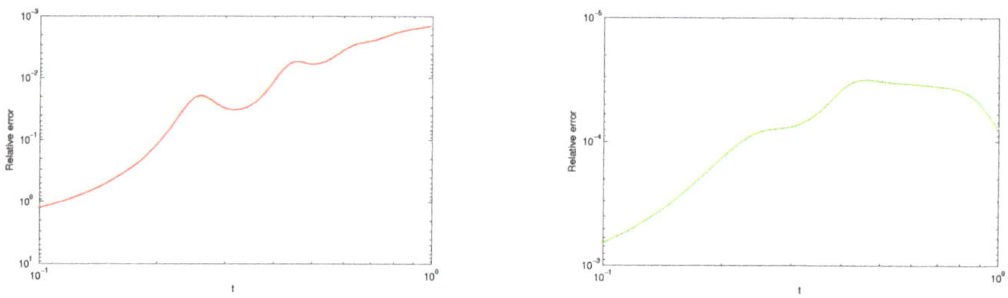

Figure 7. The relative errors for $k = 5$ (**left**) and $k = 6$ (**right**) in Example 7.

Hence, we can see that here $\|f_6 - f_7\|_\infty$ bounds the error $\|e_6\|_\infty$.

We also analyzed the L_∞-error at different values of k, i.e., $k = 5, 6, 7, 8$ for examples which clearly exclude that we achieve the best results at $k = 6$, as shown in Figures 8–11.

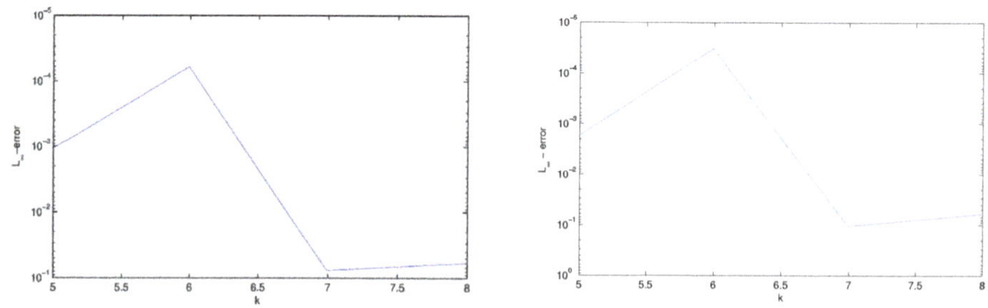

Figure 8. The $L_\infty - error$ for Example 1 (**left**) and Example 2 (**right**) for different values of k.

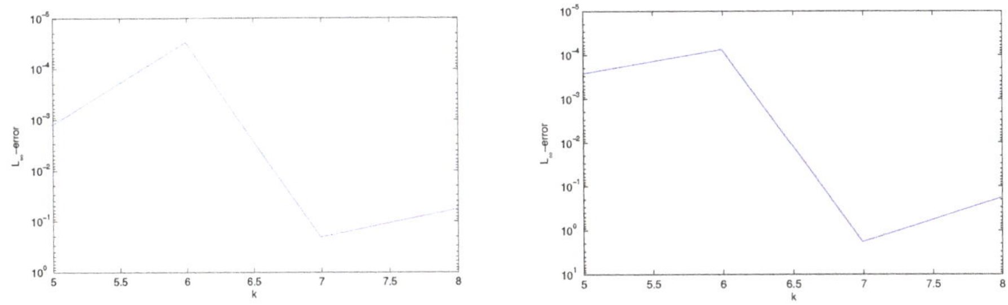

Figure 9. The $L_\infty - error$ for Example 3 (**left**) and Example 4 (**right**) for different values of k.

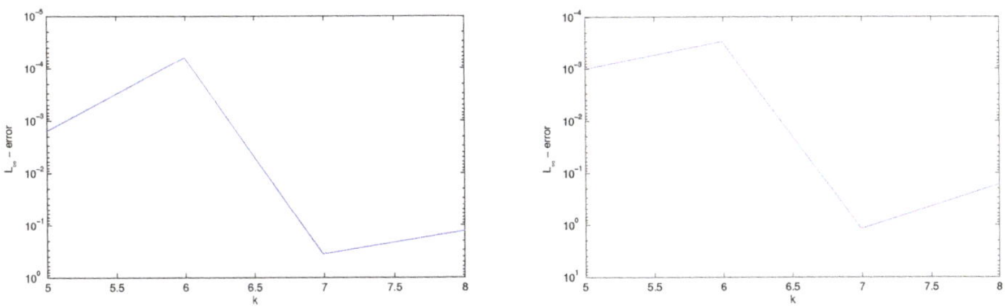

Figure 10. The $L_\infty - error$ for $f_1(t)$ (**left**) and $f_2(t)$ (**right**) at different values of k for Example 5.

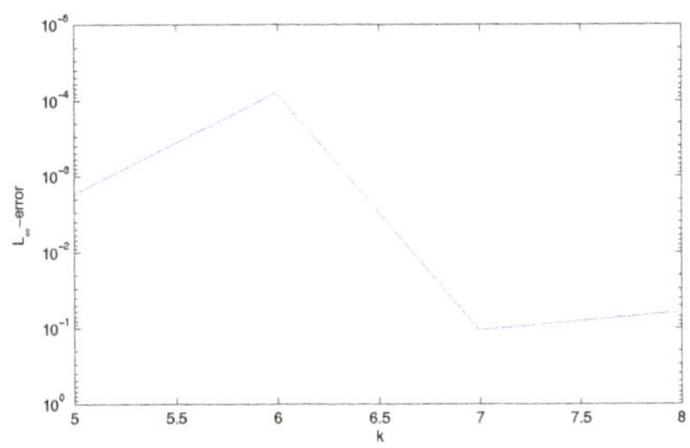

Figure 11. The $L_\infty - error$ for at different values of k for Example 7.

7. Conclusions

Enormous efforts and advances have been conducted to obtain the numerical solutions of fractional differential equations. An operational matrix of orthonormal Bernstein polynomials is derived to find the inverse Laplace transform in [50]. Here, the practical use of our proposed method is discussed for finding the solutions of some fractional order ordinary differential equations including the mathematical model of instrument (MEMS) and partial differential equations (particularly wave equation) that converts the problem to system of linear algebraic equations. The accuracy of the method is illustrated while comparing the solutions with some existing methods like VIM, FFDM, cubic spline and IMQ-RBF. We have also combined our method with Laplace adomian decomposition method that is advantageous to solve nonlinear fractional differential equations. Finally, we have analyzed the solution of each illustrative example at different values of k, i.e., at $k = 5, 6, 7, 8$ and with the help of graphs relative errors are shown. It is also observed from the plots of supremum norm error that at $k = 6$, we achieve the best accurate result compared to others.

Author Contributions: The Authors have equally contributed to this paper.

Funding: This research received no external funding.

Conflicts of Interest: The authors declare no conflict of interest.

References

1. Kilbas, A.A.; Srivastava, H.M.; Trujillo, J.J. *Theory and Applications of Fractional Differential Equations*; Elsevier: San Diego, CA, USA, 2006.
2. Podlubny, I. *Fractional Differential Equations: An Introduction to Fractional Derivatives, Fractional Differential Equations to Methods of Their Solution and Some of Their Applications*; Academic Press: New York, NY, USA, 1999.
3. Debnath, L.; Bhatta, D. *Integral Transforms and their Applications*, 2nd ed.; Chapman and Hall/CRC: Boca Raton, FL, USA, 2007.
4. Miller, K.S.; Ross, B. *An Introduction to the Fractional Calculus and Fractional Differential Equations*; John Wiley and Sons, Inc.: New York, NY, USA, 1993.
5. Oldham, K.B.; Spanier, J. *The Fractional Calculus*; Academic Press: New York, NY, USA, 1974.
6. Xiao-Jun, Y. *General Fractional Derivatives: Theory, Methods and Applications*; CRC Press: Boca Raton, FL, USA, 2018.
7. Li, Y.; Sun, N. Numerical solution of fractional differential equations using the generalized block pulse operational matrix. *Comput. Math. Appl.* **2011**, *62*, 1046–1054. [CrossRef]
8. Saadatmandi, A.; Dehghan, M. A new operational matrix for solving fractional-order differential equations. *Comput. Math. Appl.* **2010**, *59*, 1326–1336. [CrossRef]
9. Saadatmandi, A.; Dehghan, M. A tau approach for solution of the space fractional diffusion equation. *Comput. Math. Appl.* **2011**, *62*, 1135–1142. [CrossRef]
10. Doha, E.H.; Bhrawy, A.H.; Ezz-Eldien, S.S. A new Jacobi operational matrix: An application for solving fractional differential equations. *Appl. Math. Model.* **2012**, *36*, 4931–4943. [CrossRef]
11. Kazem, S.; Abbasbandy, S.; Kumar, S. Fractional-order Legendre functions for solving fractional-order differential equations. *Appl. Math. Model.* **2013**, *37*, 5498–5510. [CrossRef]
12. Mokhtary, P.; Ghoreishi, F.; Srivastava, H.M. The Muntz-Legendre Tau method for fractional differential equations. *Appl. Math. Model.* **2016**, *40*, 671–684. [CrossRef]
13. Keshavarz, E.; Ordokhani, Y.; Razzaghi, M. Bernoulli wavelet operational matrix of fractional order integration and its applications in solving the fractional order differential equations. *Appl. Math. Model.* **2014**, *38*, 6038–6051. [CrossRef]
14. Sabermahani, S.; Ordokhani, Y.; Yousefi, S.A. Numerical approach based on fractional-order Lagrange polynomials for solving a class of fractional differential equations. *Comput. Appl. Math.* **2017**, *1*, 1–23.
15. Albadarneh, R.B.; Zerqat, M.; Batiha, I.M. Numerical solutions for linear and non-linear fractional differential equations. *Int. J. Pure Appl. Math.* **2016**, *106*, 859–871. [CrossRef]
16. Garrappa, R. Numerical solution of fractional differential equations: A survey and a software tutorial. *Mathematics* **2018**, *6*, 16. [CrossRef]
17. Dehghan, M.; Manafian, J.; Saadatmandi, A. The solution of the linear fractional partial differential equations using the homotopy analysis method. *Z. Naturforsch.* **2010**, *65a*, 935–949.
18. Vanani, S.K.; Aminataei, A. On the numerical solution of fractional partial differential equations. *Math. Comput. Appl.* **2012**, *17*, 140–151. [CrossRef]
19. Rehman, M.U.; Khan, R.A. Numerical solutions to initial and boundary value problems for linear fractional partial differential equations. *Appl. Math. Model.* **2013**, *37*, 5233–5244. [CrossRef]
20. Li, W.; Bai, L.; Chen, Y.; Santos, S.D.; Li, B. Solution of linear fractional partial differential equations based on the operator matrix of fractional Bernstein polynomials and error correction. *Inter. J. Innov. Comput. Inf. Control* **2018**, *14*, 211–226.

21. Xiao-Jun, Y.; Gao, F.; Ju, Y.; Zhou, H.W. Fundamental solutions of the general fractional-order diffusion equations. *Math. Methods Appl. Sci.* **2018**, *41*, 9312–9320.
22. Xiao-Jun, Y.; Gao, F.; Srivastava, H.M. A new computational approach for solving nonlinear local fractional PDEs. *J. Comput. Appl. Math.* **2018**, *339*, 285–296.
23. Cesarano, C. Generalized special functions in the description of fractional diffusive equations. *Commun. Appl. Ind. Math.* **2019**, *10*, 31–40. [CrossRef]
24. Babolian, E.; Shamloo, A.S. Numerical solution of Volterra integral and integro-differential equations of convolution type by using operational matrices of piecewise constant orthogonal functions. *J. Comput. Appl. Math* **2008**, *214*, 495–508. [CrossRef]
25. Maleknejad, K.; Nouri, M. A direct method to solve integral and integro-differential equations of convolution type by using improved operational matrix. *Inter. J. Syst. Sci.* **2012**, *2012*, 1–8. [CrossRef]
26. Murli, A.; Rizzardi, M. Algorithm 682 Talbot's method for the Laplace inversion problem. *ACM Trans. Math. Softw.* **1990**, *16*, 158–168. [CrossRef]
27. Massouros, P.G.; Genin, G.M. Algebraic inversion of the Laplace transform. *Comput. Math. Appl.* **2005**, *50*, 179–185. [CrossRef]
28. Lee, J.; Sheen, D. An accurate numerical inversion of Laplace transforms based on the location of their poles. *Comput. Math. Appl.* **2004**, *48*, 1415–1423. [CrossRef]
29. Matsuura, T.; Saitoh, S. Real inversion formulas and numerical experiments of the Laplace transform by using the theory of reproducing kernels. *Procedia Soc. Behav. Sci.* **2010**, *2*, 111–119. [CrossRef]
30. Hsiao, C.H. Numerical inversion of Laplace transform via wavelet in ordinary differential equations. *Comput. Methods Diff. Equ.* **2014**, *2*, 186–194.
31. Iqbal, M. On comparison of spline regularization with exponential sampling method for Laplace transform inversion. *Comput. Phys.Commun.* **1995**, *88*, 43–50. [CrossRef]
32. Cuomo, S.; D'Amore, L.; Murli, A.; Rizzardi, M. Computation of the inverse Laplace transform based on a collocation method which uses only real values. *J. Comput. Appl. Math.* **2007**, *198*, 98–115. [CrossRef]
33. Dubner, H.; Abate, J. Numerical inversion of Laplace transforms by relating them to the finite Fourier cosine transform. *J. Association Comput. Mach.* **1968**, *15*, 115–123. [CrossRef]
34. Durbin, F. Numerical inversion of Laplace transforms: an efficient improvement to Dubner and Abate's method. *Comput. J.* **1974**, *17*, 371–376. [CrossRef]
35. Davis, B.; Martin, B. Numerical inversion of Laplace transform: A survey and comparison of methods. *J. Comput. Phys.* **1979**, *33*, 1–32. [CrossRef]
36. Cohen, A.M. *Numerical methods for Laplace transform inversion*; Springer: New Your, NY, USA, 2007.
37. Sastre, J.; Defez, E.; Jódar, L. Application of Laguerre matrix polynomials to the numerical inversion of Laplace transforms of matrix functions. *Appl. Math. Lett.* **2011**, *24*, 1527–1532. [CrossRef]
38. Aznam, S.M.; Hussin, A. Numerical method for inverse Laplace transform with Haar Wavelet operational matrix. *Malays. J. Fund. Appl. Sci.* **2012**, *8*, 182–188. [CrossRef]
39. Chen, C.F.; Tsay, Y.T.; Wu, T.T. Walsh operational matrices for fractional calculus and their application to distributed parameter systems. *J. Frankl. Inst.* **1977**, *503*, 267–284. [CrossRef]
40. Wu, J.L.; Chen, C.F.; Chen, C.F. Numerical inversion of Laplace transform using Haar wavlet operational matrices. *IEEE Trans. Circuit Syst.-I: Fundam. Theory Appl.* **2001**, *48*, 120–122.
41. Shamloo, A.S.; Hosseingholizadeh, R.; Nouri, M. Numerical solution of nonlinear Volterra integral equations of the first kind with convolution kernel. *World Appl. Program.* **2014**, *4*, 172–180.
42. Chen, C.F.; Haiso, C.H. Haar wavlet method for solving lumped and distributed-parameter systems. *IEEE Control Theory Appl.* **1997**, *144*, 87–94. [CrossRef]
43. Bhatti, M.I.; Bracken, P. Solutions of differential equations in a Bernstein polynomial basis. *J. Comput. Appl. Math.* **2007**, *205*, 272–280. [CrossRef]
44. Maleknejad, K.; Basirat, B.; Hashemizadeh, E. A Bernstein operational matrix approach for solving a system of high order linear Volterra-Fredholm integro-differential equations. *Math. Comput. Model.* **2012**, *55*, 1363–1372. [CrossRef]
45. Singh, A.K.; Singh, V.K.; Singh, O.P. The Bernstein operational matrix of integration. *Appl. Math. Sci.* **2009**, *3*, 2427–2436.
46. Quain, W.; Riedel, M.D.; Rosenberg, I. Uniform approximation and Bernstein polynomial with coefficients in the unit interval. *Eur. J. Comb.* **2011**, *32*, 448–463.

47. Bataineh, A.S. Bernstein polynomials method and its error analysis for solving nonlinear problems in the calculus of variations:convergnece analysis via residual function. *Filomat* **2018**, *32*, 1379–1393. [CrossRef]
48. Rostamy, D.; Alipour, M.; Jafari, H.; Baleanu, D. Solving multi-term orders fractional differential equations by operational matrices of BPs with convergence analysis. *Roman. Rep. Phys.* **2013**, *65*, 334–349.
49. Alshbool, M.H.T.; Bataineh, A.S.; Hashim, I.; Isik, O.R. Solution of fractional-order differential equations based on the operational matrices of new fractional Bernstein functions. *J. King Saud Univ. Sci.* **2017**, *29*, 1–18. [CrossRef]
50. Rani, D.; Mishra, V.; Cattani, C. Numerical inversion of Laplace transform based on Bernstein operational matrix. *Math. Methods Appl. Sci.* **2018**, *41*, 9231–9243. [CrossRef]
51. Khuri, S.A. A Laplace decomposition algorithm applied to a class of nonlinear differential equation. *J. Appl. Math.* **2001**, *1*, 141–155. [CrossRef]
52. Dattoli, G.; Lorenzutta, S.; Cesarano, C. Bernestein polynomials and operational methods. *J. Comput. Anal. Appl.* **2006**, *8*, 369–377.
53. Kurkcu, O.K.; Aslan, E.; Sezera, M.E. A numerical method for solving some model problems arising in science and convergence analysis based on residual function. *Appl. Numer. Math.* **2017**, *121*, 134–148. [CrossRef]
54. Zahra, W.K.; Elkholy, S.M. The use of cubic splines in the numerical solution of fractional differential equations. *Int. J. Math. Math. Sci.* **2012**, *2012*, 1–16. [CrossRef]
55. Momani, S.; Odibatb, Z. Numerical approach to differential equations of fractional order. *J. Comput. Appl. Math.* **2007**, *207*, 96–110. [CrossRef]

symmetry

MDPI

Article

An Efficient Class of Weighted-Newton Multiple Root Solvers with Seventh Order Convergence

Janak Raj Sharma [1], Deepak Kumar [1] and Carlo Cattani [2,3,*]

1 Department of Mathematics, Sant Longowal Institute of Engineering & Technology, Longowal, Sangrur 148106, India
2 Engineering School (DEIM), University of Tuscia, 01100 Viterbo, Italy
3 Ton Duc Thang University, Ho Chi Minh City (HCMC) 758307, Vietnam
* Correspondence: cattani@unitus.it

Received: 27 June 2019; Accepted: 13 August 2019; Published: 16 August 2019

Abstract: In this work, we construct a family of seventh order iterative methods for finding multiple roots of a nonlinear function. The scheme consists of three steps, of which the first is Newton's step and last two are the weighted-Newton steps. Hence, the name of the scheme is 'weighted-Newton methods'. Theoretical results are studied exhaustively along with the main theorem describing convergence analysis. Stability and convergence domain of the proposed class are also demonstrated by means of using a graphical technique, namely, basins of attraction. Boundaries of these basins are fractal like shapes through which basins are symmetric. Efficacy is demonstrated through numerical experimentation on variety of different functions that illustrates good convergence behavior. Moreover, the theoretical result concerning computational efficiency is verified by computing the elapsed CPU time. The overall comparison of numerical results including accuracy and CPU-time shows that the new methods are strong competitors for the existing methods.

Keywords: nonlinear equations; multiple roots; higher order methods; attraction basins

MSC: 65H05; 41A25; 49M15

---

1. Introduction

Finding numerically a root of an equation is an interesting and challenging problem. It is also very important in many diverse areas such as Mathematical Biology, Physics, Chemistry, Economics and Engineering, to name a few [1–4]. This is due to the fact that many problems from these disciplines are ultimately reduced to finding the root of an equation. Researchers are using iterative methods for approximating root since closed form solutions cannot be obtained in general. In particular, here we consider the problem of computing multiple roots of equation $f(x) = 0$ by iterative methods. A root (say, α) of $f(x) = 0$ is called multiple root with multiplicity m, if $f^{(j)}(\alpha) = 0,\ j = 0, 1, 2, \ldots, m-1$ and $f^{(m)}(\alpha) \neq 0$.

A basic and widely used iterative method is the well-known modified Newton's method

$$x_{n+1} = x_n - m\frac{f(x_n)}{f'(x_n)} \quad \forall\ n = 0, 1, 2, \ldots. \tag{1}$$

This method efficiently locates the required multiple root with quadratic order of convergence provided that the initial value x_0 is sufficiently close to root [5]. In terms of Traub's classification (see [1]), Newton's method (1) is called one-point method. Some other important methods that belong to this class have been developed in [6–9].

Recently, numerous higher order methods, either independent or based on the modified Newton's method (1), have been proposed and analyzed in the literature, see e.g., [10–23] and references cited therein. Such methods belong to the category of multipoint methods [1]. Multipoint iterative methods compute new approximations to root α by sampling the function $f(x)$, and its derivatives at several points of the independent variable, per each step. These methods have the strategy similar to Runge–Kutta methods for solving differential equations and Gaussian quadrature integration rules in the sense that they possess free parameters which can be used to ensure that the convergence speed is of a certain order, and that the sampling is done at some suitable points.

In particular, Geum et al. in [22,23] have proposed two- and three-point Newton-like methods with convergence order six for finding multiple roots. The two-point method [22], applicable for $m > 1$, is given as

$$\begin{aligned} y_n &= x_n - m\frac{f(x_n)}{f'(x_n)} \\ x_{n+1} &= y_n - Q(u,s)\frac{f(y_n)}{f'(y_n)} \end{aligned} \tag{2}$$

where $u = \left(\frac{f(y_n)}{f(x_n)}\right)^{\frac{1}{m}}$ and $s = \left(\frac{f'(y_n)}{f'(x_n)}\right)^{\frac{1}{m-1}}$ and $Q : \mathbb{C}^2 \to \mathbb{C}$ is a holomorphic function in some neighborhood of origin $(0,0)$. The three-point method [23] for $m \geq 1$ is given as

$$\begin{aligned} y_n &= x_n - m\frac{f(x_n)}{f'(x_n)} \\ z_n &= x_n - mQ_f(u)\frac{f(x_n)}{f'(x_n)} \\ x_{n+1} &= x_n - mK_f(u,v)\frac{f(x_n)}{f'(x_n)} \end{aligned} \tag{3}$$

wherein $u = \left(\frac{f(y_n)}{f(x_n)}\right)^{\frac{1}{m}}$ and $v = \left(\frac{f(z_n)}{f(x_n)}\right)^{\frac{1}{m}}$. The function $Q_f : \mathbb{C} \to \mathbb{C}$ is analytic in a neighborhood of 0 and $K_f : \mathbb{C}^2 \to \mathbb{C}$ is holomorphic in a neighborhood of $(0,0)$. Both schemes (2) and (3) require four function evaluations to obtain sixth order convergence with the efficiency index (see [24]), $6^{1/4} \approx 1.565$.

The goal and motivation in constructing iterative methods is to attain convergence of order as high as possible by using function evaluations as small as possible. With these considerations, here we propose a family of three-point methods that attain seventh order of convergence for locating multiple roots. The methodology is based on Newton's and weighted-Newton iterations. The algorithm requires four evaluations of function per iteration and, therefore, possesses the efficiency index $7^{1/4} \approx 1.627$. This shows that the proposed methods have better efficiency (1.627) than the efficiency (1.565) of existing methods (2) and (3). Theoretical results concerning convergence order and computational efficiency are verified by performing numerical tests. In the comparison of numerical results with existing techniques, the proposed methods are observed computationally more efficient since they require less computing time (CPU-time) to achieve the solution of required accuracy.

Contents of the article are summarized as follows. In Section 2, we describe the approach to develop new methods and prove their seventh order convergence. In Section 3, stability of the methods is checked by means of using a graphical technique called basins of attraction. In Section 4, some numerical tests are performed to verify the theoretical results by implementing the methods on some examples. Concluding remarks are reported in Section 5.

2. Formulation of Method

Let $m \geq 1$ be the multiplicity of a root of the equation $f(x) = 0$. To compute the root let us consider the following three-step iterative scheme:

$$\begin{cases} y_n = x_n - m\frac{f(x_n)}{f'(x_n)} \\ z_n = y_n - muH(u)\frac{f(x_n)}{f'(x_n)} \\ x_{n+1} = z_n - mvG(u,w)\frac{f(x_n)}{f'(x_n)} \end{cases} \tag{4}$$

where $u = \left(\frac{f(y_n)}{f(x_n)}\right)^{\frac{1}{m}}$, $v = \left(\frac{f(z_n)}{f(x_n)}\right)^{\frac{1}{m}}$, $w = \left(\frac{f(z_n)}{f(y_n)}\right)^{\frac{1}{m}}$, and the function $H : \mathbb{C} \to \mathbb{C}$ is analytic in some neighborhood of 0 and $G : \mathbb{C}^2 \to \mathbb{C}$ is holomorphic in a neighborhood of $(0,0)$. Notice that the first step is Newton iteration (1) whereas second and third steps are weighted by employing the factors $H(u)$ and $G(u,w)$, and so we call the algorithm (4) by the name weighted-Newton method. Factors H and G are called weight factors or more appropriately weight functions.

In the sequel we shall find conditions under which the algorithm (4) achieves high convergence order. Thus, the following theorem is stated and proved:

Theorem 1. *Assume that $f : \mathbb{C} \to \mathbb{C}$ is an analytic function in a domain enclosing a root α with multiplicity m. Suppose that initial point x_0 is closer enough to the root α, then the iterative formula defined by (4) has seventh order of convergence, if the functions $H(u)$ and $G(u,w)$ verify the conditions: $H(0) = 1$, $H'(0) = 2$, $H''(0) = -2$, $G(0,0) = 1$, $G_{10}(0,0) = 2$, $G_{01}(0,0) = 1$, $G_{20}(0,0) = 0$, $|H'''(0)| < \infty$ and $|G_{11}(0,0)| < \infty$, where $G_{ij}(0,0) = \frac{\partial^{i+j}}{\partial u^i \partial w^j} G(u,w)|_{(0,0)}$.*

Proof. Let $e_n = x_n - \alpha$ be the error at n-th iteration. Taking into account that $f^{(j)}(\alpha) = 0$, j = $0, 1, 2, \ldots, m-1$, we have by the Taylor's expansion of $f(x_n)$ about α

$$\begin{aligned} f(x_n) = {} & \frac{f^{(m)}(\alpha)}{m!}e_n^m + \frac{f^{(m+1)}(\alpha)}{(m+1)!}e_n^{m+1} + \frac{f^{(m+2)}(\alpha)}{(m+2)!}e_n^{m+2} + \frac{f^{(m+3)}(\alpha)}{(m+3)!}e_n^{m+3} + \frac{f^{(m+4)}(\alpha)}{(m+4)!}e_n^{m+4} \\ & + \frac{f^{(m+5)}(\alpha)}{(m+5)!}e_n^{m+5} + \frac{f^{(m+6)}(\alpha)}{(m+6)!}e_n^{m+6} + \frac{f^{(m+7)}(\alpha)}{(m+7)!}e_n^{m+7} + O(e_n^{m+8}) \end{aligned}$$

or:

$$f(x_n) = \frac{f^{(m)}(\alpha)}{m!}e_n^m\left(1 + C_1e_n + C_2e_n^2 + C_3e_n^3 + C_4e_n^4 + C_5e_n^5 + C_6e_n^6 + C_7e_n^7 + O(e_n^8)\right) \tag{5}$$

where $C_k = \dfrac{m!}{(m+k)!}\dfrac{f^{(m+k)}(\alpha)}{f^{(m)}(\alpha)}$ for $k \in \mathbb{N}$.

also

$$\begin{aligned} f'(x_n) = {} & \frac{f^{(m)}(\alpha)}{m!}e_n^{m-1}\big(m + C_1(m+1)e_n \\ & + C_2(m+2)e_n^2 + C_3(m+3)e_n^3 + C_4(m+4)e_n^4 \\ & + C_5(m+5)e_n^5 + C_6(m+6)e_n^6 + C_7(m+7)e_n^7 + O(e_n^8)\big), \end{aligned} \tag{6}$$

where $C_k = \frac{m!}{(m+k)!}\frac{f^{(m+k)}(\alpha)}{f^{(m)}(\alpha)}$ for $k \in \mathbb{N}$.

Using (5) and (6) in first step of (4), it follows that

$$y_n - \alpha = \frac{C_1}{m}e_n^2 + \sum_{i=1}^{5}\omega_i e_n^{i+2} + O(e_n^8), \tag{7}$$

where $\omega_i = \omega_i(m, C_1, C_2, \ldots, C_7)$ are given in terms of $m, C_1, C_2, \ldots, C_7$ with explicitly written two coefficients $\omega_1 = \frac{2mC_2-(m+1)C_1^2}{m^2}$, $\omega_2 = \frac{1}{m^3}\left(3m^2C_3 + (m+1)^2C_1^3 - m(4+3m)C_1C_2\right)$. Here, rest of the expressions of ω_i are not being produced explicitly since they are very lengthy.

Expansion of $f(y_n)$ about α yields

$$f(y_n) = \frac{f^{(m)}(\alpha)}{m!}\left(\frac{C_1}{m}\right)^m e_n^{2m}\Big(1 + \frac{2C_2 m - C_1^2(m+1)}{C_1}e_n + \frac{1}{2mC_1^2}((3+3m+3m^2+m^3)C_1^4 - 2m(2+3m+2m^2)C_1^2C_2 + 4(-1+m)m^2C_2^2 + 6m^2C_1C_3)e_n^2 + \sum_{i=1}^{4}\bar{\omega}_i e_n^{i+2} + O(e_n^8)\Big), \tag{8}$$

where $\bar{\omega}_i = \bar{\omega}_i(m, C_1, C_2, \ldots, C_7)$.

Using (5) and (8) in the expression of u, it follows that

$$u = \frac{C_1}{m}e_n + \frac{2C_2 m - C_1^2(m+2)}{m^2}e_n^2 + \sum_{i=1}^{5}\eta_i e_n^{i+2} + O(e_n^8), \tag{9}$$

where $\eta_i = \eta_i(m, C_1, C_2, \ldots, C_7)$ with explicitly written one coefficient $\eta_1 = \frac{1}{2m^3}\big(C_1^3(2m^2+7m+7) + 6C_3m^2 - 2C_2C_1m(3m+7)\big)$.

Developing weight function $H(u)$ in neighborhood 0,

$$H(u) \approx\ H(0) + uH'(0) + \frac{1}{2!}u^2H''(0) + \frac{1}{2!}u^3H'''(0). \tag{10}$$

Inserting Equations (5), (8) and (10) in the second step of (4), after some simplifications we have that

$$z_n - \alpha = -\frac{A}{m}C_1e_n^2 + \frac{1}{m^2}\big(-2mAC_2 + C_1^2(-1+mA+3H(0)-H'(0))\big)e_n^3 + \frac{1}{2m^3}\big(-6Am^2C_3 + 2mC_1C_2(-4+3Am+11H(0)-4H'(0)) + C_1^3(2-2Am^2-13H(0)+10H'(0) + m(4-11H(0)+4H'(0)) - H''(0))\big)e_n^4 + \sum_{i=1}^{3}\gamma_i e_n^{i+4} + O(e_n^8), \tag{11}$$

where $A = -1 + H(0)$ and $\gamma_i = \gamma_i(m, H(0), H'(0), H''(0), H'''(0), C_1, C_2, \ldots, C_7)$.

In order to accelerate convergence, the coefficients of e_n^2 and e_n^3 should be equal to zero. That is possible only if we have

$$H(0) = 1, \quad H'(0) = 2. \tag{12}$$

By using the above values in (11), we obtain that

$$z_n - \alpha = \frac{-2mC_1C_2 + C_1^3(9+m-H''(0))}{2m^3}e_n^4 + \sum_{i=1}^{3}\gamma_i e_n^{i+4} + O(e_n^8). \tag{13}$$

Expansion of $f(z_n)$ about α yields

$$f(z_n) = \frac{f^{(m)}(\alpha)}{m!}(z_n-\alpha)^m\big(1 + C_1(z_n-\alpha) + C_2(z_n-\alpha)^2 + O((z_n-\alpha)^3)\big). \tag{14}$$

From (5), (8) and (14), we obtain forms of v and w as

$$v = \frac{(9+m)C_1^3 - 2mC_1C_2}{2m^3}e_n^3 + \sum_{i=1}^{4}\tau_i e_n^{i+3} + O(e_n^8), \tag{15}$$

where $\tau_i = \tau_i(m, H''(0), H'''(0), C_1, C_2, \ldots, C_7)$ and

$$w = \frac{(9+m-H''(0))C_1^2 - 2mC_2}{2m^3}e_n^2 + \sum_{i=1}^{5} \varsigma_i e_n^{i+2} + O(e_n^8), \tag{16}$$

where $\varsigma_i = \varsigma_i(m, H''(0), H'''(0), C_1, C_2, \ldots, C_7)$.

Expanding $G(u,w)$ in neighborhood of origin $(0,0)$ by Taylor series, it follows that

$$G(u,w) \approx G_{00}(0,0) + uG_{10}(0,0) + \frac{1}{2}u^2G_{20}(0,0) + w(G_{01}(0,0) + uG_{11}(0,0)), \tag{17}$$

where $G_{ij}(0,0) = \frac{\partial^{i+j}}{\partial u^i \partial w^j} G(u,w)|_{(0,0)}$.

Then by substituting (5), (6), (15)–(17) into the last step of scheme (4), we obtain that

$$\begin{aligned} e_{n+1} = & \frac{1}{2m^3}((-1+G_{00}(0,0))C_1(2mC_1 - (9+m-H''(0))C_1^2))e_n^4 \\ & + \sum_{i=1}^{3} \xi_i e_n^{i+4} + O(e_n^8), \end{aligned} \tag{18}$$

where $\xi_i = \xi_i(m, H''(0), H'''(0), G_{00}(0,0), G_{10}(0,0), G_{20}(0,0), G_{01}(0,0), G_{11}(0,0), C_1, C_2, \ldots, C_7)$.

From Equation (18) it is clear that we can obtain at least fifth order convergence when $G_{00}(0,0) = 1$. In addition, using this value in $\xi_1 = 0$, we will obtain that

$$G_{10}(0,0) = 2. \tag{19}$$

By using $G_{00} = 1$ and (19) in $\xi_2 = 0$, the following equation is obtained

$$\begin{aligned} & C_1\Big(2mC_2 - C_1^2(9+m-H''(0))\Big) \\ & \Big(-2mC_2(-1+G_{01}(0,0)) + C_1^2(-11+m(-1+G_{01}(0,0)) \\ & -(-9+H''(0))G_{01}(0,0) + G_{20}(0,0))\Big) \\ & = 0, \end{aligned} \tag{20}$$

which further yields

$$G_{01}(0,0) = 1, \quad G_{20}(0,0) = 0 \quad \text{and} \quad H''(0) = -2. \tag{21}$$

Using the above values in (18), we obtain the error equation

$$\begin{aligned} e_{n+1} = & \frac{1}{360m^6}\Big(360m^3((39+5m)C_2^3 - 6mC_3^2 - 10mC_2C_4) + 120m^3C_1((515+78m)C_2C_3 \\ & - 12mC_5) - 60m^2C_1^3C_3\Big(1383 + 845m + 78m^2 + 12H'''(0)\Big) + 10mC_1^4C_2(21571 \\ & + 8183m^2 + 558m^3 + 515H'''(0) + 324G_{11}(0,0) + 36m(667 + 6H'''(0) + G_{11}(0,0))) \\ & - 60m^2C_1^2\Big(-6m(55+9m)C_4 + C_2^2(2619 + 1546m + 135m^2 + 24H'''(0) + 6G_{11}(0,0))\Big) \\ & - C_1^6\Big(55017 + 17005m + 978m^4 + 2775H'''(0) + 7290G_{11}(0,0) + 15m^2(4463 + 40H'''(0) \\ & + 6G_{11}(0,0)) + 5m(21571 + 515H'''(0) + 324G_{11}(0,0))\Big)\Big)e_n^7 + O(e_n^8). \end{aligned} \tag{22}$$

Thus, the seventh order convergence is established. □

Based on the conditions on $H(u)$ and $G(u,w)$ as shown in Theorem 1, we can generate numerous methods of the family (4). However, we restrict to the following simple forms:

2.1. Some Concrete Forms of $H(u)$

Case 1. *Considering $H(u)$ a polynomial function, i.e.,*

$$H(u) = A_0 + A_1 u + A_2 u^2.$$

Using the conditions of Theorem 1, we get $A_0 = 1$, $A_1 = 2$ and $A_2 = -1$. Then

$$H(u) = 1 + 2u - u^2.$$

Case 2. *When $H(u)$ is a rational function, i.e.,*

$$H(u) = \frac{1 + A_0 u}{A_1 + A_2 u}.$$

Using the conditions of Theorem 1, we get that $A_0 = \frac{5}{2}$, $A_1 = 1$ and $A_2 = \frac{1}{2}$. So

$$H(u) = \frac{2 + 5u}{2 + u}.$$

Case 3. *Consider $H(u)$ as another rational weight function, e.g.,*

$$H(u) = \frac{1 + A_0 u + A_1 u^2}{1 + A_2 u}.$$

Using the conditions of Theorem 1, we obtain $A_0 = 3$, $A_1 = 2$ and $A_2 = 1$. Then $H(u)$ becomes

$$H(u) = \frac{1 + 3u + u^2}{1 + u}.$$

Case 4. *When $H(u)$ is a yet another rational function of the form*

$$H(u) = \frac{1 + A_0 u}{1 + A_1 u + A_2 u^2}.$$

Using the conditions of Theorem 1, we have $A_0 = 1$, $A_1 = -1$ and $A_2 = 1$. Then

$$H(u) = \frac{1 + u}{1 - u + 3u^2}.$$

2.2. Some Concrete Forms of $G(u, w)$

Case 5. *Considering $G(u, w)$ a polynomial function, e.g.,*

$$G(u, w) = A_0 + A_1 u + A_2 u^2 + A_3 w.$$

From the conditions of Theorem 1, we get $A_0 = 1$, $A_1 = 2$, $A_2 = 0$ and $A_3 = 1$. So

$$G(u, w) = 1 + 2u + w.$$

Case 6. *Considering $G(u, w)$ a sum of two rational functions, that is*

$$G(u, w) = \frac{A_0 + 2u}{1 + A_1 u} + \frac{B_0}{1 + B_1 w}.$$

By using the conditions of Theorem 1, we find that $A_0 = 0$, $A_1 = 0$, $B_0 = 1$ *and* $B_1 = -1$. $G(u, w)$ *becomes*

$$G(u,w) = 2u + \frac{1}{1-w}.$$

Case 7. *When* $G(u, w)$ *is a product of two rational functions, that is*

$$G(u,w) = \frac{1 + A_0 u}{1 + A_1 u} \times \frac{B_0}{1 + B_1 w}.$$

Then the conditions of Theorem 1 yield $A_0 = 2$, $A_1 = 0$, $B_0 = 1$ *and* $B_1 = -1$. *So*

$$G(u,w) = \frac{1 + 2u}{1 - w}.$$

3. Complex Dynamics of Methods

Here we analyze the complex dynamics of new methods based on a graphical tool 'the basins of attraction' of the roots of a polynomial $p(z)$ in Argand plane. Analysis of the basins gives an important information about the stability and convergence region of iterative methods. Wider is the convergence region (i.e., basin), better is the stability. The idea of complex dynamics was introduced initially by Vrscay and Gilbert [25]. In recent times, many authors have used this concept in their work, see, for example [26,27] and references therein. We consider some of the cases corresponding to the previously obtained forms of $H(u)$ and $G(u, w)$) of family (4) to assess the basins of attraction. Let us select the combinations: cases 1 and 2 of $H(u)$ with the cases 5, 6 and 7 of $G(u, w)$) in the scheme (4), and denote the corresponding methods by NM-i(j), i = I, II and j = a, b, c.

To start with we take the initial point z_0 in a rectangular region $R \in \mathbb{C}$ that contains all the roots of a polynomial $p(z)$. The iterative method when starts from point z_0 in a rectangle either converges to the root $P(z)$ or eventually diverges. The stopping criterion for convergence is considered to be 10^{-3} up to a maximum of 25 iterations. If the required accuracy is not achieved in 25 iterations, we conclude that the method with initial point z_0 does not converge to any root. The strategy adopted is as follows: A color is allocated to each initial point z_0 lying in the basin of attraction of a root. If the iteration initiating at z_0 converges, it represents the attraction basin painted with assigned color to it, otherwise, the non-convergent cases are painted by the black color.

To view the geometry in complex plane, we characterize attraction basins associated with the methods NM-I(a–c) and NM-II(a–c) considering the following four polynomials:

Problem 1. *Consider the polynomial* $p_1(z) = (z^2 - 1)^3$, *which has roots* $\{-1, 1\}$ *with multiplicity three. We use a grid of* 400×400 *points in a rectangle* $R \in \mathbb{C}$ *of size* $[-3, 3] \times [-3, 3]$ *and assign red color to each initial point in the attraction basin of root* -1 *and green color to each point in the attraction basin of root* 1. *The basins so plotted for NM-I(a–c) and NM-II(a–c) are displayed in Figure 1. Looking at these graphics, we conclude that the method NM-II(c) possesses better stability followed by NM-I(c) and NM-II(b). Black zones in the figures show the divergent nature of a method when it starts assuming initial point from such zones.*

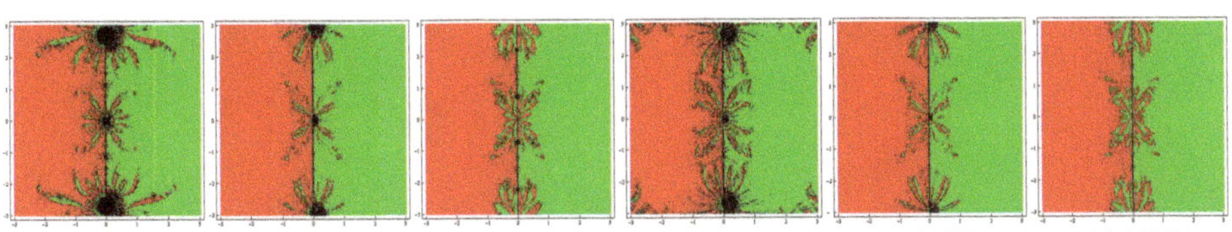

NM-I(**a**). NM-I(**b**). NM-I(**c**). NM-II(**a**). NM-II(**b**). NM-II(**c**).

Figure 1. Basins of attraction for NM-I(**a–c**) and NM-II(**a–c**) in polynomial $p_1(z)$.

Problem 2. *Let* $p_2(z) = (z^3 - 1)^2$ *that has three roots* $\{-0.5 \pm 0.866025i, 1\}$ *each with multiplicity two. To plot the graphics, we use a grid of* 400×400 *points in a rectangle* $R \in \mathbb{C}$ *of size* $[-3, 3] \times [-3, 3]$ *and assign the colors blue, green and red corresponding to each point in the basins of attraction of* 1, $-0.5 + 0.866025i$ *and* $-0.5 - 0.866025i$. *Basins drawn for the methods NM-I(a–c) and NM-II(a–c) are shown in Figure 2. As can be observed from the pictures, the method NM-I(c) and NM-II(c) possess a small number of divergent points and therefore have better convergence than the remaining methods.*

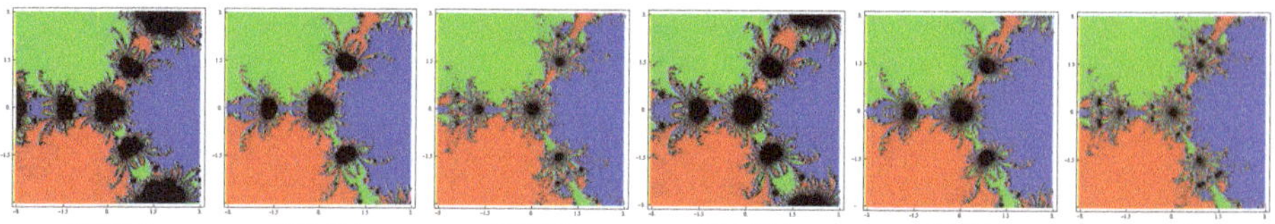

NM-I(**a**). NM-I(**b**). NM-I(**c**). NM-II(**a**). NM-II(**b**). NM-II(**c**).

Figure 2. Basins of attraction for NM-I(**a–c**) and NM-II(**a–c**) in polynomial $p_2(z)$.

Problem 3. *Let* $p_3(z) = (z^6 - 1)^3$ *with six roots* $\{\pm 1, -0.5 \pm 0.866025i, 0.5 \pm 0.866025i\}$ *each with multiplicity* $m = 3$. *Basins obtained for the considered methods are presented in Figure 3. To draw the pictures, the red, blue, green, pink, cyan and magenta colors have been assigned to the attraction basins of the six roots. We observe from the graphics that the method NM-I(c) and NM-II(c) have better convergence behavior since they have lesser number of divergent points. On the other hand NM-I(a) and NM-II(a) contain large black regions followed by NM-I(b) and NM-II(b) indicating that the methods do not converge in 25 iterations starting at those points.*

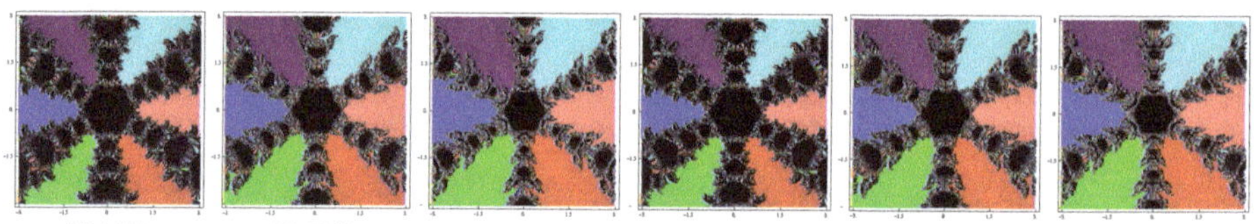

NM-I(**a**). NM-I(**b**). NM-I(**c**). NM-II(**a**). NM-II(**b**). NM-II(**c**).

Figure 3. Basins of attraction for NM-I(**a–c**) and NM-II(**a–c**) in polynomial $p_3(z)$.

Problem 4. *Consider the polynomial* $p_4(z) = z^4 - 6z^2 + 8$ *that has four simple roots* $\{\pm 2, \pm 1.414\ldots\}$. *In this case also, we use a grid of* 400×400 *points in a rectangle* $R \in \mathbb{C}$ *of size* $[-3, 3] \times [-3, 3]$ *and allocate the red, blue, green and yellow colors to the basins of attraction of these four roots. Basins obtained for the methods are shown in Figure 4. Observing the basins, we conclude that the method NM-II(c) possesses better stability followed by NM-I(c). Remaining methods show chaotic nature along the boundaries of the attraction basins.*

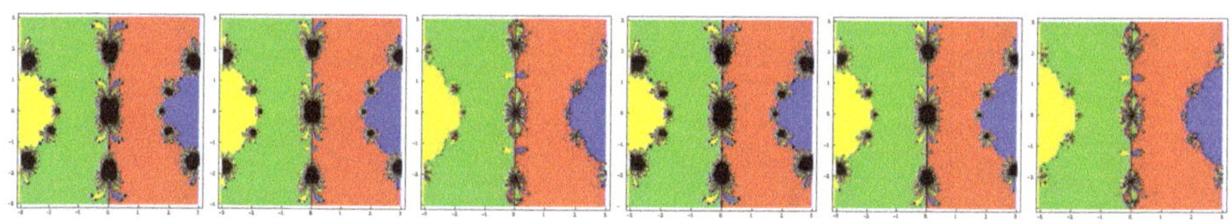

NM-I(**a**). NM-I(**b**). NM-I(**c**). NM-II(**a**). NM-II(**b**). NM-II(**c**).

Figure 4. Basins of attraction for NM-I(**a–c**) and NM-II(**a–c**) in polynomial $p_4(z)$.

Looking at the graphics, one can easily judge the stable behavior and so the better convergence of any method. We reach to a root, if we start the iteration choosing z_0 anywhere in the basin of that root. However, if we choose an initial guess z_0 in a region wherein different basins of attraction meet each other, it is difficult to predict which root is going to be attained by the iterative method that starts from z_0. So, the choice of z_0 in such a region is not a good one. Both black regions and the regions with different colors are not suitable to assume the initial guess as z_0 when we are required to achieve a particular root. The most intricate geometry is between the basins of attraction, which corresponds to the cases where the method is more demanding with respect to the initial point. From the basins, one can conclude that the method NM-II(c) possesses better stability followed by NM-I(c) than the remaining methods.

4. Numerical Tests

In this section, we apply the special cases NM-i(j), i = I, II and j = a, b, c of the scheme (4), corresponding to the combinations of $H(u)$: cases 1 and 2 with that of $G(u,w)$): cases 5, 6 and 7, to solve some nonlinear equations for validation of the theoretical results that we have derived. The theoretical seventh order of convergence is verified by calculating the computational order of convergence (COC)

$$\text{COC} = \frac{\ln|(x_{n+1}-\alpha)/(x_n-\alpha)|}{\ln|(x_n-\alpha)/(x_{n-1}-\alpha)|},$$

which is given in (see [28]). Comparison of performance is also done with some existing methods such as the sixth order methods by Geum et al. [22,23], which are already expressed by (2) and (3). To represent $Q_f(u,s)$, we choose the following four special cases in the formula (2) and denote the respective methods by GKN-I(j), j = a, b, c, d:

(a) $Q_f(u,s) = m(1+2(m-1)(u-s)-4us+s^2)$.
(b) $Q_f(u,s) = m(1+2(m-1)(u-s)-u^2-2us)$.
(c) $Q_f(u,s) = \frac{m+au}{1+bu+cs+dus}$, where $a = \frac{2m}{m-1}$, $b = 2-2m$, $c = \frac{2(2-2m+m^2)}{m-1}$, $d = -2m(m-1)$.
(d) $Q_f(u,s) = \frac{m+a_1u}{1+b_1u+c_1u^2}\frac{1}{1+d_1s}$, where $a_1 = \frac{2m(4m^4-16m^3+31m^2-30m+13}{(m-1)(4m^2-8m+7)}$, $b_1 = \frac{4(2m^2-4m+3)}{(m-1)(4m^2-8m+7)}$, $c_1 = -\frac{4m^2-8m+3}{4m^2-8m+7}$, $d_1 = 2(m-1)$.

For the formula (3), considering the following four combinations of the functions $Q_f(u)$ and $K_f(u,v)$, and denoting the corresponding methods by GKN-II(j), j = a, b, c, d:

(a) $Q_f(u) = \frac{1+u^2}{1-u}$, $K_f(u,v) = \frac{1+u^2-v}{1-u+(u-2)v}$.
(b) $Q_f(u) = 1+u+2u^2$, $K_f(u,v) = 1+u+2u^2+(1+2u)v$.
(c) $Q_f(u) = \frac{1+u^2}{1-u}$, $K_f(u,v) = 1+u+2u^2+2u^3+2u^4+(2u+1)v$.
(d) $Q_f(u) = \frac{(2u-1)(4u-1)}{1-7u+13u^2}$, $K_f(u,v) = \frac{(2u-1)(4u-1)}{1-7u+13u^2-(1-6u)v}$.

Computational work is compiled in the programming package of Mathematica software using multiple-precision arithmetic. Numerical results as displayed in Tables 1–5 contain: (i) number of iterations (n) needed to converge to desired solution, (ii) last three successive errors $e_n = |x_{n+1}-x_n|$, (iii) computational order of convergence (COC) and (iv) CPU-time (CPU-t) in seconds elapsed during the execution of a program. Required iteration (n) and elapsed CPU-time are computed by selecting $|x_{n+1}-x_n| + |f(x_n)| < 10^{-350}$ as the stopping condition.

For numerical tests we select seven problems. The first four problems are of practical interest where as last three are of academic interest. In the problems we need not to calculate the root multiplicity m and it is set a priori, before running the algorithm.

Example 1 (Eigen value problem). *Finding Eigen values of a large sparse square matrix is a challenging task in applied mathematics and engineering sciences. Calculating the roots of a characteristic equation of matrix of order larger than 4 is even a big job. We consider the following 9×9 matrix.*

$$A = \frac{1}{8}\begin{bmatrix} -12 & 0 & 0 & 19 & -19 & 76 & -19 & 18 & 437 \\ -64 & 24 & 0 & -24 & 24 & 64 & -8 & 32 & 376 \\ -16 & 0 & 24 & 4 & -4 & 16 & -4 & 8 & 92 \\ -40 & 0 & 0 & -10 & 50 & 40 & 2 & 20 & 242 \\ -4 & 0 & 0 & -1 & 41 & 4 & 1 & 0 & 25 \\ -40 & 0 & 0 & 18 & -18 & 104 & -18 & 20 & 462 \\ -84 & 0 & 0 & -29 & 29 & 84 & 21 & 42 & 501 \\ 16 & 0 & 0 & -4 & 4 & -16 & 4 & 16 & -92 \\ 0 & 0 & 0 & 0 & 0 & 0 & 0 & 0 & 24 \end{bmatrix}.$$

We calculate the characteristic polynomial of the matrix (A) as

$$f_1(x) = x^9 - 29x^8 + 349x^7 - 2261x^6 + 8455x^5 - 17663x^4 + 15927x^3 + 6993x^2 - 24732x + 12960.$$

This function has a multiple root $\alpha = 3$ with multiplicity 4. We select initial value $x_0 = 2.25$ and obtain the numerical results as shown in Table 1.

Table 1. Comparison of the numerical results for Example 1.

Methods	n	$\lvert e_{n-3}\rvert$	$\lvert e_{n-2}\rvert$	$\lvert e_{n-1}\rvert$	COC	CPU-t (s)
GKN-I(a)	4	1.06×10^{-9}	3.86×10^{-56}	9.03×10^{-335}	6.0000	0.1567
GKN-I(b)	4	1.06×10^{-9}	3.91×10^{-56}	9.85×10^{-335}	6.0000	0.1583
GKN-I(c)	4	1.06×10^{-9}	4.34×10^{-56}	2.02×10^{-334}	6.0000	0.1525
GKN-I(d)	4	1.07×10^{-9}	1.17×10^{-55}	2.02×10^{-331}	6.0000	0.1600
GKN-II(a)	4	1.19×10^{-6}	5.39×10^{-38}	4.56×10^{-226}	5.9999	0.1835
GKN-II(b)	4	1.20×10^{-6}	1.61×10^{-37}	9.49×10^{-223}	5.9999	0.1640
GKN-II(c)	4	1.20×10^{-6}	1.12×10^{-37}	7.51×10^{-224}	5.9999	0.1718
GKN-II(d)	4	1.20×10^{-6}	1.87×10^{-37}	2.76×10^{-222}	5.9999	0.1680
NM-I(a)	3	9.83×10^{-8}	4.34×10^{-51}	0	7.0000	0.1562
NM-I(b)	3	1.16×10^{-9}	1.38×10^{-64}	0	7.0000	0.1170
NM-I(c)	3	6.30×10^{-10}	7.75×10^{-67}	0	7.0000	0.1485
NM-II(a)	3	9.83×10^{-8}	4.41×10^{-51}	0	7.0000	0.1367
NM-II(b)	3	1.16×10^{-9}	1.40×10^{-64}	0	7.0000	0.1562
NM-II(c)	3	6.30×10^{-10}	8.07×10^{-67}	0	7.0000	0.1405

Example 2 (Manning equation for fluid dynamics). *Next, the problem of isentropic supersonic flow around a sharp expansion corner is chosen (see [2]). Relation among the Mach number before the corner (say M_1) and after the corner (say M_2) is given by*

$$\delta = b^{1/2}\left(\tan^{-1}\left(\frac{M_2^2-1}{b}\right)^{1/2} - \tan^{-1}\left(\frac{M_1^2-1}{b}\right)^{1/2}\right) - \left(\tan^{-1}(M_2^2-1)^{1/2} - \tan^{-1}(M_1^2-1)^{1/2}\right),$$

where $b = \frac{\gamma+1}{\gamma-1}$ and γ is the specific heat ratio of gas.

For a specific case, the above equation is solved for for M_2, given that $M_1 = 1.5$, $\gamma = 1.4$ and $\delta = 10^0$. Then, we have that

$$\tan^{-1}\left(\frac{\sqrt{5}}{2}\right) - \tan^{-1}(\sqrt{x^2-1}) + \sqrt{6}\left(\tan^{-1}\left(\sqrt{\frac{x^2-1}{6}}\right) - \tan^{-1}\left(\frac{1}{2}\sqrt{\frac{5}{6}}\right)\right) - \frac{11}{63} = 0$$

where $x = M_2$.

Let us consider this particular case seven times using same values of the involved parameters and then obtain the nonlinear function

$$f_2(x) = \left[\tan^{-1}\left(\frac{\sqrt{5}}{2}\right) - \tan^{-1}(\sqrt{x^2-1}) + \sqrt{6}\left(\tan^{-1}\left(\sqrt{\frac{x^2-1}{6}}\right) - \tan^{-1}\left(\frac{1}{2}\sqrt{\frac{5}{6}}\right)\right) - \frac{11}{63}\right]^7.$$

The above function has one root at $\alpha = 1.8411027704\ldots$ *of multiplicity 7 with initial approximations* $x_0 = 1.50$. *Computed numerical results are shown in Table 2.*

Table 2. Comparison of the numerical results for Example 2.

Methods	n	$\|e_{n-3}\|$	$\|e_{n-2}\|$	$\|e_{n-1}\|$	COC	CPU-t (s)
GKN-I(a)	4	2.17×10^{-8}	4.61×10^{-25}	1.01×10^{-152}	6.0000	1.4218
GKN-I(b)	4	2.17×10^{-8}	4.60×10^{-25}	2.27×10^{-151}	6.0000	1.4923
GKN-I(c)	4	2.11×10^{-8}	4.21×10^{-25}	1.03×10^{-150}	6.0000	1.4532
GKN-I(d)	4	1.77×10^{-8}	2.48×10^{-25}	2.68×10^{-151}	6.0000	1.4960
GKN-II(a)	4	4.83×10^{-7}	1.36×10^{-41}	6.84×10^{-249}	6.0000	1.3867
GKN-II(b)	4	4.90×10^{-7}	2.89×10^{-41}	1.21×10^{-246}	6.0000	1.3790
GKN-II(c)	4	4.88×10^{-7}	2.22×10^{-41}	1.98×10^{-247}	6.0000	1.4110
GKN-II(d)	4	4.89×10^{-7}	3.22×10^{-41}	2.62×10^{-246}	6.0000	1.3982
NM-I(a)	3	1.65×10^{-8}	2.82×10^{-58}	0	7.0000	1.1367
NM-I(b)	3	7.69×10^{-9}	1.35×10^{-60}	0	7.0000	1.1915
NM-I(c)	3	3.65×10^{-9}	3.19×10^{-63}	0	7.0000	1.1407
NM-II(a)	3	1.65×10^{-9}	2.86×10^{-58}	0	7.0000	1.1290
NM-II(b)	3	7.69×10^{-9}	1.36×10^{-60}	0	7.0000	1.2540
NM-II(c)	3	3.65×10^{-9}	3.27×10^{-63}	0	7.0000	1.1445

Example 3 (Beam designing model). *We consider the problem of beam positioning (see [4]) where a beam of length r unit leans against the edge of a cubical box of sides 1 unit distance each, such that one end of the beam touches the wall and the other end touches the floor, as depicted in Figure 5.*

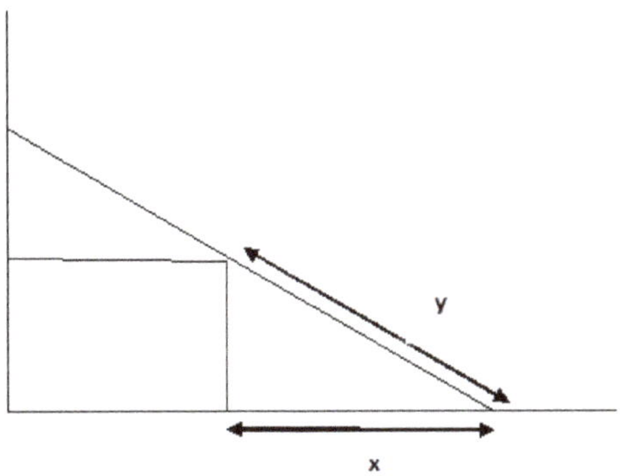

Figure 5. Beam positioning problem.

The problem is: What will be the distance alongside the floor from the base of wall to the bottom of beam? Suppose that y is distance along the beam from the floor to the edge of the box and x is the distance from the bottom of box to the bottom of beam. For a given r, we can obtain the equation

$$f_3(x) = x^4 + 4x^3 - 24x^2 + 16x + 16 = 0.$$

One of the roots of this equation is the double root $x = 2$. We select the initial guess $x_0 = 3$ to find the root. Numerical results by various methods are shown in Table 3.

Table 3. Comparison of numerical results for Example 3.

Methods	n	$\lvert e_{n-3}\rvert$	$\lvert e_{n-2}\rvert$	$\lvert e_{n-1}\rvert$	COC	CPU-t (s)
GKN-I(a)	4	1.29×10^{-3}	5.18×10^{-20}	2.19×10^{-118}	6.0000	0.0313
GKN-I(b)	4	1.48×10^{-3}	1.63×10^{-19}	2.19×10^{-115}	5.9998	0.0390
GKN-I(c)	4	1.45×10^{-3}	1.76×10^{-19}	5.56×10^{-115}	5.9997	0.0352
GKN-I(d)	4	1.97×10^{-3}	1.80×10^{-18}	1.07×10^{-108}	5.9996	0.0428
GKN-II(a)	4	5.67×10^{-4}	1.20×10^{-22}	1.06×10^{-134}	5.9999	0.0314
GKN-II(b)	4	2.39×10^{-3}	5.78×10^{-18}	1.16×10^{-105}	5.9996	0.0396
GKN-II(c)	4	1.70×10^{-3}	4.26×10^{-19}	1.08×10^{-112}	5.9997	0.0392
GKN-II(d)	4	1.55×10^{-2}	5.18×10^{-13}	7.23×10^{-76}	6.0000	0.0354
NM-I(a)	4	1.13×10^{-4}	6.52×10^{-23}	1.41×10^{-157}	6.9998	0.0275
NM-I(b)	4	9.26×10^{-4}	1.63×10^{-23}	8.75×10^{-162}	6.9998	0.0313
NM-I(c)	4	4.64×10^{-4}	4.44×10^{-26}	3.23×10^{-180}	6.9998	0.0275
NM-II(a)	4	1.13×10^{-4}	6.83×10^{-23}	2.00×10^{-157}	6.9998	0.0316
NM-II(b)	4	9.33×10^{-4}	1.77×10^{-23}	1.58×10^{-161}	6.9998	0.0275
NM-II(c)	4	4.78×10^{-4}	5.86×10^{-26}	2.43×10^{-179}	6.9998	0.0354

Example 4 (van der Waals equation). *Consider the Van der Waals equation*

$$\left(P + \frac{a_1 n^2}{V^2}\right)(V - na_2) = nRT,$$

that describes nature of a real gas by adding in the ideal gas equation two parameters, a_1 and a_2, which are specific for each gas. To find the volume V in terms of rest of the parameters one requires to solve the equation

$$PV^3 - (na_2P + nRT)V^2 + a_1n^2V = a_1a_2n^3.$$

Given a set of values of a_1 and a_2 of a particular gas, one can find values for n, P and T, so that this equation has three real roots. Using a particular set of values (see [3]), we have the equation

$$f_4(x) = x^3 - 5.22x^2 + 9.0825x - 5.2675 = 0,$$

where $x = V$. This equation has a multiple root $\alpha = 1.75$ with multiplicity 2. The initial guess chosen to obtain the root 1.75 is $x_0 = 2$. Numerical results are shown in Table 4.

Table 4. Comparison of numerical results for Example 4.

Methods	n	$\lvert e_{n-3}\rvert$	$\lvert e_{n-2}\rvert$	$\lvert e_{n-1}\rvert$	COC	CPU-t (s)
GKN-I(a)	5	1.90×10^{-5}	9.03×10^{-22}	1.05×10^{-119}	6.0000	0.0471
GKN-I(b)	5	2.31×10^{-5}	3.69×10^{-21}	6.14×10^{-116}	6.0000	0.0472
GKN-I(c)	5	2.18×10^{-5}	3.18×10^{-21}	3.14×10^{-116}	6.0000	0.0465
GKN-I(d)	5	3.58×10^{-5}	1.01×10^{-19}	5.02×10^{-107}	6.0000	0.0483
GKN-II(a)	5	3.00×10^{-6}	4.91×10^{-27}	9.51×10^{-152}	6.0000	0.0474
GKN-II(b)	5	4.78×10^{-5}	5.42×10^{-19}	1.17×10^{-102}	6.0000	0.0472
GKN-II(c)	5	2.51×10^{-5}	6.82×10^{-21}	2.75×10^{-114}	6.0000	0.0481
GKN-II(d)	7	3.85×10^{-11}	1.78×10^{-55}	1.75×10^{-321}	6.0000	0.0625
NM-I(a)	5	1.06×10^{-5}	4.09×10^{-26}	5.33×10^{-169}	7.0000	0.0368
NM-I(b)	5	5.10×10^{-6}	2.51×10^{-28}	1.73×10^{-184}	7.0000	0.0322
NM-I(c)	5	1.15×10^{-6}	2.55×10^{-33}	6.75×10^{-220}	7.0000	0.0327
NM-II(a)	5	1.05×10^{-5}	4.13×10^{-23}	5.89×10^{-169}	7.0000	0.0316
NM-II(b)	5	5.16×10^{-6}	2.76×10^{-23}	3.48×10^{-184}	7.0000	0.0323
NM-II(c)	5	1.20×10^{-6}	3.65×10^{-26}	9.09×10^{-219}	7.0000	0.0314

Example 5. *Consider now the standard nonlinear test function (see [23])*

$$f_5(x) = \left(9 - 2x - 2x^4 + \cos 2x\right)\left(5 - x - x^4 - \sin^2 x\right).$$

The root $\alpha = 1.29173329244360\ldots$ *of multiplicity 2 is computed with initial guess* $x_0 = 1.5$. *Numerical results are displayed in Table 5.*

Table 5. Comparison of the numerical results for Example 5.

Methods	n	$\lvert e_{n-3}\rvert$	$\lvert e_{n-2}\rvert$	$\lvert e_{n-1}\rvert$	COC	CPU-t (s)
GKN-I(a)	4	1.12×10^{-4}	5.78×10^{-24}	1.10×10^{-139}	6.0000	0.2772
GKN-I(b)	4	1.55×10^{-4}	7.30×10^{-23}	8.07×10^{-133}	6.0000	0.2462
GKN-I(c)	4	1.39×10^{-4}	4.40×10^{-23}	4.43×10^{-134}	6.0000	0.2497
GKN-I(d)	4	2.32×10^{-4}	1.95×10^{-21}	6.85×10^{-124}	6.0000	0.2812
GKN-II(a)	4	3.36×10^{-5}	8.72×10^{-28}	2.66×10^{-163}	6.0000	0.3397
GKN-II(b)	4	3.39×10^{-5}	2.19×10^{-20}	1.57×10^{-117}	6.0000	0.2695
GKN-II(c)	4	2.16×10^{-5}	7.70×10^{-22}	1.58×10^{-126}	6.0000	0.2460
GKN-II(d)	4	3.51×10^{-3}	3.25×10^{-14}	2.03×10^{-80}	6.0000	0.2342
NM-I(a)	4	1.52×10^{-4}	8.45×10^{-26}	1.41×10^{-174}	6.9999	0.1445
NM-I(b)	4	1.25×10^{-4}	2.22×10^{-26}	1.23×10^{-178}	6.9999	0.1522
NM-I(c)	4	5.26×10^{-4}	1.58×10^{-29}	3.54×10^{-201}	6.9999	0.1640
NM-II(a)	4	1.52×10^{-4}	9.05×10^{-26}	2.36×10^{-174}	6.9999	0.1482
NM-II(b)	4	1.27×10^{-4}	2.49×10^{-26}	2.84×10^{-178}	6.9999	0.1492
NM-II(c)	4	5.54×10^{-4}	2.51×10^{-29}	9.82×10^{-200}	6.9999	0.1642

Example 6. *Let us assume another nonlinear test function given as (see [22])*

$$f_6(x) = \left(x - \sqrt{3}x^3 \cos\left(\frac{\pi x}{6}\right) + \frac{1}{x^2+1} - \frac{11}{5} + 4\sqrt{3}\right)(x-2)^4.$$

The root $\alpha = 2$ *of this function is of multiplicity 5. This root is calculated assuming the initial approximation* $x_0 = 1.5$. *Results so obtained are shown in Table 6.*

Table 6. Comparison of the numerical results for Example 6.

Methods	n	$\lvert e_{n-3}\rvert$	$\lvert e_{n-2}\rvert$	$\lvert e_{n-1}\rvert$	COC	CPU-t (s)
GKN-I(a)	4	1.20×10^{-5}	6.82×10^{-31}	2.31×10^{-182}	6.0000	0.6797
GKN-I(b)	4	1.20×10^{-5}	6.86×10^{-31}	2.40×10^{-182}	6.0000	0.6680
GKN-I(c)	4	1.21×10^{-5}	7.72×10^{-31}	5.18×10^{-182}	6.0000	0.6992
GKN-I(d)	4	1.58×10^{-5}	1.00×10^{-29}	6.51×10^{-175}	6.0000	0.6720
GKN-II(a)	4	3.17×10^{-5}	1.64×10^{-28}	3.21×10^{-168}	6.0000	0.8047
GKN-II(b)	4	3.50×10^{-5}	6.90×10^{-28}	4.05×10^{-164}	6.0000	0.8280
GKN-II(c)	4	3.41×10^{-5}	4.42×10^{-28}	2.09×10^{-165}	6.0000	0.7967
GKN-II(d)	4	3.54×10^{-5}	8.45×10^{-28}	1.56×10^{-163}	6.0000	0.8242
NM-I(a)	4	5.14×10^{-6}	4.35×10^{-38}	1.35×10^{-262}	7.0000	0.5625
NM-I(b)	4	3.45×10^{-6}	2.68×10^{-39}	4.53×10^{-271}	7.0000	0.5782
NM-I(c)	4	2.05×10^{-6}	2.95×10^{-41}	3.76×10^{-285}	7.0000	0.5277
NM-II(a)	4	5.14×10^{-6}	4.42×10^{-38}	1.53×10^{-262}	7.0000	0.4805
NM-II(b)	4	3.45×10^{-6}	2.73×10^{-39}	5.24×10^{-271}	7.0000	0.4725
NM-II(c)	4	2.05×10^{-6}	3.07×10^{-41}	5.17×10^{-285}	7.0000	0.4610

Example 7. *Lastly, consider the test function*

$$f_7(x) = \left(x^2+1\right)\left(2xe^{x^2+1} + x^3 - x\right)\cosh^2\left(\frac{\pi x}{2}\right)$$

The function has multiple root $\alpha = i$ of multiplicity 4. We choose the initial approximations $x_0 = 1.25i$ for obtaining the root of the function. The results computed by various methods are shown in Table 7.

Table 7. Comparison of the numerical results for Example 7.

Methods	n	$\lvert e_{n-3}\rvert$	$\lvert e_{n-2}\rvert$	$\lvert e_{n-1}\rvert$	COC	CPU-t (s)
GKN-I(a)	4	2.53×10^{-6}	3.79×10^{-35}	4.32×10^{-208}	6.0000	1.1564
GKN-I(b)	4	2.53×10^{-6}	3.92×10^{-35}	5.33×10^{-208}	6.0000	1.1577
GKN-I(c)	4	2.68×10^{-6}	6.07×10^{-35}	8.23×10^{-207}	6.0000	1.1415
GKN-I(d)	4	4.80×10^{-6}	5.34×10^{-33}	1.01×10^{-194}	6.0000	1.0473
GKN-II(a)	4	5.04×10^{-6}	1.82×10^{-33}	4.04×10^{-198}	6.0000	1.0212
GKN-II(b)	4	7.15×10^{-6}	4.23×10^{-32}	1.81×10^{-189}	6.0000	1.1215
GKN-II(c)	4	6.39×10^{-6}	1.51×10^{-32}	2.64×10^{-192}	6.0000	1.2035
GKN-II(d)	4	8.22×10^{-6}	1.41×10^{-31}	8.09×10^{-187}	6.0000	1.1416
NM-I(a)	4	1.08×10^{-6}	6.96×10^{-43}	3.13×10^{-296}	7.0000	0.5787
NM-I(b)	4	9.01×10^{-7}	1.91×10^{-43}	3.71×10^{-300}	7.0000	0.5632
NM-I(c)	4	4.64×10^{-7}	7.44×10^{-46}	2.01×10^{-317}	7.0000	0.5586
NM-II(a)	4	1.09×10^{-6}	7.21×10^{-43}	4.10×10^{-296}	7.0000	0.5478
NM-II(b)	4	9.04×10^{-7}	2.00×10^{-43}	5.10×10^{-300}	7.0000	0.5946
NM-II(c)	4	4.68×10^{-7}	8.21×10^{-46}	4.20×10^{-317}	7.0000	0.5644

From the numerical values of errors we observe the increasing accuracy in the values of successive approximations as the iteration proceed, which points to the stable nature of the methods. Like the existing methods, the convergence behavior of new methods is also consistent. At the stage when stopping criterion $|x_{n+1} - x_n| + |f(x_n)| < 10^{-350}$ has been satisfied we display the value '0' of $|x_{n+1} - x_n|$. From the calculation of computational order of convergence shown in the penultimate column in each table, we verify the theoretical convergence of seventh order. The entries of last column in each table show that the new methods consume less CPU-time during the execution of program than the time taken by existing methods. This confirms the computationally more efficient nature of the new methods. Among the new methods, the better performers (in terms of accuracy) are NM-I(c) and NM-II(c) since they produce approximations of the root with small error. However, this is not true when execution time is taken into account because if one method is better in some situations, then the other is better in some other situation. The main purpose of implementing the new methods for solving different type of nonlinear equations is purely to illustrate the better accuracy of the computed solution and the better computational efficiency than existing techniques. Similar numerical experimentation, performed on a variety of numerical problems of different kinds, confirmed the above remarks to a large extent.

5. Conclusions

In the present work, we have constructed a class of seventh order methods for solving nonlinear equations containing multiple roots. Analysis of the convergence has been carried out, which proves the seventh order of convergence under standard conditions of the function whose zeros we are looking for. Some particular cases of the family are presented. The stability of these cases are tested by means of visual display of the basins of attraction when the methods are applied on different polynomials. The methods are also implemented to solve nonlinear equations including those arising in practical problems. The performance is compared with existing methods in numerical testing. Superiority of proposed methods over the known techniques is endorsed by the numerical tests including the elapsed CPU-time in execution of program.

Author Contributions: Methodology, J.R.S.; Formal analysis, J.R.S.; Investigation, D.K.; Data Curation, D.K.; Conceptualization, C.C.; Writing—review & editing, C.C.

Funding: This research received no external funding.

Conflicts of Interest: The authors declare no conflict of interest.

References

1. Traub, J.F. *Iterative Methods for the Solution of Equations*; Chelsea Publishing Company: New York, NY, USA, 1982.
2. Hoffman, J.D. *Numerical Methods for Engineers and Scientists*; McGraw-Hill Book Company: New York, NY, USA, 1992.
3. Constantinides, A.; Mostoufi, N. *Numerical Methods for Chemical Engineers with MATLAB Applications*; Prentice Hall PTR: Upper Saddle River, NJ, USA, 1999.
4. Zachary, J.L. *Introduction to Scientific Programming: Computational Problem Solving Using Maple and C*; Springer: New York, NY, USA, 2012.
5. Schröder, E. Über unendlich viele Algorithmen zur Auflösung der Gleichungen. *Math. Ann.* **1870**, *2*, 317–365. [CrossRef]
6. Chun C.; Neta, B. A third order modification of Newton's method for multiple roots. *Appl. Math. Comput.* **2009**, *211*, 474–479. [CrossRef]
7. Hansen, E.; Patrick, M. A family of root finding methods. *Numer. Math.* **1977**, *27*, 257–269. [CrossRef]
8. Neta, B. New third order nonlinear solvers for multiple roots. *App. Math. Comput.* **2008**, *202*, 162–170 [CrossRef]
9. Osada, N. An optimal multiple root-finding method of order three. *J. Comput. Appl. Math.* **1994**, *51*, 131–133. [CrossRef]
10. Li, S.G.; Cheng L.Z.; Neta, B. Some fourth-order nonlinear solvers with closed formulae for multiple roots. *Comput. Math. Appl.* **2010**, *59*, 126–135. [CrossRef]
11. Liu, B.; Zhou, X. A new family of fourth-order methods for multiple roots of nonlinear equations. *Non. Anal. Model. Cont.* **2013**, *18*, 143–152.
12. Li, S.; Liao X.; Cheng, L. A new fourth-order iterative method for finding multiple roots of nonlinear equations. *Appl. Math. Comput.* **2009**, 215, 1288–1292.
13. Sharifi, M.; Babajee, D.K.R.; Soleymani, F. Finding the solution of nonlinear equations by a class of optimal methods. *Comput. Math. Appl.* **2012**, *63*, 764–774. [CrossRef]
14. Sharma, J.R.; Sharma, R. Modified Jarratt method for computing multiple roots. *Appl. Math. Comput.* **2010**, *217*, 878–881. [CrossRef]
15. Soleymani, F.; Babajee, D.K.R.; Lotfi, T. On a numerical technique for finding multiple zeros and its dynamics. *J. Egypt. Math. Soc.* **2013**, *21*, 346–353. [CrossRef]
16. Victory, H.D.; Neta, B. A higher order method for multiple zeros of nonlinear functions. *Int. J. Comput. Math.* **1983**, *12*, 329–335. [CrossRef]
17. Zhou, X.; Chen, X.; Song, Y. Constructing higher-order methods for obtaining the multiple roots of nonlinear equations. *J. Comput. Math. Appl.* **2011**, *235*, 4199–4206. [CrossRef]
18. Soleymani, F.; Babajee, D.K.R. Computing multiple zeros using a class of quartically convergent methods. *Alex. Eng. J.* **2013**, *52*, 531–541. [CrossRef]
19. Zhou, X.; Chen, X.; Song, Y. Families of third and fourth order methods for multiple roots of nonlinear equations. *Appl. Math. Comput.* **2013**, *219*, 6030–6038. [CrossRef]
20. Thukral, R. A new family of fourth-order iterative methods for solving nonlinear equations with multiple roots. *J. Numer. Math. Stoch.* **2014**, *6*, 37–44.
21. Hueso, J.L.; Martínez, E.; Teruel, C. Determination of multiple roots of nonlinear equations and applications. *J. Math. Chem.* **2015**, *53*, 880–892. [CrossRef]
22. Geum, Y.H.; Kim, Y.I.; Neta, B. A class of two-point sixth-order multiple-zero finders of modified double-Newton type and their dynamics. *Appl. Math. Comput.* **2015**, *270*, 387–400. [CrossRef]
23. Geum, Y.H.; Kim, Y.I.; Neta, B. A sixth–order family of three–point modified Newton–like multiple–root finders and the dynamics behind their extraneous fixed points. *Appl. Math. Comput.* **2016**, *283*, 120–140. [CrossRef]
24. Ostrowski, A.M. *Solution of Equations and Systems of Equations*; Academic Press: New York, NY, USA, 1966.

25. Vrscay, E.R.; Gilbert, W.J. Extraneous fixed points, basin boundaries and chaotic dynamics for Schröder and König rational iteration functions. *Numer. Math.* **1988**, *52*, 1–16. [CrossRef]
26. Varona, J.L. Graphic and numerical comparison between iterative methods. *Math. Intell.* **2002**, *24*, 37–46. [CrossRef]
27. Scott, M.; Neta B.; Chun, C. Basin attractors for various methods. *Appl. Math. Comput.* **2011**, *218*, 2584–2599. [CrossRef]
28. Weerakoon, S.; Fernando, T.G.I. A variant of Newton's method with accelerated third-order convergence. *Appl. Math. Lett.* **2000**, *13*, 87–93. [CrossRef]

symmetry

MDPI

Article

Nondifferentiable G-Mond–Weir Type Multiobjective Symmetric Fractional Problem and Their Duality Theorems under Generalized Assumptions

Ramu Dubey [1,†], Lakshmi Narayan Mishra [2,3,†] and Luis Manuel Sánchez Ruiz [4,*,†]

1 Department of Mathematics, J C Bose University of Science and Technology, YMCA, Faridabad 121 006, Haryana, India; rdubeyjiya@gmail.com
2 Department of Mathematics, School of Advanced Sciences, Vellore Institute of Technology (VIT) University, Vellore 632 014, Tamil Nadu, India; lakshminarayanmishra04@gmail.com
3 L. 1627 Awadh Puri Colony, Beniganj, Phase III, Opposite Industrial Training Institute (I.T.I.), Ayodhya main road, Faizabad 224 001, Uttar Pradesh, India
4 ETSID- Departamento de Matemática Aplicada & CITG, Universitat Politecnica de Valencia, E-46022 Valencia, Spain
* Correspondence: lmsr@mat.upv.es
† These authors contributed equally to this work.

Received: 29 September 2019; Accepted: 20 October 2019; Published: 1 November 2019

Abstract: In this article, a pair of nondifferentiable second-order symmetric fractional primal-dual model (G-Mond–Weir type model) in vector optimization problem is formulated over arbitrary cones. In addition, we construct a nontrivial numerical example, which helps to understand the existence of such type of functions. Finally, we prove weak, strong and converse duality theorems under aforesaid assumptions.

Keywords: multiobjective; symmetric duality; second-order; nondifferentiable; fractional programming; support function; G_f-bonvexity/G_f-pseudobonvexity

1. Introduction

In multiobjective programming problems, convexity plays an important role in deriving optimality conditions and duality results. To relax convexity assumptions involved in sufficient optimality conditions and duality theorems, various generalized convexity notions have been proposed. Multiobjective type programming problem [1] is common in mathematical modeling of realistic phenomenon with a wide spectrum of utilization. Symmetric duality in nonlinear programming deals with the situation where dual of the dual is primal. Special dual problems of optimization are applied to many types of optimization problems. They are used for the proof of optimality of solutions, for designing and a theoretical justification of optimization algorithms, and for physical or economic interpretation of received solutions. Quite often dual problems introduce new meaning to modeled problems. For many interesting applications and developments of multiobjective optimization, we refer to the work of A. Chinchuluun and P.M. Pardalos [2] and the references cited therein.

In economics, we often come across a case where we have to maximize the efficiency of an economic system resulting optimization problems whose objective function is a ratio. Mangasarian [3] proposed the idea of second-order duality for nonlinear optimization problems. The perusal of second-order duality is important due to the computer simulation benefit over the first-order duality since this one supplies narrow ranges for the cost of the objectives when estimations are applied. Suneja et al. [1] and Kim et al. [4] extended the concept of symmetric duality to arbitrary cones.

Suneja et al. [5] considered a pair of multiobjective second order symmetric dual problems of Mond–Weir type without non-negativity constraints and established duality results under η-bonvexity and η-pseudobonvexity assumptions. Later, Khurana [6] defined cone-pseudoinvex and strongly cone-pseudoinvex functions and proved duality theorems for a pair of Mond–Weir type symmetric dual multiobjective programs over arbitrary cones. For more information on fractional programming, readers are advised to see [7–13].

The purpose of the present work is to study second order multiobjective fractional symmetric duality over arbitrary cones for nondifferentiable G-Mond–Weir type program under G_f-bonvexity/G_f-pseudobonvexity assumptions. The paper is organized as follows. In Section 2, we present some relevant preliminaries. In Section 3, we consider a pair of G-Mond–Weir type nondifferentiable multiobjective second order fractional symmetric dual problems with cone constraints and establish appropriate duality theorems under aforesaid assumptions followed by conclusions.

2. Preliminaries and Definitions

Throughout this paper, R^n stands for the n-dimensional Euclidean space and R^n_+ for its non-negative orthant. Consider the following vector minimization problem:

(MP) $$\text{Minimize } f(x) = \left\{f_1(x), f_2(x), f_3(x), ..., f_k(x)\right\}^T$$

$$\text{Subject to } X^0 = \{x \in X \subset R^n : g_j(x) \leq 0,\ j = 1, 2, ..., m\}$$

where $f = \{f_1, f_2, ..., f_k\} : X \to R^k$ and $g = \{g_1, g_2, ..., g_m\} : X \to R^m$ are differentiable functions defined on X.

Definition 1. *A point $\bar{x} \in X^0$ is said to be an efficient solution of (MP) if there exists no other $x \in X^0$ such that $f_r(x) < f_r(\bar{x})$, for some $r = 1, 2, ..., k$ and $f_i(x) \leq f_i(\bar{x})$, for all $i = 1, 2, ..., k$.*

Definition 2. *The positive polar cone S^* of a cone $S \subseteq R^s$ is defined by*

$$S^* = \{y \in R^s : x^T y \geq 0, \text{ for all } x \in S\}.$$

Let $C_1 \subseteq R^n$ and $C_2 \subseteq R^m$ be closed convex cones with non-empty interiors and S_1 and S_2 be non-empty open sets in R^n and R^m, respectively, such that $C_1 \times C_2 \subseteq S_1 \times S_2$. Suppose $f = (f_1, f_2, ..., f_k) : S_1 \times S_2 \to R^k$ is a vector- valued differentiable function.

Definition 3. *The function f is said to be invex at $u \in S_1$ (with respect to η, where $\eta : S_1 \times S_2 \to R^n$), if $\forall\, x \in S_1$ and for fixed $v \in S_2$, we have*

$$f_i(x, v) - f_i(u, v) \geq \eta^T(x, u) \nabla_x f_i(u, v), \text{ for all } i = 1, 2, ..., k,$$

If the above inequality sign changes to $\leq$, then f is called incave at $u \in S_1$ with respect to η.

Definition 4. *The function f is said to be pseudoinvex at $u \in S_1$ (with respect to η, where $\eta : S_1 \times S_2 \to R^n$), if $\forall\, x \in S_1$ and for fixed $v \in S_2$, we have*

$$\eta^T(x, u) \nabla_x f_i(u, v) \geq 0 \Rightarrow f_i(x, v) - f_i(u, v) \geq 0, \text{ for all } i = 1, 2, ..., k.$$

If the above inequality sign changes to $\leq$, then f is called pseudoincave at $u \in X$ with respect to η.

Definition 5. *The function f is said to be G_f-invex at $u \in S_1$ (with respect to η), if there exists a differentiable function $G_f = (G_{f_1}, G_{f_2}, ..., G_{f_k}) : R \to R^k$ such that each component $G_{f_i} : I_{f_i}(S_1 \times S_2) \to R$, where $I_{f_i}(S_1 \times S_2)$, $i = 1,2,3,...,k$ is the range of f_i, is strictly increasing on its domain and $\eta : S_1 \times S_2 \to R^n$, so that $\forall\ x \in S_1$, for fixed $v \in S_2$, we have*

$$G_{f_i}(f_i(x,v)) - G_{f_i}(f_i(u,v)) \geq \eta^T(x,u) G'_{f_i}(f_i(u,v)) \nabla_x f_i(u,v), \text{ for all } i = 1,2,...,k,$$

If the above inequality sign changes to $\leq$, then f is called G_f-incave at $u \in S_1$ with respect to η.

Definition 6. *The function f is said to be G_f-pseudoinvex at $u \in S_1$ (with respect to η), if there exists a differentiable function $G_f = (G_{f_1}, G_{f_2}, ..., G_{f_k}) : R \to R^k$ such that each component $G_{f_i} : I_{f_i}(S_1 \times S_2) \to R$, where $I_{f_i}(S_1 \times S_2)$, $i = 1,2,3,...,k$ is the range of f_i, is strictly increasing on its domain and $\eta : S_1 \times S_2 \to R^n$, so that $\forall\ x \in S_1$, for fixed $v \in S_2$, we have*

$$\eta^T(x,u) G'_{f_i}(f_i(u,v)) \nabla_x f_i(u,v) \geq 0 \Rightarrow G_{f_i}(f_i(x,v)) - G_{f_i}(f_i(u,v)) \geq 0, \text{ for all } i = 1,2,...,k.$$

If the above inequality sign changes to $\leq$, then f is called G_f-pseudoincave at $u \in X$ with respect to η.

Definition 7. *The function f is said to be G_f-bonvex at $u \in S_1$ (with respect to η), if there exists a differentiable function $G_f = (G_{f_1}, G_{f_2}, ..., G_{f_k}) : R \to R^k$ such that each component $G_{f_i} : I_{f_i}(S_1 \times S_2) \to R$, where $I_{f_i}(S_1 \times S_2)$, $i = 1,2,3,...,k$ is the range of f_i, is strictly increasing on its domain and $\eta : S_1 \times S_2 \to R^n$, so that $\forall\ x \in S_1$, for fixed $v \in S_2$ and $p_i \in R^n$, we have*

$$\begin{aligned} G_{f_i}(f_i(x,v)) - G_{f_i}(f_i(u,v)) &\geq \eta^T(x,u)[G'_{f_i}(f_i(u,v)) \nabla_x f_i(u,v) + \{G''_{f_i}(f_i(u,v)) \nabla_x f_i(u,v) (\nabla_x f_i(u,v))^T \\ &+ G'_{f_i}(f_i(u,v)) \nabla_x f_i(u,v)\} p_i] - \frac{1}{2} p_i^T [G''_{f_i}(f_i(u,v)) \nabla_x f_i(u,v) (\nabla_x f_i(u,v))^T \\ &+ G'_{f_i}(f_i(u,v)) \nabla_x f_i(u,v)] p_i, \text{ for all } i = 1,2,...,k. \end{aligned}$$

If the above inequality sign changes to $\leq$, then f is called G_f-boncave at $u \in S_1$ with respect to η.

Definition 8. *The function f is said to be G_f-pseudobonvex at $u \in S_1$ (with respect to η), if there exists a differentiable function $G_f = (G_{f_1}, G_{f_2}, ..., G_{f_k}) : R \to R^k$ such that each component $G_{f_i} : I_{f_i}(S_1 \times S_2) \to R$, where $I_{f_i}(S_1 \times S_2)$, $i = 1,2,3,...,k$ is the range of f_i, is strictly increasing on its domain and $\eta : S_1 \times S_2 \to R^n$, so that $\forall\ x \in S_1$, for fixed $v \in S_2$ and $p_i \in R^n$, $\eta^T(x,u)[G'_{f_i}(f_i(u,v)) \nabla_x f_i(u,v) + \{G''_{f_i}(f_i(u,v)) \nabla_x f_i(u,v) (\nabla_x f_i(u,v))^T + G'_{f_i}(f_i(u,v))$*

$$\begin{aligned} \nabla_{xx} f_i(u,v)\} p_i] \geq 0 \Rightarrow G_{f_i}(f_i(x,v)) - G_{f_i}(f_i(u,v)) &+ \frac{1}{2} p_i^T [G''_{f_i}(f_i(u,v)) \nabla_x f_i(u,v) (\nabla_x f_i(u,v))^T \\ &+ G'_{f_i}(f_i(u,v)) \nabla_{xx} f_i(u,v)] p_i \geq 0, \text{ for all } i = 1,2,...,k. \end{aligned}$$

If the above inequality sign changes to $\leq$, then f is called G_f-pseudoboncave at $u \in S_1$ with respect to η.

We now give an example of G_f-bonvexity with respect to η, but not η-bonvex.

Example 1. *Let $k = 4, n = 1, S_1 = S_2 = \left[-\frac{\pi}{6}, \frac{\pi}{6} \right]$, $C_1 = C_2 = \left[\frac{-\pi}{6}, \frac{\pi}{6} \right]$.*

Let $f : \left[-\frac{\pi}{6}, \frac{\pi}{6} \right] \times \left[-\frac{\pi}{6}, \frac{\pi}{6} \right] \to R^4$ be defined as

$$f(x,y) = \{f_1(x,y), f_2(x,y), f_3(x,y), f_4(x,y)\},$$

where $f_1(x,y) = e^y$, $f_2(x,y) = xe^y$, $f_3(x,y) = x^2 sin^2 y$, $f_4(x,y) = y^2$ *and* $G_f = \{G_{f_1}, G_{f_2}, G_{f_3}, G_{f_4}\} : R \to R^4$ *be defined as:*

$$G_{f_1}(t) = t,\ G_{f_2}(t) = t^4,\ G_{f_3}(t) = t,\ G_{f_4}(t) = t^2.$$

Let $\eta : \left[-\frac{\pi}{6}, \frac{\pi}{6}\right] \times \left[-\frac{\pi}{6}, \frac{\pi}{6}\right] \to R$ be given as:

$$\eta(x,u) = xu.$$

To show that f is G_f-bonvex at $u = 0$ with respect to η, we have to claim that

$$\pi_i = G_{f_i}(f_i(x,v)) - G_{f_i}(f_i(u,v)) - \eta^T(x,u)[G'_{f_i}(f_i(u,v))\nabla_x f_i(u,v) + \{G''_{f_i}(f_i(u,v))\nabla_x f_i(u,v)$$
$$(\nabla_x f_i(u,v))^T\} + G'_{f_i}(f_i(u,v))\nabla_{xx} f_i(u,v)\}p_i + \frac{1}{2}p_i^T[G''_{f_i}(f_i(u,v))\nabla_x f_i(u,v)$$
$$(\nabla_x f_i(u,v))^T + G'_{f_i}(f_i(u,v))\nabla_{xx} f_i(u,v)]p_i \geq 0,\ \ i = 1,2,3,4.$$

Putting the values of f_1, f_2, f_3, f_4, G_{f_1}, G_{f_2}, G_{f_3}, G_{f_4} and $u = 0$ in the above expressions, we have

$$\pi_1 = 0,\ \forall\ p,\ \forall\ x,\ v \in \left[-\frac{\pi}{6}, \frac{\pi}{6}\right],$$

$$\pi_2 = x^4 e^{4v},\ \forall\ p,\ \forall\ x,\ v \in \left[-\frac{\pi}{6}, \frac{\pi}{6}\right],$$

$$\pi_3 = x^2 sin^2 v,\ \forall\ p,\ \forall\ x,\ v \in \left[-\frac{\pi}{6}, \frac{\pi}{6}\right],$$

and

$$\pi_4 = 0,\ \forall\ p,\ \forall\ x,\ v \in \left[-\frac{\pi}{6}, \frac{\pi}{6}\right].$$

Hence, $\pi_1 \geq 0,\ \pi_2 \geq 0$ (from Figure 1), $\pi_3 \geq 0$ (in Figure 2) and $\pi_4 \geq 0,\ \forall\ x,\ v \in \left[-\frac{\pi}{6}, \frac{\pi}{6}\right]$ and $\forall\ p$.

Therefore, f is G_f-bonvex at $u = 0$ with respect to η and p.

Next, we claim that function f is not η-bonvex. For this, it is sufficient to prove that at least one $f_i's$ is not η-bonvex.

Let

$$\xi = f_3(x) - f_3(u) - \eta^T(x,u)[\nabla_x f_3(u) - \nabla_{xx} f_3(u)p_3] + \frac{1}{2}p_3^T[\nabla_{xx} f_3(u)]p_3$$

or

$$\xi = xe^v - ue^v - 0,\ \forall\ p,\ \forall\ x,\ v \in \left[-\frac{\pi}{6}, \frac{\pi}{6}\right],$$

$$\xi = xe^v \ \text{ at } u = 0 \in \left[-\frac{\pi}{6}, \frac{\pi}{6}\right].$$

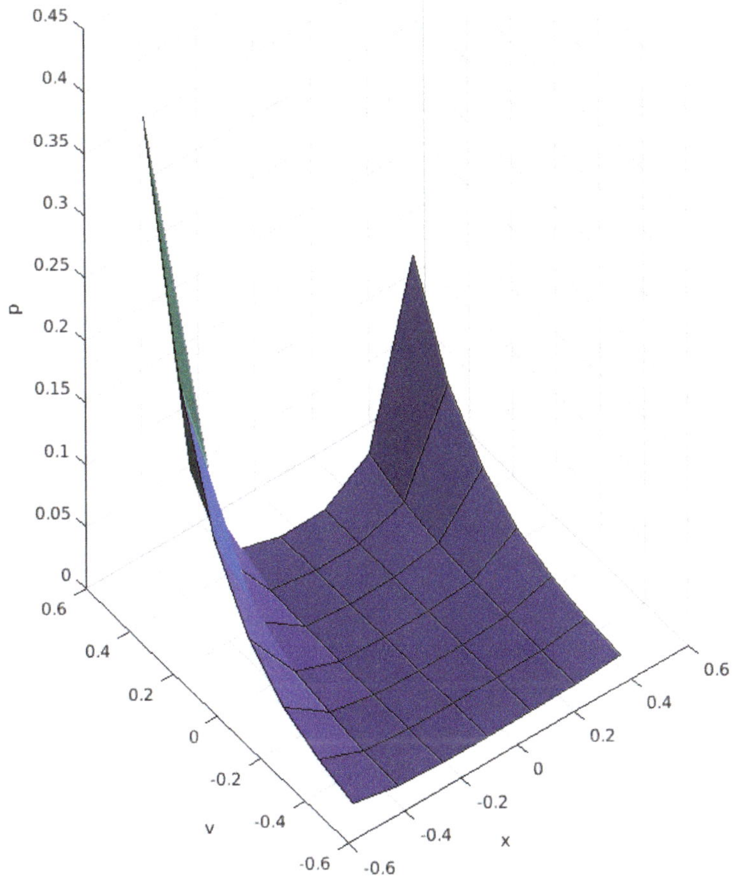

Figure 1. The function $\pi_2 = x^4e^{4v}$, $\forall\, p$, $\forall\, x,\, v \in \left[-\frac{\pi}{6}, \frac{\pi}{6}\right]$ is non-negative.

It follows that $\xi \ngeq 0$, $u \in \left[-\frac{\pi}{6}, \frac{\pi}{6}\right]$ and $\forall\, p$ (in Figure 3). Therefore, f_3 is not η-bonvex at $u = 0$ with respect to p_3. Hence, $f = (f_1,\, f_2,\, f_3,\, f_4)$ is not η-bonvex at $u = 0$ with respect to p.

Definition 9. *Let C be a compact convex set in R^n. The support function of C is defined by*

$$s\,(x|C) = max\{x^T y : y \in C\}.$$

The subdifferential of $s(x|C)$ is given by

$$\partial s(x|C) = \{z \in C : z^T x = s(x|C)\}.$$

For any convex set $S \subset R^n$, the normal cone to S at a point $x \in S$ is defined by

$$N_S(x) = \{y \in R^n : y^T(z - x) \leq 0 \text{ for all } z \in S\}.$$

It is readily verified that for a compact convex set S, y is in $N_S(x)$ if and only if

$$s(y|S) = x^T y.$$

Suppose that $S_1 \subseteq R^n$ and $S_2 \subseteq R^m$ are open sets such that $C_1 \times C_2 \subset S_1 \times S_2$.

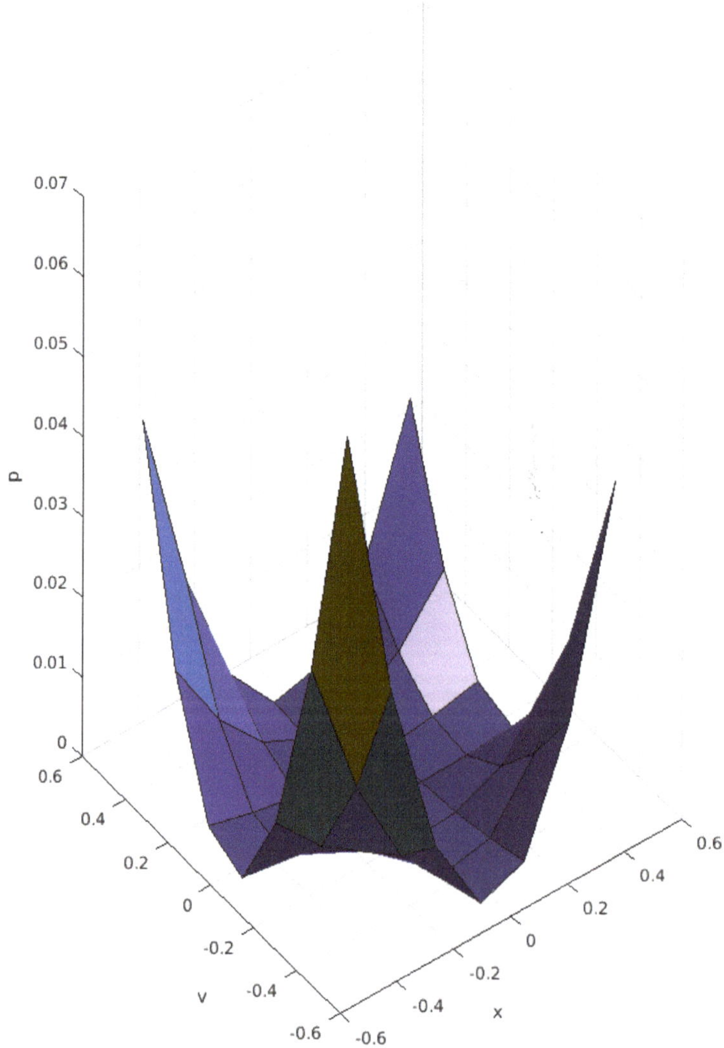

Figure 2. The function $\pi_3 = x^2 sin^2 v$, $\forall\ p$, $\forall\ x$, $v \in \left[-\frac{\pi}{6}, \frac{\pi}{6}\right]$ is non-negative.

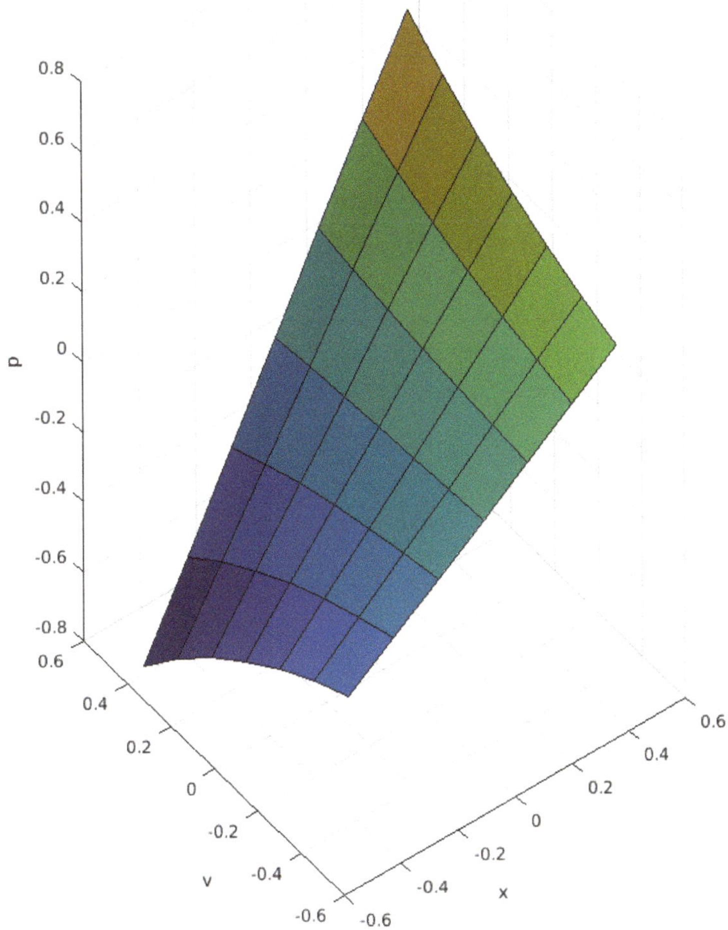

Figure 3. The function $\xi = xe^{v}$ becomes negative at some $x, v \in \left[-\frac{\pi}{6}, \frac{\pi}{6}\right]$.

3. Second-Order Nondifferentiable Multiobjective Symmetric Fractional Programming Problem Over Arbitrary Cones

Now, we consider the following pair of a nondifferentiable multiobjective second-order fractional symmetric dual program over arbitrary cones

(GMFP) Minimize $U(x,y,z,r,p) = (U_1(x,y,z_1,r_1,p_1), U_2(x,y,z_2,r_2,p_2), \ldots, U_k(x,y,z_k,r_k,p_k))^T$

subject to

$$-\sum_{i=1}^{k}\lambda_i[G'_{f_i}(f_i(x,y)) \nabla_y f_i(x,y) - z_i + \{G''_{f_i}(f_i(x,y)) \nabla_y f_i(x,y)(\nabla_y f_i(x,y))^T + G'_{f_i}(f_i(x,y))$$

$$\nabla_{yy} f_i(x,y)\}p_i - U_i(x,y,p_i)\{G'_{g_i}(g_i(x,y)) \nabla_y g_i(x,y) + r_i + \{G''_{g_i}(g_i(x,y)) \nabla_y g_i(x,y)\}$$
$$(\nabla_y g_i(x,y))^T + G'_{g_i}(g_i(x,y)) \nabla_{yy} g_i(x,y)\}p_i] \in C_2^*,$$

$$y^T\Bigg[\sum_{i=1}^{k}\lambda_i\Big\{G'_{f_i}(f_i(x,y)) \nabla_y f_i(x,y) - z_i + \{G''_{f_i}(f_i(x,y)) \nabla_y f_i(x,y)(\nabla_y f_i(x,y))^T + G'_{f_i}(f_i(x,y))$$
$$\nabla_{yy} f_i(x,y)\}p_i - U_i(x,y,p_i)\{G'_{g_i}(g_i(x,y)) \nabla_y g_i(x,y) + r_i + \{G''_{g_i}(g_i(x,y)) \nabla_y g_i(x,y)\}$$
$$(\nabla_y g_i(x,y))^T + G'_{g_i}(g_i(x,y)) \nabla_{yy} g_i(x,y)\}p_i\Big\}\Bigg] \geq 0,$$

$$x \in C_1,\ \lambda > 0,\ z_i \in D_i,\ r_i \in F_i,\ i = 1,2,...,k.$$

(GMFD) Maximize $T(u,v,w,t,q) = (T_1(u,v,w_1,t_1,q_1),(T_2(u,v,w_2,t_2,q_2),\ldots,T_k(u,v,w_k,t_k,q_k))^T$

subject to

$$\sum_{i=1}^{k}\lambda_i[G'_{f_i}(f_i(u,v)) \nabla_x f_i(u,v) + w_i + G''_{f_i}(f_i(u,v)) \nabla_x f_i(u,v)(\nabla_x f_i(u,v))^T + G'_{f_i}(f_i(u,v))$$
$$\nabla_{xx} f_i(u,v)]q_i - T_i(u,v,q_i)[G'_{g_i}(g_i(u,v)) \nabla_x g_i(u,v) - t_i + \{G''_{g_i}(g_i(u,v)) \nabla_x g_i(u,v)$$
$$(\nabla_x g_i(u,v))^T + G'_{g_i}(g_i(u,v)) \nabla_{xx} g_i(u,v)\}q_i] \in C_1^*,$$

$$u^T\Bigg[\sum_{i=1}^{k}\lambda_i\Big\{G'_{f_i}(f_i(u,v)) \nabla_x f_i(u,v) - w_i + G''_{f_i}(f_i(u,v)) \nabla_x f_i(u,v)(\nabla_x f_i(u,v))^T + G'_{f_i}(f_i(u,v))$$
$$\nabla_{xx} f_i(u,v)]q_i - T_i(u,v,q_i)[G'_{g_i}(g_i(u,v)) \nabla_x g_i(u,v) - t_i + \{G''_{g_i}(g_i(u,v))$$
$$\nabla_x g_i(u,v)(\nabla_x g_i(u,v))^T + G'_{g_i}(g_i(u,v)) \nabla_{xx} g_i(u,v)\}q_i\Big\}\Bigg] \leq 0,$$

$$v \in C_1,\ \lambda > 0,\ w_i \in Q_i,\ t_i \in E_i, i = 1,2,...,k.$$

where

$$U_i(x,y,z_i,r_i,p_i) = \frac{G_{f_i}(f_i(x,y)) + s(x|Q_i) - y^T z_i - \frac{1}{2}p_i^T[G''_{f_i}(f_i(x,y)) \nabla_y f_i(x,y)(\nabla_y f_i(x,y))^T + G'_{f_i}(f_i(x,y)) \nabla_{yy} f_i(x,y)]p_i}{G_{g_i}(g_i(x,y)) - s(x|E_i) + y^T r_i - \frac{1}{2}p_i^T[G''_{g_i}(g_i(x,y)) \nabla_y g_i(x,y)(\nabla_y g_i(x,y))^T + G'_{g_i}(g_i(x,y)) \nabla_{yy} g_i(x,y)]p_i}$$

and

$$T_i(u,v,w_i,t_i,q_i) = \frac{G_{f_i}(f_i(u,v)) - s(v|D_i) + u^T w_i - \frac{1}{2}q_i^T[G''_{f_i}(f_i(u,v)) \nabla_x f_i(u,v)(\nabla_x f_i(u,v))^T + G'_{f_i}(f_i(u,v)) \nabla_{xx} f_i(u,v)]q_i}{G_{g_i}(g_i(u,v)) + s(v|F_i) - u^T t_i - \frac{1}{2}q_i^T[G''_{g_i}(g_i(u,v)) \nabla_x g_i(u,v)(\nabla_x g_i(u,v))^T + G'_{g_i}(g_i(u,v)) \nabla_{xx} g_i(u,v)]q_i};$$

and

$S_1 \subseteq R^n$ and $S_2 \subseteq R^m$; C_1 and C_2 are arbitrary cones in R^n and R^m, respectively, such that $C_1 \times C_2 \subseteq S_1 \times S_2$; $f_i : S_1 \times S_2 \to R$ and $g_i : S_1 \times S_2 \to R$ are differentiable functions; $G_{f_i} : I_{f_i} \to R$ and $G_{g_i} : I_{g_i} \to R$ are differentiable strictly increasing functions on their domains; Q_i, E_i are compact convex sets in R^n; and D_i, F_i are compact convex sets in R^m, $i = 1,2,3,...,k$. C_1^* and C_2^* are positive polar cones of C_1 and C_2, respectively. It is assumed that in the feasible regions, the numerators are nonnegative and denominators are positive. p_i and q_i are vectors in R^m and R^n, respectively, $\lambda \in R^k$.

Equivalently, the above problem is reduced in the given form:

(EGMFP) Min $R(x,y,z,r,p) = (R_1(x,y,z_1,r_1,p_1), R_2(x,y,z_2,r_2,p_2), \ldots, R_k(x,y,z_k,r_k,p_k))$

subject to

$$G_{f_i}(f_i(x,y)) + s(x|Q_i) - y^T z_i - \tfrac{1}{2}p_i^T[G''_{f_i}(f_i(x,y)) \bigtriangledown_y f_i(x,y)(\bigtriangledown_y f_i(x,y))^T + G'_{f_i}(f_i(x,y))$$
$$\bigtriangledown_{yy} f_i(x,y)]p_i - R_i(x,y,z_i,r_i,p_i)[G_{g_i}(g_i(u,v)) - s(x|E_i) + y^T r_i - \frac{1}{2}q_i^T[G''_{g_i}(g_i(u,v))$$
$$\bigtriangledown_x g_i(u,v)(\bigtriangledown_x g_i(u,v))^T + G'_{g_i}(g_i(u,v)) \bigtriangledown_{xx} g_i(u,v)]q_i] = 0,\ i = 1,2,\ldots,k, \tag{1}$$

$$-\sum_{i=1}^{k}\lambda_i[G'_{f_i}(f_i(x,y)) \bigtriangledown_y f_i(x,y) - z_i + [G''_{f_i}(f_i(x,y)) \bigtriangledown_y f_i(x,y)(\bigtriangledown_y f_i(x,y))^T +$$
$$G'_{f_i}(f_i(x,y)) \bigtriangledown_{yy} f_i(x,y)]p_i - R_i(x,y,z_i,r_i,p_i)\{G'_{g_i}(g_i(x,y)) + r_i \bigtriangledown_y g_i(x,y) + (G''_{g_i}(g_i(x,y))$$
$$\bigtriangledown_y g_i(x,y)(\bigtriangledown_y g_i(x,y))^T + G'_{g_i}(g_i(x,y)) \bigtriangledown_{yy} g_i(x,y))p_i\}] \in C_2^*, \tag{2}$$

$$y^T\sum_{i=1}^{k}\lambda_i[G'_{f_i}(f_i(x,y)) \bigtriangledown_y f_i(x,y) - z_i + [G''_{f_i}(f_i(x,y)) \bigtriangledown_y f_i(x,y)(\bigtriangledown_y f_i(x,y))^T +$$
$$G'_{f_i}(f_i(x,y)) \bigtriangledown_{yy} f_i(x,y)]p_i - R_i(x,y,z_i,r_i,p_i)\{G'_{g_i}(g_i(x,y)) \bigtriangledown_y g_i(x,y) + r_i + (G''_{g_i}(g_i(x,y))$$
$$\bigtriangledown_y g_i(x,y)(\bigtriangledown_y g_i(x,y))^T + G'_{g_i}(g_i(x,y)) \bigtriangledown_{yy} g_i(x,y))p_i\}] \geq 0, \tag{3}$$

$$x \in C_1,\ \lambda > 0,\ z_i \in D_i,\ r_i \in F_i,\ i = 1,2,...,k. \tag{4}$$

(EGMFD) Maximize $S(u,v,w,t,q) = [S_1(u,v,w_1,t_1,q_1), S_2(u,v,w_2,t_2,q_2), \ldots, S_k(u,v,w_k,t_k,q_k)]$

subject to

$$G_{f_i'}(f_i(u,v)) - s(v|D_i) + u^T w_i - \tfrac{1}{2}q_i^T\{G''_{f_i}(f_i(u,v)) \bigtriangledown_x f_i(u,v)(\bigtriangledown_x f_i(u,v))^T + G'_{f_i}(f_i(u,v))$$
$$\bigtriangledown_{xx} f_i(u,v)\}q_i - S_i(u,v,w_i,t_i,q_i)[G'_{g_i}(g_i(u,v)) + s(v|F_i) - u^T t_i - \tfrac{1}{2}q_i^T\{G''_{g_i}(g_i(u,v)) \bigtriangledown_x g_i(u,v)$$
$$(\bigtriangledown_x g_i(u,v))^T + G_{g_i'}(g_i(u,v)) \bigtriangledown_{xx} g_i(u,v)q_i\}] = 0,\ i = 1,2,\ldots,k. \tag{5}$$

$$\sum_{i=1}^{k}\lambda_i[G'_{f_i}(f_i(u,v)) \bigtriangledown_x f_i(u,v) + w_i + \{G''_{f_i}(f_i(u,v)) \bigtriangledown_x f_i(u,v)(\bigtriangledown_x f_i(u,v))^T +$$
$$G'_{f_i}(f_i(u,v)) \bigtriangledown_{xx} f_i(u,v)\}q_i - T_i(u,v,w_i,t_i,q_i)\{G'_{g_i}(g_i(u,v)) \bigtriangledown_x g_i(u,v) - t_i + G''_{g_i}(g_i(u,v))$$
$$\bigtriangledown_x g_i(u,v)(\bigtriangledown_x g_i(u,v))^T + G'_{g_i}(g_i(u,v)) \bigtriangledown_{xx} g_i(u,v)\}q_i] \in C_1^*, \tag{6}$$

$$u^T\sum_{i=1}^{k}\lambda_i[G'_{f_i}(f_i(u,v))\nabla_x f_i(u,v)+w_i+\{G''_{f_i}(f_i(u,v))\nabla_x f_i(u,v)(\nabla_x f_i(u,v))^T+$$
$$G'_{f_i}(f_i(u,v))\nabla_{xx} f_i(u,v)\}q_i-T_i(u,v,w_i,t_i,q_i)\{G'_{g_i}(g_i(u,v))\nabla_x g_i(u,v)-t_i+G''_{g_i}(g_i(u,v))$$
$$\nabla_x g_i(u,v)(\nabla_x g_i(u,v))^T+G'_{g_i}(g_i(u,v))\nabla_{xx} g_i(u,v)\}q_i]\leq 0, \tag{7}$$

$$v\in C_1,\ \lambda>0,\ w_i\in Bi,\ t_i\in E_i, i=1,2,...,k. \tag{8}$$

Let Z^0 and W^0 be the sets of feasible solutions of (EGMFP) and (EGMFD), respectively. Next, we prove duality theorems for (EGMFP) and (EGMFD), which equally apply to (GMFP) and (GMFD), respectively. Let $z=(z_1,z_2,...,z_k)$, $r=(r_1,r_2,...,r_k)$, $w=(w_1,w_2,...,w_k)$, $t=(t_1,t_2,...,t_k)$ and $\lambda=(\lambda_1,\lambda_2,...,\lambda_k)$.

Theorem 1. *(Weak Duality). Let $(x,y,R,z,r,\lambda,p)\in Z^0$ and $(u,v,S,w,t,\lambda,q)\in W^0$. Assume that for $i=1,2,3,\ldots,k$:*

(i) $f_i(.,v)$ is G_{f_i}- bonvex and $(.)^T w_i$ is invex at u for fixed v with respective to η_1.
(ii) $g_i(.,v)$ is a G_{g_i}- boncave and $(.)^T t_i$ is invex at u for fixed v with respective to η_1.
(iii) $f_i(x,.)$ is a G_{f_i}- boncave and $(.)^T z_i$ is invex at y for fixed x with respective to η_2.
(iv) $g_i(x,.)$ is a G_{g_i}- bonvex and $(.)^T r_i$ is invex at y for fixed x with respective to η_2.
(v) $\eta_1(x,u)+u\in C_1$ and $\eta_2(v,y)+y\in C_2$.
(vi) $G_{g_i}((x,v))+v^T r_i-x^T t_i>0$.

Then, the following can not hold simultaneously:

$$R_i\leq S_i, \textit{for all } i=1,2,3,...,k \textit{ and } R_j<S_j, \textit{for some } j=1,2,3,...,m.$$

Proof. From Assumption (v) and Equation (6), we get

$$(\eta_1(x,u)+u)^T\sum_{i=1}^{k}\lambda_i[G'_{f_i}(f_i(u,v))\nabla_x f_i(u,v)+w_i+[G''_{f_i}(f_i(u,v))\nabla_x f_i(u,v)(\nabla_x f_i(u,v))^T$$
$$+G'_{f_i}(f_i(u,v))\nabla_{xx} f_i(u,v)]q_i-T_i(u,v,q_i)\{G'_{g_i}(g_i(u,v))\nabla_x g_i(u,v)-t_i$$
$$+(G''_{g_i}(g_i(u,v))\nabla_x g_i(u,v)(\nabla_x g_i(u,v))^T+G'_{g_i}(g_i(u,v))\nabla_{xx} g_i(u,v))q_i\}]\geq 0. \tag{9}$$

Using Equations (7) and (9), we obtain,

$$\eta_1^T(x,u)\left[\sum_{i=1}^{k}\lambda_i(G'_{f_i}(f_i(u,v))\nabla_x f_i(u,v)+w_i+[G''_{f_i}(f_i(u,v))\nabla_x f_i(u,v)(\nabla_x f_i(u,v))^T\right.$$
$$+G'_{f_i}(f_i(u,v))\nabla_{xx} f_i(u,v)]q_i-T_i(u,v,q_i)\{G'_{g_i}(g_i(u,v))\nabla_x g_i(u,v)-t_i$$
$$\left.+(G''_{g_i}(g_i(u,v))\nabla_x g_i(u,v)(\nabla_x g_i(u,v))^T+G'_{g_i}(g_i(u,v))\nabla_{xx} g_i(u,v))q_i\})\right]\geq 0. \tag{10}$$

From Assumption (i), we have

$$G_{f_i}(f_i(x,v))-G_{f_i}(f_i(u,v))\geq n_1^T(x,u)G'_{f_i}(f_i(u,v))\nabla_x f_i(u,v)+[G''_{f_i}(f_i(u,v))\nabla_x f_i(u,v)$$
$$(\nabla_x f_i(u,v))^T+G'_{f_i}(f_i(u,v))\nabla_{xx} f_i(u,v)]p_i-\tfrac{1}{2}p_i{}^T[G''_{f_i}(f_i(u,v))\nabla_x f_i(u,v)$$
$$(\nabla_x f_i(u,v))^T+G'_{f_i}(f_i(u,v))\nabla_{xx} f_i(u,v)p_i],\ i=1,2,...,k. \tag{11}$$

and

$$x^T w_i - u^T w_i \geq \eta_1^T(x,u) w_i, \;\; i = 1,2,...,k. \tag{12}$$

Since $\lambda > 0$ and combining above inequalities, it follows that

$$\sum_{i=1}^{k} [G_{f_i}(f_i(x,v)) + x^T w_i - G_{f_i}(f_i(u,v)) - u^T w_i] \geq n_1^T(x,u) \sum_{i=1}^{k} \lambda_i [G'_{f_i}(f_i(u,v)) \bigtriangledown_x f_i(u,v) + w_i$$
$$+ [G''_{f_i}(f_i(u,v)) \bigtriangledown_x f_i(u,v)(\bigtriangledown_x f_i(u,v))^T + G'_{f_i}(f_i(u,v)) \bigtriangledown_{xx} f_i(u,v)]p_i$$
$$-\frac{1}{2} p_i{}^T \{G''_{f_i}(f_i(u,v)) \bigtriangledown_x f_i(u,v)(\bigtriangledown_x f_i(u,v))^T + G'_{f_i}(f_i(u,v)) \bigtriangledown_{xx} f_i(u,v)\} p_i]. \tag{13}$$

Similarly, from Assumption (ii), we get

$$-G_{g_i}(g_i(x,v)) + G_{g_i}(g_i(u,v)) \geq -\eta_1^T(x,u)[G'_{g_i}(g_i(u,v)) \bigtriangledown_x g_i(u,v) + [G''_{g_i}(g_i(u,v))$$
$$\bigtriangledown_x g_i(u,v)(\bigtriangledown_x g_i(u,v))^T + G'_{g_i}(g_i(u,v)) \bigtriangledown_{xx} g_i(u,v)]p_i + \tfrac{1}{2} p_i{}^T \{G''_{g_i}(g_i(u,v)) \bigtriangledown_x g_i(u,v)$$
$$(\bigtriangledown_x g_i(u,v))^T + G'_{g_i}(g_i(u,v)) \bigtriangledown_x g_i(u,v)\} p_i], \;\; i = 1,2,...,k, \tag{14}$$

and

$$x^T t_i - u^T t_i \geq \eta_1^T(x,u) t_i, \; i = 1,2,...,k. \tag{15}$$

Multiplying by $\lambda_i T_i$ in above inequalities and taking summation over $i = 1,2,3,...,k$, it follows that

$$\sum_{i=1}^{\hat{k}} \lambda_i T_i [-G_{g_i}(g_i(x,v)) + x^T t_i + G_{g_i}(g_i(u,v)) - u^T t_i] \geq -\eta_1^T(x,u) \sum_{i=1}^{\hat{k}} \lambda_i T_i [[G'_{g_i}(g_i(u,v)) - t_i +$$
$$\bigtriangledown_x g_i(u,v)[G''_{g_i}(g_i(u,v)) \bigtriangledown_x g_i(u,v)(\bigtriangledown_x g_i(u,v))^T + G'_{g_i}(g_i(u,v)) \bigtriangledown_{xx} g_i(u,v)]p_i$$
$$-\frac{1}{2} \bar{p}_i{}^T \{G''_{g_i}(g_i(u,v)) \bigtriangledown_x g_i(u,v)(\bigtriangledown_x g_i(u,v))^T + G'_{g_i}(g_i(u,v)) \bigtriangledown_x g_i(u,v) p_i\}]. \tag{16}$$

Adding the inequalities in Equations (13) and (16), we get

$$\sum_{i=1}^{\hat{k}} \lambda_i [G_{f_i}(f_i(x,v)) - G_{f_i}(f_i(u,v)) - T_i(G_{g_i}(g_i(x,v)) - G_{g_i}(g_i(u,v)))]$$
$$\geq -\sum_{i=1}^{k} \frac{\lambda_i q_i^T}{2} [G'_{f_i}(f_i(u,v)) \bigtriangledown_{xx} f_i(u,v) + G''_{f_i}(f_i(u,v)) \bigtriangledown_x f_i(u,v)(\bigtriangledown_x f_i(u,v))^T$$
$$-T_i\{G'_{g_i}(g_i(u,v)) \bigtriangledown_{xx} g_i(u,v) + G''_{g_i}(g_i(u,v)) \bigtriangledown_x g_i(u,v)(\bigtriangledown_x g_i(u,v))^T\} g_i]. \tag{17}$$

Since $v^T r_i \leq s(v|F_i)$, from Equations (17) and (5), we get

$$\sum_{i=1}^{k} \lambda_i [G_{f_i}(f_i(x,v)) + x^T w_i - s(v|D_i) + T_i(x^T t_i - v^T r_i - G_{g_i}(g_i(x,v)))] \geq 0. \tag{18}$$

Similarly, using Hypotheses (iii)–(v) and the primal constraints in Equations (1)–(4), we have

$$\sum_{i=1}^{k} \lambda_i [-G_{f_i}(f_i(x,v)) + v^T z_i - s(x|Q_i) + R_i(-x^T t_i + v^T r_i + G_{g_i}(g_i(x,v)))] \geq 0. \tag{19}$$

On adding the inequalities in Equations (18) and (19), we get

$$\sum_{i=1}^{k}\lambda_i[v^Tz_i - s(v|D_i) + x^Tw_i - G_{f_i}(f_i(x,v)) - s(x|Q_i)$$

$$+(R_i - S_i)(-x^Tt_i + v^Tr_i + G_{g_i}(g_i(x,v)))] \geq 0. \tag{20}$$

Since $\lambda_i > 0$, $v^Tz_i - s(v|Di) + x^Tw_i - s(x|Ci) \leq 0, i = 1,2,3,...,k$, it yields

$$\sum_{i=1}^{k}\lambda_i(R_i - T_i)(G_{g_i}(g_i(x,v)) + v^Tr_i - x^Tt_i) \geq 0.$$

From Assumption (vi), we have, $G_{g_i}((x,v)) + v^Tr_i - x^Tt_i >,\ \ i = 1,2,3,...,k$. Since $\lambda > 0$, it follows that $R \nleq S$, hence the result. □

Remark 1. *Since every convex function is pseudoconvex, the above weak duality theorem for the symmetric dual pair (EGMFP) and (EGMFD) can also be obtained under pseudobonvexity assumptions.*

Theorem 2. *(Weak Duality). Let $(x,y,R,z,r,\lambda,p) \in Z^0$ and $(u,v,S,w,t,\lambda,q) \in W^0$. Assume that for $i = 1,2,3,\ldots,k$:*

(i) $f_i(.,v)$ is G_{f_i}- pseudobonvex and $(.)^Tw_i$ is pseudoinvex at u for fixed v with respective to η_1.
(ii) $g_i(.,v)$ is a G_{g_i}- pseudoboncave and $(.)^Tt_i$ is pseudoinvex at u for fixed v with respective to η_1.
(iii) $f_i(x,.)$ is a G_{f_i}- pseudoboncave and $(.)^Tz_i$ is pseudoinvex at y for fixed x with respective to η_2.
(iv) $g_i(x,.)$ is a G_{g_i}- pseudobonvex and $(.)^Tr_i$ is pseudoinvex at y for fixed x with respective to η_2.
(v) $\eta_1(x,u) + u \in C_1$ and $\eta_2(v,y) + y \in C_2$.
(vi) $G_{g_i}((x,v)) + v^Tr_i - x^Tt_i > 0$.

Then, the following cannot hold simultaneously:

$$R_i \leq S_i, \text{for all } i = 1,2,3,...,k \ \text{ and } \ R_j < S_j, \text{for some } j = 1,2,3,...,m.$$

Proof. The proof follows on the lines of Theorem 1. □

Theorem 3. *(Strong Duality). Let $(\bar{x},\bar{y},\bar{R},\bar{z},\bar{r},\bar{\lambda},\bar{p})$ be an efficient solution to (EGMFP), fix $\lambda = \bar{\lambda}$ in (EGMFD). Further, assume that*

(i) $\{G'_{f_i}(f_i(\bar{x},\bar{y})) \bigtriangledown_{yy} f_i(\bar{x},\bar{y}) + G''_{f_i}(f_i(\bar{x},\bar{y})) \bigtriangledown_y f_i(\bar{x},y)(\bigtriangledown_y f_i(\bar{x},\bar{y}))^T - \bar{R}_i\{G'_{g_i}(g_i(\bar{x},\bar{y})) \bigtriangledown_{yy} g_i(\bar{x},\bar{y})$

$+G''_{g_i}(g_i(\bar{x},\bar{y})) \bigtriangledown_y g_i(\bar{x},\bar{y})(\bigtriangledown_y g_i(\bar{x},\bar{y}))^T\}]$ *is positive definite*

and

$p_i^T[G'_{f_i}(f_i(\bar{x},\bar{y})) \bigtriangledown_{yy} f_i(\bar{x},\bar{y}) + [G''_{f_i}(f_i(\bar{x},\bar{y})) \bigtriangledown_y f_i(\bar{x},y)(\bigtriangledown_y f_i(\bar{x},\bar{y}))^T - \bar{R}_i[G'_{g_i}(g_i(\bar{x},\bar{y}))$

$\bigtriangledown_{yy} g_i(\bar{x},\bar{y}) + G''_{g_i}(g_i(\bar{x},\bar{y})) \bigtriangledown_y g_i(\bar{x},\bar{y})(\bigtriangledown_y g_i(\bar{x},\bar{y}))^T \geq 0$, *for all* $i = 1,2,3,\ldots,k$.

(ii) The matrix $\Big\{G'_{f_i}(f_i(\bar{x},\bar{y})) \bigtriangledown_{yy} f_i(\bar{x},\bar{y}) + G''_{f_i}(f_i(\bar{x},\bar{y})) \bigtriangledown_y f_i(\bar{x},y)(\bigtriangledown_y f_i(\bar{x},\bar{y}))^T - \bar{R}_i[G'_{g_i}(g_i(\bar{x},\bar{y}))$

$\bigtriangledown_{yy} g_i(\bar{x},\bar{y}) + G''_{g_i}(g_i(\bar{x},\bar{y})) \bigtriangledown_y g_i(\bar{x},\bar{y})(\bigtriangledown_y g_i(\bar{x},\bar{y}))^T\Big\}$ *is positive definite for* $i = 1,2,3,\ldots,k$.

(iii) For $\beta > 0$ and $\bar{p}_i \in \mathbb{R}^m$, $\bar{p}_i \neq 0, i = 1,2,\ldots,k$ implies that

$$\sum_{i=1}^{k} \beta_i \bar{p}_i [G'_{f_i}(f_i(\bar{x},\bar{y})) \bigtriangledown_{yy} f_i(\bar{x},\bar{y}) + [G''_{f_i}(f_i(\bar{x},\bar{y})) \bigtriangledown_y f_i(\bar{x},y)(\bigtriangledown_y f_i(\bar{x},\bar{y}))^T - \bar{R}_i[G'_{g_i}(g_i(\bar{x},\bar{y}))$$
$$\bigtriangledown_{yy} g_i(\bar{x},\bar{y}) + G''_{g_i}(g_i(\bar{x},\bar{y})) \bigtriangledown_y g_i(\bar{x},y)(\bigtriangledown_y g_i(\bar{x},\bar{y}))^T \neq 0.$$

(iv) $[G'_{f_i}(f_i(\bar{x},\bar{y})) \bigtriangledown_{yy} f_i(\bar{x},\bar{y}) + \{G''_{f_i}(f_i(\bar{x},\bar{y})) \bigtriangledown_y f_i(\bar{x},\bar{y})(\bigtriangledown_y f_i(\bar{x},\bar{y}))^T - \bar{R}_i(G'_{g_i}(g_i(\bar{x},\bar{y})) \bigtriangledown_{yy} g_i(\bar{x},\bar{y})$
$+G''_{g_i}(g_i(\bar{x},\bar{y})) \bigtriangledown_y g_i(\bar{x},y)(\bigtriangledown_y g_i(\bar{x},y))^T)\}]_{i=1}^{k}$ *is linearly independent.*

(iv) $\bar{R}_i > 0,\ i = 1,2,3,...,k.$

Then, there exist $\bar{w}_i \in Q$ and $\bar{t}_i \in E_i$, $i = 1,2,3,...,k$ such that $(\bar{x},\bar{y},\bar{R},\bar{w},\bar{\lambda},\bar{t},\bar{q} = 0)$ is feasible for (EGMFD). Furthermore, if the assumptions of Theorem 1 or Theorem 2 are satisfied, then $(\bar{x},\bar{y},\bar{R},\bar{w},\bar{\lambda},\bar{t},\bar{q} = 0)$ is an efficient solution to (EGMFD).

Proof. Since $(\bar{x},\bar{y},\bar{R},\bar{w},\bar{\lambda},\bar{t},\bar{q} = 0)$ is an efficient solution of (EMFP), by Fritz John necessary conditions [14], there exists $\alpha \in \mathbb{R}^k$, $\beta \in \mathbb{R}_+, \gamma \in \mathbb{C}_{\neq}$, $\delta \in \mathbb{R}$ and $\xi \in R^k$ such that

$$(x-\bar{x})^T \sum_{i=1}^{k} \beta_i \Big[G'_{f_i}(f_i(\bar{x},\bar{y}))\nabla_x f_i(\bar{x},\bar{y}) + \bar{w}_i - \frac{1}{2}\bar{p}_i^T \nabla_x[\{G''_{f_i}(f_i(\bar{x},\bar{y}))\nabla_y f_i(\bar{x},\bar{y})(\nabla_y f_i(\bar{x},\bar{y}))^T$$
$$+ G'_{f_i}(f_i(\bar{x},\bar{y}))\nabla_{yy} f_i(\bar{x},\bar{y})\}]\bar{p}_i - \bar{R}_i\Big(G'_{g_i}(g_i(\bar{x},\bar{y}))\nabla_x g_i(\bar{x},\bar{y}) + \bar{t}_i - \frac{1}{2}\bar{p}_i^T \nabla_x[\{G''_{g_i}(g_i(\bar{x},\bar{y}))\nabla_y g_i(\bar{x},\bar{y})$$
$$(\nabla_y g_i(\bar{x},\bar{y}))^T + G'_{g_i}(g_i(\bar{x},\bar{y}))\nabla_{yy} g_i(\bar{x},\bar{y})\}]\bar{p}_i\Big)\Big] + (\gamma - \delta\bar{y})^T \sum_{i=1}^{k} \bar{\lambda}_i \Big[G''_{f_i}(f_i(\bar{x},\bar{y}))\nabla_x f_i(\bar{x},\bar{y})\nabla_y f_i(\bar{x},\bar{y})$$
$$+ G'_{f_i}(f_i(\bar{x},\bar{y}))\nabla_{xy} f_i(\bar{x},\bar{y}) + \nabla_x[\{G''_{f_i}(f_i(\bar{x},\bar{y}))\nabla_y f_i(\bar{x},\bar{y})(\nabla_y f_i(\bar{x},\bar{y}))^T + G'_{f_i}(f_i(\bar{x},\bar{y}))$$
$$\nabla_{yy} f_i(\bar{x},\bar{y})\}\bar{p}_i] - \bar{R}_i\Big(G''_{g_i}(g_i(\bar{x},\bar{y}))\nabla_x g_i(\bar{x},\bar{y})\nabla_y g_i(\bar{x},\bar{y}) + G'_{g_i}(g_i(\bar{x},\bar{y}))\nabla_{xy} g_i(\bar{x},\bar{y})$$
$$+\nabla_x[\{G''_{g_i}(g_i(\bar{x},\bar{y}))\nabla_y g_i(\bar{x},\bar{y})(\nabla_y g_i(\bar{x},\bar{y}))^T + G'_{g_i}(g_i(\bar{x},\bar{y}))\nabla_{yy} g_i(\bar{x},\bar{y})\}\bar{p}_i]\Big)\Big] \geq 0,\ \forall x \in C_1, \quad (21)$$

$$\sum_{i=1}^{k} \Big[(\beta_i - \delta\bar{\lambda}_i)\{(G'_{f_i}(f_i(\bar{x},\bar{y})) \bigtriangledown_y f_i(\bar{x},\bar{y}) - \bar{z}_i + (G'_{f_i}(f_i(\bar{x},\bar{y})) \bigtriangledown_{yy} f_i(\bar{x},\bar{y}) + (G''_{f_i}(f_i(\bar{x},\bar{y}))$$
$$\bigtriangledown_y f_i(\bar{x},\bar{y})(\bigtriangledown_y f_i(\bar{x},\bar{y}))^T)\bar{p}_i) - \bar{R}_i((G'_{g_i}(g_i(\bar{x},\bar{y})) \bigtriangledown_y g_i(\bar{x},\bar{y}) + \bar{r}_i + (G'_{g_i}(g_i(\bar{x},\bar{y})) \bigtriangledown_{yy} g_i(\bar{x},\bar{y})$$
$$+(G''_{g_i}(g_i(\bar{x},\bar{y})) \bigtriangledown_y g_i(\bar{x},\bar{y})(\bigtriangledown_y g_i(\bar{x},\bar{y}))^T)\bar{p}_i)\}) + ((\gamma - \delta\bar{y})\bar{\lambda}_i - \beta_i\bar{p}_i)\{(G'_{f_i}(f_i(\bar{x},\bar{y}))$$
$$\bigtriangledown_{yy} f_i(\bar{x},\bar{y}) + (G''_{f_i}(f_i(\bar{x},\bar{y})) \bigtriangledown_y f_i(\bar{x},y)(\bigtriangledown_y f_i(\bar{x},\bar{y}))^T) - \bar{R}_i(G'_{g_i}(g_i(\bar{x},\bar{y})) \bigtriangledown_{yy} g_i(\bar{x},\bar{y})$$
$$+(G''_{g_i}(g_i(\bar{x},\bar{y})) \bigtriangledown_y g_i(\bar{x},y)(\bigtriangledown_y g_i(\bar{x},\bar{y}))^T)\} + \Big((\gamma - \delta\bar{y})\bar{\lambda}_i - \frac{\beta_i\bar{p}_i}{2}\Big)\{\nabla_y((G'_{f_i}(f_i(\bar{x},\bar{y})) \bigtriangledown_{yy} f_i(\bar{x},\bar{y})$$
$$+(G''_{f_i}(f_i(\bar{x},\bar{y})) \bigtriangledown_y f_i(\bar{x},\bar{y})(\bigtriangledown_y f_i(\bar{x},\bar{y}))^T)\bar{p}_i) - \bar{R}_i(\nabla_y((G'_{g_i}(g_i(\bar{x},\bar{y}))$$
$$\bigtriangledown_{yy} g_i(\bar{x},\bar{y}) + (G''_{g_i}(g_i(\bar{x},\bar{y})) \bigtriangledown_y g_i(\bar{x},y)(\bigtriangledown_y g_i(\bar{x},y)(\bar{x},\bar{y}))^T)\bar{p}_i)\}\Big] = 0, \quad (22)$$

$$(\gamma - \delta\bar{y})\{(G'_{f_i}(f_i(\bar{x},\bar{y})) \bigtriangledown_y f_i(\bar{x},\bar{y}) - \bar{z}_i + (G'_{f_i}(f_i(\bar{x},\bar{y})) \bigtriangledown_{yy} f_i(\bar{x},\bar{y}) + (G''_{f_i}(f_i(\bar{x},\bar{y}))$$
$$\bigtriangledown_y f_i(\bar{x},\bar{y})(\bigtriangledown_y f_i(\bar{x},\bar{y}))^T)\bar{p}_i) - \bar{R}_i((G'_{g_i}(g_i(\bar{x},\bar{y})) \bigtriangledown_y g_i(\bar{x},\bar{y}) + \bar{r}_i + (G'_{g_i}(g_i(\bar{x},\bar{y})) \bigtriangledown_{yy} g_i(\bar{x},\bar{y})$$
$$g_i(\bar{x},\bar{y}) + (G''_{g_i}(g_i(\bar{x},\bar{y})) \bigtriangledown_y g_i(\bar{x},\bar{y})(\bigtriangledown_y g_i(\bar{x},\bar{y}))^T)\bar{p}_i)) = 0, i = 1,2,3,\ldots,k, \tag{23}$$

$$(\bar{\lambda}_i(\gamma - \delta\bar{y}) - \beta_i\bar{p}_i)^T[G'_{f_i}(f_i(\bar{x},\bar{y})) \bigtriangledown_{yy} f_i(\bar{x},\bar{y}) + G''_{f_i}(f_i(\bar{x},\bar{y})(\bigtriangledown_y f_i(\bar{x},\bar{y}))$$
$$(\bigtriangledown_y f_i(\bar{x},\bar{y}))^T - \bar{R}_i[G'_{g_i}(g_i(\bar{x},\bar{y})) \bigtriangledown_{yy} g_i(\bar{x},\bar{y}) + G''_{g_i}(g_i(\bar{x},\bar{y}))(\bigtriangledown_y g_i(\bar{x},\bar{y}))$$
$$(\bigtriangledown_y g_i(\bar{x},\bar{y}))^T]] = 0, \ i = 1,2,3,\ldots,k. \tag{24}$$

$$\alpha_i - \beta_i[G_{g_i}(g_i(\bar{x},\bar{y})) - s(\bar{x}|E_i) + \bar{y}^T\bar{r}_i - \tfrac{1}{2}\bar{p}_i{}^T[G'_{g_i}(g_i(\bar{x},\bar{y})) \bigtriangledown_{yy} g_i(\bar{x},\bar{y}) + G''_{g_i}(g_i(\bar{x},\bar{y}))$$
$$(\bigtriangledown_y g_i(\bar{x},\bar{y}))(\bigtriangledown_y g_i(\bar{x},\bar{y}))^T)p_i] - (\gamma - \delta\bar{y})[\bar{\lambda}_i(G'_{g_i}(g_i(\bar{x},\bar{y})) \bigtriangledown_y g_i(\bar{x},\bar{y}) + \bar{r}_i +$$
$$(G'_{g_i}(g_i(\bar{x},\bar{y})) \bigtriangledown_{yy} g_i(\bar{x},\bar{y}) + G''_{g_i}(g_i(\bar{x},\bar{y}))(\bigtriangledown_y g_i(\bar{x},\bar{y}))(\bigtriangledown_y g_i(\bar{x},\bar{y}))^T))\bar{p}_i] = 0, \ i = 1,2,\ldots,k. \tag{25}$$

$$\beta_i\bar{y} + (\gamma - \delta\bar{y})\bar{\lambda}_i \in N_{D_i}(\bar{z}_i), i = 1,2,...,K, \tag{26}$$

$$\beta_i\bar{R}_i\bar{y} + (\gamma - \delta\bar{y})\bar{R}_i\bar{\lambda}_i \in N_{F_i}(\bar{r}_i), \ i = 1,2,3,...,k, \tag{27}$$

$$\bar{y}^T \sum_{i=1}^{k} \bar{\lambda}_i[G'_{f_i}(f_i(\bar{x},\bar{y})) \bigtriangledown_y f_i(\bar{x},\bar{y}) - \bar{z}_i + (G'_{f_i}(f_i(\bar{x},\bar{y})) \bigtriangledown_{yy} f_i(\bar{x},\bar{y}) + G''_{f_i}(f_i(\bar{x},\bar{y}))$$
$$(\bigtriangledown_y f_i(\bar{x},\bar{y}))(\bigtriangledown_y f_i(\bar{x},\bar{y}))^T)\bar{p}_i - \bar{R}_i[G'_{f_i}(f_i(\bar{x},\bar{y})) \bigtriangledown_y f_i(\bar{x},\bar{y}) + \bar{r}_i + \{G'_{g_i}(g_i(\bar{x},\bar{y}))$$
$$\bigtriangledown_{yy} g_i(\bar{x},\bar{y}) + G''_{g_i}(g_i(\bar{x},\bar{y}))(\bigtriangledown_y g_i(\bar{x},\bar{y}))(\bigtriangledown_y g_i(\bar{x},\bar{y}))^T)\bar{p}_i]] = 0. \tag{28}$$

$$\delta\bar{y}^T \sum_{i=1}^{k} \bar{\lambda}_i[G'_{f_i}(f_i(\bar{x},\bar{y})) \bigtriangledown_y f_i(\bar{x},\bar{y}) - \bar{z}_i + (G'_{f_i}(f_i(\bar{x},\bar{y})) \bigtriangledown_{yy} f_i(\bar{x},\bar{y}) + G''_{f_i}(f_i(\bar{x},\bar{y}))$$
$$(\bigtriangledown_y f_i(\bar{x},\bar{y}))(\bigtriangledown_y f_i(\bar{x},\bar{y}))^T)\bar{p}_i - \bar{R}_i[G'_{f_i}(f_i(\bar{x},\bar{y})) \bigtriangledown_y f_i(\bar{x},\bar{y}) + \bar{r}_i + \{G'_{g_i}(g_i(\bar{x},\bar{y}))$$
$$\bigtriangledown_{yy} g_i(\bar{x},\bar{y}) + G''_{g_i}(g_i(\bar{x},\bar{y}))..(\bigtriangledown_y g_i(\bar{x},\bar{y}))(\bigtriangledown_y g_i(\bar{x},\bar{y}))^T)\bar{p}_i]] = 0. \tag{29}$$

$$\bar{\lambda}^T\xi = 0, \tag{30}$$

$$\bar{w}_i \in Q_i, \bar{t}_i \in E_i, \bar{x}^T\bar{t}_i = S(\bar{x}|E_i), \bar{x}^T\bar{w}_i = S(\bar{x}|Q_i), \ i = 1,2,3,...,k, \tag{31}$$

$$(\alpha,\delta,\xi) \geq 0, \ (\alpha,\beta,\gamma,\delta,\xi) \neq 0. \tag{32}$$

From Assumption (i) and Equation (24), we have

$$\gamma\bar{\lambda}_i - \beta_i\bar{p}_i - \bar{\lambda}_i\delta\bar{y} = 0. \tag{33}$$

We claim that $\beta_i \neq 0, \forall i$. The proof is by contradiction. Let $\beta_i = 0$ for some i. Since $\bar{\lambda} > 0$, the relation in Equation (33) yields

$$\gamma = \delta \bar{y}. \tag{34}$$

From the relation in Equations (22), (33) and (34), we obtain

$$\sum_{i=1}^{k} (\beta_i - \delta\bar{\lambda}_i)[G'_{f_i}(f_i(\bar{x},\bar{y})) \nabla_y f_i(\bar{x},\bar{y}) - \bar{z}_i + (G'_{f_i}(f_i(\bar{x},\bar{y})) \nabla_{yy} f_i(\bar{x},\bar{y})+$$
$$G''_{f_i}(f_i(\bar{x},\bar{y}))(\nabla_y f_i(\bar{x},\bar{y}))(\nabla_y f_i(\bar{x},\bar{y}))^T)\bar{p}_i - \bar{R}_i[G'_{g_i}(g_i(\bar{x},\bar{y})) \nabla_y g_i(\bar{x},\bar{y}) + \bar{r}_i$$
$$+\{G''_{g_i}(g_i(\bar{x},\bar{y}))(\nabla_y g_i(\bar{x},\bar{y}))(\nabla_y g_i(\bar{x},\bar{y}))^T) + G'_{g_i}(\nabla_{yy} g_i(\bar{x},\bar{y}))\}\bar{p}_i)] = 0. \tag{35}$$

On using Asumption (iv), this gives

$$\beta_i - \delta\bar{\lambda}_i = 0, i = 1,2,\ldots,k. \tag{36}$$

Since $\beta_i = 0$, we obtain $\delta\bar{\lambda}_i = 0$ but $\bar{\lambda}_i > 0,\ i = 1,2,\ldots,k$ and thus the relation in Equation (36) implies $\delta = 0$. Thus, from the relation in Equations (25), (34) and (36), we get $\alpha_i = 0,\ i = 1,2,\ldots,k$. In addition, from the relation in Equation (34), we get $\gamma = 0$, which is a contradiction, since $(\alpha,\beta,\gamma,\delta) \neq 0$. Hence, we get $\beta_i \neq 0,\ i = 1,2,\ldots,k$.

Since $\bar{\lambda} > 0$, using Equations (22) and (33), we get

$$\sum_{i=1}^{k} \beta_i \bar{p}_i [G'_{f_i}(f_i(\bar{x},\bar{y})) \nabla_y f_i(\bar{x},\bar{y}) + \bar{w}_i + (G'_{f_i}(f_i(\bar{x},\bar{y})) \nabla_{yy} f_i(\bar{x},\bar{y})+$$
$$G''_{f_i}(f_i(\bar{x},\bar{y}))(\nabla_y f_i(\bar{x},\bar{y}))(\nabla_y f_i(\bar{x},\bar{y}))^T)\bar{p}_i - \bar{R}_i[G'_{g_i}(g_i(\bar{x},\bar{y})) \nabla_y g_i(\bar{x},\bar{y}) - \bar{t}_i$$
$$+\{G''_{g_i}(g_i(\bar{x},\bar{y}))(\nabla_y g_i(\bar{x},\bar{y}))(\nabla_y g_i(\bar{x},\bar{y}))^T) + G'_{g_i}(\nabla_{yy} g_i(\bar{x},\bar{y}))\}\bar{p}_i)] = 0. \tag{37}$$

Hence, from Assumption (iii), we get $\bar{p}_i = 0,\ i = 1,2,\ldots,k$. From the relation in Equation (33), $\bar{p}_i = 0,\ i = 1,2,\ldots,k$ and $\bar{\lambda} > 0$, we have $\gamma = \delta\bar{y}$, from Equations (21) and (22), we have

$$\sum_{i=1}^{k} \bar{\lambda}_i [G'_{f_i}(f_i(\bar{x},\bar{y})) \nabla_x f_i(\bar{x},\bar{y}) + \bar{w}_i - \bar{R}_i[G'_{g_i}(g_i(\bar{x},\bar{y})) \nabla_x g_i(\bar{x},\bar{y}) - \bar{t}_i]] = 0. \tag{38}$$

$$\sum_{i=1}^{k} (\beta_i - \delta\bar{\lambda}_i)[G'_{f_i}(f_i(\bar{x},\bar{y})) \nabla_y f_i(\bar{x},\bar{y}) + \bar{w}_i - \bar{R}_i G'_{g_i} \nabla_y g_i(\bar{x},\bar{y}) - \bar{t}_i] = 0. \tag{39}$$

By Assumptions (i) and (iii), we have

$$\beta_i = \delta\bar{\lambda}_i,\ i = 1,2,\ldots,k. \tag{40}$$

Since $\beta_i > 0$ and $\bar{\lambda}_i > 0,\ i = 1,2,\ldots,k$, the relation in Equation (40) implies that $\delta > 0$, and the relation in Equation (38) reduces to

$$(x-\bar{x})^T \sum_{i=1}^{k} \bar{\lambda}_i [G'_{f_i}(f_i(\bar{x},\bar{y})) \nabla_x f_i(\bar{x},\bar{y}) + \bar{w}_i - \bar{R}_i (G'_{g_i}(g_i(\bar{x},\bar{y})) \nabla_x g_i(\bar{x},\bar{y}) - \bar{t}_i)] \geq 0,\ \forall x \in C_1. \tag{41}$$

Let $x \in C_1$. Then, $x + \bar{x} \in C_1$ as C_1 is a closed convex cone. On substituting $x + \bar{x}$ into the place of x in Equation (41), we get

$$x^T \sum_{i=1}^{k} \bar{\lambda}_i [(G'_{f_i}(f_i(\bar{x},\bar{y}))\nabla_x f_i(\bar{x},\bar{y}) + \bar{w}_i) - \bar{R}_i (G'_{g_i}(g_i(\bar{x},\bar{y}))\nabla_x g_i(\bar{x},\bar{y}) - \bar{t}_i)] \geq 0.$$

Hence,

$$\sum_{i=1}^{k} \bar{\lambda}_i [(G'_{f_i}(f_i(\bar{x},\bar{y}))\nabla_x f_i(\bar{x},\bar{y}) + \bar{w}_i) - \bar{R}_i (G'_{g_i}(g_i(\bar{x},\bar{y}))\nabla_x g_i(\bar{x},\bar{y}) - \bar{t}_i)] \in C_1^*. \tag{42}$$

In addition, by letting $x = 0$ and $x = 2\bar{x}$ simultaneously in Equation (41), we have

$$\bar{x}^T \sum_{i=1}^{k} \bar{\lambda}_i [(G'_{f_i}(f_i(\bar{x},\bar{y}))\nabla_x f_i(\bar{x},\bar{y}) + \bar{w}_i) - \bar{R}_i (G'_{g_i}(g_i(\bar{x},\bar{y}))\nabla_x g_i(\bar{x},\bar{y}) - \bar{t}_i)] = 0. \tag{43}$$

Since $\gamma = \delta\bar{y}$ and $\delta > 0$, we have

$$\bar{y} = \frac{\gamma}{\delta} \in C_2. \tag{44}$$

From Equations (26) and (34) and using $\beta > 0$, we get $\bar{y} \in N_{D_i}(\bar{z}_i),\ i = 1,2,3,...,k$ This implies

$$\bar{y}^T \bar{z}_i = S(\bar{y}|D_i),\ i = 1,2,3,...,k. \tag{45}$$

Similarly, by Equation (27) and Assumption (iii), $\bar{y} \in N_{F_i}(\bar{r}_i),\ i = 1,2,3,...,k,$ we obtain

$$\bar{y}^T \bar{r}_i = S(\bar{y}|F_i),\ i = 1,2,3,...,k. \tag{46}$$

Combining Equations (31), (45), (46) and (31), it follows that

$$(G_{f_i}(f_i(\bar{x},\bar{y})) - S(\bar{y}|D_i) + \bar{x}^T\bar{w}_i) - \bar{R}_i(G_{g_i}(g_i(\bar{x},\bar{y})) + S(\bar{y}|F_i) - \bar{x}^T\bar{t}_i) = 0,\ i = 1,2,3,...,k. \tag{47}$$

This together with Equations (42), (43) and (47) shows that $(\bar{x},\bar{y},\bar{R},\bar{\lambda},\bar{w},\bar{t}) \in W^0$. Now, let $(\bar{x},\bar{y},\bar{R},\bar{\lambda},\bar{w},\bar{t})$ be not an efficient solution of (EGMFD). Then, there exists other $(u,v,R,\lambda,w,t) \in W^0$ such that $\bar{R}_i \leq S_i, \forall\, i = 1,2,...,k$ and $\bar{R}_j < S_j$, for some $j = 1,2,...,m$. This contradicts the result of the Theorems 1 and 2. Hence, the proof is complete. □

Remark 2. *In the case of symmetric programming problem, the proof of converse duality theorem remains same as Theorem 3.*

Theorem 4. *(Converse duality theorem). Let $(\bar{u},\bar{v},\bar{S},\bar{t},\bar{w},\bar{\lambda},\bar{q})$ be an efficient solution to (EGMFD), fix $\lambda = \bar{\lambda}$ in (EGMFP). Further, assume that*

(i) $\{G'_{f_i}(f_i(\bar{u},\bar{v}))\nabla_{xx} f_i(\bar{u},\bar{v}) + G''_{f_i}(f_i(\bar{u},\bar{v}))\nabla_x f_i(\bar{u},\bar{v})(\nabla_x f_i(\bar{u},\bar{v}))^T - \bar{S}_i\{G'_{g_i}(g_i(\bar{u},\bar{v}))\nabla_{xx} g_i(\bar{u},\bar{v})$

$+G''_{g_i}(g_i(\bar{u},\bar{v}))\nabla_x g_i(\bar{u},\bar{v})(\nabla_x g_i(\bar{u},\bar{v}))^T\}]$ *is positive definite and*

$q_i^T[G'_{f_i}(f_i(\bar{u},\bar{v}))\nabla_{xx} f_i(\bar{u},\bar{v}) + [G''_{f_i}(f_i(\bar{u},\bar{v}))\nabla_x f_i(\bar{u},\bar{v})(\nabla_x f_i(\bar{u},\bar{v}))^T - \bar{S}_i[G'_{g_i}(g_i(\bar{u},\bar{v}))$

$\nabla_{xx} g_i(\bar{u},\bar{v}) + G''_{g_i}(g_i(\bar{u},\bar{v}))\nabla_x g_i(\bar{u},\bar{v})(\nabla_x g_i(\bar{u},\bar{v}))^T \geq 0,$ *for all* $i = 1,2,3,\ldots,k.$

(ii) The matrix $\Big\{G'_{f_i}(f_i(\bar{u},\bar{v}))\nabla_{xx} f_i(\bar{u},\bar{v}) + [G''_{f_i}(f_i(\bar{u},\bar{v}))\nabla_x f_i(\bar{u},\bar{v})(\nabla_x f_i(\bar{u},\bar{v}))^T - \bar{S}_i[G'_{g_i}(g_i(\bar{u},\bar{v}))$

$$\nabla_{xx}g_i(\bar{u},\bar{v}) + G''_{g_i}(g_i(\bar{u},\bar{v})) \nabla_x g_i(\bar{u},\bar{v})(\nabla_x g_i(\bar{u},\bar{v}))^T \Big\} \text{ is positive definite for } i = 1,2,3,\ldots,k.$$

(iii) For $\beta > 0$ and $\bar{q}_i \in \mathbb{R}^n$, $\bar{q}_i \neq 0, i = 1,2,\ldots,k$ implies that

$$\sum_{i=1}^{k} \beta_i \bar{q}_i [G'_{f_i}(f_i(\bar{u},\bar{v})) \nabla_{xx} f_i(\bar{u},\bar{v}) + [G''_{f_i}(f_i(\bar{u},\bar{v})) \nabla_x f_i(\bar{u},\bar{v})(\nabla_x f_i(\bar{u},\bar{v}))^T - \bar{S}_i[G'_{g_i}(g_i(\bar{u},\bar{v}))$$
$$\nabla_{xx}g_i(\bar{u},\bar{v}) + G''_{g_i}(g_i(\bar{u},\bar{v})) \nabla_x g_i(\bar{u},\bar{v})(\nabla_x g_i(\bar{u},\bar{v}))^T \neq 0,$$

(iv) $[G'_{f_i}(f_i(\bar{u},\bar{v})) \nabla_{xx} f_i(\bar{u},\bar{v}) + \{G''_{f_i}(f_i(\bar{u},\bar{v})) \nabla_x f_i(\bar{u},\bar{v})(\nabla_x f_i(\bar{u},\bar{v}))^T - \bar{S}_i(G'_{g_i}(g_i(\bar{u},\bar{v}))$

$\nabla_{xx}g_i(\bar{u},\bar{v}) + G''_{g_i}(g_i(\bar{u},\bar{v})) \nabla_x g_i(\bar{u},\bar{v})(\nabla_x(g_i(\bar{u},\bar{v}))^T)\}]_{i=1}^{k}$ *is linearly independent.*

(v) $\bar{S}_i > 0$, $i = 1,2,3,...,k$. Then, there exist $\bar{z}_i \in D_i$ and $\bar{r}_i \in E_i$, $i = 1,2,3,...,k$ such that $(\bar{u},\bar{v},\bar{S},\bar{z},\bar{\lambda},\bar{r},\bar{p} = 0)$ is feasible for (EGMFP). Furthermore, if the assumptions of Theorem 1 or Theorem 2 are satisfied, then $(\bar{u},\bar{v},\bar{S},\bar{z},\bar{\lambda},\bar{r},\bar{p} = 0)$ is an efficient solution to (EGMFP).

Proof. The results can be obtained on the lines of Theorem 3. □

4. Conclusions

In this paper, we use the concept of G_f- bonvex/G_f-pseudobonvex functions to establish duality results for G- Mond–Weir type dual model related to multiobjective nondifferentiable second-order symmetric fractional programming problem over arbitrary cones. Numerical examples are also illustrated to justify the existence of such type of functions. The present work can be further extended to nondifferentiable higher-order symmetric fractional programming over cones. This will orient the future task for the researcher working in this area.

Author Contributions: All authors contributed equally in writing this article. All authors read and approved the final manuscript.

Acknowledgments: The authors wish to thank the referees for useful comments.

Conflicts of Interest: The authors declare no conflict of interest.

References

1. Suneja, S.K.; Aggarwal, S.; Davar, S. Multiobjective symmetric duality involving cones. *Eur. J. Oper. Res.* **2002**, *141*, 471–479. [CrossRef]
2. Chinchuluun, A.; Pardalos, P.M. A survey of multiobjective optimization. *Ann. Oper. Res.* **2007**, *154*, 29–50. [CrossRef]
3. Mangasarian, O.L. Second and higher-order duality in nonlinear programming. *J. Math. Anal. Appl.* **1975**, *51*, 607–620. [CrossRef]
4. Kim, D.S.; Yun, Y.B.; Lee, W.J. Multiobjective symmetric duality with cone constraints. *Eur. J. Oper. Res.* **1998**, *107*, 686–691. [CrossRef]
5. Suneja, S.K.; Lalitha, C.S.; Khurana, S. Second order symmetric duality in multiobjective programming. *Eur. J. Oper. Res.* **2003**, *144*, 492–500. [CrossRef]
6. Khurana, S. Symmetric duality in multiobjective programming involving generalized cone-invex functions. *Eur. J. Oper. Res.* **2005**, *165*, 592–597. [CrossRef]
7. Antcza, T. New optimality conditions and duality results of G-type in differentiable mathematical programming. *Nonlinear Anal.* **2007**, *66*, 1617–1632. [CrossRef]
8. Dubey, R.; Mishra, L.N.; Mishra, V.N. Duality relations for a class of a multiobjective fractional programming problem involving support functions. *Am. J. Oper. Res.* **2018**, *8*, 294–311.
9. Dubey, R.; Mishra, V.N. Symmetric duality results for a nondifferentiable multiobjective programming problem with support function under strongly assumptions. *RAIRO-Operation Res.* **2019**, *53*, 539–558. [CrossRef]

10. Dubey, R.; Mishra, L.N.; Ali, R. Special class of second-order non-differentiable symmetric duality problems with (G, α_f)-pseudobonvexity assumptions. *Mathematics* **2019**, *7*, 763. [CrossRef]
11. Antczak, T. On G-invex multiobjective programming. *Part I. Optim. J. Glob. Optim.* **2009**, *43*, 97–109. [CrossRef]
12. Kang, Y.M.; Kim, D.S.; Kim, M.H. Optimality conditions of G-type in locally Lipchitz multiobjective programming. *Viet. J. Math.* **2014**, *40*, 275–285.
13. Gao, X. Sufficiency in multiobjective programming under second-order $B-(p,r)-V$-type Ifunctions. *J. Interdiscip. Math.* **2014**, *17*, 385–402. [CrossRef]
14. Brumelle, S. Duality for multiple objective convex programs. *Math. Oper. Res.* **1981**, *6*, 159–172. [CrossRef]

symmetry

MDPI

Article

Improved Hydrodynamic Analysis of 3-D Hydrofoil and Marine Propeller Using the Potential Panel Method Based on B-Spline Scheme

Chen-Wei Chen [1,*] and Ming Li [1,2]

1 Institute of Marine Structures and Naval Architecture, Ocean College, Zhejiang University, Zhoushan 316000, China; ming_lihk@yahoo.com
2 Shanghai Key Laboratory of Multidimensional Information Processing, No. 500, Dong-Chuan Road, East China Normal University, Shanghai 200241, China
* Correspondence: cwchen@zju.edu.cn

Received: 7 January 2019; Accepted: 2 February 2019; Published: 11 February 2019

Abstract: In this paper, the hydrodynamic performance of lift-body marine propellers and hydrofoils is analyzed using a B-spline potential-based panel method. The potential panel method, based on a combination of two singularity elements, is proposed, and a B-spline curve interpolation method is integrated with the interpolation of the corner points and collocation points to ensure accuracy and continuity of the interpolation points. The B-spline interpolation is used for the distribution of the singularity elements on a complex surface to ensure continuity of the results for the intensity of the singular points and to reduce the possibility of abrupt changes in the surface velocity potential to a certain extent. A conventional cubic spline method is also implemented as a comparison of the proposed method. The surface pressure coefficient and lift the performance of 2-D and 3-D hydrofoils of sweepback and dihedral type with different aspect ratios are analyzed to verify the rationality and feasibility of the present method. The surface pressure distribution and coefficients of thrust and torque are calculated for different marine propellers and compared with the experimental data. A parametric study on the propeller wake model was carried out. The validated results show that it is practical to improve the accuracy of hydrodynamic performance prediction using the improved potential panel method proposed.

Keywords: B-spline scheme; dihedral hydrofoil; hydrodynamics; marine propeller; propeller wake; sweptback hydrofoil; surface panel method

1. Introduction

The panel method has been widely used as a powerful tool in the aerodynamic and hydrodynamic fields since the successful work presented by Hess and Smith in 1967 [1]. The velocity potential method transforms the Laplace equation using the Green formula, which transforms an integral problem into a surface integral problem of the object, and the velocity potential is used as the unknown quantity. Brandner [2] and Baltazar et al. [3] first proposed this basic principle and compared the results obtained using the panel method and the RANS equation method. Moreover, applications of Green's function method to the frontier science field and engineering problems include the analysis of the stress gradient of nanobeams in nanotechnology with a possible loss of symmetry of Green's function for nonlocal boundary conditions [4], to predict accurately dynamic behavior of vibroacoustic systems [5], and to solve transmission loss problems [5,6], radiation, diffraction problems [7,8], etc.

For three-dimensional objects that generate lift, such as hydrofoils and propeller blades, apart from the basic objective of providing lift, we considered that there should be some attached vortices on the blades and that the additional wake vortices are assumed at the wake line, hence, all the vortices form a closed interval or the equal strength of an infinitely long vortex line. The panel method based on the velocity potential normally requires the Dirichlet boundary condition (B.C.) to solve the governing equation matrix, and the Dirichlet boundary condition requires the virtual velocity potential in the interior region of the object to be constant. Based on the results reported by Belibassakis et al. [9], it was assumed in the present research that the internal velocity potential of a three-dimensional object is equal to the perturbation velocity potential at infinity, hence, the intensity distribution of the source on the surface of the object can be determined, and then the governing equation, i.e., the Laplace equation, for each collocation point can be solved.

For hydrofoils with a continuous slope, there is no suitable additional condition. Because the slope at the trailing edge of a two-dimensional hydrofoil is discontinuous, it is possible to find a definite loop in which the velocity at the trailing edge of the hydrofoil is infinite. This additional constraint is the Kutta condition [10], i.e., zero velocity at the trailing edge. However, this condition is not directly suitable for performing numerical calculations. To meet the Kutta condition, Kinnas and Hsin [11] applied a simplex iterative method. This method can meet the equal pressure at the trailing edge of the hydrofoil, but it requires several iterative operations to reach convergence. Another approach, the Newton–Raphson method [12], has been proposed to satisfy the Kutta condition; however, it is difficult to reach convergence by this method when the pressures at the trailing edge are equal. Therefore, Srivastava and Roychowdhury [13] presented an improved Newton–Raphson iteration method for enhanced stability and reliability of the Kutta condition in numerical calculations. With this method, the computation efficiency is greatly improved.

For a specific problem, the panel method should be used according to the specific conditions. Mantia and Dabnichki [14] and Eça and Vaz [15] studied two-dimensional wings using the panel method based on velocity and based on velocity potential, respectively. They found that the panel method based on velocity and/or velocity potential can be applied to estimate the irrotational potential flow around 3-D hydrodynamic shapes with a certain thickness. However, for thin objects, the velocity potential panel method can provide more accurate results. The influence of the source and doublet combination elements on the velocity potential of the collocation point leads to a lower-order singularity than the velocity influence coefficient.

For a continuous geometric surface, Kanemaru and Ando [16] applied flat rectangular panels that are simple and easy to calculate; however, there will always be gaps between adjacent rectangular panels. Tarafder and Suzuki [17] applied surface modelling with discretized hyperboloid panels, where the gap between the panels can be narrow. The accuracy of the discretized form directly influences the accuracy of the calculations. If the difference in the strengths of the adjacent source and doublet is relatively large, a discontinuous situation will arise, leading to inaccuracy in the velocity potential derivative.

Based on this consideration, the grid meshing method based on the B-spline scheme was adopted in this study to make the spatial interpolation of the singular point of the panel elements more accurate and generate a reasonable surface mesh. The B-spline scheme based on the Bezier scheme [18] uses fewer control vertices to generate the high-order curves and surfaces and enables curve control over local deformation without affecting the global deformation. Hence, it is considered one of the most important geometric modelling methods [19]. The B-spline scheme is widely applied in vehicle design [20,21], aerodynamic optimization [22], submarine guidance systems with path planning [23], etc.

Prediction of hydrodynamic performance of hydrofoil and marine propeller using B-spline high-order panel method has been studied in the literature [24–27]. The lower number of panels and high computational efficiency were experienced in the high-order panel methods. However, the effect of the singularity war [28,29] on numerical instability is still difficult to overcome in the high-order methods whereas the low-order panel method [2,9,11,28,30,31], i.e., the constant potential strength with an equal singularity density on a panel, is more stable to numerical computation for the analysis of a smooth hydrodynamic shape [25]. To improve the accuracy of high- or low-order panel method, the integrated trailing wake surface shall be an improved wake modelling [24,30,32].

In this study, we propose a method by integrating the B-spline geometry representation with the low-order panel method so as to obtain a smoother and continuous discretized surface integrated with an analyzed trailing wake model for the purpose of enhancing the accuracy and numerical stability in the hydrodynamic analysis. Extending the earlier research work [21], the surface pressure coefficient and lift forces on a hydrofoil and propeller were calculated and analyzed using the cubic B-spline low-order panel method.

The potential panel method based on a combination of two singularity elements was proposed and a B-spline curve interpolation method was integrated with the interpolation of the corner points and collocation points to ensure the accuracy and continuity of the interpolation points. The B-spline interpolation was used for distribution of the singularity elements on a complex surface to ensure continuity of the results of the intensity of the singular points and to reduce the possibility of abrupt changes in the surface velocity potential to a certain extent.

Control polygon and control points of the B-spline interpolation were inversely calculated by the basis function for fitting and smoothing 3-D hydrofoil and propeller discreted panels integrated with a low-order panel method for efficient and effective hydrodynamic prediction improved in this study. The hydrodynamic performance of the 3-D hydrofoil, sweepback hydrofoil, and dihedral hydrofoil with different aspect ratios was analyzed to verify the rationality and feasibility of the method. Finally, the results of pressure distribution and the coefficients of thrust and torque for different 3-D screw propellers were obtained and compared with the experimental data. The comparison results were found to be satisfactory. National Advisory Committee for Aeronautics (NACA) airfoils [33–35] and David Taylor Model Basin (DTMB) propellers [2,36–38] have been extensively adopted for the validation of the numerical schemes thanks to the availability of the experimental data for several different foil profiles and propeller models.

2. Methodology

2.1. Surface Panel Method

Panel methods involve numerical models based on simplified assumptions regarding inviscid and incompressible flow problems. In principle, if a problem can be solved by distributing the unknown quantities on the boundary surface surrounding a foil under simplified potential flow, the characteristic coefficients of the foil can be obtained. Under these assumptions, the velocity vector that describes the flow field can be represented as the gradient of a scalar velocity potential. A statement of conservation of mass in the flow field leads to Laplace's equation as the governing equation for the velocity potential. If the flow in the fluid region is considered to be incompressible and irrotational, the continuity equation can be expressed as follows [1]:

$$\nabla^2 \varphi = 0 \tag{1}$$

where ∇^2 is the Laplacian and ϕ is the velocity potential. The solution of Equation (1) is based on the Green formula, a well-known function method, which can be derived from Gauss's law and the divergence theorem. For a finite fluid domain V, the Green formula can be obtained using the divergence theorem [9]:

$$\iint_S \left(\phi \frac{\partial \varphi}{\partial n} - \varphi \frac{\partial \phi}{\partial n}\right) dS = \iiint_V \left(\phi \nabla^2 \varphi - \varphi \nabla^2 \phi\right) dV \tag{2}$$

where $\varphi(x,y,z)$ and $\phi(x,y,z)$ can represent any function in the finite field that has a continuous first-order partial derivative. The velocity field function in the fluid domain is represented in the calculation of the surface element, where S represents the outer boundary of the volume V and n is the outer normal vector of the boundary. By Green's third formula for hydrofoil and propeller problems, the velocity potential at point $P(x, y, z)$ can be expressed as [32]:

$$4\pi E\phi(P) = \iint\limits_{S} \left[\phi(Q)\frac{\partial}{\partial n}\left(\frac{1}{R}\right) - \frac{1}{R}\frac{\partial\phi(Q)}{\partial n}\right]dS, E = \begin{cases} 0 & P \notin V \\ 1/2 & P \in S \\ 1 & P \in V \end{cases} \quad (3)$$

where Q represents the singularity point in the field and R is the distance between the collocation point P and the singularity point Q. Hence, the basic assumptions of the potential panel method for a lifting body can be illustrated as in Figure 1.

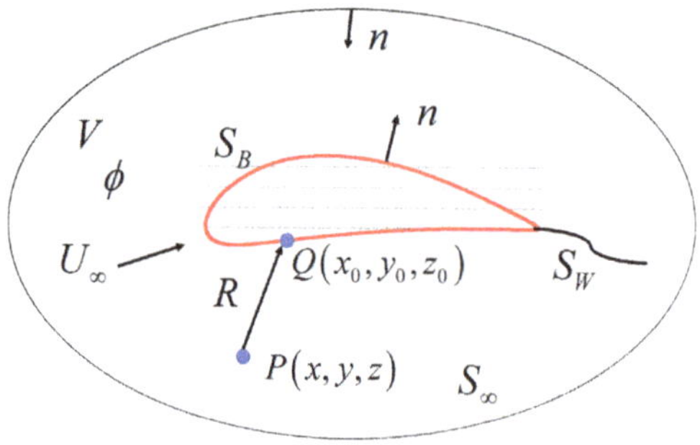

Figure 1. Potential flow over a lifting body.

In the nonpenetrating and perturbation velocity potential attenuation conditions, the integral equation for the surface of the object can be expressed as [28,29]

$$\begin{aligned} &2\pi\phi(P) - \iint\limits_{S_B} \phi(Q)\tfrac{\partial}{\partial n}\left(\tfrac{1}{R}\right)dS - \iint\limits_{S_W} \Delta\phi(Q)\tfrac{\partial}{\partial n}\left(\tfrac{1}{R}\right)dS \\ &= -\iint\limits_{S_B} \tfrac{\partial\phi(Q)}{\partial n}\left(\tfrac{1}{R}\right)dS = -\iint\limits_{S_B} \tfrac{n\cdot\Delta\phi(Q)}{R}dS \end{aligned} \quad (4)$$

where $\Delta\varphi$ represents the distribution of the doublet of the object's wake surface S_W and $n \cdot \nabla\varphi(q)$ represents the distribution of the source of the object surface S_B. Equation (4) is the Fredholm integral equation of the second kind, and the velocity potential φ can be obtained from this equation. The doublet strength on the wake surface can be determined by the Kutta condition and the wake surface, and the perturbation velocity potential of the object surface can then be obtained. The perturbation velocity can be obtained by differentiating the perturbation velocity potential, and then the hydrodynamic performance can be obtained. So far, the main symbolic relationship of the above-mentioned potential flow theory is shown in Figure 1.

2.2. B-Spline Model Scheme

The motivation for using B-splines is to ensure that the B-spline curved panels have at least C^2 continuity and the B-splines can be provided with local control, i.e., changing a single point does not affect the entire curve. While natural cubic splines provide that level of continuity, they are also subject to global control, i.e., changing a single point affects the entire curve.

The B-spline curve method inherits the advantages of the Bezier curve method, preserving the basic modelling features such as defining the vertices of curves and surfaces. The B-spline curve uses a set of basic functions that is different from the one used in the Bezier curve.

The B-spline curve is expressed as follows [19]:

$$p(u) = \sum_{i=0}^{n} d_i N_{i,k}(u) \tag{5}$$

where $d_i(i = 0,\ 1,\ 2,\ \cdots, n)$ is the de Boor point, which is the control point of the curve. The polygon defined by all of the control vertices is defined as the control polygon, for instance, a control polygon covering the NACA4410 hydrofoil geometry [33] can be calculated to generate a discreted cubic B-Spline curve or surface as shown in Figure 2.

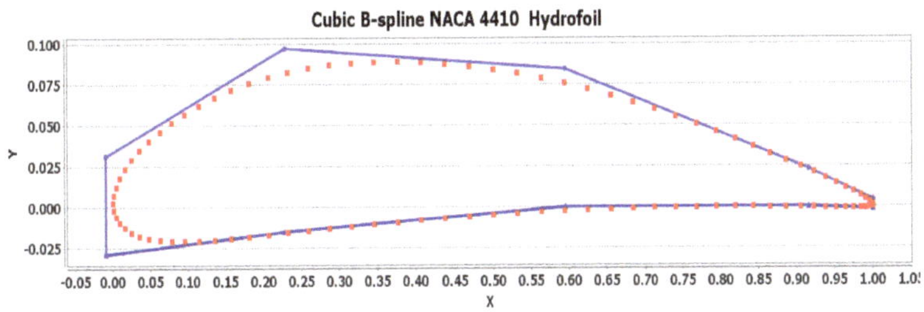

Figure 2. The control polygon for generating the discreted cubic B-Spline NACA4410 curve.

In Equation (5), $N_{i,k}(u)(i = 0,\ 1,\ 2,\ \cdots, m)$ is a base function of order k. Every base function is a polynomial of order k, consisting of increasing vector nodes u. The base function can be expressed using the de Boor–Cox recursive formula [19,24] as follows:

$$N_{i,0}(u) = \begin{cases} 1, & u_i \le u \le u_{i+1} \\ 0, & else \end{cases} \tag{6}$$

$$N_{i,k}(u) = \frac{u - u_i}{u_{i+k-1}} N_{i,k-1}(u) + \frac{u_{i+k+1} - u}{u_{i+k+1} - u_{i+1}} N_{i+1,k-1}(u) \tag{7}$$

$$\frac{d}{du} N_{i,k}(u) = k\left[\frac{N_{i,k-1}(u)}{u_{i+k} - u_i} - \frac{N_{i+1,k-1}(u)}{u_{i+k+1} - u_{i+1}}\right] \tag{8}$$

In practice, the calculation of the entire curve is rarely performed in the application of the B-spline curve. To calculate a series of normalized B-spline basis function from Equations (6)–(8), a B-spline function of a specific order can be obtained; some of the curves are shown in Figure 3.

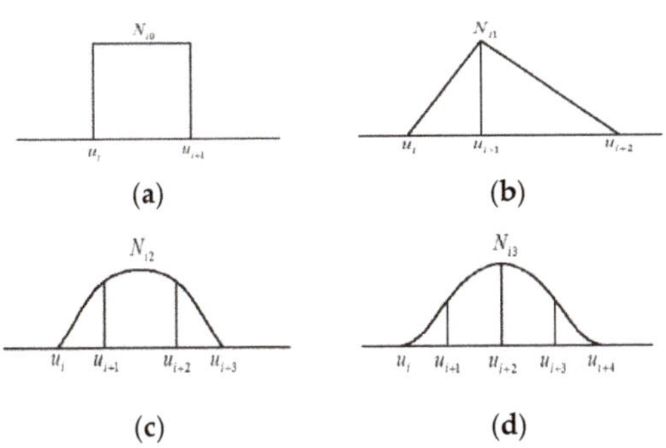

Figure 3. The B-spline first- to fourth-order base function curves: (**a**) first-order base function; (**b**) second-order base function; (**c**) third-order base function; (**d**) fourth-order base function.

The number of control vertices $d_{i,j}(i = 0, 1, 2, \cdots, m; j = 0, 1, 2, \cdots, n)$ is $(m+1) \times (n+1)$, and all the control vertices constitute the control surface grid. The order of u is k, and that of v is l. The corresponding node vectors are $U = \begin{bmatrix} u_0, & u_1, & \cdots & , u_{m+k+1} \end{bmatrix}$ and $V = \begin{bmatrix} v_0, & v_1, & \cdots, & v_{n+l+1} \end{bmatrix}$. The $k \times l$ order tensor product B-spline surface can be expressed as [19]:

$$\begin{aligned} p(u,v) &= \sum_{i=0}^{m} \sum_{j=0}^{n} d_{i,j} N_{i,k}(u) N_{j,l}(v) \\ u_k &\leq u \leq u_{m+1}, v_l \leq v \leq u_{n+1} \end{aligned} \tag{9}$$

where the basic functions $N_{i,k}(u)(i = 0, 1, 2, \cdots, m)$ and $N_{j,l}(v)(i = 0, 1, 2, \cdots, m)$ are determined by the de Boor–Cox recursion formula corresponding to the node vectors $U = \begin{bmatrix} u_0, & u_1, & \cdots, & u_{m+k+1} \end{bmatrix}$ and $V = \begin{bmatrix} v_0, & v_1, & \cdots, & v_{n+l+1} \end{bmatrix}$. For the surface problem, the inverse solution process involves calculating the inverse of the tensor product. Therefore, the inverse of the surface can be expressed by 2 inverse curves. The B-spline surface equation that needs to be inversed can be expressed as follows [19]:

$$p(u,\, v) = \sum_{i=0}^{m} \left[\sum_{j=0}^{n} d_{i,j} N_{j,l}(v) \right] N_{i,k}(u) \tag{10}$$

where an equation similar to the curve equation [19] is constructed as given below:

$$p(u,\, v) = \sum_{i=0}^{m} c_i(v) N_{i,k}(u) \tag{11}$$

The inverse of the applied base function is used to obtain the control vertices on the isoparametric lines. The B-spline curve is defined by the resulting control vertices. The control vertices of the required curves can be solved using the control vertices obtained as the initial points.

2.3. Numerical Representation of Surface Panel Method

The numerical calculation process for a surface panel method is shown in Figure 4. First, the hydrodynamic surface is defined to generate grid coordinates and points to discretize the hydrodynamic surface into N panels. Second, the number and type of singularities (solutions of the Laplace equation), such as the source, doublet, and vortex, are chosen. Third, the collocation points are placed on the curved panels to satisfy the surface-tangency condition by using the cubic B-spline interpolation scheme. Fourth, the Kutta condition is imposed and the influence of all of the singularities and the freestream at a fixed collocation point is summed. Fifth, solving the singularity strength from the system of equations generated by step 4, i.e., the zero normal flow boundary condition on each of the collocation points resulting in the set of algebraic equations as shown in Equations (12)–(14), is carried out. Finally, the estimated performance to be improved or satisfied is determined, including pressure, velocity, and load. The related numerical representation of the panel method can be referenced in Appendix A.

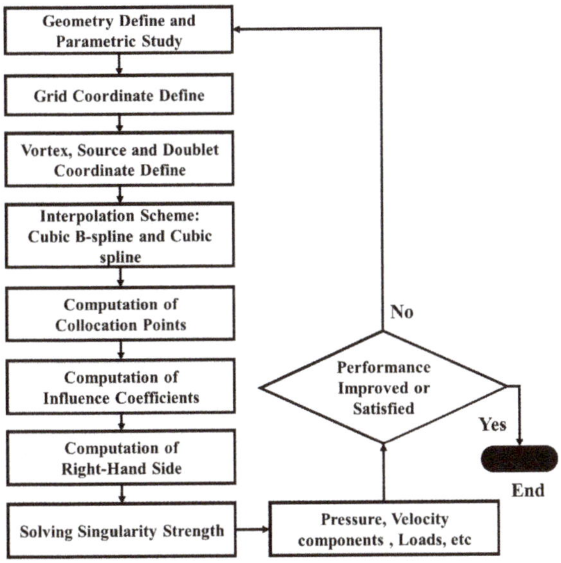

Figure 4. The numerical calculation process for the surface panel method.

Modelling of 3-D Hydrofoil

The Dirichlet boundary condition is used for a hydrofoil with a certain thickness. The governing equation can be expressed as follows (the internal disturbance potential of the object is equal to zero) [28]:

$$\sum_{k=1}^{N} c_k\mu_k + \sum_{l=1}^{N_W} c_l\mu_l + \sum_{k=1}^{N} b_k\sigma_k = 0 \tag{12}$$

Equation (12) is applied to every collocation point P. In the equation, l and k are the count numbers of the singularity elements. The geometry of the hydrofoil is shown in Figure 5. The influence coefficients of the doublets c_k, the wake doublets c_l, and the source b_k can be calculated. The Kutta condition is expressed by the relationship between the strength of the doublets on the upper and lower surfaces and the doublet strength on the wake surface [28,32]:

$$(\mu_1 - \mu_N) + \mu_W = 0 \tag{13}$$

The matrix equation under the Kutta condition can be expressed as follows:

$$\sum_{i=1}^{N=1}\sum_{j=1}^{N=1} c_{ij}\mu_j = \begin{bmatrix} c_{11} & c_{12} & \cdots & \cdots & c_{1N} & c_{1W} \\ c_{21} & c_{22} & \cdots & \cdots & c_{2N} & c_{2W} \\ \cdots & \cdots & \cdots & \cdots & \cdots & \cdots \\ c_{N1} & c_{N2} & \cdots & \cdots & c_{NN} & c_{NW} \\ 1 & 0 & 0 & \cdots & -1 & 1 \end{bmatrix} \begin{bmatrix} \mu_1 \\ \mu_2 \\ \cdots \\ \mu_N \\ \mu_W \end{bmatrix} \tag{14}$$

In this case, μ_w is replaced with $\mu_N - \mu_1$. Therefore, the order will be reduced to N. Only the first and the Nth columns will change because of the term $\pm c_{iw}$. Hence, the influence of the doublet can be expressed as follows:

$$\begin{cases} a_{ij} = c_{ij} & j \neq 1,\ N \\ a_{i1} = c_{i1} - c_{iW} & j = 1 \\ a_{iN} = c_{iN} + c_{iW} & j = N \end{cases} \tag{15}$$

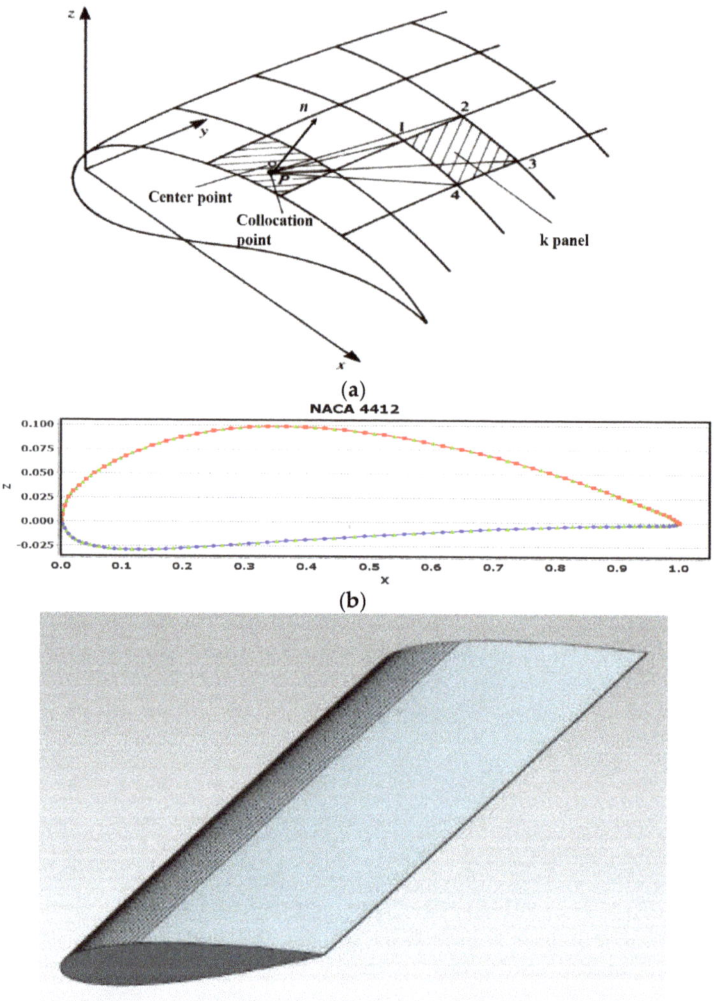

Figure 5. The discrete panels on the hydrofoil surface: (**a**) definitions of collocation point P and panel k for calculation of the influence coefficient of panel k on collocation point P; (**b**) cosine distribution of singularity on the upper surface (red line denoted with square points) and on the lower surface (blue line with circular points) and collocation point (green triangle points); (**c**) resolution of panels.

After the known matrix multiplication is moved to the right-hand side (RHS) of this equation, the RHS vector can be obtained. Based on the theory of the panel method shown above, a program based on MATLAB was developed to calculate the strength of the singularities. Then, the coefficient of pressure on the surface of the foil can be obtained. The external potential ϕ_u can be represented as the internal potential ϕ_i plus the doublet strength μ. Then, the local external tangential velocity component at each collocation point can be expressed by the following equations:

$$\phi_u = \phi_i + \mu \tag{16}$$

$$Q_t = \frac{\partial \phi_u}{\partial l} \tag{17}$$

$$C_{pj} = 1 - \frac{Q_{tj}^2}{Q_\infty} \quad (18)$$

where l is the distance between 2 adjacent collocation points.

2.4. Numerical Modelling of 3-D Propeller

For geometric boundary discretization and the motion of the propeller, 2 sets of coordinate systems are established to describe the propeller geometry accurately. The rectangular coordinate system and the cylindrical coordinate system are defined as shown in Figure 6. The conversion between the 2 sets of coordinate systems is as follows [32]:

$$\begin{cases} x = x \\ y = r\cos\theta \\ z = r\sin\theta \end{cases}, \quad (19)$$

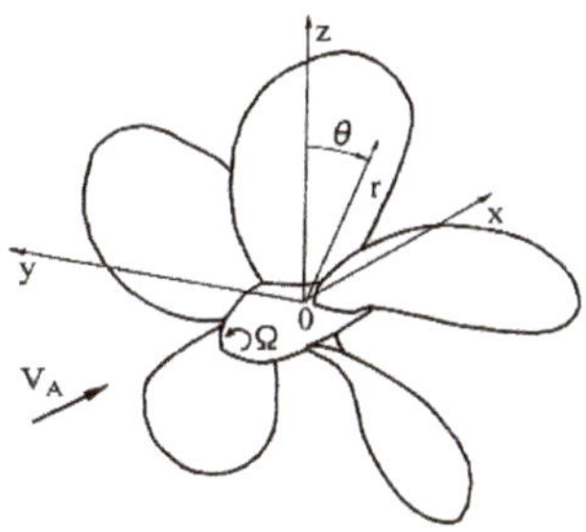

Figure 6. The cartesian coordinate system and the cylindrical coordinate system.

The key point in this step is the definition of the propeller coordinates. The Cartesian and cylindrical coordinates of a point on the camber surface with radial coordinate r and chordwise coordinate s can be expressed in terms of the skew (θ_s), rake (x_m), chord (c), and camber (f) [29]. With this definition, the coordinates of a point on the camber surface of the key blade can be written as [29,36]:

$$\begin{cases} x_c = x_m + c(s - 1/2)\sin\phi - f\cos\phi \\ \theta_c = \theta_m + c(s - 1/2)\frac{\cos\phi}{r} + f\frac{\sin\phi}{r} \\ y_c = r\cos\theta_c \\ z_c = r\sin\theta_c \end{cases}, \quad (20)$$

The above equation expresses the coordinates of any point on the camber surface of the propeller blade. If we use $f \pm 1/2t$ to replace f in the equation, we can obtain the numerical expression for the upper and lower surfaces of the propeller blade. Figure 7 shows the 3-D propeller generation process for the surface panel method.

The surface of the propeller blade is discretized using hyperbolic quadrilateral panels. Each propeller blade is divided into $N_R \times 2M_C$ panels; N_R is the number of radial panels and M_C is the number of spanwise panels. This method of discretization is called the cosine-clustering method. The coordinates of the corner points of the panels can be expressed as [39]:

$$\begin{cases} r_m = \frac{1}{2}\left[\left(1 + r_h - \frac{\delta r_1}{4}\right) - \left(1 - r_h - \frac{\delta r_1}{4}\right)\frac{\cos(\alpha+\beta_m)}{\cos\alpha}\right] \\ \alpha = \frac{\arcsin\left(\frac{\delta r_1}{\delta r_{\max}}\right)N_R - \frac{\pi}{2}}{N_R} \\ \beta = \frac{(\pi-2\alpha)(m-1)}{N_R}, m = 1,2,3\ldots N_R + 1 \end{cases}, \quad (21)$$

where $\delta r_{\max}$ and δr_1 are the radial distribution of the panel with a maximum radial dimension and the first panel, respectively. At any given radius r, the propeller blade is discretized in the chordwise direction using the cosine-clustering method. From the leading edge to the trailing edge, the coordinates of the corner points are expressed as [39]:

$$\begin{cases} s_n = \frac{1}{2}\left(1 - \frac{\cos(\alpha+\beta_m)}{\cos\alpha}\right) \\ \alpha = \frac{\arcsin\left(\frac{\delta s_1}{\delta s_{\max}}\right)(M_C+1)-\frac{\pi}{2}}{M_C+1} \\ \beta = \frac{(\pi-2\alpha)(m-1)}{M_C}, m = 1,2,3\ldots M_C+1 \end{cases}, \tag{22}$$

where $\delta s_{\max}$ and δs_1 are the spanwise dimensions of the panel with the maximum spanwise dimension and the first panel, respectively. The cosine-clustering method was adopted to optimize the sparse density of the discrete surfaces of the propeller panels.

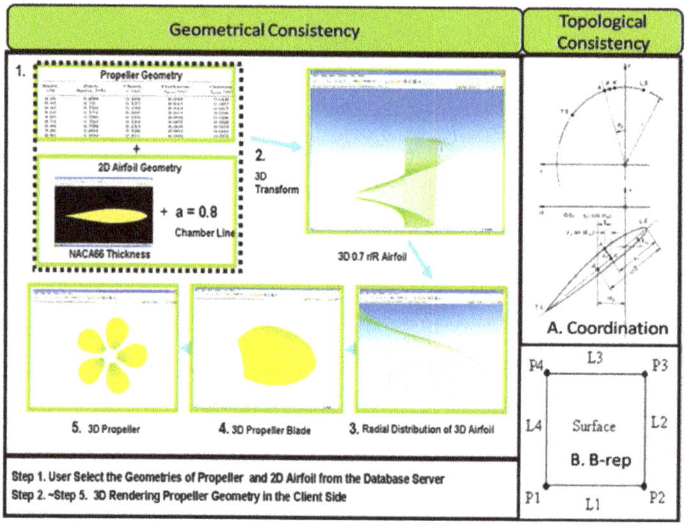

Figure 7. The three-dimensional propeller generation process for surface panel method.

Propeller Wake Model

The propeller wake model has an influence on the perturbation velocity potential flow of the propeller. The helical spiral surface can be used to model the wake surface of a propeller. The propeller pitch angle can be expressed by a linear combination of the blade pitch angle and that based on lift-line theory. Greeley and Kerwin [36] proposed a model of radial shrinkage in the wake region and achieved good results. In this wake model, the radius of the wake surface decreases and eventually shrinks into 2 vortices, the hub vortex and the tip vortex. Kanemaru and Ando [16] improved the geometry of the wake surface by making it more refined, combining Gaschler's experimental results [40] on the wake surface into the modelling.

In this study, a wake shrinkage model was applied to improve the prediction of the performance of a propeller in which the pitch of the wake surface changes with shrinkage (Figure 8). To obtain a more reasonable panel distribution on the wake surface, the discrete coordinates of the wake surface can be calculated using Equations (21)–(25). Specifically, when a downstream coordinate is given, the region from the blade to the downstream coordinate θ_w is the near-wake region, and the region farther than the downstream coordinate θ_w is the far-wake region. The distance from the angular coordinate θ_{TE} of the trailing edge to the downstream coordinate θ_w is defined by s, and the radius of the wake is contracted to r_w at $s = 1$. It is assumed that the wake radius of the central part changes smoothly, and that the pitch angle of the wake surface changes smoothly in the direction of the blade radius and wake flow. The radius of the far-wake region at $s > 1$ does not shrink, while the pitch angle does not change in the direction of the wake flow. The geometric parameters of the wake surface are defined as follows:

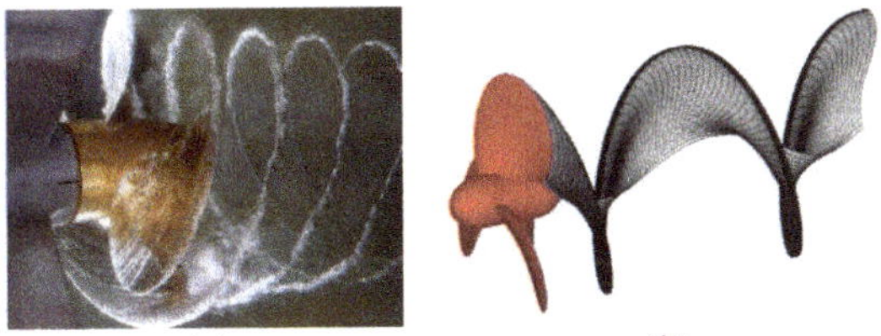

Figure 8. The wake surface model of a propeller: (**a**) experimental results [38], (**b**) numerical model.

(1) Tip vortex radius [39]:

$$r_{tip}(s) = \begin{cases} r_{TE} + (r_w - r_{TE}) \cdot (3s - 3s^2 + s^3) & 0 \le s \le 1 \\ r_{rw} & s > 1 \end{cases} \tag{23}$$

(2) Wake radius at other locations [41]:

$$r(s) = \begin{cases} \sqrt{r_h^2 + (r_{TE}^2 - r_h^2) \cdot \left(\frac{r_{tip}^2(s) - r_h^2}{r_{tip}^2(s) - r_h^2} \right)} & 0 \le s \le 1 \\ C_{rw} = 0.625 + 0.05\theta_s/90^\circ & s > 1 \end{cases} \tag{24}$$

where r_{TE} represents the blade radius at the trailing edge, r_h represents the radius of the propeller hub, and r_w represents the radius of the shrinking tip vortex and satisfies the following conditions [29,31]:

$$\begin{cases} r_w = C_{rw} + \frac{J}{P_{0.7R}/D} \left(0.475 - 0.255 \frac{J}{P_{0.7R}/D} \right) \\ C_{rw} = 0.625 + 0.05\theta_s/90^\circ \end{cases} \tag{25}$$

The pitch angle of the wake surface changes as the following equation [39,42]:

$$\beta_w(s) = \begin{cases} \beta_w(0) + (\beta_w(1) - \beta_w(0)) \cdot (3s - 3s^2 + s^3) & 0 \le s \le 1 \\ \beta_w(1) = \beta_G(r) & s > 1 \end{cases} \tag{26}$$

Figure 9 shows discrete panels on a propeller and wake surface for the surface panel method. The geometric model of the wake surface can be constructed accurately using the characteristic quantities of the wake surface. Because the influence of the panels at the trailing edge is greater, the cosine-clustering method was applied for the wake surface at a given radial profile. The corner point coordinates of the panels in the wake transition region were determined using the following equation [37]:

$$\begin{cases} \theta(s) = \theta_{TE} + s(i) \cdot (\theta_w - \theta_{TE}) \\ s(i) = \frac{1}{2}\left(1 - \cos \frac{i-1}{N_w}\pi\right),\ i = 1,\ 2,\ \cdots,\ N_w + 1 \end{cases} \tag{27}$$

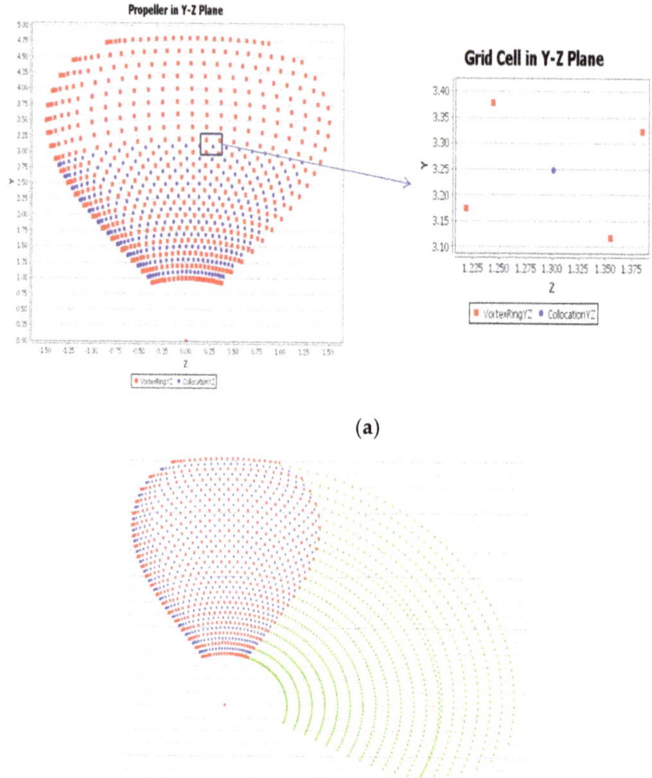

(a)

(b)

Figure 9. The discrete panels on the DTMB4381 propeller surface: (**a**) definitions of collocation point and panel for calculation of the influence coefficient of the panel on collocation point P; (**b**) cosine distribution of singularity, collocation point, and propeller wake chordwise and spanwise.

3. Results

3.1. Validation of the Panel Method Using 2-D Hydrofoil Performance

To validate the two-dimensional hydrofoil panel method, the hydrofoil profiles NACA0012, NACA2410, NACA6412, and NACA4418 [33] were selected for calculation. The lift coefficient of NACA0012 was calculated at different angles of attack using the two-dimensional hydrofoil panel method and compared with the experimental data; the results are shown in Figure 10a–d.

Figure 10a–d show that the calculated results are in good agreement with the experimental data [33], and the average error using the B-spline panel method is within 6.3% whereas the average error was 7.1% using the cubic spline panel method. The results for NACA0012 and NACA2410 have the highest degree of agreement. There are some differences between the calculated results and experimental data for NACA6412 and NACA4418, but the trend of the pressure coefficient curve obtained through calculation is in good agreement with the experimental data [33]. The higher the degree of the camber line, the greater the effect on the calculation accuracy; the number of panels also has some influence on the calculation results. For the two-dimensional hydrofoil, the number of panels should be controlled in the range of 40–80. A lower grid density would lead to less accuracy. Figure 11 shows a comparison of the calculated and experimental results of lift coefficient for NACA0012 at different angles of attack using the B-spline and cubic spline panel methods. The calculated results using the B-spline panel method show a higher degree of agreement with the experimental data.

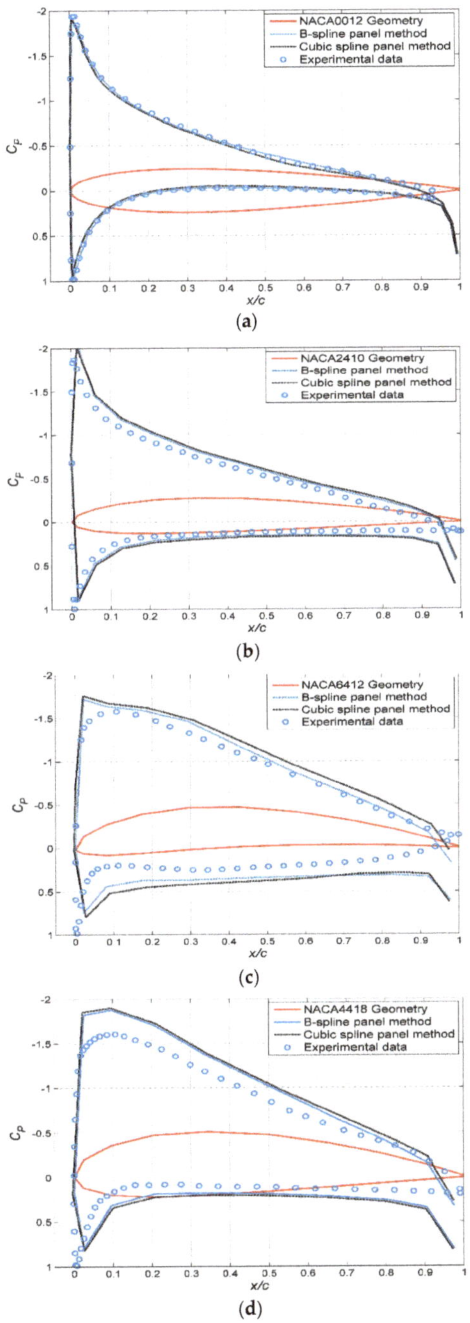

Figure 10. The comparison of calculated and experimental results of the surface pressure coefficient at a 5° angle of attack for (**a**) NACA0012, (**b**) NACA2410, (**c**) NACA6412, (**d**) NACA4418.

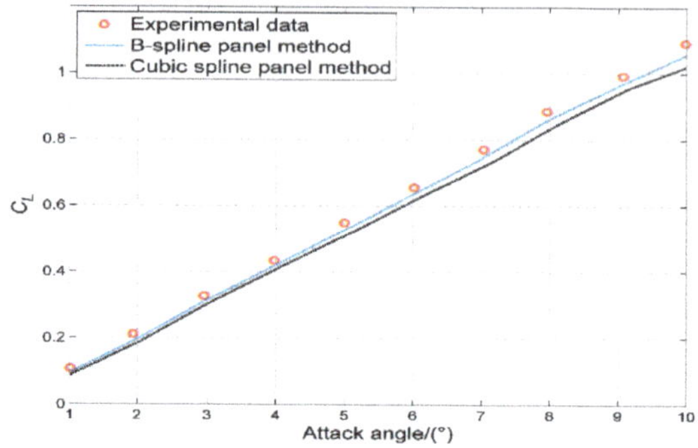

Figure 11. The comparison of calculated and experimental results of the lift coefficient for NACA0012 at different angles of attack.

3.2. Validation of the Panel Method Using 3-D Hydrofoil Performance

In this study, a three-dimensional rectangular hydrofoil was selected for calculation using the panel method. The hydrofoil type selected was NACA0012 [33]. The aspect ratio was set to 10, the angle of attack was set to 5°, and the number of panels was 180. The surface pressure distribution on the hydrofoil is shown in Figure 12.

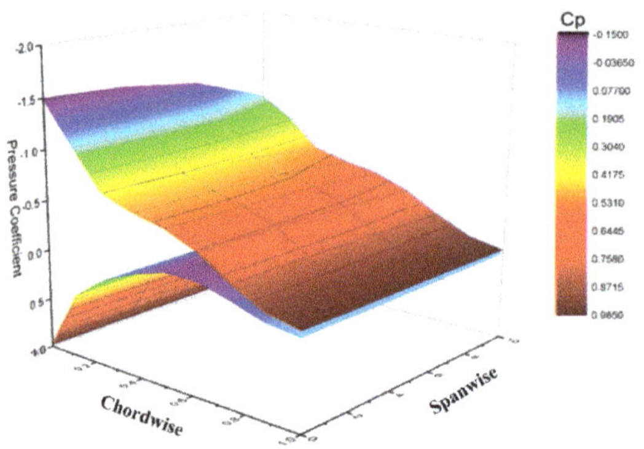

Figure 12. The NACA0012 3-D hydrofoil surface pressure coefficient distribution.

The surface pressure distribution of the hydrofoil can be directly compared with the experimental data reported by Lee and Jang [34]. In the present study, the number of chordwise panels per section is 18, and the number of spanwise panels per section is 10. With this discretization method, it is possible that some regions of the panel shape become relatively slender, resulting in a lower calculation accuracy. It is necessary to develop some algorithm to judge the size of the panels to ensure that they are not very slender. The results obtained using the hydrofoil panel method are compared with the experimental results for the sections defined by a span fraction (s) to the wing span (S) as s/S = 0.25, 0.5, 0.65, and 0.85 as shown in Figure 13a–d, respectively.

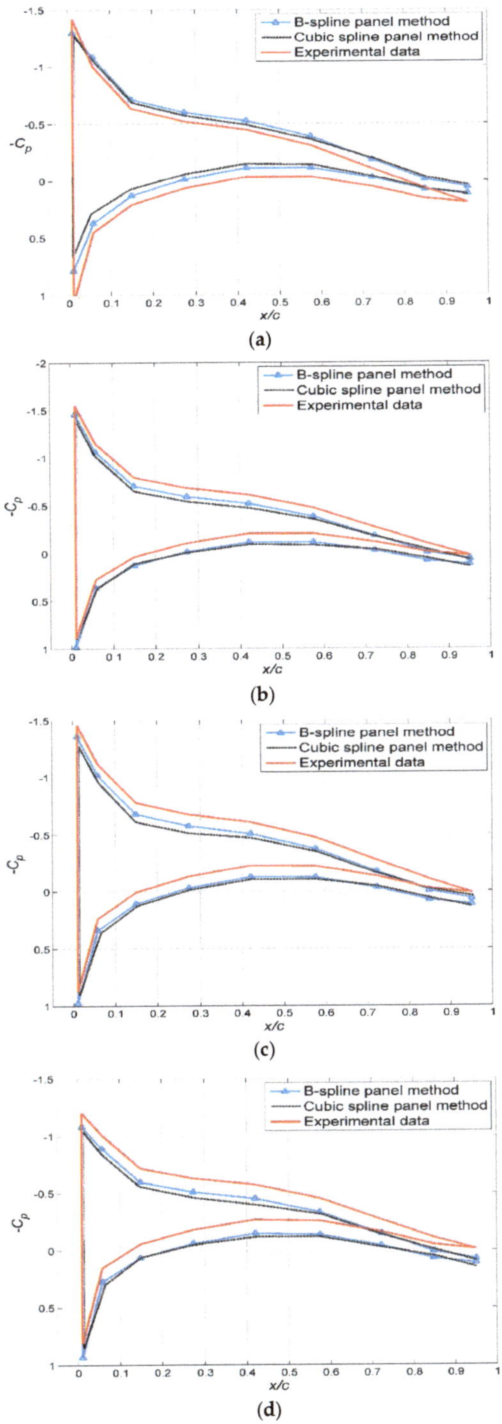

Figure 13. The chordwise distribution of the surface pressure coefficient of 3-D hydrofoil at different spanwise fractions: (**a**) s/S = 0.25, (**b**) s/S = 0.5, (**c**) s/S = 0.65, (**d**) s/S = 0.85.

3.3. Hydrodynamic Analysis of Sweptback and Sweptforward Hydrofoils

Compared with the straight rectangular hydrofoil, the sweptback hydrofoil can effectively reduce the impact force of the flow and improve the hydrodynamic performance. To test the adaptability and stability of the panel method, rectangular hydrofoils with different sweptback and sweptforward angles were selected for analysis. The hydrodynamic performance was calculated with different sweepback angles at a limited angle of attack using the panel method. Discretization of the hydrofoil was performed using the cosine-clustering method. The aspect ratio was set to 6 and the angle of attack was set to 5°. A comparison of the calculation results and the experimental data of the sweptback hydrofoil performance [41] is given in Figure 14. The experimental data considered the ground effect on the sweptback hydrofoils whereas the prediction of the sweptback hydrofoil performance in the present panel method did not consider the ground effect, hence, the value of the experimental lift coefficient at the zero sweptback angle is greater than those of the panel methods as shown in Figure 14a.

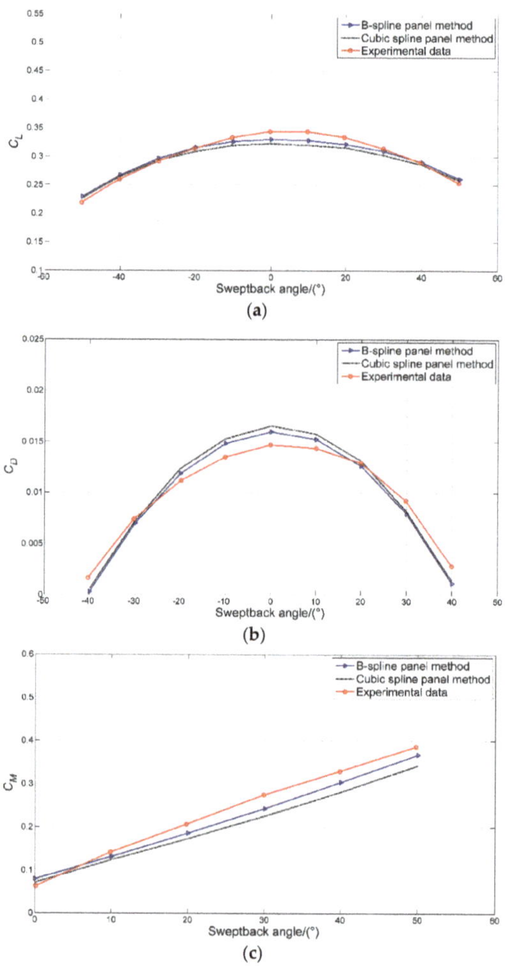

Figure 14. The comparison of the B-spline and cubic spline results and experimental data for the hydrodynamic performance of a sweptback wing in a range of sweptback angles: (**a**) lift coefficients, (**b**) drag coefficients, (**c**) pitch moment coefficients.

Figure 14a–c show that the lift coefficients, induced drag coefficients, and moment coefficients calculated by the B-spline and cubic spline panel methods are in good agreement with the experimental data. The average error based on the cubic B-spline method is within 4% whereas the cubic spline one is 5.9%. The calculated results using the B-spline panel method show a higher degree of agreement with the experimental data. Hence, a better superiority was experienced in the cubic B-spline panel method compared to the cubic spline one. The lift coefficient and induced drag coefficient of the hydrofoil reach the maximum value when the sweptback angle is near zero and decrease with the increased sweep angle. A comparison of the calculation results of hydrofoil performance with different aspect ratios based on the cubic B-spline panel method is given in Figures 13 and 15.

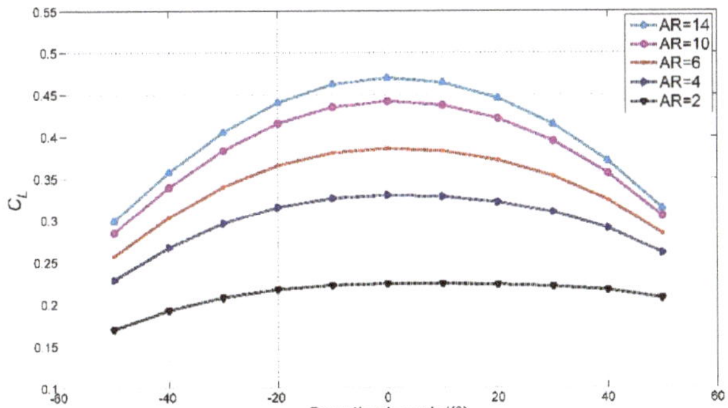

(a)

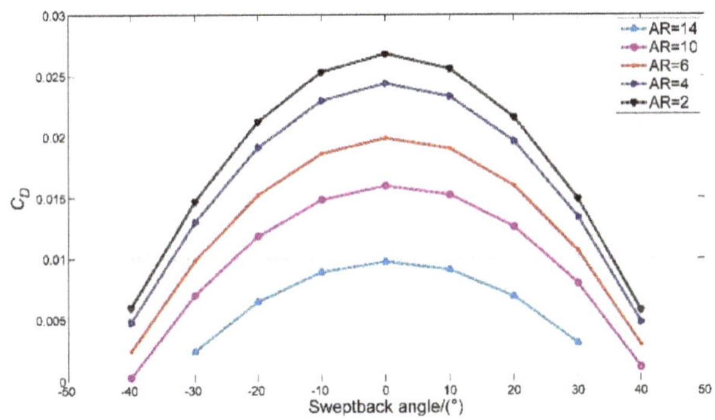

(b)

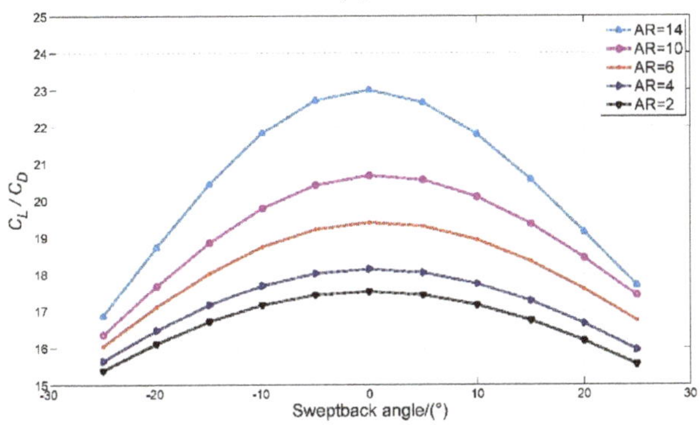

(c)

Figure 15. *Cont.*

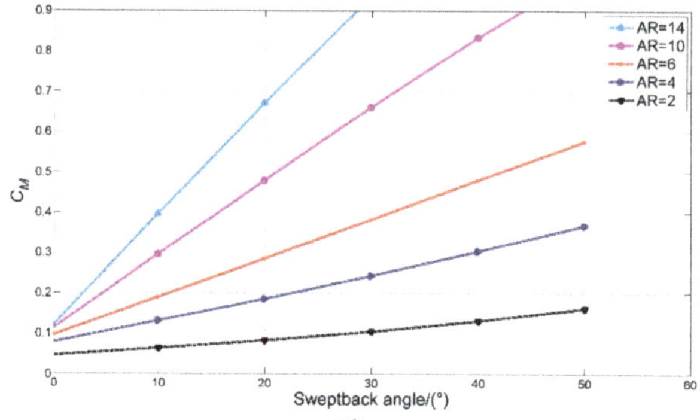

(d)

Figure 15. The comparison of sweptback hydrofoil with different aspect ratios and sweptback angles: (**a**) lift coefficient, (**b**) drag coefficient, (**c**) lift-to-drag ratio, (**d**) pitch moment coefficient.

Figure 15a–d show that the lift coefficient and the lift-to-drag ratio increase with the increased aspect ratio of the sweptback hydrofoil, while the induced drag coefficient and moment coefficient decrease.

3.4. Hydrodynamic Analysis of Dihedral Hydrofoils

The stability of the hydrofoil can be improved by using a dihedral hydrofoil as shown in Figure 16. The calculation results of the hydrodynamic performance of the dihedral hydrofoil with different aspect ratios based on the cubic B-spline panel method are shown in Figure 17.

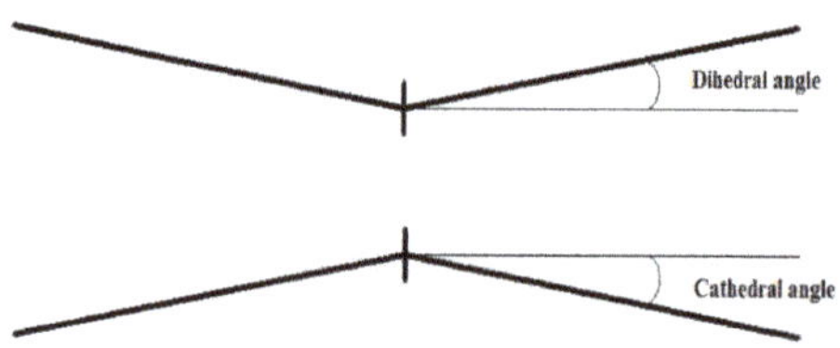

Figure 16. Dihedral hydrofoil geometry.

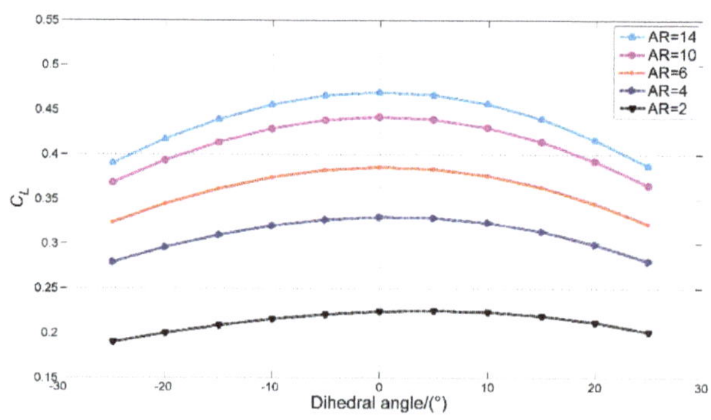

(a)

Figure 17. *Cont.*

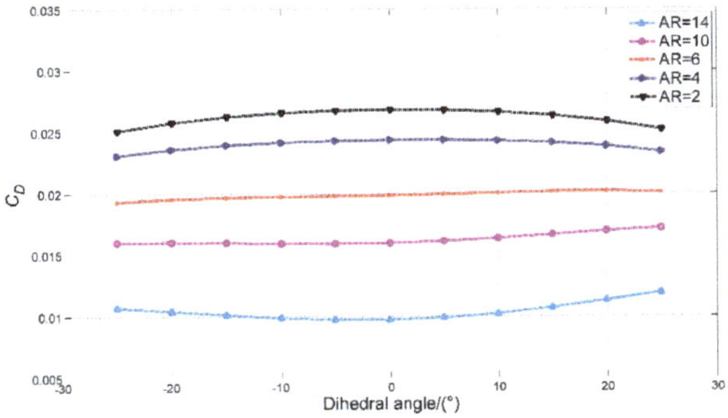

(b)

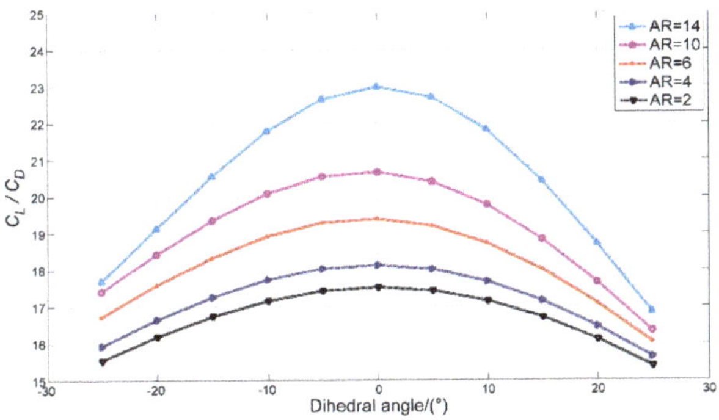

(c)

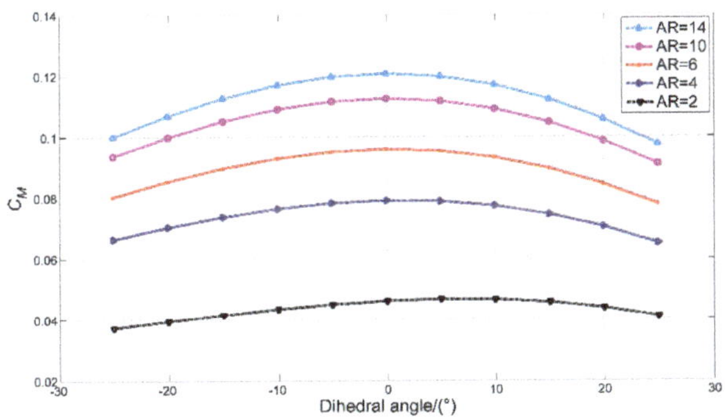

(d)

Figure 17. The comparison of dihedral hydrofoil with different aspect ratios and dihedral angles: (**a**) lift coefficient, (**b**) induced drag coefficient, (**c**) lift-to-drag ratio, (**d**) pitch moment coefficient.

Figure 17a–d show that with the increased dihedral angle, the lift coefficient, lift-to-drag ratio, and moment coefficient decrease, and the induced drag coefficient changes with the aspect ratio.

For low-speed hydrofoils, the hydrodynamic performance can be improved by increasing the lift coefficient and the lift-to-drag ratio; in this case, it is possible to increase the hydrofoil aspect ratio, reduce the sweptback angle, and keep the hydrofoil as straight as possible. At higher speeds, the hydrofoil must not only have a good hydrodynamic performance, but it may also be resistant to stress fatigue; hence, it is not appropriate to reduce the aspect ratio and increase the sweptback angle.

3.5. Results of Propeller Performance

In this study, the DTMB series propellers with different values of rake and skew were selected for calculation of hydrodynamic performance using the panel methods with the present method and the conventional cubic spline scheme. The selected propellers were DTMB4119, DTMB4497, DTMB4498, DTMB4381, DTMB4382, DTMB4383, and DTMB4384.

3.5.1. Validation of the Panel Method Using DTMB4119 Propeller Performance

DTMB4119 was approved by the 20th International Towing Tank Conference as a standard propeller to verify the accuracy of the panel method. The geometric parameters of the propeller and the experimental data pertaining to DTMB4119 were taken from the conference reports [2,37], and the corresponding details for the other propellers were taken from previously published documents [29,36,43].

The propeller blade surface was discretized in the chordwise direction using the cosine-clustering method. The DTMB4119 propeller panel discretization model is shown below (Figure 18) and the relative characteristics of the DTMB4119 are shown in Table 1 [37].

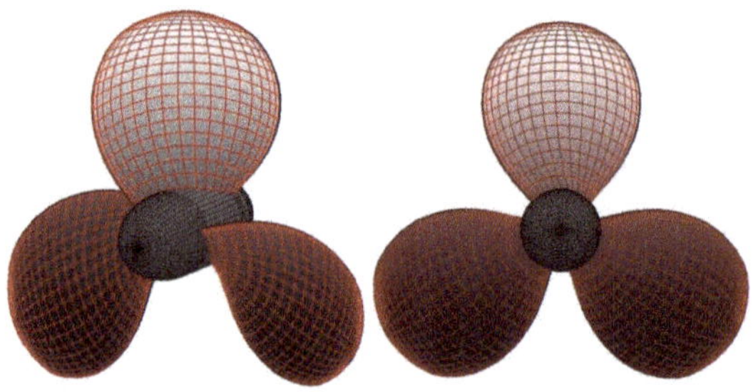

Figure 18. The DTMB4119 propeller panel discretization model.

Table 1. The geometric parameters of propeller DTMB4119 [37].

Number of blades, Z: 3
Hub diameter ratio: 0.2
Expanded area ratio: 0.6
Section mean line: NACA a = 0.8
Section thickness distribution: NACA66 (modified)
Design advance coefficient, J = 0.833

r/R	c/D	P/D	θ_s	x_m/D	t/D	f_0/c
0.200	0.321	1.105	0	0	0.200	0.014
0.250	0.343	1.103	0	0	0.180	0.019
0.300	0.361	1.102	0	0	0.161	0.023
0.400	0.405	1.098	0	0	0.113	0.023
0.500	0.443	1.093	0	0	0.093	0.021
0.600	0.463	1.087	0	0	0.074	0.020
0.700	0.465	1.083	0	0	0.052	0.020
0.800	0.436	1.081	0	0	0.043	0.019
0.900	0.363	1.078	0	0	0.032	0.018
0.950	0.286	1.077	0	0	0.034	0.016
1.000	0.031	1.075	0	0	0.001	0

The literature [2,37] offers the experimental data of the pressure coefficient distribution for DTMB4119 at different sections at a speed factor of J = 0.833. A comparison of the pressure coefficients obtained from the experimental data and the calculated results is shown in Figure 19a–d.

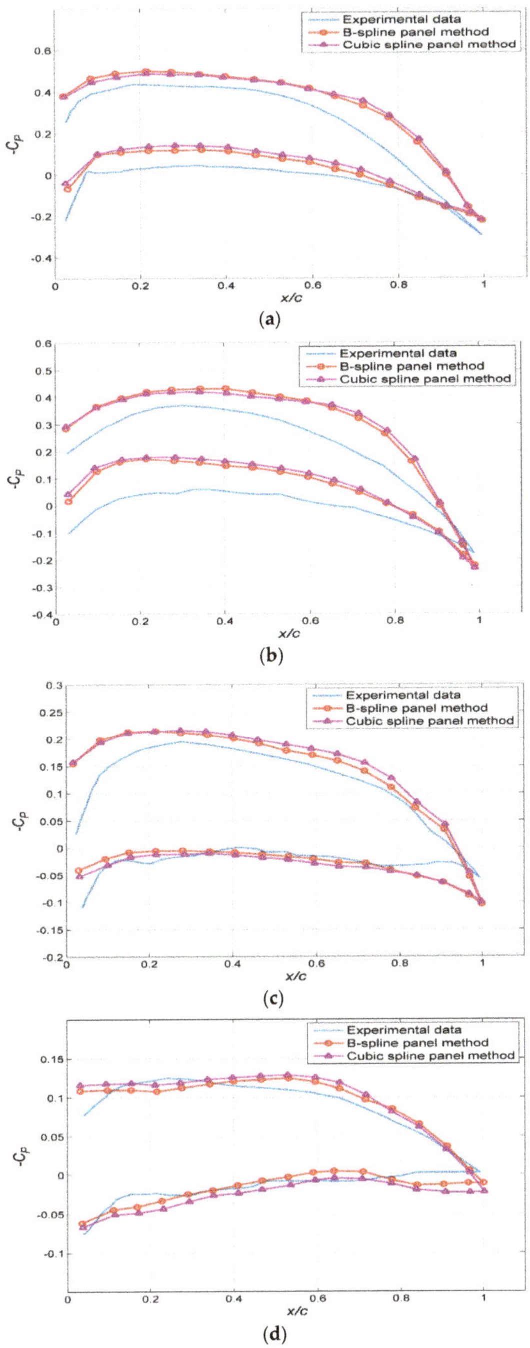

Figure 19. The comparison of the calculated and experimental results of the surface pressure coefficient of DTMB4119. The propeller at different radius fractions using the cubic B-spline and cubic spline panel methods: (**a**) r/R = 0.3, (**b**) r/R = 0.5, (**c**) r/R = 0.7, (**d**) r/R = 0.9.

In general, the calculated results of the surface pressure coefficient of the DTMB4119 propeller blade are in agreement with the experimental results [2,37], and the average error is within 7.3%. The selection of the wake model has some influence on the calculation results. The model selected in this study is the empirical wake model, which is different from the actual wake. Therefore, the wake model is not the main factor in the calculation results.

3.5.2. Results of Propeller Thrust and Toque

A comparison of the hydrodynamic coefficients for the propellers obtained from the calculation results and experimental results [2,36,38] is shown in Figure 20a–f. The number of panels is 280. Based on the cosine clustering method together with cubic B-spline interpolation, these panels were clustered more tightly and smoothly near leading and trailing edges of 3-D propeller hydrofoils placed on the propeller helix lines for the best results.

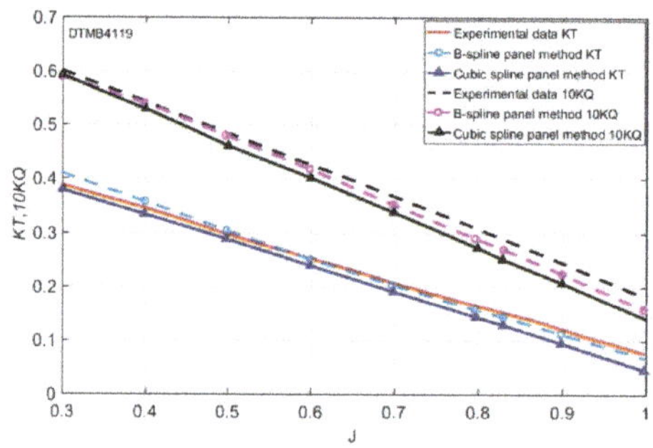

(a)

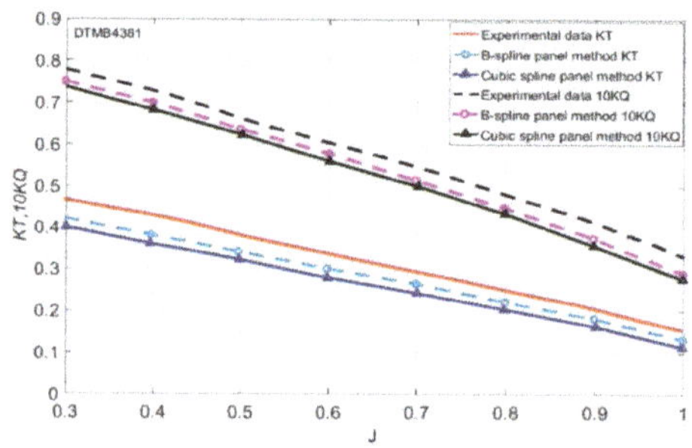

(b)

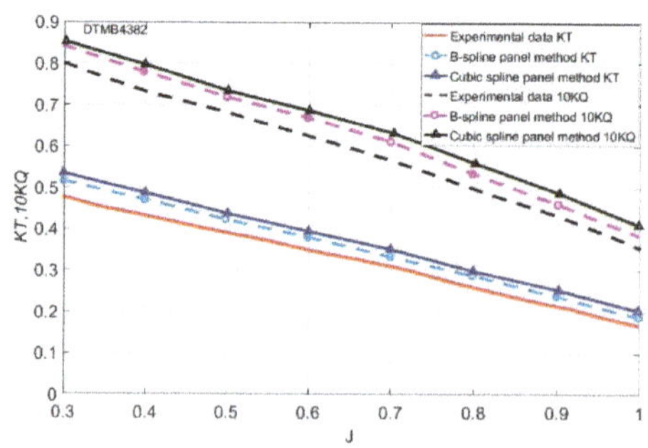

(c)

Figure 20. *Cont.*

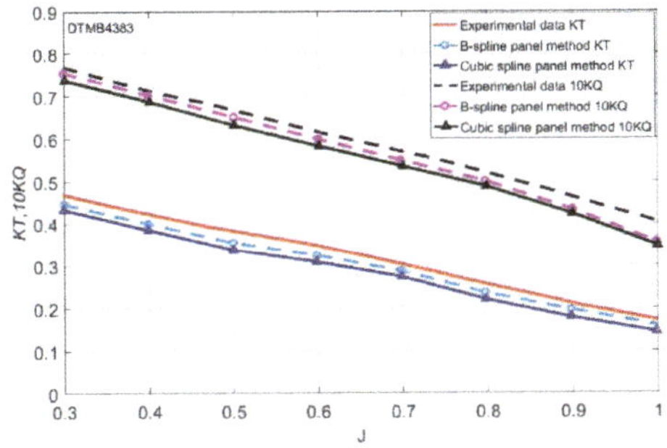

(**d**)

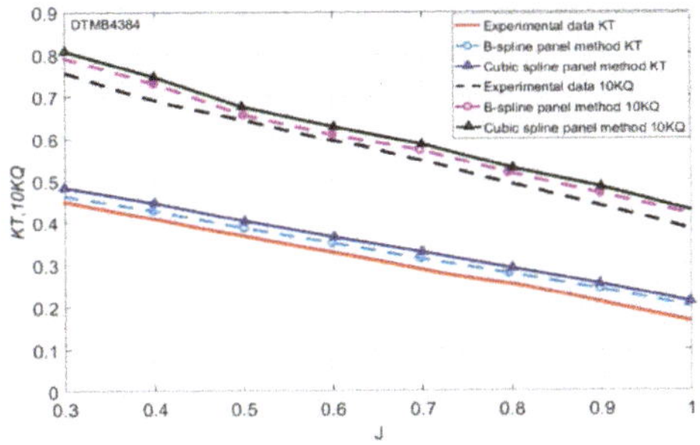

(**e**)

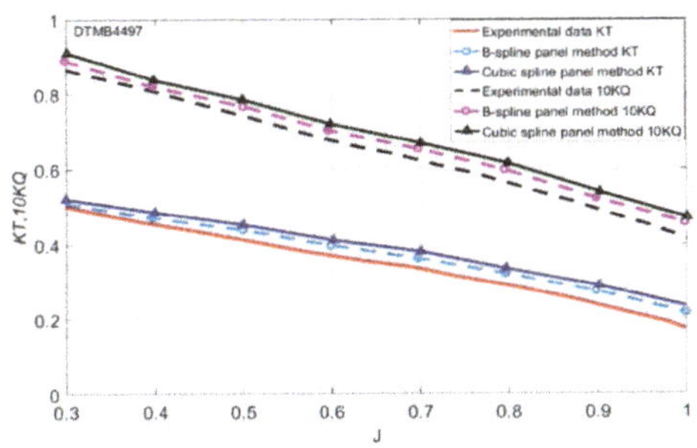

(**f**)

Figure 20. The comparison of the calculated and experimental results of hydrodynamic coefficients for different marine propeller types using the cubic B-spline and cubic spline panel methods: (**a**) DTMB4119 propeller, (**b**) DTMB4381 propeller, (**c**) DTMB4382 propeller, (**d**) DTMB4383 propeller, (**e**) DTMB4384 propeller, (**f**) DTMB4497 propeller.

Figure 20a–f show that the hydrodynamic performance of the propellers is in good agreement with the experimental data, and the average error is within 7.3% based on the cubic B-spline panel method whereas the average error 8.2% using the conventional cubic spline panel method. Especially for the DTMB4119 propeller, the accuracy of the hydrodynamic performance calculation is high, and the average error is within 6% based on the proposed method.

4. Conclusions

A computational technique using the B-spline panel method and the cubic spline panel method to predict the performance of a hydrofoil and propeller is presented in this paper. The surface pressure coefficient and lift performance of a two-dimensional hydrofoil are calculated and analyzed by the above numerical methods. The results are compared with the experimental data. In addition, the hydrodynamic performance of a three-dimensional hydrofoil and of a sweepback hydrofoil and a dihedral hydrofoil with different aspect ratios are analyzed, and the average error is within 4% in the present method, whereas it is 5.9% of the cubic spline method. The results confirm the rationality and feasibility of the method. The surface pressure distribution and the coefficients of thrust and torque of different propellers are calculated and compared with the experimental data, and the average error is within 7.3% in the present method whereas it is 8.2% using the conventional method. The comparison results are found satisfactory. In conclusion, the validity of all the calculated results from the cases studied in this paper was proved by comparing them to the available studies in the literature. Additionally, the calculated results using the proposed B-spline panel method show a higher degree of agreement with the experimental data than the cubic spline one. In the future, the panel method can be used for the inverse design based on the required load and modelling predictions of hydrofoil cavitation.

Author Contributions: Conceptualization, C.-W.C.; Data curation, M.L.; Formal analysis, M.L.; Funding acquisition, C.-W.C. and M.L.; Resources, C.-W.C.; Validation, C.-W.C.; Writing–original draft, C.-W.C.; Writing–review & editing, M.L. The revision was done by C.-W.C. under the direction and supervision by M.L.

Acknowledgments: The authors wish to thank the National Natural Science Foundation of China and Zhejiang Zhoushan Science and Technology Project for financial support for this research under the project grant numbers 51409230 and 2018C81041. Ming Li acknowledges the National Natural Science Foundation of China under the project grant numbers 61672238 and 61272402. We are grateful for the anonymous reviewers for their valuable comments in improving the paper. The authors highly appreciate the National Advisory Committee for Aeronautics (NACA), David Taylor Model Basin (DTMB) centers, and Greeley, D.S., Kerwin, J.E., Abbott H., Yang W., Ping. N. and Lee S.J. for their real experimental data.

Conflicts of Interest: The authors declare no conflict of interest.

Appendix A

A1. Numerical Representation of Panel Method

Numerical Modelling of the Integral Equation for Propellers

The doublet and source are distributed on the panels, and the surface integral equation (Equation (4)) is satisfied at the collocation point of every panel. The second type of Fredholm integral equation can be discretized as [28]

$$\sum_{\substack{j=1 \\ i \neq j}}^{N} (\delta_{ij} - C_{ij})\phi_j - \sum_{j=1}^{N_R} W_{ij}\Delta\phi_j = \sum_{j=1}^{N} B_{ij}(-V_I \cdot n_j) \tag{A1}$$

where δ_{ij} represents the Kronecker number and N represents the total number of panels on a single propeller blade. B_{ij}, C_{ij}, and W_{ij} are the influence coefficients of the singularity points, which can be calculated using the following equation [1]:

$$\begin{cases} C_{ij} = \sum_{K=1}^{K} \left[\frac{1}{2\pi} \iint_{S_j} \frac{\partial}{\partial n_j} \left(\frac{1}{R_{ijk}} \right) dS_j \right] \\ B_{ij} = \sum_{K=1}^{K} \left[-\frac{1}{2\pi} \iint_{S_j} \frac{\partial}{\partial n_j} \left(\frac{1}{R_{ijk}} \right) dS_j \right] \\ W_{ij} = \sum_{K=1}^{K} \sum_{L=1}^{L} \left[-\frac{1}{2\pi} \iint_{S_l} \frac{\partial}{\partial n_l} \left(\frac{1}{R_{ilk}} \right) dS_l \right] \end{cases} \tag{A2}$$

where R_{ilk} and R_{ijk} are the distances between the collocation points and panels S_l and S_j, respectively. The influence coefficient of the singularity element on the collocation point is the key part of the calculation in the panel method. The doublet and source are distributed at the corners of the hyperboloid panels, and then the integral of the influence coefficient can be calculated after discretization. Calculating the influence coefficient requires establishing the panel coordinate system

(ξ_1, ξ_2). By considering the coordinates of the four corner points of the panel as $Q_i(i = 1 \sim 4)$, any vector in the panel coordinate system can be expressed as follows [43]:

$$\mathbf{Q}(\xi_1, \xi_2) = \omega_1 Q_1 + \omega_2 Q_2 + \omega_3 Q_3 + \omega_4 Q_4 \tag{A3}$$

where $\omega_i(i = 1 \sim 4)$ represents the interpolation function, and the values of the interpolation function are given as follows [43]:

$$\begin{cases} \omega_1 = \frac{1}{4}(1+\xi_1)(1+\xi_2) \\ \omega_2 = \frac{1}{4}(1+\xi_1)(1-\xi_2) \\ \omega_3 = \frac{1}{4}(1-\xi_1)(1+\xi_2) \\ \omega_4 = \frac{1}{4}(1-\xi_1)(1-\xi_2) \end{cases} \tag{A4}$$

The tangent vector of the panels can be expressed as

$$\begin{cases} \mathbf{a}_1(\xi_1, \xi_2) = \frac{\partial \mathbf{Q}}{\partial \xi_1} \\ \mathbf{a}_2(\xi_1, \xi_2) = \frac{\partial \mathbf{Q}}{\partial \xi_2} \end{cases} \tag{A5}$$

The unit normal vector of the panels can be expressed as

$$\mathbf{n}(\xi_1, \xi_2) = \frac{\mathbf{a}_1 \times \mathbf{a}_2}{|\mathbf{a}_1 \times \mathbf{a}_2|} \tag{A6}$$

The surface of the panels can be expressed as follows:

$$dS_j = |\mathbf{a}_1 \times \mathbf{a}_2| d\xi_1 d\xi_2 \tag{A7}$$

The integral of the influence coefficient can be expressed as follows [24]:

$$I = \int_{-1}^{1} \int_{-1}^{1} f(\xi_1, \xi_2) d\xi_1 d\xi_2 \tag{A8}$$

The integral function $\mathbf{f}(\xi_1, \xi_2)$ in the above equation can be expressed as

$$\mathbf{f}(\xi_1, \xi_2) = \frac{\partial^2 \mathbf{F}(\xi_1, \xi_2)}{\partial \xi_1 \partial \xi_2} \tag{A9}$$

The following equation is obtained as follows [24]:

$$I = F(1,1) - F(1,-1) - F(-1,1) + F(-1,-1) \tag{A10}$$

The influence coefficients C_{ij} and B_{ij} can be expressed as follows [25]:

$$\begin{cases} C_{ij} = I_D(1,1) - I_D(1,-1) - I_D(-1,1) + I_D - (1,-1) \\ B_{ij} = I_S(1,1) - I_S(1,-1) - I_S(-1,1) + I_S - (1,-1) \end{cases} \tag{A11}$$

where the influence coefficients can be expressed as follows [25]:

$$\begin{cases} I_D(\xi,\eta) = \frac{1}{2\pi} \tan^{-1}\left(\frac{(\mathbf{R}\times\mathbf{a}_1)\cdot(\mathbf{R}\times\mathbf{a}_2)}{|\mathbf{R}|\mathbf{R}\cdot(\mathbf{a}_1\times\mathbf{a}_2)}\right) \\ I_S(\xi,\eta) = -\frac{1}{2\pi}\left\{ \begin{array}{c} -\frac{(\mathbf{R}\times\mathbf{a}_1)\cdot\mathbf{n}_C}{|\mathbf{a}_1|}\sinh^{-1}\left(\frac{\mathbf{R}\times\mathbf{a}_1}{|\mathbf{R}\times\mathbf{a}_1|}\right) + \frac{(\mathbf{R}\times\mathbf{a}_2)\cdot\mathbf{n}_C}{|\mathbf{a}_2|} + \\ \mathbf{R}\cdot\mathbf{n}_C \tan^{-1}\left(\frac{(\mathbf{R}\times\mathbf{a}_1)\cdot(\mathbf{R}\times\mathbf{a}_2)}{|\mathbf{R}|\mathbf{R}\cdot(\mathbf{a}_1\times\mathbf{a}_2)}\right) \end{array} \right\}, \\ \mathbf{R} = \mathbf{Q}(\xi,\eta) - \mathbf{P} \end{cases} \tag{A12}$$

where $\mathbf{n}_C$ is a normal vector on the collocation point and **P** is the coordinate of the collocation point. Thus, the governing equation can be solved.

A2. The Geometric Parameters of the Propeller for Surface Panel Code (see Tables A1–A5)

Table A1. The geometric parameters of propeller DTMB4381 [36].

Number of blades, Z: 5
Hub diameter ratio: 0.2
Expanded area ratio: 0.6
Section mean line: NACA a = 0.8
Section thickness distribution: NACA66 (modified)
Design advance coefficient, J = 0.833

r/R	c/D	P/D	θ_s	x_m/D	t/D	f_0/c
0.200	0.174	1.332	0	0	0.043	0.035
0.250	0.202	1.338	0	0	0.039	0.036
0.300	0.229	1.345	0	0	0.035	0.036
0.400	0.275	1.358	0	0	0.029	0.034
0.500	0.312	1.336	0	0	0.024	0.030
0.600	0.337	1.280	0	0	0.019	0.024
0.700	0.347	1.210	0	0	0.014	0.019
0.800	0.334	1.137	0	0	0.010	0.014
0.900	0.280	1.066	0	0	0.006	0.012
0.950	0.210	1.031	0	0	0.004	0.007
1.000	0	0.995	0	0	0.003	0

Table A2. The geometric parameters of propeller DTMB4382 [36].

Number of blades, Z: 5
Hub diameter ratio: 0.2
Expanded area ratio: 0.6
Section mean line: NACA a = 0.8
Section thickness distribution: NACA66 (modified)
Design advance coefficient, J = 0.833

r/R	c/D	P/D	θ_s	x_m/D	t/D	f_0/c
0.200	0.167	1.451	0	0	0.041	0.046
0.250	0.205	1.441	2.328	0.008	0.033	0.039
0.300	0.221	1.436	4.655	0.025	0.031	0.035
0.400	0.286	1.423	9.363	0.038	0.024	0.032
0.500	0.314	1.352	13.948	0.048	0.021	0.030
0.600	0.334	1.281	18.378	0.067	0.017	0.025
0.700	0.352	1.185	22.747	0.078	0.012	0.018
0.800	0.334	1.112	27.145	0.081	0.010	0.016
0.900	0.282	1.024	31.575	0.090	0.007	0.014
0.950	0.216	0.973	33.788	0.093	0.005	0.012
1.000	0	0.938	35.000	0.094	0.004	0

Table A3. The geometric parameters of propeller DTMB4383 [36].

Number of blades, Z: 5
Hub diameter ratio: 0.2
Expanded area ratio: 0.6
Section mean line: NACA a = 0.8
Section thickness distribution: NACA66 (modified)
Design advance coefficient, J = 0.833

r/R	c/D	P/D	θ_s	x_m/D	t/D	f_0/c
0.200	0.174	1.566	0	0	0.043	0.040
0.250	0.202	1.539	4.647	0.019	0.039	0.040
0.300	0.229	1.512	9.293	0.039	0.035	0.041
0.400	0.275	1.459	18.816	0.076	0.029	0.038
0.500	0.312	1.386	27.991	0.107	0.024	0.034
0.600	0.337	1.296	36.770	0.132	0.019	0.028
0.700	0.347	1.198	45.453	0.151	0.014	0.023
0.800	0.334	1.096	54.254	0.165	0.010	0.018
0.900	0.280	0.996	63.102	0.174	0.006	0.015
0.950	0.210	0.945	67.531	0.177	0.004	0.016
1.000	0	0.895	72.000	0.179	0.003	0

Table A4. The geometric parameters of propeller DTMB4384 [36].

Number of blades, Z: 5
Hub diameter ratio: 0.2
Expanded area ratio: 0.6
Section mean line: NACA a = 0.8
Section thickness distribution: NACA66 (modified)
Design advance coefficient, J = 0.833

r/R	c/D	P/D	θ_s	x_m/D	t/D	f_0/c
0.200	0.174	1.675	0	0	0.043	0.054
0.250	0.202	1.629	6.961	0.031	0.039	0.050
0.300	0.229	1.584	13.921	0.061	0.035	0.047
0.400	0.275	1.496	28.426	0.118	0.029	0.045
0.500	0.312	1.406	42.152	0.164	0.024	0.040
0.600	0.337	1.305	55.199	0.200	0.019	0.033
0.700	0.347	1.199	68.098	0.226	0.014	0.027
0.800	0.334	1.086	81.283	0.245	0.010	0.023
0.900	0.280	0.973	94.624	0.255	0.006	0.019
0.950	0.210	0.916	101.300	0.257	0.004	0.020
1.000	0	0.859	108.000	0.257	0.003	0

Table A5. The geometric parameters of propeller DTMB4497 [44].

Number of blades, Z: 5
Hub diameter ratio: 0.2
Expanded area ratio: 0.6
Section mean line: NACA a = 0.8
Section thickness distribution: NACA66 (modified)
Design advance coefficient, J = 0.833

r/R	c/D	P/D	θ_s	x_m/D	t/D	f_0/c
0.200	0.178	1.450	0	0	0.043	0.042
0.250	0.217	1.445	2.272	0	0.036	0.037
0.300	0.242	1.432	4.675	0	0.032	0.034
0.400	0.298	1.427	9.312	0	0.024	0.031
0.500	0.314	1.365	13.941	0	0.021	0.029
0.600	0.332	1.291	18.732	0	0.016	0.022
0.700	0.351	1.182	22.578	0	0.014	0.017
0.800	0.329	1.119	27.533	0	0.012	0.015
0.900	0.279	1.015	31.534	0	0.009	0.012
0.950	0.209	0.969	34.783	0	0.007	0.010
1.000	0	0.917	36.000	0	0.005	0

References

1. Hess, J.L.; Smith AM, O. Calculation of potential flow about arbitrary bodies. *Prog. Aerosp. Sci.* **1967**, *8*, 1–138. [CrossRef]
2. Brandner, P. Calculation Results for the 22nd ITTC Propulsor Committee Workshop on Propeller RANS/PANEL Methods Steady Panel Method Analysis of DTMB 4119 Propeller. Available online: https://www.researchgate.net/publication/237730350_Calculation_Results_for_the_22nd_ITTC_Propulsor_Committee_Workshop_on_Propeller_RANSPANEL_Methods_Steady_Panel_Method_Analysis_of_DTMB_4119_Propeller (accessed on 8 February 2019).
3. Baltazar, J.; Campos, J.; Bosschers, J. A Comparison of Panel Method and RANS Calculations for a Ducted Propeller System in Open-Water. In Proceedings of the International Symposium on Marine Propulsors, Launceston, Tasmania, Australia, 5–8 May 2013; pp. 334–343.
4. Hozhabrossadati, S.M.; Challamel, N.; Rezaiee-Pajand, M.; Sani, A.A. Application of Green's function method to bending of stress gradient nanobeams. *Int. J. Solids Struct.* **2018**, *143*, 209–217. [CrossRef]
5. Kesour, K.; Atalla, N. A hybrid patch transfer-Green functions method to solve transmission loss problems of flat single and double walls with attached sound packages. *J. Sound Vib.* **2018**, *429*, 1–17. [CrossRef]
6. Esfahani, J.A.; Vahidhosseini, S.M.; Barati, E. Three-dimensional analytical solution for transport problem during convection drying using Green's function method (GFM). *Appl. Therm. Eng.* **2015**, *85*, 264–277. [CrossRef]
7. Datta, R.; Sen, D. A B-spline-based method for radiation and diffraction problems. *Ocean Eng.* **2006**, *33*, 2240–2259. [CrossRef]
8. Datta, R.; Rodrigues, J.M. Guedes Soares C. Study of the motions of fishing vessels by a time domain panel method. *Ocean Eng.* **2011**, *38*, 782–792. [CrossRef]
9. Belibassakis, K.A.; Politis, G.K.; Thens, A. Analysis of Unsteady Propeller Performance by a Surface Vorticity Panel Method. *Ship Technol. Res.* **2002**, *35*, 342–355.
10. Katz, J.; Weihs, D. Wake Rollup and the Kutta Condition for Airfoils Oscillating at High Frequency. *AIAA J.* **2015**, *19*, 1604–1606. [CrossRef]
11. Kinnas, S.A.; Hsin, C. Boundary element method for the analysis of the unsteady flow around extreme propeller geometries. *AIAA J.* **1992**, *30*, 688–696. [CrossRef]

12. Hartmann, S. A remark on the application of the Newton-Raphson method in non-linear finite element analysis. *Comput. Mech.* **2005**, *36*, 100–116. [CrossRef]
13. Srivastava, S.; Roychowdhury, J. Independent and Interdependent Latch Setup/Hold Time Characterization via Newton–Raphson Solution and Euler Curve Tracking of State-Transition Equations. *IEEE Trans. Comput. Aided Des. Integr. Circ. Syst.* **2008**, *27*, 817–830. [CrossRef]
14. Mantia, M.L.; Dabnichki, P. Unsteady panel method for flapping foil. *Eng. Anal. Bound. Elem.* **2009**, *33*, 572–580. [CrossRef]
15. Eça, L.; Vaz, G.B.; Campos, J.F.D. Verification Study of Low and Higher-Order Potential Based Panel Methods for 2D Foils. In Proceedings of the AIAA Fluid Dynamics Conference and Exhibit, St. Louis, MI, USA, 24–26 June 2002; pp. 197–208.
16. Kanemaru, T.; Ando, J. Numerical analysis of cavitating propeller and pressure fluctuation on ship stern using a simple surface panel method "SQCM". *J. Mar. Sci. Technol.* **2013**, *18*, 294–309. [CrossRef]
17. Tarafder, M.S.; Suzuki, K. Numerical calculation of free-surface potential flow around a ship using the modified Rankine source panel method. *Ocean Eng.* **2008**, *35*, 536–544. [CrossRef]
18. Huang, Q.; Huang, Y.; Hu, W. Bezier Interpolation for 3-D Freehand Ultrasound. *IEEE Trans. Hum. Mach. Syst.* **2015**, *45*, 385–392. [CrossRef]
19. Catmull, E.; Clark, J. Recursively generated B-spline surfaces on arbitrary topological meshes. *Comput. Aided Des.* **2010**, *10*, 350–355. [CrossRef]
20. Chen, C.W.; Kang, D.D.; Leng, J.X.; Lin, H.T.; Wang, J.; Jiao, L. Numerical analysis on the wake field of fast container ship stern with novel propeller duct. In Proceedings of the International Symposium on Fluid Machinery and Fluid Engineering—IET, Wuhan, China, 22 October 2014; pp. 1–7.
21. Chen, C.W.; Ning, P. Prediction and analysis of 3D hydrofoil and propeller under potential flow using panel method. *EDP Sciences* **2016**, *77*, 01013. [CrossRef]
22. Lee, C.; Koo, D.; Zingg, D.W. Comparison of B-spline surface and free-form deformation geometry control for aerodynamic optimization. *AIAA J.* **2017**, *55*, 228–240. [CrossRef]
23. Chen, C.W.; Tsai, J.S.K.J.F. Modeling and Simulation of an AUV Simulator With Guidance System. *IEEE J. Ocean. Eng.* **2013**, *38*, 211–225. [CrossRef]
24. Kim, G.-D.; Lee, C.-S.; Kerwin, J.E. A B-spline higher order panel method for analysis of steady flow around marine propellers. *Ocean Eng.* **2007**, *34*, 2045–2060. [CrossRef]
25. Kim, G.; Ahn, B.; Kim, J.; Lee, C. Improved hydrodynamic analysis of marine propellers using a B-spline-based higher-order panel method. *J. Mar. Sci. Technol.* **2015**, *20*, 670–678. [CrossRef]
26. Lee, C.-S.; Kerwin, J.E. A B-spline higher order panel method applied to two-dimensional lifting problem. *J. Ship Res.* **2003**, *47*, 290–298.
27. Hsin, C.Y.; Kerwin, J.E.; Newman, J.N. A higher-order panel method based on B-splines. In Proceedings of the 6th International Conference on Numerical Ship Hydrodynamics, Iowa City, IA, USA, 2–5 August 1993.
28. Plotkin, A.; Katz, J. *Low-Speed Aerodynamics*; Cambridge University Press: Cambridge, UK, 2001.
29. Kerwin, J.E.; Hadler, J.B. *Principles of Naval Architecture Series: Propulsion*; The Society of Naval Architects and Marine Engineers (SNAME): Alexandria, VA, USA, 2010.
30. Su, Y.; Kim, S.; Kinnas, S.A. Prediction of propeller-induced hull pressure fluctuations via a potential-based method: Study of the effects of different wake alignment methods and of the rudder. *J. Mar. Sci. Eng.* **2018**, *6*, 52. [CrossRef]
31. Kanemaru, T.; Ando, J. Calculation of Propeller Cavitation and Pressure Fluctuation on Ship Stern Using a Simple Surface Panel Method. *J. Soc. Naval Arch. Japan* **2009**, *10*, 1–10.
32. Politis, G.K. Simulation of unsteady motion of a propeller in a fluid including free wake modeling. *Eng. Anal. Bound. Elem.* **2004**, *28*, 633–653. [CrossRef]
33. von Doenhoff, H.A.A. *Theory of Wing Sections: Including a Summary of Airfoil Data*; Dover Publications: New York, NY, USA, 1959.
34. Lee, S.J.; Jang, Y.G. Control of flow around a NACA 0012 airfoil with a micro-riblet film. *J. Fluids Struct.* **2005**, *20*, 659–672. [CrossRef]
35. Wang Yang, Wu Weiwei, Li Zhiguo, Investigation on Aerodynamic Characteristics of Forward/Backward Swept Wing in Ground Effect. *Adv. Aeronaut. Sci. Eng.* **2015**, *6*, 412–418.
36. Greeley, D.S.; Kerwin, J.E. Numerical methods for propeller design and analysis in steady flow. *Trans. SNAME* **1982**, *90*, 415–453.

37. The Propulsion Committee. *Final Report and Recommendations to the 20th ITTC*; ITTC: San Francisco, CA, USA, 1993; Volume 5, pp. 111–142.
38. Jessup, S.D. An Experimental Investigation of Viscous Aspects of Propeller Blade Flow. Ph.D. Thesis, The Catholic University of America, Washington, DC, USA, 1989.
39. Tachmindji, A.J. *The Potential Problem of the Optimum Propeller with Finite Number of Blades Operating in a Cylindrical Duct*; Navy Department, David Taylor Model Basin: Bethesda, MD, USA, 1958.
40. Gaschler, M.; Abdel-Maksoud, M. Computation of hydrodynamic mass and damping coefficients for a cavitating marine propeller flow using a panel method. *J. Fluids Struct.* **2014**, *49*, 574–593. [CrossRef]
41. Koyama, K. Relation between the lifting surface theory and the lifting line theory in the design of an optimum screw propeller. *J. Mar. Sci. Technol.* **2013**, *18*, 145–165. [CrossRef]
42. Pyo, S.; Kinnas, S.A. Propeller wake sheet roll-up modeling in three dimensions. *J. Ship Res.* **1997**, *41*, 81–92.
43. Carmichael, R.; Erickson, L. PAN AIR—A higher order panel method for predicting subsonic or supersonic linear potential flows about arbitrary configurations. In Proceedings of the 14th Fluid and Plasma Dynamics Conference, Palo Alto, CA, USA, 23–25 June 1981.
44. Ramsey, W.D. Boundary Integral Methods for Lifting Bodies with Vortex Wakes. Ph.D. Thesis, Department of Ocean Engineering, MIT, Cambridge, MA, USA, 1995.

symmetry

MDPI

Article

Enhanced Membrane Computing Algorithm for SAT Problems Based on the Splitting Rule

Le Hao [1,*] and Jun Liu [2]

1 School of Mathematics, Southwest Jiaotong University, Chengdu 611756, China
2 School of Computing, Ulster University, Northern Ireland BT37 0QB, UK; j.liu@ulster.ac.uk
* Correspondence: haole2018@outlook.com; Tel.: +86-028-15708418678

Received: 13 October 2019; Accepted: 13 November 2019; Published: 15 November 2019

Abstract: Boolean propositional satisfiability (SAT) problem is one of the most widely studied NP-complete problems and plays an outstanding role in many domains. Membrane computing is a branch of natural computing which has been proven to solve NP problems in polynomial time with a parallel compute mode. This paper proposes a new algorithm for SAT problem which combines the traditional membrane computing algorithm of SAT problem with a classic simplification rule, the splitting rule, which can divide a clause set into two axisymmetric subsets, deal with them respectively and simultaneously, and obtain the solution of the original clause set with the symmetry of their solutions. The new algorithm is shown to be able to reduce the space complexity by distributing clauses with the splitting rule repeatedly, and also reduce both time and space complexity by executing one-literal rule and pure-literal rule as many times as possible.

Keywords: SAT problem; membrane computing; P system; splitting rule

1. Introduction

Boolean satisfiability problem, namely, SAT problem, is one of the most important problems of theoretical computer science. Its range of application includes multiple significance areas [1], such as mathematics, artificial intelligence, data mining, circuit design, etc., and it has attracted much attention since it was put forward. Since the 1960s, research has produced several models of SAT solvers, such as conflict driven clause learning [2], CDCL for short, in which its basic structure comes from the DPLL(Davis-Putnam-Logemann-Loveland) algorithm [3]. This model has made improvements to conflict analysis, clause learning, and some other aspects and has occupied the main battleground of SAT competitions. Representative solvers of this model include Mini-SAT [4], Lingeling [5], Chaff [6], and Glucose [7] all of which have received many achievements in international SAT competitions. With improvements of the SAT algorithm performance, many application examples can obtain satisfactory solutions in a given period of time, and SAT solvers are gradually being applied to more and more actual fields, such as circuit design verification [8,9] and cryptanalysis [10], however, as the first problem which has been proven as NP problem [11], the primary research direction of SAT problem is to reduce its computational complexity. Meanwhile, NP problems can transfer between each other in polynomial time, and therefore previous, current and future studies are also efforts to solve NP problems.

In recently decades, the research of SAT solvers has been mainly focused on three directions, complete solution algorithm, incomplete solution algorithm, and parallel solution algorithm. The membrane computing algorithm of SAT problems, which we are going to discuss, is a kind of parallel algorithm, which can solve any SAT problem in polynomial time but with exponential space occupation [12]. Among the three types of SAT solvers mentioned above, the complete solution algorithm is certain to obtain the solution of a given SAT problem, but it takes an unacceptable amount of time. By comparison, the incomplete solution algorithm uses less time, but it is not able to ensure results. The disadvantages of these two algorithms can both be solved by the algorithm of membrane computing.

Membrane computing is a branch of natural computing, which was proposed in 1998 by professor Gheorghe Păun when he visited Finland [13]. Five years later, the Institute of Scientific Information listed it in the fast-growing frontier area of computational science. The membrane computing system, also known as P system, is a kind of distributed parallel computational model, with good computational performance by referring to and simulating the way cells, tissues, organs, or other biological structures process chemical substances, which has been provn to have the computing power of Turing machine, and a computing power to solve NP problems in polynomial time [14–18]. This characteristic has attracted significant attention from the scientific community who have promoted its development tremendously. To be specific, because membrane computation is performed at the cellular level, biochemical reactions and material transfer at the cellular level can be understood as computational processes. The cell membrane divides the cell into compartments, each compartment synchronously processes multiple resets of objects (corresponding to evolving compounds in the cell), the objects permeate through membranes, the membranes are dissolved, split, and produced, and their penetrability can also be changed. A series of transfers of a system is called a computation, and the calculation result is defined as the objects that appear in a particular membrane (also known as the output membrane) at termination. As a typical type of NP problem, solving SAT problems with membrane computing has a long history [19–23]. Although membrane computing has strong computational power, it is predicted that there is still an upper limit, and therefore developing a new model of membrane computing for solving SAT problems that simplifies algorithms' structure is also important, which is the aim of this paper.

In this paper, the traditional membrane computing algorithm of SAT problems is combined with a typical classic simplification rule, the splitting rule, to improve the algorithm's structure, from assigning values to all clauses in a membrane to dividing the given clauses into two parts, and deals with them respectively and simultaneously. Because this divide operation can be executed as many times as needed, it significantly reduces the space occupation of the algorithm. Meanwhile, it is obvious that a clause set dealt by the splitting rule must not include tautologies, one literals, and pure literals, and therefore it is indispensable to operate the following three simplified rules beforehand: tautology rule, one-literal rule, and pure-literal rule. The tautology rule needs only to be done once, at the beginning, and it does not make contributions to the simplification of the algorithm, however, the one-literal rule and pure-literal rule are two rules which can be used repeatedly and intensively, they can decrease the number of membrane divisions during the assignment process of the clause set, and can assign more than one value at one time, and therefore both space and time complexity of the algorithm is reduced.

The remainder of the paper is organized as follows: Section 2 provides some preliminary and review about P system and the traditional membrane computing algorithm for SAT, Section 3 proposes a new algorithm for SAT problem by combining the traditional membrane computing algorithm for SAT with splitting rule, in Section 4 an example is provided to illustrate the proposed algorithm, and the paper is concluded in Section 5.

2. Preliminaries

This section provides some preliminaries to be used in the present work.

2.1. Membrane Computing System (P System)

A membrane computing system [24] can be defined as $\prod = (O,\mu,w_i,R_i,i_0)$, of which:
O is a finite and nonempty alphabet of objects;
μ is a membrane structure made up by several membranes;
w_i denotes the character string inside the nth membrane in the initial state;
R_i is a finite set of the evolutionary rules which are carried out inside the ith membrane;

i_0 is the membrane which stores the final result.

2.2. Tradtional Membrane Computing Algorithm of SAT Problems

Using membrane computing system to solve SAT problem has decades of history. Figure 1 shows the process diagram of the traditional membrane computing algorithm of SAT problems [24].

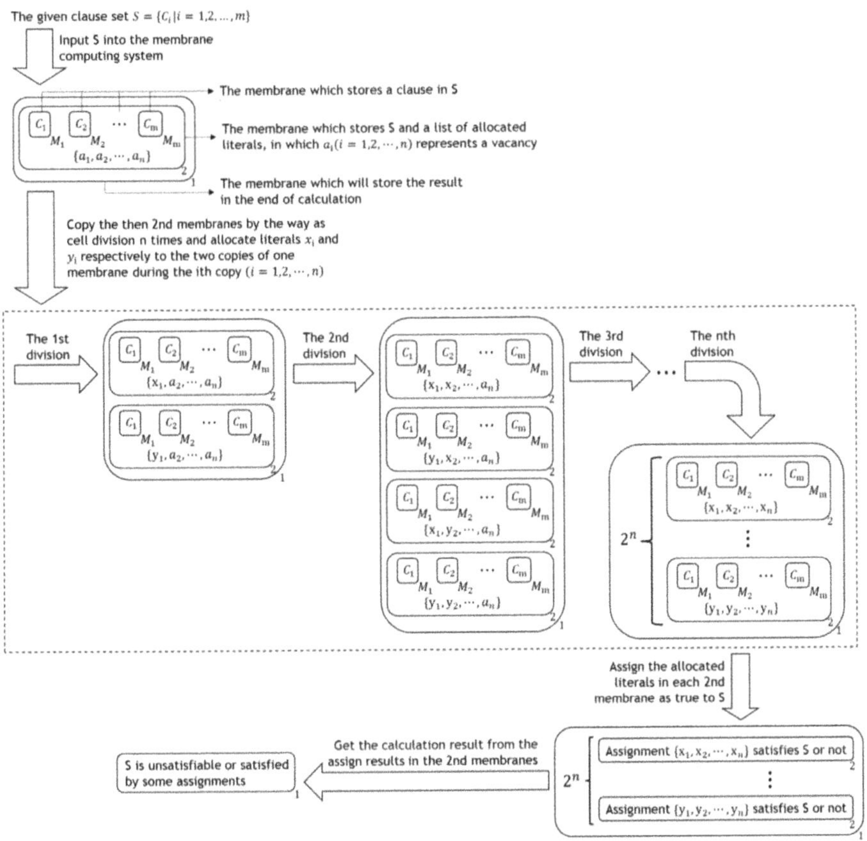

Figure 1. Process diagram of the traditional membrane computing algorithm of SAT problems, of which m is the number of given clauses, n is the number of the given clause set's atoms.

From the above diagram, shown in Figure 1, it is clear that this algorithm consists of the following two steps, except the input and output steps:

Step 1: Copy the, then, 2nd membranes by cell division n times and allocate literals x_i and y_i, respectively to the two copies of one membrane during the ith copy $(ir[1,n]\cap,)$, its time complexity is $O(1)$, and its once execution makes the space occupation of this algorithm twice as much as before.

Step 2: Assign the allocated literals in each 2nd membrane as true to the given clause set, its time complexity is $O(1)$, and since it is completed with sending each allocated literal into all membrane M_is which is inside the same 2nd membrane, the largest increase in space occupation of this algorithm during its execution process is $O(2^n n(m-1))$, and the space occupation of this algorithm will decrease to $O(2^n)$ after its execution process.

To sum up, the time complexity of this algorithm is O(n), and since the biggest initial space occupation of this algorithm is O(nm), the space complexity of this algorithm is $O(2^n nm) + O(2^n n(m-1)) = O(2^n nm)$.

2.3. Simplification Rules

Splitting Rule [25]: Assume a clause set S can be arranged into a form such as $[(A_1 \vee L) \wedge \cdots \wedge (A_n \vee L)] \wedge [(B_1 \vee \neg L) \wedge \cdots \wedge (B_m \vee \neg L)] \wedge R$, of which $A_i(i = 1, 2, \cdots, n)$ and $b_i(i = 1, 2, \cdots, m)$, R are clauses which exclude literals L and ¬L, then S is unsatisfiable if and only if $A_1 \wedge \cdots \wedge A_n \wedge R$ and $B_1 \wedge \cdots \wedge B_m \wedge R$ are unsatisfiable, A is an assignment which satisfies $A_1 \wedge \cdots \wedge A_n \wedge R$ ($B_1 \wedge \cdots \wedge B_m \wedge R$) if and only if $A \cup \{\neg L = 1\}(A \cup \{L = 1\})$ is an assignment which satisfies S.

Tautology Rule [25]: Delete tautologies, namely the clauses which include complemental pairs of literals, from a clause set does not change if an assignment satisfies this clause set.

One-literal Rule [25]: Assume a clause set S includes a clause which only contains one literal L, then L is a single literal of S. Since one-literal clause can only be satisfied by assignments which assign its single literal as true, if S is empty after deleting the clauses contain S's single literals from S, S can be satisfied by any assignments which assign its single literals as true, or else S is satisfied by assignment A after deleting the clauses contain S's single literals from it and deleting literal the negations of S's single literals from all of its clauses if and only if S is satisfied by an assignment $A \cup \{\text{S's sin gle literals are all true}\}$.

Pure-literal Rule [25]: Assume a clause set S includes literal L but excludes literal ¬L, then L is a pure literal of S. Since the clause which contains pure literals can only be satisfied by assignments which assign the pure literals contained by it as true, if S is empty after deleting the clauses contain S's pure literals from S, S can be satisfied by any assignments which assign its pure literals as true, or else S is satisfied by assignment A after deleting the clauses contain S's pure literals from it if and only if S is satisfied by an assignment $A \cup \{\text{S's pure literals are all true}\}$.

3. Proposed New Algorithm

3.1. Definition

This part is the elements' definitions of the algorithm's membrane computing structure as follows:

(1) O: {x_i, y_i, denote the literals of given clause set;

h, denotes the number of literals in a clause;

t_i, f_i, are the literal symbols of x_i and y_i;

g, denotes the number of literals' symbols;

a_i, denote the atoms of the given formulae;

T_i, F_i, are objects in the literal list of 2nd membrane;

c_i, e_i, are the transitive symbols of t_i and f_i;

p_i, denote the transport sign of membrane M_i;

z, denotes the number of membranes $M_i(i = 1, \ldots, m)$;

t_i, f_i, denote the assignments of literals x_i and y_i;

λ, δ, are the melting symbol of objects and membrane;

$Y, N, a, b, c, d, p, q, s$, are aided symbols}.

(2) μ : the initial membrane structure is as Figure 2:

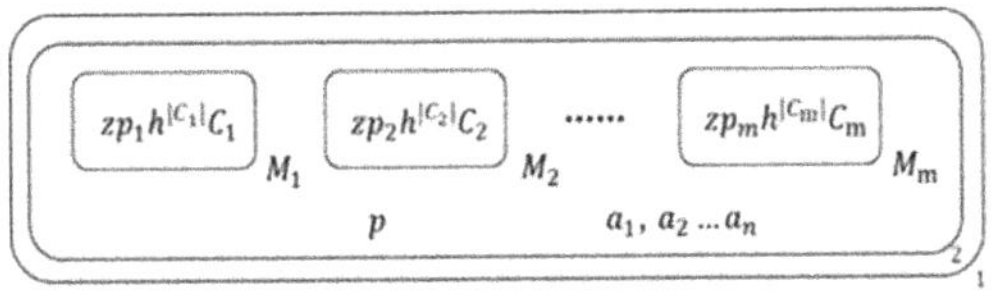

Figure 2. Initial membrane structure of the algorithm.

where C_j denote the jth clause in the given clause set, $|C_j|$ represents the number of literals in $C_j(j = 1, \ldots, m)$.

In addition, there's a literal list inside 2nd membrane which is empty in the initial status, it is used to store the literals which would be assigned as true in all assignments which satisfy the inside clause set of 2nd membrane and the given clause set.

(3) $w_1 = \lambda$;

$w_2 = p, a_i\ (i = 1, \ldots, n)$;

$w_{M_j} = z, p_j, h, x_i, y_i\ (i = 1, \ldots, n,\ j = 1, \ldots, m)$.

(4) $R_i(i = 1,\ 2, M_j)$ specific as described in the part of algorithm, and they are made up by three types of evolutionary rules:

(a) $[a]_i \to [b]_i[c]_i$, copy the membrane i into two copies when it or them contains object a, turn a in two copies into b and c respectively;

(b) $a \to b|c$ or $\neg c$, turn a into b when c is existent or inexistent;

(c) $a \to b(c, in_j \text{ or } out)$, turn a into b and c, meanwhile, send c into the membrane i or out the membrane which a was inside.

(5) $i_0 = 1$.

3.2. Compiling

As stated in the Introduction, the new membrane computing algorithm, proposed in this paper, is specifically based on the splitting rule and uses tautology rule, one-literal rule and pure-literal rule, and therefore for clarity this section compiles the simplified rules with membrane computing language, first, as follows:

(1) Tautology rule:

This rule is executed on the only 2nd membrane at the beginning of the calculation.

Pick tautologies from the clauses inside membrane M_js, and mark the membrane M_js which contain tautologies with a delete symbol, $r_1 = (zp_jh^2x_iy_i \to s, 1)$;

Delete all contents inside the membrane M_js which are marked with a delete symbol:
$r_2 = (hx_i \to \lambda|s, 1)$
$r_3 = (hy_i \to \lambda|s, 1)$;

Delete the membrane M_js which are marked with a delete symbol, $r_4 = (s \to \delta, 1)$.

(2) One-literal rule:

This rule is executed on each 2nd membrane.

Pick the one-literal clauses from the clauses inside membrane M_js, mark the literals of these selected one-literal clauses with one-literal symbols t_i, f_i, send these symbols outside their original membranes.

$$r_1 = \left(zp_j x_i \to x_i\left(zp_j gt_i, out\right)\middle|\neg h^2, 4\right)$$
$$r_2 = \left(zp_j y_i \to y_i\left(zp_j gf_i, out\right)\middle|\neg h^2, 4\right);$$
$$r_3 = \left(zp_j, out, 5\right)$$

Then, judge if its inside clause set has complementary pairs of single literals with the one-literal symbols, if so, this clause set is unsatisfiable, so what we need to do is to delete this 2nd membrane and all its contents:

$$r_4 = \left(g^2 t_i f_i \to N\middle|\neg N, 1\right);$$
$$r_5 = \left(zp_j \to \left(szp_j, inM_j\right)\middle|N, 1\right)$$
$$r_6 = \left(hx_i \to \lambda\middle|s, 1\right)$$
$$r_7 = \left(hy_i \to \lambda\middle|s, 1\right)$$
$$r_8 = (s \to \delta, 1)$$
$$r_9 = (a_i \to \lambda|N, 1)$$
$$r_{10} = \left(gt_i \to \lambda\middle|N, 1\right)$$
$$r_{11} = \left(gf_i \to \lambda\middle|N, 1\right)$$
$$r_{12} = \left(p \to \lambda\middle|N, 1\right)$$
$$r_{13} = (N \to \delta, 1)$$

If not, since all assignments which satisfy the inside clause set assign these single literals as true, add these one-literal symbols into the literal list of this 2nd membrane:

$$r_{14} = \left(g^2 t_i^2 \to gt_i, 2\right)$$
$$r_{15} = \left(g^2 f_i^2 \to gf_i, 2\right)$$
$$r_{16} = \left(ga_i t_i \to bt_i T_i\middle|\neg t_i^2, 2\right)$$
$$r_{17} = \left(ga_i f_i \to bf_i F_i\middle|\neg f_i^2, 2\right)$$

then package these one-literal symbols and copy this symbol pack into as many copies as membrane M_is inside this 2nd membrane, then send each copy into one membrane M_i:

$$r_{18} = \left(bp \to bq[\ ]_3\middle|\neg g, 1\right)$$
$$r_{19} = (bt_i \to (t_i, in_3), 1)$$
$$r_{20} = (bf_i, \to (f_i, in_3), 1)$$
$$r_{21} = \left([\ ]_3 zp_j \to [\ ]_3([\ \delta\ ]_3 zp_j, inM_j)\middle|\neg b, 1\right);$$
$$r_{22} = \left(q \to p(s, in_3)\middle|\neg z, 1\right)$$
$$r_{23} = \left(t_i \to \lambda\middle|s, 1\right)$$
$$r_{24} = \left(f_i \to \lambda\middle|s, 1\right)$$

Next execute following three operations respectively on the membrane M_is which contain different types of clauses:

For the membrane M_i whose inside clause includes some of these single literals, the inside clause can be satisfied by any assignment which assigns these single literals as true, so what needs to be done is delete it and all of its contents:

$$r_{25} = \left(t_i zp_j hx_i \to s, 1\right)$$
$$r_{26} = \left(f_i zp_j hy_i \to s, 1\right);$$

For the membrane M_i whose inside clause excludes these single literals, but includes some of their negations, the inside clause can be satisfied by an assignment which assigns these single literals as true if and only if it can still be satisfied by this assignment after deleting the negations of these single literals from it, so delete these negations:

$r_{27} = (t_i hy_i \rightarrow \lambda, 2)$, $r_{28} = (f_i hx_i \rightarrow \lambda, 2)$, for the same reason, if the inside clause is built only by the negations of single literals, the clause set inside this 2nd membrane is unsatisfiable, delete this membrane and all of its contents:

$r_{29} = (zp_j \rightarrow \delta(N, out)|\neg h, 3)$
$r_{30} = (T_i \rightarrow \lambda|N, 1)$
$r_{31} = (F_i \rightarrow \lambda|N, 1)$, or else delete the left one-literal symbols in the end:
$r_{32} = (t_i \rightarrow \lambda|h, 3$
$r_{33} = (f_i \rightarrow \lambda|h, 3$

For the membrane M_i whose inside clause excludes single literals and their negations, we can just delete the one-literal symbols from this membrane M_i.

(3) Pure-literal rule:

This rule is executed on each 2nd membrane as follows:

Send all literals in the inside clause set outside from their original membranes:
$r_1 = ([zp_j]_{M_i} \rightarrow [\,]_{M_i}[c]_{M_i}(zp_j, out), 1)$
$r_2 = (h \rightarrow \lambda|c, 1)$
$r_3 = (c \rightarrow \delta, 1)$
, then delete the repetitive literals: $r_4 = (x_i^2 \rightarrow x_i, 1)$, $r_5 = (y_i^2 \rightarrow y_i, 1)$,

finally delete the complemental pairs of literals:
$r_6 = (x_i \rightarrow gt_i|\neg x_i^2, 1)$
$r_7 = (y_i \rightarrow gf_i|\neg y_i^2, 1)$
$r_8 = (g^2 t_i f_i \rightarrow \lambda, 1)$

At this point, the left literals are all pure literals of the inside clause set, since all assignments which satisfy the inside clause set assign these pure literals as true, what needs to be done is that add the symbols of them into the literal list of this 2nd membrane: $r_9 = (ga_i t_i \rightarrow bt_i T_i, 2)$, $r_{10} = (ga_i f_i \rightarrow bf_i F_i, 2)$, then package these one-literal symbols and copy these symbols into as many copies as membrane M_is inside this 2nd membrane, then send each copy into one membrane M_i:

$r_{11} = (bp \rightarrow bq[\,]_3|\neg g, 1)$
$r_{12} = (bt_i \rightarrow (t_i, in_3), 1)$
$r_{13} = (bf_i \rightarrow (f_i, in_3), 1)$
$r_{14} = ([\,]_3 zp_j \rightarrow [\,]_3([\,\delta\,]_3 zp_j, inM_j)|\neg b, 1)$
$r_{15} = (q \rightarrow p(s, in_3)|\neg z, 1)$
$r_{16} = (hx_i \rightarrow \lambda|s, 1)$
$r_{17} = (hy_i \rightarrow \lambda|s, 1)$
$r_{18} = (t_i \rightarrow \lambda|s, 1)$
$r_{19} = (f_i \rightarrow \lambda|s, 1)$
$r_{20} = (s \rightarrow \delta, 1)$

Next execute following three operations, respectively, on the membrane M_is which contain different types of clauses:

For the membrane M_i whose inside clause includes some of these pure literals, the inside clause can be satisfied by any assignment which assigns these pure literals as true, therefore what needs to be done is delete it and all of its contents: $r_{21} = (t_i zp_j hx_i \rightarrow s, 1)$, $r_{22} = (f_i zp_j hy_i \rightarrow s, 1)$;

For the membrane M_i whose inside clause excludes single literals and their negations, we can just delete the one-literal symbols from this membrane M_i: $r_{23} = (t_i \rightarrow \lambda, 2)$, $r_{24} = (f_i \rightarrow \lambda, 2)$.

(4) Splitting rule:

This rule is executed on each 2nd membrane whose inside clause set has no one-literal clause and pure-literal clause.

Splitting rule is supposed to be distributing the clauses in the given clause set into two parts which excludes a specified literal and its negation separately, and putting them into two new second membranes, but this operation is too complicated for the membrane computing program language, so we transpose the order of processes:

First, we copy the 2nd membrane and its contents into two copies and allocate a specified literal and its negation to two copies, respectively, the allocated literal of each copy is the literal which its negation and the clauses contain it should be excluded by the inside clause set of this copy, since all assignments which satisfy the inside clause set of one of two copies and the inside clause of the original 2nd membrane assign the allocated literal of this copy as true, add the symbol of the allocated literal into the literal list of each copy: $\begin{array}{l} r_1 = \left(zp_j, out, 4\right) \\ r_2 = \left(\left[a_i z^j\right]_2 \to \left[t_i^j T_i\right]_2 \left[f_i^j F_i\right]_2 \middle| j \in Q^+, 1\right) \end{array}$, then send the literal symbols into each membrane M_i which inside two copies respectively:

$r_3 = (t_i p_j \to (t_i z p_j, inM_j), 1;$

$r_4 = (f_i p_j \to (f_i z p_j, inM_j), 1$

Next execute following three operations, respectively, on the membrane M_is which contain different types of clause.

For the membrane M_i whose inside clause includes the allocated literal of the 2nd membrane which contains it, this inside clause should be excluded by the inside clause set of this 2nd membrane, so what needs to be done is delete it and all of its contents: $\begin{array}{l} r_5 = \left(t_i z p_j h x_i \to s, 1\right) \\ r_6 = \left(f_i z p_j h y_i \to s, 1\right) \\ r_7 = \left(h x_i \to \lambda \middle| s, 1\right) \\ r_8 = \left(h y_i \to \lambda \middle| s, 1\right) \\ r_9 = \left(s \to \delta, 1\right) \end{array}$;

For the membrane M_i whose inside clause includes the negation of the allocated literal of the 2nd membrane which contains it, this negation should be excluded by the inside clause, so delete this negation: $\begin{array}{l} r_{10} = \left(t_i h y_i \to \lambda, 2\right) \\ r_{11} = \left(f_i h x_i \to \lambda, 2\right) \end{array}$;

For the membrane M_i whose inside clause excludes the allocated literal of the 2nd membrane which contains this membrane M_i and its negation, we can just delete the literal symbol from this membrane M_i: $\begin{array}{l} r_{12} = \left(t_i \to \lambda, 3\right) \\ r_{13} = \left(f_i \to \lambda, 3\right) \end{array}$;

Since these four program modules have a lot of overlap, we can concordance them to realize the algorithm of this paper. The flowing chart of the new algorithm is shown in Figure 3.

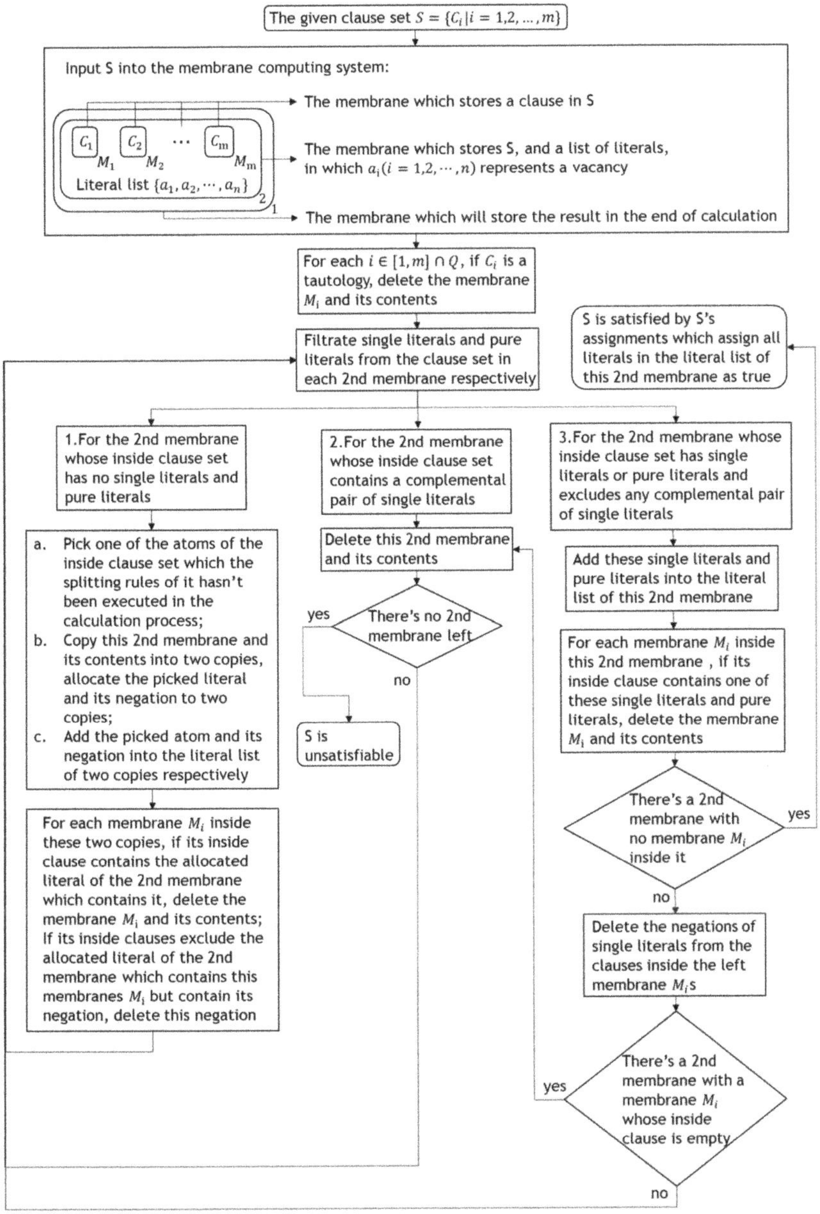

Figure 3. Flow chart of the algorithm.

From the above diagram (Figure 3), it is clear that this algorithm is completed by repeat executing the following two steps, in the ith (1th < n) repeat:

Step 1: Filtrate single literals and pure literals from the clause set in each 2nd membrane simultaneously, its time complexity is $O(1)$, the largest increase in space occupation of this algorithm during its execution process is no more than $O\left(2^{i-1}(m+1-i)(n+1-i)\right)$, and this increase will decrease to no more than $O\left(2^{i-1}(n+1-i)\right)$ after its execution process.

Step 2: Execute one of the following three substeps on each then 2nd membrane simultaneously based on the type of the 2nd membrane's inside clause set:

For the 2nd membrane whose inside clause set has no single literals and pure literals, copy this 2nd membrane into two copies and allocate an atom of this inside clause set and its negation, respectively, to the two copies, then in each copy, delete the clauses which contains its allocated literal from its inside clause set and the negation of its allocated literal from the clauses inside its inside clause set;

For the 2nd membrane whose inside clause set contains a complemental pair of single literals, delete this 2nd membrane and its contents;

For the 2nd membrane whose inside clause set has single literals or pure literals and excludes any complemental pair of single literals, assign these single literals and pure literals as true to its inside clause set.

The time complexity of this step is $O(1)$, and if the ith repeat is not the finial repeat, in the worst case, namely, in the case that all 2nd membranes' inside clause sets exclude single literals and pure literals, the largest increase in space occupation of this algorithm during the execution process of this step is no more than $O\left(2^{i-1}(m+1-i)(n+1-i)\right)+O\left(2^{i}(m-i)\right)$, and the space occupation of this algorithm will decrease to no more than $O\left(2^{i}(m-i)(n-i)\right)$ after its execution process; if not, all 2nd membranes' inside clause set has single literals or pure literals and excludes any complemental pair of single literals, the largest increase in space occupation of this algorithm during the execution process of this step is no more than $O\left(2^{i-1}(m+1-i)\right)$, and the space occupation of this algorithm will decrease to 0 after its execution process.

Since the upper bound of this algorithm's repeat time is Min(n,m), the time complexity of this algorithm is $O(\mathrm{Min}(n,m))$, the space complexity of this algorithm is:

$$\begin{aligned}&\mathrm{Max}_{ia[1,\mathrm{Min}(n,m)]\cap,}(\mathrm{Max}(O(2^{i-1}(m+1-i)(n+1-i))+\ O(2^{i-1}(m+1-i)(n+1-i))\\&+O(2^{i}(m-i))),O(2^{i-1}(m+1-i)(n+1-i))+O(2^{i-1}(m+1-i))(n+1-i)),\\&O(2^{i-1}(m+1-i)(n+1-i))+O(2^{i-1}(n+1-i))+O(2^{i-1}(m+1-i)))\\&=\mathrm{Max}_{ia[1,\mathrm{Min}(n,m)]\cap,}\left(O\left(2^{i}(m-i)(n-i)\right)\right)<O(2^{n}nm).\end{aligned}$$

It is clear that the new algorithm is more efficient than the traditional algorithm. To be noted, the new algorithm's computing complexity is worked out based on its worst case, which almost never happens, such as assuming that no pure-literal and one-literal clause exist before the final repeat, but the traditional algorithm's computing complexity is worked out based on the general change of the time and space which it occupies, and therefore, in practice, the difference between the two algorithm will be much more significant.

In order to describe the difference between the new algorithm's computing complexity and its general time and space occupation, work out the three situations of the time and space occupation of the step 2's execution on a 2nd membrane in the ith repeat, of which the ith repeat is not the finial repeat as follows:

If the 2nd membrane whose inside clause set has no single literals and pure literals, the time complexity is $O(1)$, the largest increase in space occupation of this 2nd membrane, during the execution process, is no more than $O((m+1-i)(n+1-i))+O(2(m-i))$, and the space occupation of this 2nd membrane will decrease to no more than $O(2(m-i)(n-i))$ after the execution process, the new algorithm's computing complexity is worked out based on this situation;

If the 2nd membrane whose inside clause set contains a complemental pair of single literals, the time complexity is $O(1)$, the largest increase in space occupation of this 2nd membrane during the execution process is 0, and the space occupation of this 2nd membrane will decrease to 0 after the execution process;

If the 2nd membrane whose inside clause set has single literals or pure literals and excludes any complemental pair of single literals, the time complexity is $O(1)$, the largest increase in space occupation of this 2nd membrane during the execution process is no more than $O((m-i)(n+1-i))$, and the space occupation of this 2nd membrane will decrease to no more than $O((m-i)(n-i))$ after the execution process. In this situation, assume the inside clause set of this 2nd membrane has k $(k \le n+1-i)$ single literals and pure literals, then k atoms of it have been assigned values simultaneously by this step.

Since the upper bound of the new algorithm's rest repeat time is no more than the biggest number of unassigned atoms of the then 2nd membranes' inside clause sets, the above step makes a contribution to reducing the repeat time of the new algorithm, namely reducing the time complexity of the new algorithm in the second and the third situations.

According to the above flow chart and program modules, we get the program of the object algorithm:

$$
\begin{aligned}
R_{M_i}:\ & r_1 = (zp_j h^2 x_i y_i \to s, 1) \\
& r_2 = (t_i zp_j h x_i \to s, 1) \\
& r_3 = (f_i zp_j h y_i \to s, 1) \\
& r_4 = (hx_i \to \lambda | s, 1) \\
& r_5 = (hy_i \to \lambda | s, 1) \\
& r_6 = (t_i \to \lambda | s, 1) \\
& r_7 = (f_i \to \lambda | s, 1) \\
& r_8 = (c \to \lambda | s, 1) \\
& r_9 = (s \to \delta, 1) \\
& r_{10} = (t_i h y_i \to \lambda, 2) \\
& r_{11} = (f_i h x_i \to \lambda, 2) \\
& r_{12} = (t_i \to \lambda | h, 3) \\
& r_{13} = (f_i \to \lambda | h, 3) \\
& r_{14} = (zp_j \to \delta(N, out) | \neg h, 3) \\
& r_{15} = (h \to \lambda | c, 4) \\
& r_{16} = (c \to \delta | \neg s, 4) \\
& r_{17} = (x_i \to x_i(gt_i, out) | \neg h^2, 5) \\
& r_{18} = (y_i \to y_i(gf_i, out) | \neg h^2, 5)
\end{aligned}
$$

$$
\begin{aligned}
R_2:\ & r_{19} = (g^2 t_i f_i \to N | \neg N, 1) \\
& r_{20} = (t_i p_j \to (t_i zp_j, inM_j) | \neg z, 1) \\
& r_{21} = (f_i p_j \to (f_i zp_j, inM_j) | \neg z, 1) \\
& r_{22} = (bp \to bq[\ ]_3 | \neg g, 1) \\
& r_{23} = (bt_i \to (t_i, in_3), 1) \\
& r_{24} = (bf_i \to (f_i, in_3), 1) \\
& r_{25} = ([\ ]_3 zp_j \to [\ ]_3([\ \delta\ ]_3 zp_j, inM_j) | \neg b, 1) \\
& r_{26} = (q \to p(s, in_3) | \neg z, 1) \\
& r_{27} = (x_i \to c_i | \neg c_i, 1) \\
& r_{28} = (y_i \to e_i | \neg e_i, 1) \\
& r_{29} = (x_i \to \lambda | c_i, 1) \\
& r_{30} = (y_i \to \lambda | e_i, 1) \\
& r_{31} = (c_i e_i \to \lambda, 1) \\
& r_{32} = (c_i \to gt_i | \neg e_i, 1) \\
& r_{33} = (e_i \to gf_i | \neg c_i, 1) \\
& r_{34} = (zp_j \to (szp_j, inM_j) | N, 1) \\
& r_{35} = (a_i \to \lambda | N, 1) \\
& r_{36} = (gt_i \to \lambda | N, 1) \\
& r_{37} = (gf_i \to \lambda | N, 1) \\
& r_{38} = (p \to \lambda | N, 1) \\
& r_{39} = (T_i \to \lambda | N, 1) \\
& r_{40} = (F_i \to \lambda | N, 1) \\
& r_{41} = (N \to \delta, 1) \\
& r_{42} = (a_i \to \lambda | Y, 1) \\
& r_{43} = (Y \to \delta, 1) \\
& r_{44} = (gt_i \to \lambda | T_i, 1) \\
& r_{45} = (gf_i \to \lambda | F_i, 1) \\
& r_{46} = (ga_i t_i \to bt_i T_i | \neg T_i, 2) \\
& r_{47} = (ga_i f_i \to bf_i F_i | \neg F_i, 2) \\
& r_{48} = ([zp_j]_{M_i} \to [\ ]_{M_i} [c]_{M_i} (zp_j, out), 2) \\
& r_{49} = ([a_i z^j]_2 \to [t_i^j T_i]_2 [f_i^j F_i]_2 | j \in Q^+, 3) \\
& r_{50} = (p \to q(d, out), 4)
\end{aligned}
$$

$$
\begin{aligned}
R_1:\ & r_{51} = (d \to a | \neg a, 1) \\
& r_{52} = (d \to \lambda | a, 1) \\
& r_{53} = ([p]_2 \to [N]_2 | Y, 1) \\
& r_{54} = ([q]_2 \to [N]_2 | Y, 1) \\
& r_{55} = (a[q]_2 \to Y[Y]_2, 2) \\
& r_{56} = (Y \to \lambda, 3)
\end{aligned}
$$

The following is the explanation:

First, delete the membrane M_is which contain tautologies and their content with a delete symbol s: $r_1 = (zp_j h^2 x_i y_i \to s | \neg s, 1)$, this is how delete symbol s works in this algorithm:

$$
\begin{aligned}
& r_4 = (hx_i \to \lambda | s, 1) \\
& r_5 = (hy_i \to \lambda | s, 1) \\
& r_6 = (t_i \to \lambda | s, 1) \\
& r_7 = (f_i \to \lambda | s, 1) \\
& r_8 = (c \to \lambda | s, 1) \\
& r_9 = (s \to \delta, 1)
\end{aligned}
;
$$

Then filtrate single literals and pure literals from the clause set in each 2nd membrane respectively. First, execute the 22th rule to copy each membrane M_i and its contents into two copies, $\mathrm{r}_{48} = \left(\left[\mathrm{zp_j}\right]_{M_i} \rightarrow [\,]_{M_i}[c]_{M_i}\left(\mathrm{zp_j}, out\right), 2\right)$, then for copy with mark c, release all literals of its inside clause to the 2nd membrane which contains it and generate literal symbols of all pure literals of this 2nd membrane's inside clause set with these released literals:

$$\begin{aligned}
\mathrm{r}_{15} &= \left(h \rightarrow \lambda\middle|\mathrm{c}, 4\right)\\
r_{16} &= (\mathrm{c} \rightarrow \delta|\neg \mathrm{s}, 4)\\
\mathrm{r}_{27} &= (\mathrm{x_i} \rightarrow \mathrm{c_i}|\neg \mathrm{c_i}, 1)\\
\mathrm{r}_{28} &= \left(\mathrm{y_i} \rightarrow \mathrm{e_i}\middle|\neg \mathrm{e_i}, 1\right)\\
\mathrm{r}_{29} &= \left(\mathrm{x_i} \rightarrow \lambda\middle|\mathrm{c_i}, 1\right)\\
\mathrm{r}_{30} &= \left(\mathrm{y_i} \rightarrow \lambda\middle|\mathrm{e_i}, 1\right)\\
\mathrm{r}_{31} &= (\mathrm{c_i e_i} \rightarrow \lambda, 1)\\
\mathrm{r}_{32} &= \left(\mathrm{c_i} \rightarrow \mathrm{gt_i}\middle|\neg \mathrm{e_i}, 1\right)\\
\mathrm{r}_{33} &= \left(\mathrm{e_i} \rightarrow \mathrm{gf_i}\middle|\neg \mathrm{c_i}, 1\right)
\end{aligned}$$

, and for the copy without mark c, if its inside clause is a one-literal clause, mark the literal of this inside clause with one-literal symbol and send the one-literal symbol outside:

$$\begin{aligned}
\mathrm{r}_{17} &= \left(\mathrm{x_i} \rightarrow \mathrm{x_i}\left(\mathrm{gt_i}, \mathrm{out}\right)\middle|\neg \mathrm{h}^2, 5\right)\\
\mathrm{r}_{18} &= \left(\mathrm{y_i} \rightarrow \mathrm{y_i}\left(\mathrm{gf_i}, \mathrm{out}\right)\middle|\neg \mathrm{h}^2, 5\right)
\end{aligned};$$

Now execute following three operations on the 2nd membranes which contain different types of clause sets respectively:

(a) For the 2nd membrane whose inside clause set has no single literals and pure literals, pick one of the atoms of the inside clause set which the splitting rules of it hasn't been executed in the calculation process, and copy this 2nd membrane and its contents into two copies, allocate the picked literal and its negation to two copies, then add the picked atom and its negation into the literal list of two copies respectively, $\mathrm{r}_{49} = \left(\left[a_i z^j\right]_2 \rightarrow \left[\mathrm{t_i^j T_i}\right]_2\left[\mathrm{f_i^j F_i}\right]_2\middle| j \in Q^+, 3\right)$, then for each membrane M_i inside these two copies, judge the type of its inside clause:

$$\begin{aligned}
\mathrm{r}_{20} &= \left(\mathrm{t_i p_j} \rightarrow \left(\mathrm{t_i zp_j}, \mathrm{inM_j}\right)\middle|\neg z, 1\right)\\
\mathrm{r}_{21} &= \left(\mathrm{f_i p_j} \rightarrow \left(\mathrm{f_i zp_j}, \mathrm{inM_j}\right)\middle|\neg z, 1\right)
\end{aligned},$$

if its inside clause contains the allocated literal of the 2nd membrane which contains it, delete the membrane M_i and its contents, if its inside clauses exclude the allocated literal of the 2nd membrane which contains this membranes M_i, but contain its negation, delete this negation:

$$\begin{aligned}
\mathrm{r}_2 &= \left(\mathrm{t_i zp_j h x_i} \rightarrow \mathrm{s}, 1\right)\\
\mathrm{r}_3 &= \left(\mathrm{f_i zp_j h y_i} \rightarrow \mathrm{s}, 1\right)\\
\mathrm{r}_{10} &= \left(\mathrm{t_i h y_i} \rightarrow \lambda, 2\right)\\
\mathrm{r}_{11} &= (\mathrm{f_i h x_i} \rightarrow \lambda, 2)\\
\mathrm{r}_{12} &= \left(\mathrm{t_i} \rightarrow \lambda\middle|\mathrm{h}, 3\right)\\
\mathrm{r}_{13} &= \left(\mathrm{f_i} \rightarrow \lambda\middle|\mathrm{h}, 3\right)
\end{aligned};$$

(b) For the 2nd membrane whose inside clause set contains a complemental pair of single literals, delete this 2nd membrane and its contents with a unsatisfiable symbol N: $\mathrm{r}_{19} = \left(\mathrm{g^2 t_i f_i} \rightarrow \mathrm{N}\middle|\neg \mathrm{N}, 1\right)$, this is how unsatisfiable symbol N works in this algorithm:

$$\begin{aligned}
\mathrm{r}_{34} &= \left(\mathrm{zp_j} \rightarrow \left(\mathrm{szp_j}, \mathrm{inM_j}\right)\middle|\mathrm{N}, 1\right)\\
\mathrm{r}_{35} &= (\mathrm{a_i} \rightarrow \lambda|\mathrm{N}, 1)\\
\mathrm{r}_{36} &= \left(\mathrm{gt_i} \rightarrow \lambda\middle|\mathrm{N}, 1\right)\\
\mathrm{r}_{37} &= \left(\mathrm{gf_i} \rightarrow \lambda\middle|\mathrm{N}, 1\right)\\
\mathrm{r}_{38} &= \left(\mathrm{p} \rightarrow \lambda\middle|\mathrm{N}, 1\right)\\
\mathrm{r}_{39} &= (\mathrm{T_i} \rightarrow \lambda|\mathrm{N}, 1)\\
\mathrm{r}_{40} &= (F_\mathrm{i} \rightarrow \lambda|\mathrm{N}, 1)\\
\mathrm{r}_{41} &= (\mathrm{N} \rightarrow \delta, 1)
\end{aligned},$$

if all 2nd membranes have been deleted, there's no rule can be executed, the membrane structure has only the membrane which stores the final result left and this membrane is empty at this point, it means that S is unsatisfiable;

(c) For the 2nd membrane whose inside clause set has single literals or pure literals and excludes any complemental pair of single literals, first add these single literals and pure literals into the literal list of this 2nd membrane:
$$\begin{array}{l} r_{44} = \left(gt_i \to \lambda \middle| T_i, 1\right) \\ r_{45} = \left(gf_i \to \lambda \middle| F_i, 1\right) \\ r_{46} = \left(ga_i t_i \to bt_i T_i \middle| \neg T_i, 2\right) \\ r_{47} = \left(ga_i f_i \to bf_i F_i \middle| \neg F_i, 2\right) \end{array}$$
, then for each membrane M_i inside this 2nd membrane, judge the type of its inside clause:
$$\begin{array}{l} r_{22} = (bp \to bq[\ ]_3|\neg g, 1) \\ r_{23} = (bt_i \to (t_i, in_3), 1) \\ r_{24} = (bf_i \to (f_i, in_3), 1) \\ r_{25} = \left([\ ]_3 zp_j \to [\ ]_3([\ \delta\ ]_3 zp_j, inM_j) \middle| \neg b, 1\right) \\ r_{26} = (q \to p(s, in_3)|\neg z, 1) \end{array}$$
, if its inside clause contains one of these single literals and pure literals, delete the membrane M_i and its contents:
$$\begin{array}{l} r_2 = \left(t_i zp_j hx_i \to s, 1\right) \\ r_3 = \left(f_i zp_j hy_i \to s, 1\right) \end{array}$$
, or else delete the negations of single literals from the clauses inside the left membrane M_is:
$$\begin{array}{l} r_{10} = \left(t_i hy_i \to \lambda, 2\right) \\ r_{11} = \left(f_i hx_i \to \lambda, 2\right) \\ r_{12} = \left(t_i \to \lambda \middle| h, 3\right) \\ r_{13} = \left(f_i \to \lambda \middle| h, 3\right) \end{array}$$
, if there's a membrane M_i whose inside clause is empty at this point, delete this 2nd membrane and its contents: $r_{14} = \left(zp_j \to \delta(N, out) \middle| \neg h, 3\right)$.

If there's a 2nd membrane in which all membrane M_is have been deleted, S is satisfied by S's assignments which assign all literals in the literal list of this 2nd membrane as true, so what needs to be done is that pick a 2nd membrane like this, and delete all 2nd membrane and their contents except the literal list of the picked 2nd membrane
$$\begin{array}{l} r_{50} = (p \to q(d, out), 4) \\ r_{51} = (d \to a|\neg a, 1) \\ r_{52} = (d \to \lambda|a, 1) \\ r_{53} = \left([p]_2 \to [N]_2 \middle| Y, 1\right) \\ r_{54} = \left([q]_2 \to [N]_2 \middle| Y, 1\right) \\ r_{55} = \left(a[q]_2 \to Y[Y]_2, 2\right) \\ r_{56} = (Y \to \lambda, 3) \\ r_{42} = (a_i \to \lambda|Y, 1) \\ r_{43} = (Y \to \delta, 1) \end{array}$$
.

4. Example Illustration

In this section, we use an example to show how to solve a SAT problem with this new algorithm, we use clause set $\{x_1, y_1x_2y_3, x_3x_4, y_1y_3y_4\}$:

Figure 4 show the initial membrane structure of the given clause set.

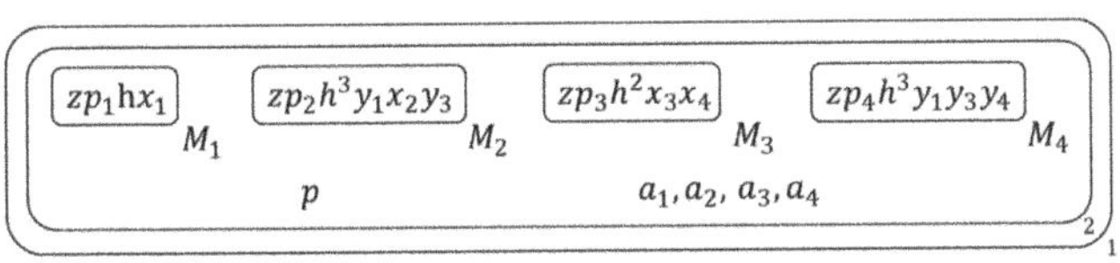

Figure 4. The initial membrane structure of the given clause set.

Delete tautologies from the given clause set (r_1): $\{x_1, y_1x_2y_3, x_3x_4, y_1y_3y_4\}$ includes no tautology.

Filtrate single literals and pure literals from the clause set in each 2nd membrane respectively:

1. Copy each membrane M_i and its contents into two copies (r_{48}), Figure 5 shows the membrane structure of this time.

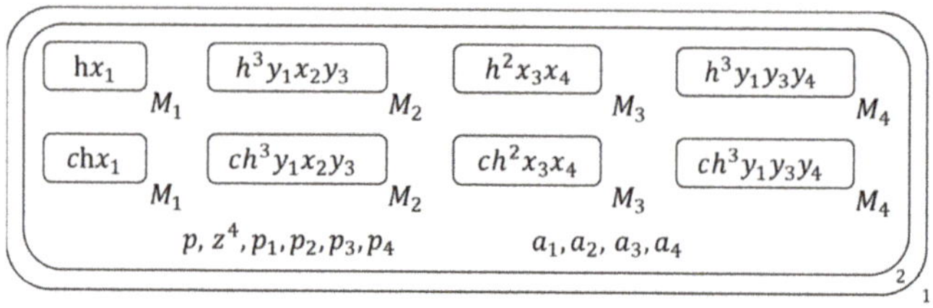

Figure 5. The membrane structure during calculation (1).

2. Then for copy with mark c, release all literals of its inside clause to the 2nd membrane which contains it and generate literal symbols of all pure literals of this 2nd membrane's inside clause set with these released literals $\left(r_{(15,16,27\text{–}33)}\right)$, and for the copy without mark c, if its inside clause is a one-literal clause, mark the literal of this inside clause with one-literal symbol and send the one-literal symbol outside $\left(r_{(17,18)}\right)$, Figure 6 shows the membrane structure of this time.

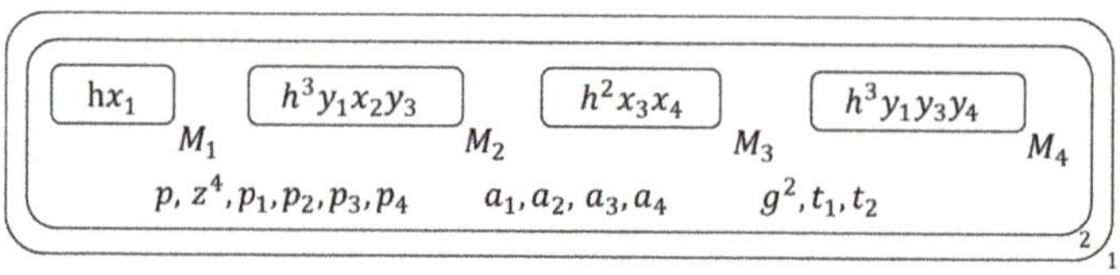

Figure 6. The membrane structure during calculation (2).

Execute the following operations on the 2nd membrane bases on the type of its inside clause set:

1. Add these single literals and pure literals into the literal list of this 2nd membrane $\left(r_{(44\text{–}47)}\right)$, Figure 7 shows the membrane structure of this time.

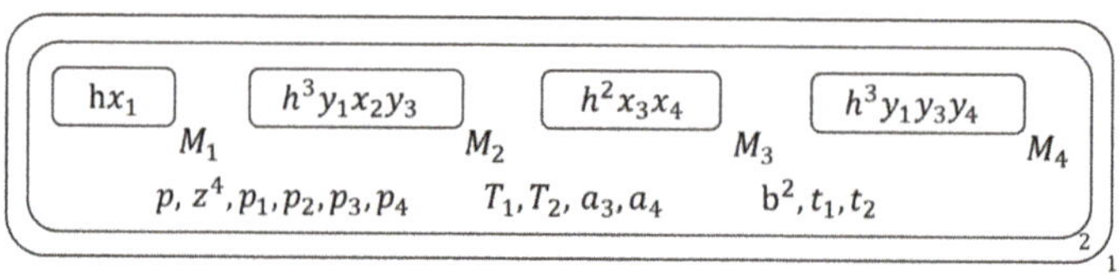

Figure 7. The membrane structure during calculation (3).

2. Then for each membrane M_i inside this 2nd membrane, judge the type of its inside clause $\left(r_{(22\text{–}26)}\right)$, Figure 8 shows the membrane structure of this time.

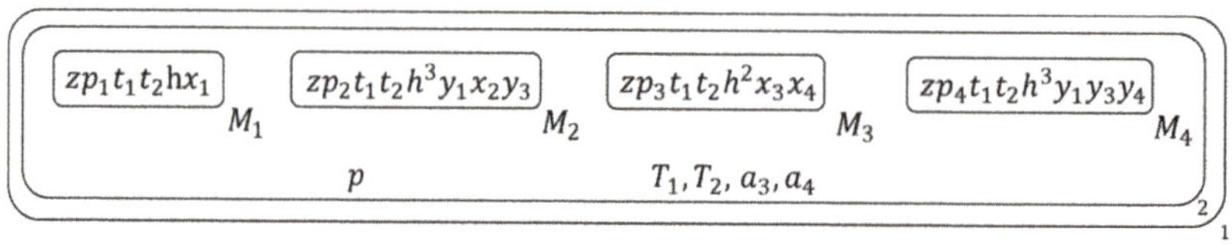

Figure 8. The membrane structure during calculation (4).

At this time, two of the S's atoms have been assigned, the membrane structures of the tradition algorithm which reach the same assign effect is as Figure 9.

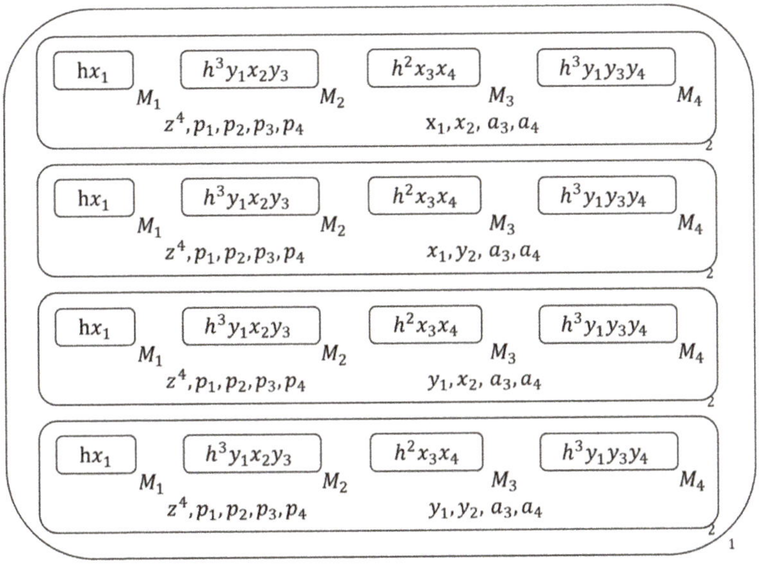

Figure 9. The membrane structure during calculation (5).

3. Next execute following three operations respectively on the membrane M_is which contain different types of clauses $\left(r_{(2,3,10\text{–}13)}\right)$, Figure 10 shows the membrane structure of this time.

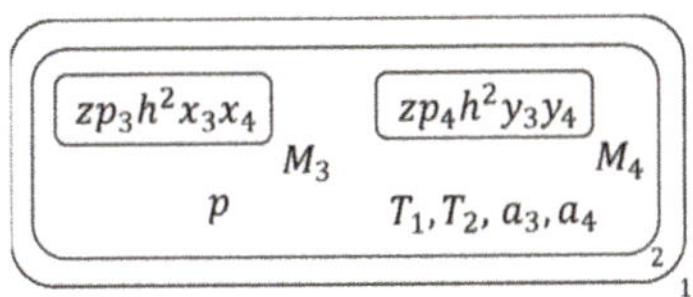

Figure 10. The membrane structure during calculation (6).

Filtrate single literals and pure literals from the clause set in each 2nd membrane respectively:

1. Copy each membrane M_i and its contents into two copies (r_{48}), Figure 11 shows the membrane structure of this time.

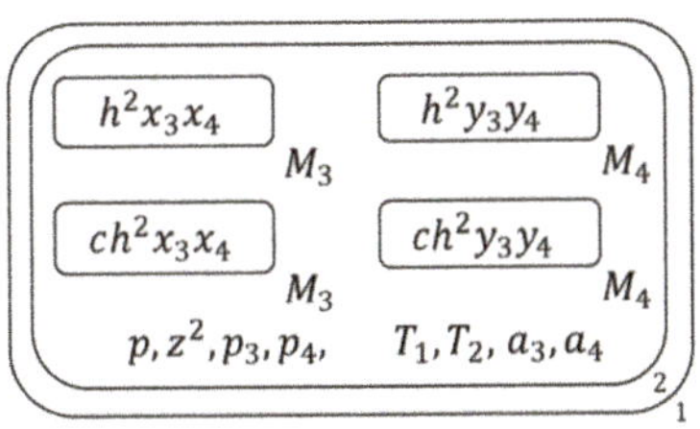

Figure 11. The membrane structure during calculation (7).

2. Then for the copy with mark c, release all literals of its inside clause to the 2nd membrane which contains it and generate literal symbols of all pure literals of this 2nd membrane's inside clause

set with these released literals $\left(r_{(15,16,27\text{–}33)}\right)$, and for the copy without mark c, if its inside clause is a one-literal clause, mark the literal of this inside clause with one-literal symbol and send the one-literal symbol outside $\left(r_{(17,18)}\right)$, Figure 12 shows the membrane structure of this time.

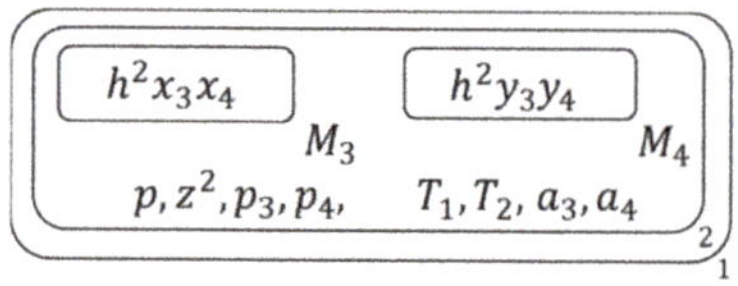

Figure 12. The membrane structure during calculation (8).

Execute the following operations on the 2nd membrane bases on the type of its inside clause set:

1. Pick one of the atoms of the inside clause set which the splitting rules of it hasn't been executed in the calculation process, and copy this 2nd membrane and its contents into two copies, allocate the picked literal and its negation to two copies, then add the picked atom and its negation into the literal list of two copies respectively (r_{49}), Figure 13 shows the membrane structure of this time.

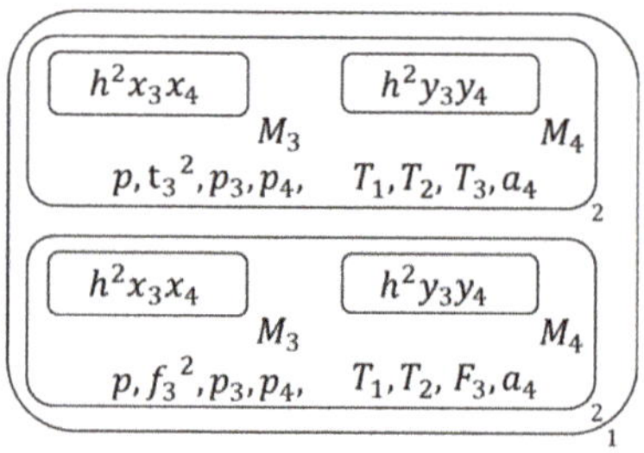

Figure 13. The membrane structure during calculation (9).

At this time, two of the S's atoms have been assigned, the membrane structures of the tradition algorithm which reach the same assign effect is as Figure 14.

2. Then for each membrane M_i inside these two copies, judge the type of its inside clause $\left(r_{(20,21)}\right)$, Figure 15 shows the membrane structure of this time.
3. Next execute following three operations respectively on the membrane M_is which contain different types of clauses $\left(r_{(2,3,10\text{–}13)}\right)$, Figure 16 shows the membrane structure of this time.

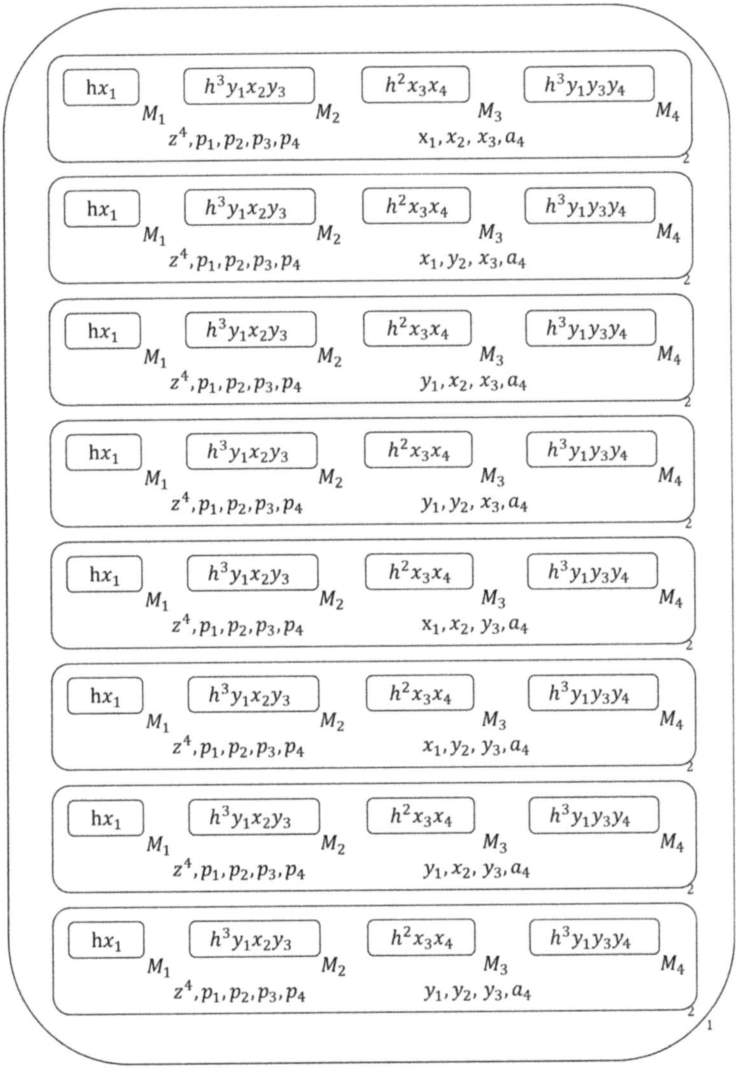

Figure 14. The membrane structure during calculation (10).

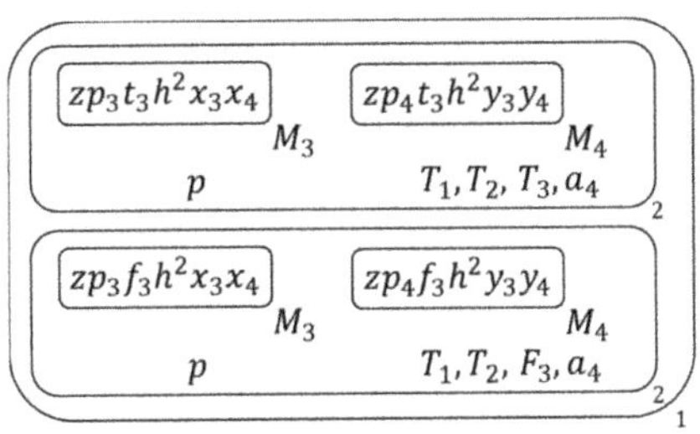

Figure 15. The membrane structure during calculation (11).

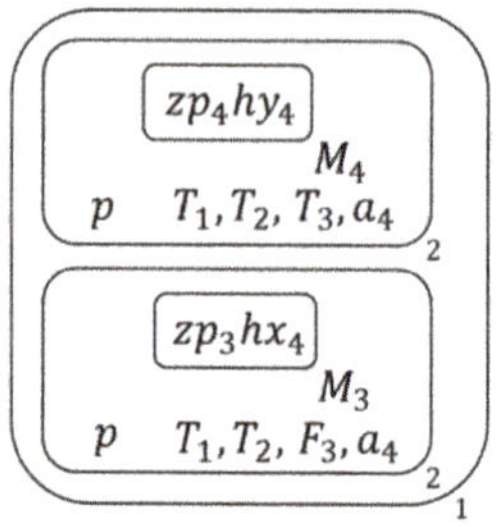

Figure 16. The membrane structure during calculation (12).

Filtrate single literals and pure literals from the clause set in each 2nd membrane respectively:

1. Copy each membrane M_i and its contents into two copies (r_{48}), Figure 17 shows the membrane structure of this time.

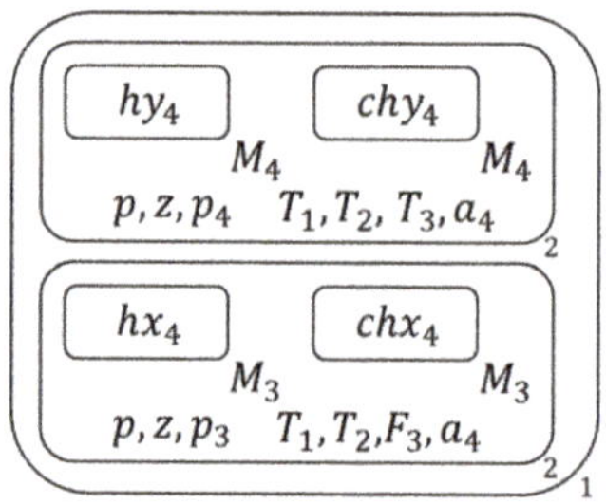

Figure 17. The membrane structure during calculation (13).

2. Then for the copy with mark c, release all literals of its inside clause to the 2nd membrane which contains it and generate literal symbols of all pure literals of this 2nd membrane's inside clause set with these released literals $\left(r_{(15,16,27\text{–}33)}\right)$, and for the copy without mark c, if its inside clause is a one-literal clause, mark the literal of this inside clause with one-literal symbol and send the one-literal symbol outside $\left(r_{(17,18)}\right)$, Figure 18 shows the membrane structure of this time.

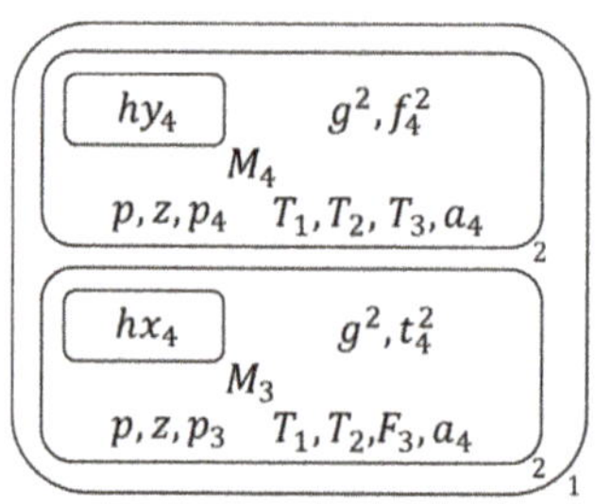

Figure 18. The membrane structure during calculation (14).

Execute the following operations on the 2nd membrane bases on the type of its inside clause set:

1. Add these single literals and pure literals into the literal list of this 2nd membrane $\left(r_{(44\text{–}47)}\right)$, then for each membrane M_i inside this 2nd membrane, judge the type of its inside clause $\left(r_{(22\text{–}26)}\right)$, Figure 19 shows the membrane structure of this time.

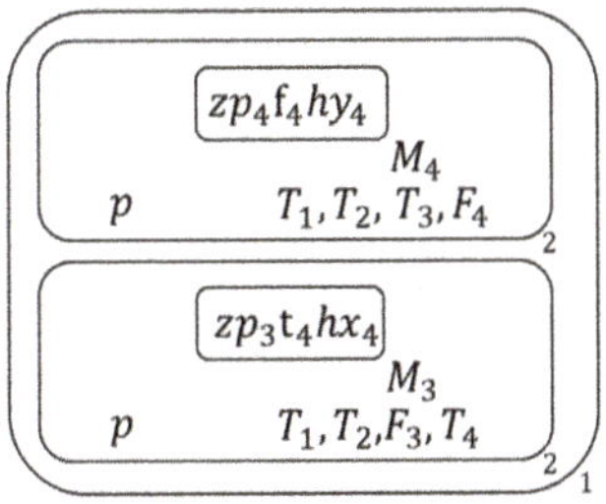

Figure 19. The membrane structure during calculation (15).

At this time, two of the S's atoms have been assigned, the membrane structures of the tradition algorithm which reach the same assign effect is as Figure 20.

Figure 20. The membrane structure during calculation (16).

2. Next execute following three operations respectively on the membrane M_is which contain different types of clauses $\left(r_{(2,3,10\text{–}13)}\right)$, Figure 21 shows the membrane structure of this time.

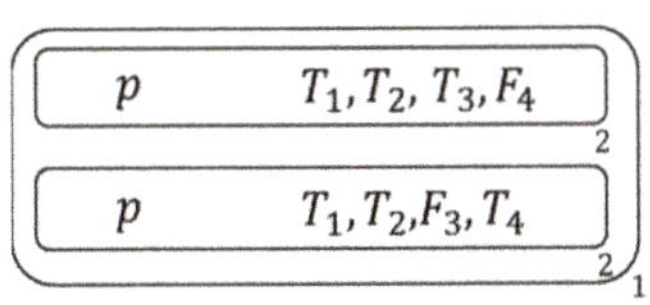

Figure 21. The membrane structure during calculation (17).

3. Pick a 2nd membrane like this, and delete all 2nd membrane and their contents except the literal list of the picked 2nd membrane $\left(r_{(50-56,42,43)}\right)$, Figure 22 shows the membrane structure of this time.

$$\left[T_1, T_2, F_3, T_4 \right]_1$$

Figure 22. The membrane structure during calculation (18).

So we get the calculation result: the given clause set $\{x_1, y_1x_2y_3, x_3x_4, y_1y_3y_4\}$ is satisfied by assignment $\{x_1 = 1; x_2 = 1; y_3 = 1; x_4 = 1\}$.

5. Conclusions

Membrane computing is a type of natural computing, similar to the neural algorithm which has been widely used, however, unlike the neural algorithm, membrane computing is still not a reality, because its infinite parallel computing mode cannot be realized under current technical conditions, Nevertheless, calculating it as the way living bodies deal with data is desired, and therefore work towards it has never stopped as it is still just a fantasy. Even if it is realized and it works as well as our body cells, simplifying the algorithms based on it is still necessary because of their unimaginable but still limited computational power and the predictable crazy growing scale and needs of data treatment in the future. The work in this paper combines the membrane computing system with the traditional SAT problem simplification rules and obtains a fairly good effect, which has certain promotion and reference significance for the research field of membrane computing model and SAT problem solving model, and therefore has research value for related research fields. Meanwhile, this research indicates a feasible direction to improve membrane computing systems for different practice problems, which is to optimize the theoretical basis of this problem based on the character of P system. In addition, with respect to this research, in order to realize the membrane computing system, the strategy could be changed. Because membrane computing is a compute process which imitates the way cells work, the use of cells to simulate the compute process of SAT problems altogether should be considered. Although this has not been considered in this paper because of a lack of biological knowledge, it is a method that should be undertaken.

Author Contributions: Conceptualization, L.H.; methodology, L.H.; validation, L.H.; formal analysis, L.H.; investigation, L.H.; data curation, L.H.; writing—original draft preparation, L.H.; writing—review and editing, J.L.

Funding: This work was supported by the National Natural Science Foundation of China(Grant No.61673320), and by the Fundamental Research Funds for the Central Universities (Grant No.2682018ZT10, No.2682018CX59).

Conflicts of Interest: The authors declare no conflict of interest.

References

1. Kautz, H.; Selman, B. The State of SAT. *Discrete Appl. Math.* **2007**, *155*, 1514–1524. [CrossRef]
2. Marques-Silva, J.; Lynce, I.; Malik, S. Conflict-driven clause learning SAT solvers. In *Handbook of Satisfiability*; Biere, A., Heule, M., Van Maaren, H., Walsh, T., Eds.; IOS Press: Amsterdam, The Netherlands, 2009; Volume 185, pp. 131–153.
3. Davis, M.; Putnam, H. A Computing Procedure for Quantification Theory. *J. ACM* **1960**, *7*, 201–215. [CrossRef]
4. Sorensson, N.; Een, N. MiniSat—A SAT solver with conflict-clause minimization. In Proceedings of the SAT-05 8th International Conference on Theory and Applications of Satisfiability Testing, St Andrews, UK, 19–23 June 2005.
5. Biere, A. MiniSat, cadical, lingeling, plingeling, treengeling and yalsat Entering the SAT Competition 2018. In Proceedings of the SAT Competition 2018: Solver and Benchmark Descriptions; 2018.
6. Moskewicz, M.W.; Madigan, C.F.; Zhao, Y.; Zhang, L.; Malik, S. Chaff: Engineering an efficient SAT solver. In Proceedings of the 38th annual Design Automation Conference, Las Vegas, NV, USA, 22 June 2001.

7. Simon, L.; Audemard, G. Predicting Learnt Clauses Quality in Modern SAT Solver. In Proceedings of the 21th International Joint Conference on Artificial Intelligence, Pasadena, CA, USA, 11–17 July 2009.
8. Juretus, K.; Savidis, I. Importance of Multi-parameter SAT Attack Exploration for Integrated Circuit Security. In Proceedings of the 2018 IEEE Asia Pacific Conference on Circuits and Systems, Chengdu, China, 26–30 October 2018.
9. Liu, C.; Zhang, L.; He, X.; Guo, Y. Analysis of SET Reconvergence and Hardening in the Combinational Circuit Using a SAT-Based Method. *IEEE Access* **2018**, *6*, 48740–48746. [CrossRef]
10. Efficient Methods for Conversion and Solution of Sparse Systems of Low-Degree Multivariate Polynomials over GF(2) Via Sat-Solvers. Available online: http://eprint.iacr.org/2007/024 (accessed on 25 January 2007).
11. Cook, S.A. The complexity of theorem-proving procedures. In Proceedings of the Third ACM Symposium on Theory of Computing, Shaker Heights, OH, USA, 3–5 May 1971.
12. Martins, R.; Manquinho, V.; Lynce, I. An overview of parallel SAT solving. *Constraints* **2012**, *17*, 304–347. [CrossRef]
13. Păun, G. Computing with Membranes. *J. Comput. Syst. Sci.* **2000**, *61*, 108–143. [CrossRef]
14. Gutiérrez-Naranjo, M.A.; Pérez-Jiménez, M.J.; Riscos-Núñez, A.; Romero-Campero, F.J. On the efficiency of cell-like and tissue-like recognizing membrane systems. *Int. J. Intell. Syst.* **2009**, *24*, 747–765. [CrossRef]
15. Gutiérrez-Naranjo, M.A.; Pérez-Jiménez, M.J.; Riscos-Núñez, A.; Romero-Campero, F.J. On the power of dissolution in P systems with active membranes. In Proceedings of the International Workshop on Membrane Computing, Vienna, Austria, 18–21 July 2005; pp. 224–240.
16. Macías-Ramos, L.F.; Pérez-Jiménez, M.J.; Riscos-Núñez, A.; Rius-Font, M. The efficiency of tissue P systems with cell separation relies on the environment. In Proceedings of the International Conference on Membrane Computing, Budapest, Hungary, 28–31 August 2012; pp. 243–256.
17. Pan, L.; Pérez-Jiménez, M.J. Computational complexity of tissue–like P systems. *J. Complex.* **2010**, *26*, 296–315. [CrossRef]
18. Pérez-Jiménez, M.J.; Riscos-Núñez, A.; Rius-Font, M.; Valencia-Cabrera, L. The relevance of the environment on the efficiency of tissue P systems. In Proceedings of the International Conference on Membrane Computing Chisinau, Chisinau, Moldova, 20–23 August 2013; pp. 308–321.
19. Manca, V. DNA and membrane algorithms for SAT. *Fund. Inform.* **2002**, *49*, 205–221.
20. Ciobanu, G.; Păun, G.; Pérez-Jiménez, M.J. *Applications of Membrane Computing*; Springer: Berlin, Germany, 2006.
21. Ishii, K.; Fujiwara, A.; Tagawa, H. Asynchronous P systems for SAT and Hamiltonian cycle problem. In Proceedings of the 2010 Second World Congress on Nature and Biologically Inspired Computing, Fukuoka, Japan, 15–17 December 2010.
22. Song, T.; Macías-Ramosb, L.F.; Pan, L.; Pérez-Jiménez, M.J. Time-free solution to SAT problem using P systems with active membranes. *Theor. Comput. Sci.* **2014**, *529*, 61–68. [CrossRef]
23. Adorna, H.N.; Pan, L.; Song, B. On Distributed Solution to SAT by Membrane Computing. *Int. J. Comp. Commun.* **2018**, *13*, 303–322. [CrossRef]
24. Păun, G. *Membrane Computing: An Introduction*; Springer: Berlin, Germany, 2002.
25. Liu, X.H. *Automatic Reasoning Based on The Resolution Method*; Science Press: Beijing, Chinese, 1994.

symmetry

MDPI

Article

A Study on Hypergraph Representations of Complex Fuzzy Information

Anam Luqman [1], Muhammad Akram [1,*], Ahmad N. Al-Kenani [2] and José Carlos R. Alcantud [3]

1 Department of Mathematics, University of the Punjab, New Campus, Lahore 54590, Pakistan; anamluqman7@yahoo.com
2 Department of Mathematics, Faculty of Science, King Abdulaziz University, P.O. Box 80219, Jeddah 21589, Saudi Arabia; aalkenani10@hotmail.com
3 BORDA Research Unit and IME, University of Salamanca, 37007 Salamanca, Spain; jcr@usal.es
* Correspondence: m.akram@pucit.edu.pk

Received: 21 October 2019; Accepted: 5 November 2019; Published: 7 November 2019

Abstract: The paradigm shift prompted by Zadeh's fuzzy sets in 1965 did not end with the fuzzy model and logic. Extensions in various lines have produced e.g., intuitionistic fuzzy sets in 1983, complex fuzzy sets in 2002, or hesitant fuzzy sets in 2010. The researcher can avail himself of graphs of various types in order to represent concepts like networks with imprecise information, whether it is fuzzy, intuitionistic, or has more general characteristics. When the relationships in the network are symmetrical, and each member can be linked with groups of members, the natural concept for a representation is a hypergraph. In this paper we develop novel generalized hypergraphs in a wide fuzzy context, namely, complex intuitionistic fuzzy hypergraphs, complex Pythagorean fuzzy hypergraphs, and complex q-rung orthopair fuzzy hypergraphs. Further, we consider the transversals and minimal transversals of complex q-rung orthopair fuzzy hypergraphs. We present some algorithms to construct the minimal transversals and certain related concepts. As an application, we describe a collaboration network model through a complex q-rung orthopair fuzzy hypergraph. We use it to find the author having the most outstanding collaboration skills using score and choice values.

Keywords: complex q-rung orthopair fuzzy set; complex q-rung orthopair fuzzy graphs; complex q-rung orthopair fuzzy hypergraphs; transversals

---

1. Introduction

In 1965, fuzzy sets (FSs) were originally defined by Zadeh [1] as a novel approach to represent uncertainty arising in various fields. The idea of "partial membership" was questioned by many researchers at that time. The extension of crisp sets to FSs, i.e., the extension of membership function $\mu(x)$ from $\{0,1\}$ to $[0,1]$, bears comparison to the generalization of $\mathbb{Q}$ to $\mathbb{R}$. Just like $\mathbb{R}$ was extended to $\mathbb{C}$ with the incorporation of imaginary quantities, FSs have been extended to complex fuzzy sets (CFSs) by Ramot et al. [2]. A CFS is characterized by a membership function $\mu(x)$ whose range is not limited to [0,1] but extends to the unit circle in the complex plane. Hence, $\mu(x)$ is a complex-valued function that assigns a grade of membership of the form $v(x)e^{i\alpha(x)}$, $i=\sqrt{-1}$ to any element x in the universe of discourse. The membership function $\mu(x)$ of CFS consists of two terms, namely, an amplitude term $v(x)$ which lies in the unit interval $[0,1]$ and a phase term (periodic term) $\alpha(x)$ which lies in the interval $[0,2\pi]$. During the last few years, many researchers have paid special attention to CFSs. Yazdanbakhsh and Dick [3] gives an updated review of the development of CFSs.

Atanassov [4] had proposed a different extension of FSs by intuitionistic fuzzy sets (IFSs). Fuzzy sets give the degree of membership of an element in a given set (the non-membership of degree equals one minus the degree of membership), while IFSs give both a degree of membership and a degree of non-membership, which are to some extent independent from each other. The truth (T) and falsity (F) membership functions are used to characterize an IFS in such a way that the sum of truth and falsity degrees should not be greater than one at any point. These figures allow for some indeterminacy in the expression of memberships. Progress on the investigation of IFSs and related extensions of the FS concept continues to be made. Liu et al. [5] introduced different types of centroid transformations of IF values. Feng et al. [6] defined various new operations for generalized IF soft sets. Recently, Shumaiza et al. [7] have proposed group decision-making based on the VIKOR method with trapezoidal bipolar fuzzy information. Akram and Arshad [8] proposed a novel trapezoidal bipolar fuzzy TOPSIS method for group decision-making. Alcantud et al [9] proposed a novel modelization of the party formation process, in which citizens' private opinions are described by means of continuous fuzzy profiles. A novel hesitant fuzzy model for group decision-making was proposed by Alcantud and Giarlotta [10].

Of particular importance are two extensions of IFSs proposed by Yager [11–13]. In these papers he introduced Pythagorean fuzzy sets (PFSs) and q-rung orthopair fuzzy sets (q-ROFSs). A q-ROFS is characterized by means of truth and falsity degrees satisfying the constraint that the sum of the qth powers of both degrees should be less than one. PFSs consist of the case where $q = 2$. Thus, q-ROFSs generalize both the notions of IFSs and PFSs so that the uncertain information can be dealt with in a more widened range. After that, Liu and Wang [14] applied certain simple weighted operators to aggregate q-ROFSs in decision-making. Intertemporal choice problems have been investigated with the help of fuzzy soft sets [15]. These problems appear in the analysis of environmental issues and sustainable development with an infinitely long horizon, project evaluations, or health care [16,17]. In multi-attribute decision making, q-ROF Heronian mean operators were defined by Wei et al. [18]. For further applications of q-ROFSs, we refer the readers to the work presented in [19,20]. Complex intuitionistic fuzzy sets (CIFSs) were introduced by Alkouri and Salleh [21] in order to generalize IFSs in the spirit of [2] by adding non-membership degree $\nu(x) = s(x)e^{i\beta(x)}$ to the CFSs subjected to the constraint $0 \leq r + s \leq 1$. The CIFSs are used to handle information about uncertainty and periodicity simultaneously. As an extension of both PFSs and CIFSs, Ullah et al. [22] proposed complex Pythagorean fuzzy sets (CPFSs) and discussed some applications.

The vagueness in the representation of various objects and the uncertain interactions between them originated the necessity of fuzzy graphs (FGs), that were first defined by Rosenfeld [23] (see also the remarks made by Bhattacharya [24]). The notion of FGs was extended to complex fuzzy graphs (CFGs) by Thirunavukarasu et al. [25]. Intuitionistic fuzzy graphs (IFGs) were defined by Parvathi and Karunambigai [26]. The energy of Pythagorean fuzzy graphs (PFGs) was discussed by Akram and Naz [27]. Akram and Habib [28] defined q-ROF competition graphs and discussed their applications. Akram et al. [29] proposed a novel description on edge-regular q-ROFGs. Yaqoob et al. [30] defined complex intuitionistic fuzzy graphs (CIFGs) and discussed an application of CIFGs in cellular networks to test the proposed model. Later on, complex neutrosophic graphs were studied by Yaqoob and Akram [31]. Recently, complex Pythagorean fuzzy graphs (CPFGs) and their applications in decision making have been put forward by Akram and Naz [32].

A hypergraph, as an extension of a crisp graph, is a powerful tool to model different practical problems in various fields, including biological sciences, computer science, sustainable development and social networks [33–35]. Co-authorship networks, an important type of social network, have been studied extensively from various angles such as degree distribution analysis, social community extraction and social entity ranking. Most of the previous studies consider the co-authorship relation between two or more authors as a collaboration using crisp hypergraphs. Han et al. [36] proposed a hypergraph analysis approach to understand the importance of collaborations in co-authorship networks. Zhang and Liu [37] proposed a hypergraph model of social tagging networks. Ouvrard et al. [38] studied the hypergraph modeling and visualization of collaboration networks.

In order to allow for uncertainty in crisp hypergraphs, fuzzy hypergraphs (FHGs) were defined by Kaufmann [39] as an extension of FGs. Lee-Kwang and Lee [40] discussed fuzzy partitions using FHGs. A valuable contribution to FGs and FHGs has been proposed by Mordeson and Nair [41]. Fuzzy transversals of FHGs were studied by Goetschel et al. [42]. Intuitionistic fuzzy hypergraphs (IFHGs) were defined by Parvathi et al. [43]. Further discussion on IFHGs can be seen in [44,45]. Akram and Luqman [46] defined bipolar neutrosophic hypergraphs with applications. Transversals and minimal transversals of m-polar FHGs were studied by Akram and Sarwar [47]. Luqman et al. [48] presented q-ROFHGs and their applications. Further, Luqman et al. [49,50] have proposed m-polar and q-rung picture fuzzy hypergraph models of granular computing.

The proposed research generalizes the concepts of CIFGs and CPFGs. These existing models can only depict the uncertainty having periodic nature occurring in pairwise relationships. The existence of various complex network models in which the relationships are more generalized rather than the pairwise relationships motivates the extension of CIFGs and CPFGs to complex intuitionistic fuzzy and complex Pythagorean fuzzy hypergraphs. Let us consider the modeling of research collaborations through CIFGs and CPFGs. The uncertainty and periodicity of the given data are dealt with with the help of phase terms and amplitude terms, respectively. Two research articles are connected through an edge if both have the same author but if more than two articles are written by the same author then CIFGs and CPFGs fail to model this situation. Thus the main objective of this study is to generalize the concepts of CIFGS and CPFGs to complex q-rung orthopair fuzzy hypergraphs. As argued above, complex q-ROF models provide more flexibility than IFSs and FSs. Therefore a complex q-rung orthopair fuzzy hypergraph model proves to be a very general framework to deal with vagueness in complex hypernetworks when the symmetrical relationships go beyond pairwise interactions. The generality of the proposed model can be observed from the reduction of complex q-rung orthopair fuzzy models to CIF and CPF models for $q = 1$ and $q = 2$, respectively. Moreover, most of the previous studies consider the co-authorship relation between two or more authors as a collaboration using crisp hypergraphs. Here we consider a complex q-rung orthopair fuzzy hypergraph model of co-authorship network to represent the collaboration relations between authors having uncertainty and vagueness of periodic nature simultaneously.

The contents of this paper are as follows. In Sections 2 and 3, we define complex intuitionistic fuzzy hypergraphs and complex Pythagorean fuzzy hypergraphs, respectively. In Section 4, complex q-ROF hypergraphs are discussed. In Section 5, we define the q-ROF transversals and minimal transversals of q-ROF hypergraphs. Section 6 illustrates an application of q-ROF hypergraphs in research collaboration networks. We also present an algorithm to select an author with powerful collaboration characteristics using the score and choice values of q-rung orthopair fuzzy hypergraphs and give a brief comparison of our proposed model with CIF and CPF models. The final Section 7 contains conclusions and future research directions.

2. Complex Intuitionistic Fuzzy Hypergraphs

In this section, we define the notion of complex intuitionistic fuzzy hypergraphs. A complex intuitionistic fuzzy hypergraph extends the concept of CIFGs. The proposed hypergraph model is used to handle the uncertain and periodic real-life situations when the relationships are analyzed between more than two objects. The main model that we use in our research design is given in the next definition:

Definition 1. *[21] A complex intuitionistic fuzzy set (CIFS) I on the universal set Y is defined as,*

$$I = \{(u, T_I(u)e^{i\phi_I(u)}, F_I(u)e^{i\psi_I(u)}) | u \in Y\},$$

where $i = \sqrt{-1}$, $T_I(u), F_I(u) \in [0,1]$, $\phi_I(u), \psi_I(u) \in [0, 2\pi]$, *and for every* $u \in Y$, $0 \leq T_I(u) + F_I(u) \leq 1$.

For every $u \in Y$, $T_I(u)$ and $F_I(u)$ are the amplitude terms for membership and non-membership of u, and $\phi_I(u)$ and $\psi_I(u)$ are the phase terms for membership and non-membership of u. CIFSs where the phase terms equal zero (for all u) reduce to ordinary IFSs. When in addition, the amplitude terms for non-membership of all elements equal zero, we obtain a FS.

The application of this concept to graphs was produced in [30]. We represent definition of complex intuitionistic fuzzy graphs as follows:

Definition 2. *A complex intuitionistic fuzzy graph (CIFG) on Y is an ordered pair $G = (A, B)$, where A is a complex intuitionistic fuzzy set on Y and B is complex intuitionistic fuzzy relation on Y such that,*

$$T_B(ab) \leq \min\{T_A(a), T_A(b)\}, \ F_B(ab) \leq \max\{F_A(a), F_A(b)\}, \text{(for amplitude terms)}$$
$$\phi_B(ab) \leq \min\{\phi_A(a), \phi_A(b)\}, \ \psi_B(ab) \leq \max\{\psi_A(a), \psi_A(b)\}, \text{(for phase terms)}$$

$0 \leq T_B(ab) + F_B(ab) \leq 1$, *and* $\phi, \psi \in [0, 2\pi]$, *for all* $a, b \in Y$.

When we apply Definition 1 to hypergraphs we obtain the following structure that generalizes Definition 2:

Definition 3. *Let Y be a non-trivial set of universe. A complex intuitionistic fuzzy hypergraph (CIFHG) is defined as an ordered pair $H = (\mathcal{C}, \mathcal{D})$, where $\mathcal{C} = \{\alpha_1, \alpha_2, \cdots, \alpha_k\}$ is a finite family of complex intuitionistic fuzzy sets on Y and $\mathcal{D}$ is a complex intuitionistic fuzzy relation on complex intuitionistic fuzzy sets α_j's such that the following conditions hold:*

(i)

$$\begin{aligned} T_{\mathcal{D}}(\{r_1, r_2, \cdots, r_l\}) &\leq \min\{T_{\alpha_j}(r_1), T_{\alpha_j}(r_2), \cdots, T_{\alpha_j}(r_l)\}, \\ F_{\mathcal{D}}(\{r_1, r_2, \cdots, r_l\}) &\leq \max\{F_{\alpha_j}(r_1), F_{\alpha_j}(r_2), \cdots, F_{\alpha_j}(r_l)\}, \text{(for amplitude terms)} \\ \phi_{\mathcal{D}}(\{r_1, r_2, \cdots, r_l\}) &\leq \min\{\phi_{\alpha_j}(r_1), \phi_{\alpha_j}(r_2), \cdots, \phi_{\alpha_j}(r_l)\}, \\ \psi_{\mathcal{D}}(\{r_1, r_2, \cdots, r_l\}) &\leq \max\{\psi_{\alpha_j}(r_1), \psi_{\alpha_j}(r_2), \cdots, \psi_{\alpha_j}(r_l)\}, \text{(for phase terms)} \end{aligned}$$

$0 \leq T_{\mathcal{D}} + F_{\mathcal{D}} \leq 1$, *and* $\phi_{\mathcal{D}}, \psi_{\mathcal{D}} \in [0, 2\pi]$, *for all* $r_1, r_2, \cdots, r_l \in Y$.

(ii) $\bigcup_j supp(\alpha_j) = Y$, *for all* $\alpha_j \in \mathcal{C}$.

Notice that $E_k = \{r_1, r_2, \cdots, r_l\}$ is the crisp hyperedge of $H = (\mathcal{C}, \mathcal{D})$.

Note that the above formula reduces to Definition 2 if we consider only two vertices in an hyperedge.

We illustrate the previous definition with a graphical example.

Example 1. *Consider a CIFHG $H = (\mathcal{C}, \mathcal{D})$ on $Y = \{v_1, v_2, v_3, v_4\}$. The CIFR is defined as, $\mathcal{D}(\{v_1, v_2, v_3, v_4\}) = (0.2e^{i0.4\pi}, 0.6e^{i0.3\pi})$, $\mathcal{D}(\{v_1, v_2\}) = (0.3e^{i0.6\pi}, 0.6e^{i0.3\pi})$, and $\mathcal{D}(\{v_3, v_4\}) = (0.2e^{i0.4\pi}, 0.5e^{i0.3\pi})$. The corresponding CIFHG is shown in Figure 1.*

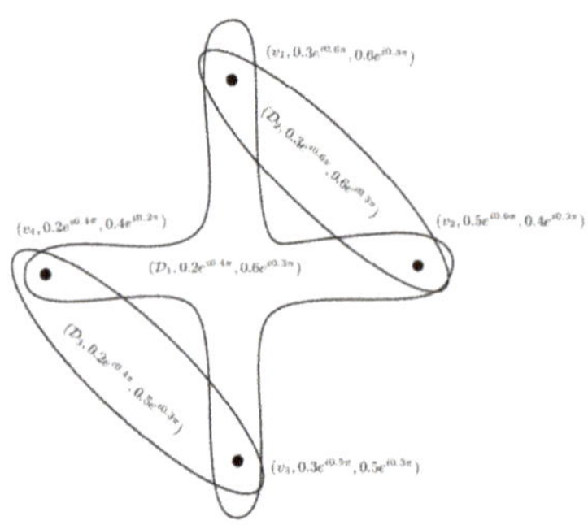

Figure 1. Complex intuitionistic fuzzy hypergraph.

Simple CIFHGs are the following special types of CIFHGs:

Definition 4. *A CIFHG $H = (\mathcal{C}, \mathcal{D})$ is simple if whenever $\mathcal{D}_j, \mathcal{D}_k \in \mathcal{D}$ and $\mathcal{D}_j \subseteq \mathcal{D}_k$, then $\mathcal{D}_j = \mathcal{D}_k$. A CIFHG $H = (\mathcal{C}, \mathcal{D})$ is support simple if whenever $\mathcal{D}_j, \mathcal{D}_k \in \mathcal{D}$, $\mathcal{D}_j \subseteq \mathcal{D}_k$, and $supp(\mathcal{D}_j) = supp(\mathcal{D}_k)$, then $\mathcal{D}_j = \mathcal{D}_k$.*

Our next notion produces a link between CIFHGs and crisp hypergraphs. The subsequent example illustrates this construction.

Definition 5. *Let $H = (\mathcal{C}, \mathcal{D})$ be a CIFHG. Suppose that $\alpha, \beta \in [0,1]$ and $\theta, \varphi \in [0, 2\pi]$ such that $0 \leq \alpha + \beta \leq 1$. The $(\alpha e^{i\theta}, \beta e^{i\varphi})-$level hypergraph of H is defined as an ordered pair $H^{(\alpha e^{i\theta}, \beta e^{i\varphi})} = (\mathcal{C}^{(\alpha e^{i\theta}, \beta e^{i\varphi})}, \mathcal{D}^{(\alpha e^{i\theta}, \beta e^{i\varphi})})$, where*

(i) $\mathcal{D}^{(\alpha e^{i\theta}, \beta e^{i\varphi})} = \{D_j^{(\alpha e^{i\theta}, \beta e^{i\varphi})} : D_j \in \mathcal{D}\}$ *and* $D_j^{(\alpha e^{i\theta}, \beta e^{i\varphi})} = \{u \in Y : T_{D_j}(u) \geq \alpha, \phi_{D_j}(u) \geq \theta$, and $F_{D_j}(u) \leq \beta, \psi_{D_j}(u) \leq \varphi\}$,
(ii) $\mathcal{C}^{(\alpha e^{i\theta}, \beta e^{i\varphi})} = \bigcup_{D_j \in \mathcal{D}} D_j^{(\alpha e^{i\theta}, \beta e^{i\varphi})}$.

Note that the $(\alpha e^{i\theta}, \beta e^{i\varphi})-$level hypergraph of H is a crisp hypergraph.

Example 2. *Consider a CIFHG $H = (\mathcal{C}, \mathcal{D})$ as shown in Figure 1. Let $\alpha = 0.2$, $\beta = 0.5$, $\theta = 0.5\pi$, and $\varphi = 0.2\pi$. The $(\alpha e^{i\theta}, \beta e^{i\varphi})-$level hypergraph of H is shown in Figure 2.*

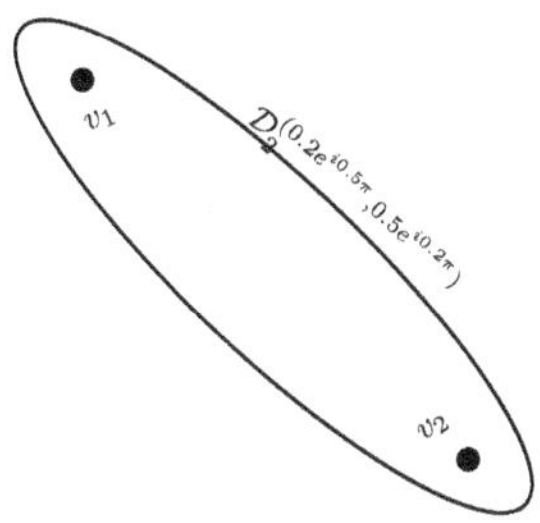

Figure 2. $(0.2e^{i(0.5)\pi}, 0.5e^{i(0.2)\pi})-$level hypergraph of H.

Definition 6. *Let $H = (\mathcal{C}, \mathcal{D})$ be a CIFHG. The complex intuitionistic fuzzy line graph of H is defined as an ordered pair $l(H) = (\mathcal{C}_l, \mathcal{D}_l)$, where $\mathcal{C}_l = \mathcal{D}$ and there exists an edge between two vertices in $l(H)$ if $|supp(D_j) \cap supp(D_k)| \geq 1$. The membership degrees of $l(H)$ are given as,*

(i) $\mathcal{C}_l(E_k) = \mathcal{D}(E_k)$,
(ii) $\mathcal{D}_l(E_jE_k) = (\min\{T_{\mathcal{D}}(E_j), T_{\mathcal{D}}(E_k)\}e^{i\min\{\phi_{\mathcal{D}}(E_j), \phi_{\mathcal{D}}(E_k)\}}, \max\{F_{\mathcal{D}}(E_j), F_{\mathcal{D}}(E_k)\}e^{i\max\{\psi_{\mathcal{D}}(E_j), \psi_{\mathcal{D}}(E_k)\}})$.

Definition 7. *A CIFHG $H = (\mathcal{C}, \mathcal{D})$ is said to be linear if for every $D_j, D_k \in \mathcal{D}$,*

(i) $supp(D_j) \subseteq supp(D_k) \Rightarrow j = k$,
(ii) $|supp(D_j) \cap supp(D_k)| \leq 1$.

Example 3. *Consider a CIFHG $H = (\mathcal{C}, \mathcal{D})$ as shown in Figure 1. By direct calculations, we have*

$$supp(\mathcal{D}_1) = \{v_1, v_2, v_3, v_4\},\ supp(\mathcal{D}_2) = \{v_1, v_2\},\ supp(\mathcal{D}_3) = \{v_3, v_4\}.$$

Note that, $supp(D_j) \subseteq supp(D_k) \Rightarrow j \neq k$ and $|supp(D_j) \cap supp(D_k)| \nleq 1$. Hence, CIFHG $H = (\mathcal{C}, \mathcal{D})$ is not linear. The corresponding CIFHG $H = (\mathcal{C}, \mathcal{D})$ and its line graph is shown in Figure 3.

Theorem 1. *A simple strong CIFG is the complex intuitionistic line graph of a linear CIFHG.*

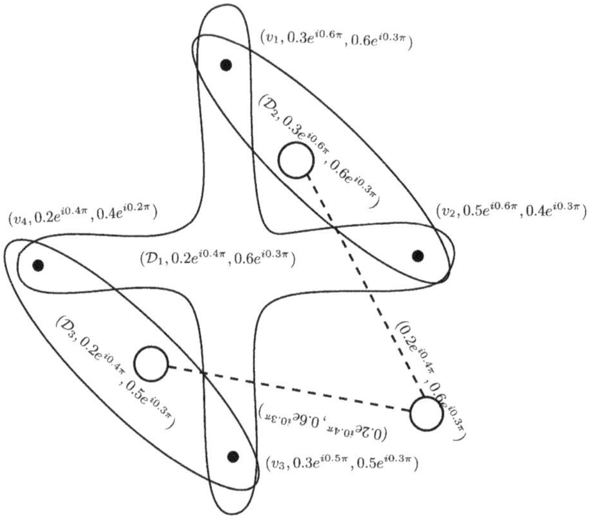

Figure 3. Complex intuitionistic fuzzy line graph of H.

Definition 8. *The 2-section $H_2 = (\mathcal{C}_2, \mathcal{D}_2)$ of a CIFHG $H = (\mathcal{C}, \mathcal{D})$ is a CIFG having same set of vertices as that of H, $\mathcal{D}_2$ is a CIFS on $\{e = u_j u_k | u_j, u_k \in E_l,\ l = 1, 2, 3, \cdots\}$, and* $\mathcal{D}_2(u_j u_k) = (\min\{\min T_{\alpha_l}(u_j), \min T_{\alpha_l}(u_k)\} e^{i\min\{\min \phi_{\alpha_l}(u_j), \min \phi_{\alpha_l}(u_k)\}}, \max\{\max F_{\alpha_l}(u_j), \max F_{\alpha_l}(u_k)\} e^{i\max\{\max \psi_{\alpha_l}(u_j), \max \psi_{\alpha_l}(u_k)\}})$ *such that* $0 \leq T_{\mathcal{D}_2}(u_j u_k) + F_{\mathcal{D}_2}(u_j u_k) \leq 1$, $\phi_{\mathcal{D}_2}, \psi_{\mathcal{D}_2} \in [0, 2\pi]$.

Example 4. *An example of a CIFHG is given in Figure 4. The 2-section of H is presented with dashed lines.*

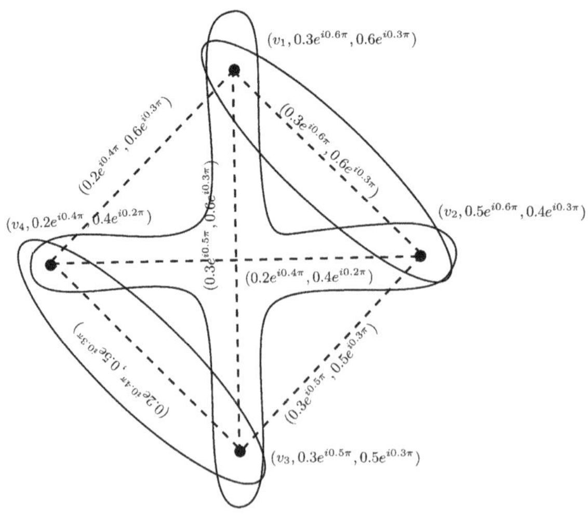

Figure 4. Two-section of complex intuitionistic fuzzy hypergraph.

Definition 9. *Let $H = (\mathcal{C}, \mathcal{D})$ be a CIFHG. A complex intuitionistic fuzzy transversal (CIFT) τ is a CIFs of Y satisfying the condition $\rho^{h(\rho)} \cap \tau^{h(\rho)} \neq \emptyset$, for all $\rho \in \mathcal{D}$, where $h(\rho)$ is the height of ρ.*

A minimal complex intuitionistic fuzzy transversal t is the CIFT of H having the property that if $\tau \subset t$, then τ is not a CIFT of H.

3. Complex Pythagorean Fuzzy Hypergraphs

We now turn our attention to the next class of hypergraphs called complex Pythagorean fuzzy hypergraphs. A complex Pythagorean fuzzy hypergraph is the generalization of CPFGs and CIFHGs. The occurrence of truth and falsity degrees whose sum is not less than one but the sum of squares does not exceed one in complex hypernetworks motivates the necessity of this proposed model.

Definition 10. [32] *A complex Pythagorean fuzzy graph (CPFG) on Y is an ordered pair $G^* = (C, D)$, where C is a CPFS on Y and D is CPFR on Y such that,*

$$T_D(ab) \leq \min\{T_C(a), T_C(b)\},\ F_D(ab) \leq \max\{F_C(a), F_C(b)\}, \text{ for amplitude terms}$$
$$\phi_D(ab) \leq \min\{\phi_C(a), \phi_C(b)\},\ \psi_D(ab) \leq \max\{\psi_C(a), \psi_C(b)\}, \text{ for phase terms}$$

$0 \leq T_D^2(ab) + F_D^2(ab) \leq 1$, *and* $\phi_D, \psi_D \in [0, 2\pi]$, *for all* $a, b \in Y$.

Definition 11. *A complex Pythagorean fuzzy hypergraph (CPFHG) on Y is defined as an ordered pair $H^* = (\mathcal{C}^*, \mathcal{D}^*)$, where $\mathcal{C}^* = \{\beta_1, \beta_2, \cdots, \beta_k\}$ is a finite family of CPFSs on Y and $\mathcal{D}^*$ is a CPFR on CPFSs β_j's such that,*

(i)

$$\begin{aligned}
T_{\mathcal{D}^*}(\{s_1, s_2, \cdots, s_l\}) &\leq \min\{T_{\beta_j}(s_1), T_{\beta_j}(s_2), \cdots, T_{\beta_j}(s_l)\},\\
F_{\mathcal{D}^*}(\{s_1, s_2, \cdots, s_l\}) &\leq \max\{F_{\beta_j}(s_1), F_{\beta_j}(s_2), \cdots, F_{\beta_j}(s_l)\}, \text{ for amplitude terms}\\
\phi_{\mathcal{D}^*}(\{s_1, s_2, \cdots, s_l\}) &\leq \min\{\phi_{\beta_j}(s_1), \phi_{\beta_j}(s_2), \cdots, \phi_{\beta_j}(s_l)\},\\
\psi_{\mathcal{D}^*}(\{s_1, s_2, \cdots, s_l\}) &\leq \max\{\psi_{\beta_j}(s_1), \psi_{\beta_j}(s_2), \cdots, \psi_{\beta_j}(s_l)\}, \text{ for phase terms}
\end{aligned}$$

$0 \leq T_{\mathcal{D}^*}^2 + F_{\mathcal{D}^*}^2 \leq 1$, *and* $\phi_{\mathcal{D}^*}, \psi_{\mathcal{D}^*} \in [0, 2\pi]$, *for all* $s_1, s_2, \cdots, s_l \in Y$.

(ii) $\bigcup_j supp(\beta_j) = Y$, *for all* $\beta_j \in \mathcal{C}^*$.

Note that, $E_k = \{s_1, s_2, \cdots, s_l\}$ *is the crisp hyperedge of* $H^* = (\mathcal{C}^*, \mathcal{D}^*)$.

Example 5. *Consider a CPFHG $H^* = (\mathcal{C}^*, \mathcal{D}^*)$ on $Y = \{s_1, s_2, s_3, s_4, s_5, s_6\}$. The CPFR is defined as, $\mathcal{D}^*(s_1, s_2, s_3) = ((0.6e^{i(0.2)\pi}, 0.5e^{i(0.9)\pi}))$, $\mathcal{D}^*(s_4, s_5, s_6) = (0.6e^{i(0.4)\pi}, 0.4e^{i(0.6)\pi})$, $\mathcal{D}^*(s_3, s_6) = (0.6e^{i(0.6)\pi}, 0.5e^{i(0.6)\pi})$, $\mathcal{D}^*(s_2, s_5) = (0.6e^{i(0.4)\pi}, 0.5e^{i(0.6)\pi})$, and $\mathcal{D}^*(s_1, s_4) = (0.6e^{i(0.2)\pi}, 0.9e^{i(0.9)\pi})$. The corresponding CPFHG is shown in Figure 5.*

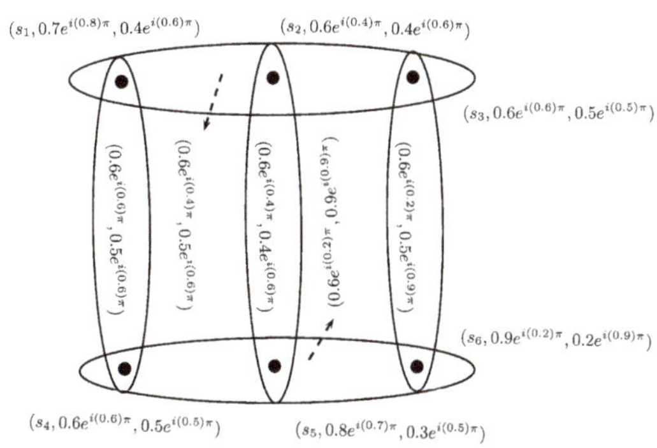

Figure 5. Complex Pythagorean fuzzy hypergraph.

Definition 12. *A CPFHG $H^* = (\mathcal{C}^*, \mathcal{D}^*)$ is simple if whenever $\mathcal{D}_j^*, \mathcal{D}_k^* \in \mathcal{D}^*$ and $\mathcal{D}_j^* \subseteq \mathcal{D}_k^*$, then $\mathcal{D}_j^* = \mathcal{D}_k^*$. A CPFHG $H^* = (\mathcal{C}^*, \mathcal{D}^*)$ is support simple if whenever $\mathcal{D}_j^*, \mathcal{D}_k^* \in \mathcal{D}^*$, $\mathcal{D}_j^* \subseteq \mathcal{D}_k^*$, and $supp(\mathcal{D}_j^*) = supp(\mathcal{D}_k^*)$, then $\mathcal{D}_j^* = \mathcal{D}_k^*$.*

Definition 13. *Let $H^* = (\mathcal{C}^*, \mathcal{D}^*)$ be a CPFHG. Suppose that $\alpha_1, \beta_1 \in [0,1]$ and $\theta, \varphi \in [0, 2\pi]$ such that $0 \leq \alpha_1^2 + \beta_1^2 \leq 1$. The $(\alpha_1 e^{i\theta}, \beta_1 e^{i\varphi})-$level hypergraph of H^* is defined as an ordered pair $H^{*(\alpha_1 e^{i\theta}, \beta_1 e^{i\varphi})} = (\mathcal{C}^{*(\alpha_1 e^{i\theta}, \beta_1 e^{i\varphi})}, \mathcal{D}^{*(\alpha_1 e^{i\theta}, \beta_1 e^{i\varphi})})$, where*

- **(i)** $\mathcal{D}^{*(\alpha_1 e^{i\theta}, \beta_1 e^{i\varphi})} = \{D_j^{*(\alpha_1 e^{i\theta}, \beta_1 e^{i\varphi})} : D_j^* \in \mathcal{D}^*\}$ *and* $D_j^{*(\alpha_1 e^{i\theta}, \beta_1 e^{i\varphi})} = \{y \in Y : T_{D_j^*}(y) \geq \alpha_1, \phi_{D_j^*}(y) \geq \theta$, and $F_{D_j^*}(y) \leq \beta_1, \psi_{D_j^*}(y) \leq \varphi\}$,
- **(ii)** $\mathcal{C}^{*(\alpha_1 e^{i\theta}, \beta_1 e^{i\varphi})} = \bigcup_{D_j^* \in \mathcal{D}^*} D_j^{*(\alpha_1 e^{i\theta}, \beta_1 e^{i\varphi})}$.

Note that, $(\alpha_1 e^{i\theta}, \beta_1 e^{i\varphi})-$level hypergraph of H^* is a crisp hypergraph.

Example 6. *Consider a CPFHG $H^* = (\mathcal{C}^*, \mathcal{D}^*)$ as shown in Figure 5. Let $\alpha_1 = 0.5$, $\beta_1 = 0.6$, $\theta = 0.3\pi$, and $\varphi = 0.7\pi$. Then, $(\alpha_1 e^{i\theta}, \beta_1 e^{i\varphi})-$level hypergraph of H^* is shown in Figure 6.*

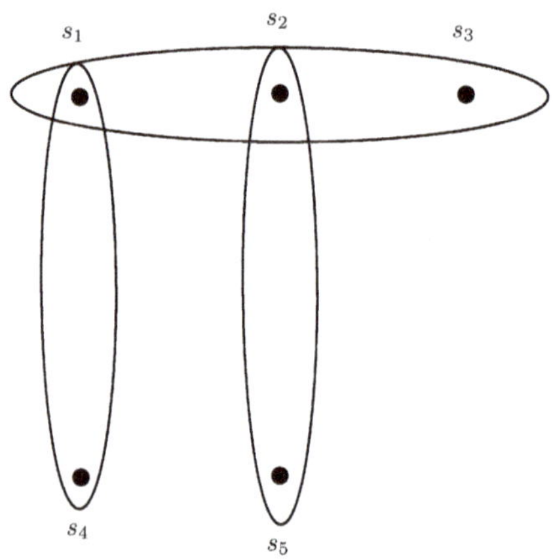

Figure 6. $(\alpha_1 e^{i\theta}, \beta_1 e^{i\varphi})-$level hypergraph of H^*.

Definition 14. *Let $H^* = (\mathcal{C}^*, \mathcal{D}^*)$ be a CPFHG. The complex Pythagorean fuzzy line graph of H^* is defined as an ordered pair $l(H^*) = (\mathcal{C}_l^*, \mathcal{D}_l^*)$, where $\mathcal{C}_l^* = \mathcal{D}^*$ and there exists an edge between two vertices in $l(H^*)$ if $|supp(D_j) \cap supp(D_k)| \geq 1$, for all $D_j, D_k \in \mathcal{D}^*$. The membership degrees of $l(H^*)$ are given as,*

- **(i)** $\mathcal{C}_l^*(E_k) = \mathcal{D}^*(E_k)$,
- **(ii)** $\mathcal{D}_l^*(E_j E_k) = (\min\{T_{\mathcal{D}^*}(E_j), T_{\mathcal{D}^*}(E_k)\} e^{i \min\{\phi_{\mathcal{D}^*}(E_j), \phi_{\mathcal{D}^*}(E_k)\}}, \max\{F_{\mathcal{D}^*}(E_j), F_{\mathcal{D}^*}(E_k)\}$ $e^{i \max\{\psi_{\mathcal{D}^*}(E_j), \psi_{\mathcal{D}^*}(E_k)\}})$.

Definition 15. *A CPFHG $H^* = (\mathcal{C}^*, \mathcal{D}^*)$ is said to be linear if for every $D_j, D_k \in \mathcal{D}^*$,*

- **(i)** $supp(D_j) \subseteq supp(D_k) \Rightarrow j = k$,
- **(ii)** $|supp(D_j) \cap supp(D_k)| \leq 1$.

Example 7. *Consider a CPFHG $H^* = (\mathcal{C}^*, \mathcal{D}^*)$ as shown in Figure 5. By direct calculations, we have*

$$\begin{aligned} supp(\mathcal{D}_1) &= \{s_1, s_2, s_3\}, \; supp(\mathcal{D}_2) = \{s_4, s_5, s_6\}, \; supp(\mathcal{D}_3) = \{s_1, s_4\}, \\ supp(\mathcal{D}_4) &= \{s_2, s_5\}, \; supp(\mathcal{D}_5) = \{s_3, s_6\}. \end{aligned}$$

Note that, $supp(D_j) \subseteq supp(D_k) \Rightarrow j = k$ and $|supp(D_j) \cap supp(D_k)| \leq 1$. Hence, CPFHG $H^* = (\mathcal{C}^*, \mathcal{D}^*)$ is linear. The corresponding CPFHG $H^* = (\mathcal{C}^*, \mathcal{D}^*)$ and its line graph is shown in Figure 7.

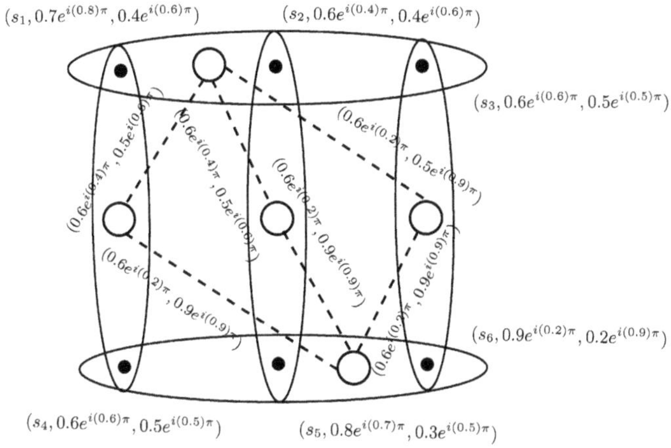

Figure 7. Line graph of complex Pythagorean fuzzy hypergraph H^*.

Theorem 2. *A simple strong CPFG is the complex Pythagorean fuzzy line graph of a linear CPFHG.*

Definition 16. *The 2-section $H_2^* = (\mathcal{C}_2^*, \mathcal{D}_2^*)$ of a CPFHG $H^* = (\mathcal{C}^*, \mathcal{D}^*)$ is a CPFG having same set of vertices as that of H^*, $\mathcal{D}_2^*$ is a CPFS on $\{e = u_j u_k | u_j, u_k \in E_l,\ l = 1, 2, 3, \cdots\}$, and $\mathcal{D}_2^*(u_j u_k) = (\min\{\min T_{\beta_l}(u_j), \min T_{\beta_l}(u_k)\} e^{i \min\{\min \phi_{\beta_l}(u_j), \min \phi_{\beta_l}(u_k)\}}, \max\{\max F_{\beta_l}(u_j), \max F_{\beta_l}(u_k)\}$ $e^{i \max\{\max \psi_{\beta_l}(u_j), \max \psi_{\beta_l}(u_k)\}})$ such that $0 \leq T^2_{\mathcal{D}_2^*}(u_j u_k) + F^2_{\mathcal{D}_2^*}(u_j u_k) \leq 1$, $\phi_{\mathcal{D}_2^*}, \psi_{\mathcal{D}_2^*} \in [0, 2\pi]$.*

Example 8. *An example of a CPFHG is given in Figure 8. The 2-section of H^* is presented with dashed lines.*

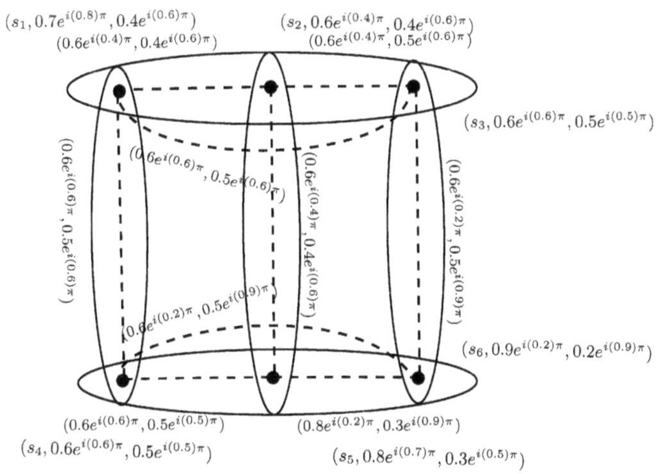

Figure 8. Two-section of complex Pythagorean fuzzy hypergraph H^*.

Definition 17. *Let $H^* = (\mathcal{C}^*, \mathcal{D}^*)$ be a CPFHG. A complex Pythagorean fuzzy transversal (CPFT) τ is a CPFS of Y satisfying the condition $\rho^{h(\rho)} \cap \tau^{h(\rho)} \neq \emptyset$, for all $\rho \in \mathcal{D}^*$, where $h(\rho)$ is the height of ρ.*

A minimal complex Pythagorean fuzzy transversal t is the CPFT of H^* having the property that if $\tau \subset t$, then τ is not a CPFT of H^*.

4. Complex q-Rung Orthopair Fuzzy Hypergraphs

This section explores the class of complex q-rung orthopair fuzzy graphs and complex q-rung orthopair fuzzy hypergraphs. Complex q-rung orthopair fuzzy hypergraphs generalize the notions of CIFHGs and CPFHGs. The class of Cq-ROFSs extends the classes of CIFSs and CPFSs. The space of Cq-ROFSs increases as the value of parameter q increases. Based on these advantages of Cq-ROFSs, we combine the theories of Cq-ROFSs and graphs to define complex q-rung orthopair fuzzy graphs and complex q-rung orthopair fuzzy hypergraphs.

Definition 18. *[13] A q-rung orthopair fuzzy set (q-ROFS) Q in the universal set Y is defined as, $Q = \{(u, T_Q(u), F_Q(u)) | u \in Y\}$, where the function $T_Q : Y \to [0,1]$ defines the truth-membership and $F_Q : Y \to [0,1]$ defines the falsity-membership of the element $u \in Y$ and for every $u \in Y$, $0 \leq T_Q^q(u) + F_Q^q(u) \leq 1$, $q \geq 1$. Furthermore, $\pi_Q(u) = \sqrt[q]{1 - T_Q^q(u) - F_Q^q(u)}$ is called the indeterminacy degree or q-ROF index of u to the set Q.*

Definition 19. *A complex q-rung orthopair fuzzy set (Cq-ROFS) S in the universal set Y is given as,*

$$S = \{(u, T_S(u)e^{i\phi_S(u)}, F_S(u)e^{i\psi_S(u)}) | u \in Y\},$$

where $i = \sqrt{-1}$, $T_S(u), F_S(u) \in [0,1]$, $\phi_S(u), \psi_S(u) \in [0, 2\pi]$, and for every $u \in Y, 0 \leq T_S^q(u) + F_S^q(u) \leq 1$, $q \geq 1$.

Remark 1.

- *When $q = 1$, C1-ROFS is called a CIFS.*
- *When $q = 2$, C2-ROFS is called a CPFS.*

Definition 20. *Let $S_1 = \{(u, T_{S_1}(u)e^{i\phi_{S_1}(u)}, F_{S_1}(u)e^{i\psi_{S_1}(u)}) | u \in Y\}$ and $S_2 = \{(u, T_{S_2}(u)e^{i\phi_{S_2}}(u), F_{S_2}(u)e^{i\psi_{S_2}}(u)) | u \in Y\}$ be two Cq-ROFSs in Y, then*

- **(i)** *$S_1 \subseteq S_2 \Leftrightarrow T_{S_1} \leq T_{S_2}(u)$, $F_{S_1}(u) \geq F_{S_2}(u)$, and $\phi_{S_1}(u) \leq \phi_{S_2}(u)$, $\psi_{S_1}(u) \geq \psi_{S_2}(u)$ for amplitudes and phase terms, respectively, for all $u \in Y$.*
- **(ii)** *$S_1 = S_2 \Leftrightarrow T_{S_1} = T_{S_2}(u)$, $F_{S_1}(u) = F_{S_2}(u)$, and $\phi_{S_1}(u) = \phi_{S_2}(u)$, $\psi_{S_1}(u) = \psi_{S_2}(u)$ for amplitudes and phase terms, respectively, for all $u \in Y$.*

Definition 21. *Let $S_1 = \{(u, T_{S_1}(u)e^{i\phi_{S_1}(u)}, F_{S_1}(u)e^{i\psi_{S_1}(u)}) | u \in Y\}$ and $S_2 = \{(u, T_{S_2}(u)e^{i\phi_{S_2}}(u), F_{S_2}(u)e^{i\psi_{S_2}}(u)) | u \in Y\}$ be two Cq-ROFSs in Y, then*

- **(i)** *$S_1 \cup S_2 = \{(u, \max\{T_{S_1}(u), T_{S_2}(u)\}e^{i\max\{\phi_{S_1}(u), \phi_{S_2}(u)\}}, \min\{F_{S_1}(u), F_{S_2}(u)\}e^{i\min\{\psi_{S_1}(u), \psi_{S_2}(u)\}})$ $|u \in Y\}$.*
- **(ii)** *$S_1 \cap S_2 = \{(u, \min\{T_{S_1}(u), T_{S_2}(u)\}e^{i\min\{\phi_{S_1}(u), \phi_{S_2}(u)\}}, \max\{F_{S_1}(u), F_{S_2}(u)\}e^{i\max\{\psi_{S_1}(u), \psi_{S_2}(u)\}})$ $|u \in Y\}$.*

Definition 22. *A complex q-rung orthopair fuzzy relation (Cq-ROFR) is a Cq-ROFS in $Y \times Y$ given as,*

$$R = \{(rs, T_R(rs)e^{i\phi_R(rs)}, F_R(rs)e^{i\psi_R(rs)}) | rs \in Y \times Y\},$$

where $i = \sqrt{-1}$, $T_R : Y \times Y \to [0,1]$, $F_R : Y \times Y \to [0,1]$ characterize the truth and falsity degrees of R, and $\phi_R(rs), \psi_R(rs) \in [0, 2\pi]$ such that for all $rs \in Y \times Y, 0 \leq T_R^q(rs) + F_R^q(rs) \leq 1, q \geq 1$.

Example 9. *Let $Y = \{b_1, b_2, b_3\}$ be the universal set and $\{b_1b_2, b_2b_3, b_1b_3\}$ be the subset of $Y \times Y$. Then, the C5-ROFR R is given as,*

$$R = \{(b_1b_2, 0.9e^{i(0.7)\pi}, 0.7e^{i(0.9)\pi}), (b_2b_3, 0.6e^{i(0.7)\pi}, 0.8e^{i(0.9)\pi}), (b_1b_3, 0.7e^{i(0.8)\pi}, 0.5e^{i(0.6)\pi})\}.$$

Note that, $0 \leq T_R^5(xy) + F_R^5(xy) \leq 1$, for all $xy \in Y \times Y$. Hence, R is a C5-ROFR on Y.

Definition 23. *A complex q-rung orthopair fuzzy graph(Cq-ROFG) on Y is an ordered pair $\mathcal{G} = (\mathcal{A}, \mathcal{B})$, where $\mathcal{A}$ is a complex q-rung orthopair fuzzy set on Y and $\mathcal{B}$ is complex q-rung orthopair fuzzy relation on Y such that,*

$$\begin{aligned}
&T_{\mathcal{B}}(ab) \leq \min\{T_{\mathcal{A}}(a), T_{\mathcal{A}}(b)\},\\
&F_{\mathcal{B}}(ab) \leq \max\{F_{\mathcal{A}}(a), F_{\mathcal{A}}(b)\}, \text{ (for amplitude terms)}\\
&\phi_{\mathcal{B}}(ab) \leq \min\{\phi_{\mathcal{A}}(a), \phi_{\mathcal{A}}(b)\},\\
&\psi_{\mathcal{B}}(ab) \leq \max\{\psi_{\mathcal{A}}(a), \psi_{\mathcal{A}}(b)\}, \text{ (for phase terms)}
\end{aligned}$$

$0 \leq T_{\mathcal{B}}^q(ab) + F_{\mathcal{B}}^q(ab) \leq 1$, $q \geq 1$, *for all $a, b \in Y$.*

Remark 2. *Note that,*

- *When $q = 1$, C1-ROFG is called a CIFG.*
- *When $q = 2$, C2-ROFG is called a CPFG.*

Example 10. *Let $\mathcal{G} = (\mathcal{A}, \mathcal{B})$ be a C6-ROFG on $Y = \{s_1, s_2, s_3, s_4\}$, where $\mathcal{A} = \{(s_1, 0.7e^{i(0.9)\pi}, 0.9e^{i(0.7)\pi}), (s_2, 0.5e^{i(0.6)\pi}, 0.6e^{i(0.5)\pi}), (s_3, 0.7e^{i(0.4)\pi}, 0.4e^{i(0.7)\pi}), (s_4, 0.8e^{i(0.5)\pi}, 0.5e^{i(0.8)\pi})\}$ and $\mathcal{B} = \{(s_1s_4, 0.7e^{i(0.7)\pi}, 0.8e^{i(0.8)\pi}), (s_2s_4, 0.5e^{i(0.5)\pi}, 0.6e^{i(0.8)\pi}), (s_3s_4, 0.7e^{i(0.4)\pi}, 0.5e^{i(0.8)\pi})\}$ are C6-ROFS and C6-ROFR on Y, respectively. The corresponding C6-ROFG $\mathcal{G}$ is shown in Figure 9.*

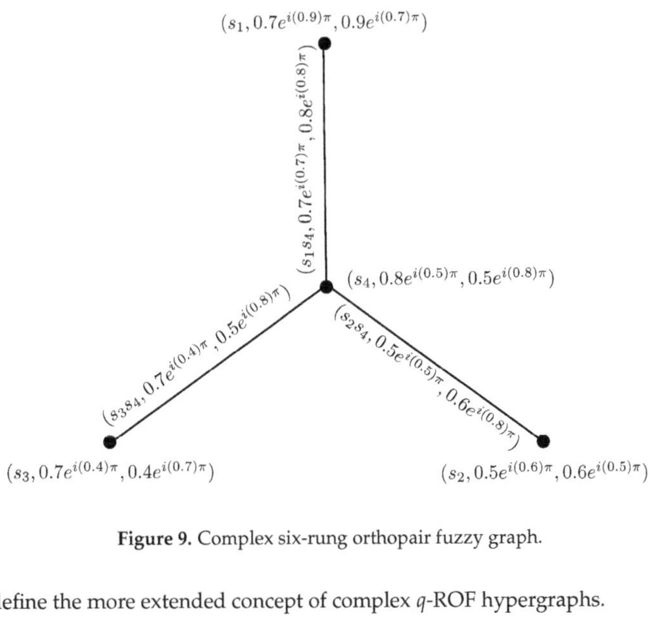

Figure 9. Complex six-rung orthopair fuzzy graph.

We now define the more extended concept of complex q-ROF hypergraphs.

Definition 24. *The support of a Cq-ROFS* $S = \{(u, T_S(u)e^{i\phi_S(u)}, F_S(u)e^{i\psi_S(u)})|u \in Y\}$ *is defined as* $supp(S) = \{u|T_S(u) \neq 0, F_S(u) \neq 1, 0 < \phi_S(u), \psi_S(u) < 2\pi\}$.. *The height of a Cq-ROFS* $S = \{(u, T_S(u)e^{i\phi_S(u)}, F_S(u)e^{i\psi_S(u)})|u \in Y\}$ *is defined as*

$$h(S) = \{\max_{u \in Y} T_S(u)e^{i \max_{u \in Y} \phi_S(u)}, \min_{u \in Y} F_S(u)e^{i \min_{u \in Y} \psi_S(u)}\}.$$

If $h(S) = (1e^{i2\pi}, 0e^{i0})$, *then S is called normal.*

Definition 25. *Let Y be a non-trivial set of universe. A complex q-rung orthopair fuzzy hypergraph (Cq-ROFHG) is defined as an ordered pair* $\mathcal{H} = (\mathcal{Q}, \eta)$, *where* $\mathcal{Q} = \{Q_1, Q_2, \cdots, Q_k\}$ *is a finite family of complex q-rung orthopair fuzzy sets on Y and* η *is a complex q-rung orthopair fuzzy relation on complex q-rung orthopair fuzzy sets* Q_j*'s such that,*

(i)

$$\begin{aligned} T_\eta(\{a_1, a_2, \cdots, a_l\}) &\leq \min\{T_{Q_j}(a_1), T_{Q_j}(a_2), \cdots, T_{Q_j}(a_l)\}, \\ F_\eta(\{a_1, a_2, \cdots, a_l\}) &\leq \max\{F_{Q_j}(a_1), F_{Q_j}(a_2), \cdots, F_{Q_j}(a_l)\}, \text{ (for amplitude terms)} \\ \phi_\eta(\{a_1, a_2, \cdots, a_l\}) &\leq \min\{\phi_{Q_j}(a_1), \phi_{Q_j}(a_2), \cdots, \phi_{Q_j}(a_l)\}, \\ \psi_\eta(\{a_1, a_2, \cdots, a_l\}) &\leq \max\{\psi_{Q_j}(a_1), \psi_{Q_j}(a_2), \cdots, \psi_{Q_j}(a_l)\}, \text{ (for phase terms)} \end{aligned}$$

$0 \leq T_\eta^q + F_\eta^q \leq 1, q \geq 1,$ *for all* $a_1, a_2, \cdots, a_l \in Y.$

(ii) $\bigcup_j supp(Q_j) = X,$ *for all* $Q_j \in \mathcal{Q}.$

Note that, $E_k = \{a_1, a_2, \cdots, a_l\}$ *is the crisp hyperedge of* $\mathcal{H} = (\mathcal{Q}, \eta)$.

Remark 3. *Note that,*

- *When* $q = 1$, *C1-ROFHG is a CIFHG.*
- *When* $q = 2$, *C2-ROFHG is a CPFHG.*

Definition 26. *Let* $\mathcal{H} = (\mathcal{Q}, \eta)$ *be a Cq-ROFHG. The height of* $\mathcal{H}$, *given as* $h(\mathcal{H})$, *is defined as* $h(\mathcal{H}) = (\max \eta_l e^{i \max \phi}, \min \eta_m e^{i \min \psi})$, *where* $\eta_l = \max T_{\rho_j}(x_k)$, $\phi = \max \phi_{\rho_j}(x_k)$, $\eta_m = \min F_{\rho_j}(x_k)$, $\psi = \min \psi_{\rho_j}(x_k)$. *Here,* $T_{\rho_j}(x_k)$ *and* $F_{\rho_j}(x_k)$ *denote the truth and falsity degrees of vertex* x_k *to hyperedge* ρ_j, *respectively.*

Definition 27. *Let* $\mathcal{H} = (\mathcal{Q}, \eta)$ *be a Cq-ROFHG. Suppose that* $\mu, \nu \in [0,1]$ *and* $\theta, \varphi \in [0, 2\pi]$ *such that* $0 \leq \mu^q + \nu^q \leq 1$. *The* $(\mu e^{i\theta}, \nu e^{i\varphi})-$*level hypergraph of* $\mathcal{H}$ *is defined as an ordered pair* $\mathcal{H}^{(\mu e^{i\theta}, \nu e^{i\varphi})} = (\mathcal{Q}^{(\mu e^{i\theta}, \nu e^{i\varphi})}, \eta^{(\mu e^{i\theta}, \nu e^{i\varphi})})$, *where*

(i) $\eta^{(\mu e^{i\theta}, \nu e^{i\varphi})} = \{\rho_j^{(\mu e^{i\theta}, \nu e^{i\varphi})} : \rho_j \in \eta\}$ *and* $\rho_j^{(\mu e^{i\theta}, \nu e^{i\varphi})} = \{u \in Y : T_{\rho_j}(u) \geq \mu, \phi_{\rho_j}(u) \geq \theta,$ and $F_{\rho_j}(u) \leq \nu, \psi_{\rho_j}(u) \leq \varphi\}$,

(ii) $\mathcal{Q}^{(\mu e^{i\theta}, \nu e^{i\varphi})} = \bigcup_{\rho_j \in \eta} \rho_j^{(\mu e^{i\theta}, \nu e^{i\varphi})}$.

Note that, $(\mu e^{i\theta}, \nu e^{i\varphi})-$*level hypergraph of* $\mathcal{H}$ *is a crisp hypergraph.*

Example 11. *Consider a C6-ROFHG* $\mathcal{H} = (\mathcal{Q}, \eta)$ *on* $Y = \{u_1, u_2, u_3, u_4, u_5, u_6\}$. *The C6-ROFR* η *is given as,* $\eta(u_1, u_2, u_3) = (0.7e^{i(0.7)\pi}, 0.8e^{i(0.8)\pi})$, $\eta(u_3, u_4, u_5) = (0.6e^{i(0.6)\pi}, 0.8e^{i(0.8)\pi})$, $\eta(u_1, u_6) = (0.8e^{i(0.8)\pi}, 0.8e^{i(0.8)\pi})$ *and* $\eta(u_4, u_6) = (0.7e^{i(0.7)\pi}, 0.8e^{i(0.8)\pi})$. *The incidence matrix of* $\mathcal{H}$ *is given in Table 1.*

Table 1. Incidence matrix of C6-ROFHG $\mathcal{H}$.

$u \in Y$	η_1	η_2	η_3	η_4
u_1	$(0.8e^{i(0.8)\pi}, 0.6e^{i(0.6)\pi})$	$(0,0)$	$(0,0)$	$(0.8e^{i(0.8)\pi}, 0.6e^{i(0.6)\pi})$
u_2	$(0.7e^{i(0.7)\pi}, 0.6e^{i(0.6)\pi})$	$(0,0)$	$(0,0)$	$(0,0)$
u_3	$(0.7e^{i(0.7)\pi}, 0.8e^{i(0.8)\pi})$	$(0.7e^{i(0.7)\pi}, 0.8e^{i(0.8)\pi})$	$(0,0)$	$(0,0)$
u_4	$(0,0)$	$(0.7e^{i(0.7)\pi}, 0.8e^{i(0.8)\pi})$	$(0.7e^{i(0.7)\pi}, 0.8e^{i(0.8)\pi})$	$(0,0)$
u_5	$(0,0)$	$(0.6e^{i(0.6)\pi}, 0.8e^{i(0.8)\pi})$	$(0,0)$	$(0,0)$
u_6	$(0,0)$	$(0,0)$	$(0.9e^{i(0.9)\pi}, 0.8e^{i(0.8)\pi})$	$(0.9e^{i(0.9)\pi}, 0.8e^{i(0.8)\pi})$

The corresponding C6-ROFHG $\mathcal{H} = (\mathcal{Q}, \eta)$ is shown in Figure 10.

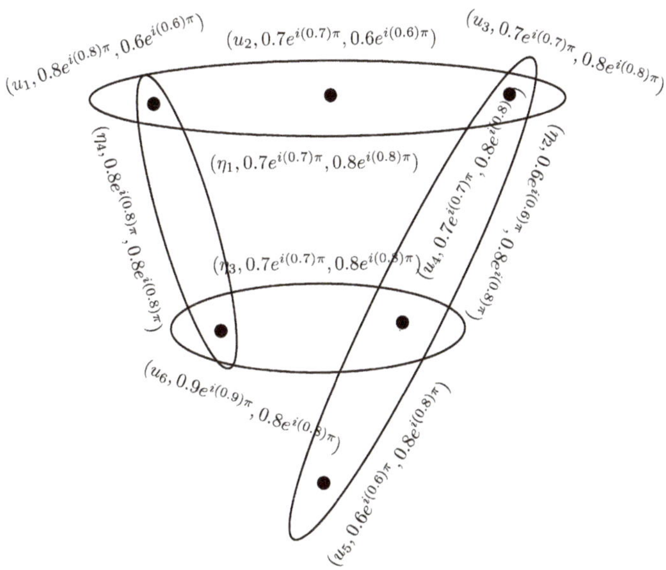

Figure 10. Complex six-rung orthopair fuzzy hypergraph.

Let $\mu = 0.7$, $\nu = 0.6$, $\theta = 0.7\pi$, and $\varphi = 0.6\pi$, then $(0.7e^{i(0.7)\pi}, 0.6e^{i(0.6)\pi})$—level hypergraph of $\mathcal{H}$ is shown in Figure 11.

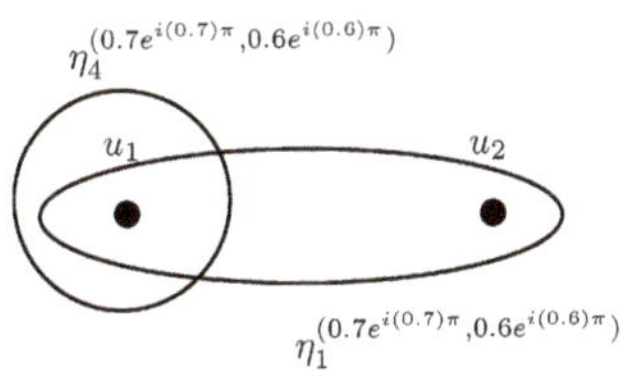

Figure 11. The $(0.7e^{i(0.7)\pi}, 0.6e^{i(0.6)\pi})$—level hypergraph of $\mathcal{H}$.

Note that,

$$\eta_1^{(0.7e^{i(0.7)\pi},0.6e^{i(0.6)\pi})} = \{u_1,u_2\},\ \eta_2^{(0.7e^{i(0.7)\pi},0.6e^{i(0.6)\pi})} = \{\emptyset\},$$
$$\eta_3^{(0.7e^{i(0.7)\pi},0.6e^{i(0.6)\pi})} = \{\emptyset\},\ \eta_4^{(0.7e^{i(0.7)\pi},0.6e^{i(0.6)\pi})} = \{u_1\}.$$

5. Transversals of Complex q-Rung Orthopair Fuzzy Hypergraphs

In this section we study transversality. Prior to the main definition we need the following auxiliary concept:

Definition 28. *Let $\mathcal{H} = (\mathcal{Q}, \eta)$ be a Cq-ROFHG and for $0 < \mu \leq T(h(\mathcal{H}))$, $\nu \geq F(h(\mathcal{H})) > 0$, $0 < \theta \leq \phi(h(\mathcal{H}))$, and $\varphi \geq \psi(h(\mathcal{H})) > 0$ let $\mathcal{H}^{(\mu e^{i\theta},\nu e^{i\varphi})} = (\mathcal{Q}^{(\mu e^{i\theta},\nu e^{i\varphi})}, \eta^{(\mu e^{i\theta},\nu e^{i\varphi})})$ be the level hypergraph of $\mathcal{H}$. The sequence of complex numbers $\{(\mu_1 e^{i\theta_1}, \nu_1 e^{i\varphi_1}), (\mu_2 e^{i\theta_2}, \nu_2 e^{i\varphi_2}), \cdots, (\mu_n e^{i\theta_n}, \nu_n e^{i\varphi_n})\}$ such that $0 < \mu_1 < \mu_2 < \cdots < \mu_n = T(h(\mathcal{H}))$, $\nu_1 > \nu_2 > \cdots > \nu_n = F(h(\mathcal{H})) > 0$, $0 < \theta_1 < \theta_2 < \cdots < \theta_n = \phi(h(\mathcal{H}))$, and $\varphi_1 > \varphi_2 > \cdots > \varphi_n = \psi(h(\mathcal{H})) > 0$ satisfying the conditions,*

- **(i)** *if $\mu_{k+1} < \alpha \leq \mu_k$, $\nu_{k+1} > \beta \geq \nu_k$, $\theta_{k+1} < \phi \leq \theta_k$, $\varphi_{k+1} > \psi \geq \varphi_k$, then $\eta^{(\alpha e^{i\phi},\beta e^{i\psi})} = \eta^{(\mu_k e^{i\theta_k},\nu_k e^{i\varphi_k})}$, and*
- **(ii)** $\eta^{(\mu_k e^{i\theta_k},\nu_k e^{i\varphi_k})} \subset \eta^{(\mu_{k+1} e^{i\theta_{k+1}},\nu_{k+1} e^{i\varphi_{k+1}})}$,

is called the fundamental sequence of $\mathcal{H} = (\mathcal{Q}, \eta)$, denoted by $\mathcal{F}_s(\mathcal{H})$. The set of $(\mu_j e^{i\theta_j}, \nu_j e^{i\varphi_j})-$level hypergraphs $\{\mathcal{H}^{(\mu_1 e^{i\theta_1},\nu_1 e^{i\varphi_1})}, \mathcal{H}^{(\mu_2 e^{i\theta_2},\nu_2 e^{i\varphi_2})}, \dots, \mathcal{H}^{(\mu_n e^{i\theta_n},\nu_n e^{i\varphi_n})}\}$ is called the set of core hypergraphs or the core set of $\mathcal{H}$, denoted by $cor(\mathcal{H})$.

Now we are ready to define:

Definition 29. *Let $\mathcal{H} = (\mathcal{Q}, \eta)$ be a Cq-ROFHG. A complex q-rung orthopair fuzzy transversal (Cq-ROFT) τ is a Cq-ROFs of Y satisfying the condition $\rho^{h(\rho)} \cap \tau^{h(\rho)} \neq \emptyset$, for all $\rho \in \eta$, where $h(\rho)$ is the height of ρ.*

A minimal complex q-rung orthopair fuzzy transversal t is the Cq-ROFT of $\mathcal{H}$ having the property that if $\tau \subset t$, then τ is not a Cq-ROFT of $\mathcal{H}$.

Let us denote the family of minimal Cq-ROFTs of $\mathcal{H}$ by $t_r(\mathcal{H})$.

Example 12. *Consider a C5-ROFHG $\mathcal{H} = (\mathcal{Q}, \eta)$ on $Y = \{a_1, a_2, a_3, a_4, a_5\}$. The C5-ROFR η is given as, $\eta(\{a_1 a_3, a_4\}) = (0.6e^{i(0.6)\pi}, 0.9e^{i(0.9)\pi})$, $\eta(\{a_2, a_3, a_5\}) = (0.7e^{i(0.7)\pi}, 0.9e^{i(0.9)\pi})$, and $\eta(\{a_1, a_2, a_4\}) = (0.6e^{i(0.6)\pi}, 0.9e^{i(0.9)\pi})$. The incidence matrix of $\mathcal{H}$ is given in Table 2.*

Table 2. Incidence matrix of C5-ROFHG $\mathcal{H}$.

$a \in Y$	η_1	η_2	η_3
a_1	$(0.8e^{i(0.8)\pi}, 0.6e^{i(0.6)\pi})$	$(0.8e^{i(0.8)\pi}, 0.6e^{i(0.6)\pi})$	$(0,0)$
a_2	$(0.7e^{i(0.7)\pi}, 0.9e^{i(0.9)\pi})$	$(0,0)$	$(0.7e^{i(0.7)\pi}, 0.9e^{i(0.9)\pi})$
a_3	$(0,0)$	$(0.8e^{i(0.8)\pi}, 0.5e^{i(0.5)\pi})$	$(0.8e^{i(0.8)\pi}, 0.5e^{i(0.5)\pi})$
a_4	$(0.6e^{i(0.6)\pi}, 0.8e^{i(0.8)\pi})$	$(0.6e^{i(0.6)\pi}, 0.8e^{i(0.8)\pi})$	$(0,0)$
a_5	$(0,0)$	$(0,0)$	$(0.7e^{i(0.7)\pi}, 0.5e^{i(0.5)\pi})$

The corresponding C5-ROFHG is shown in Figure 12.

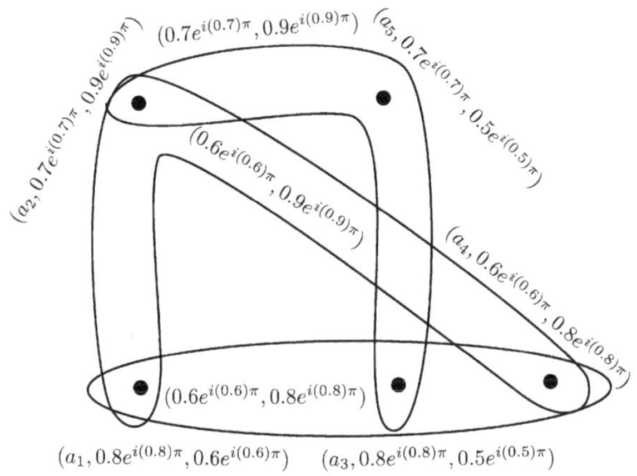

Figure 12. Complex five-rung orthopair fuzzy hypergraph.

By routine calculations, we have $h(\eta_1) = (0.8e^{i(0.8)\pi}, 0.6e^{i(0.6)\pi})$, $h(\eta_2) = (0.8e^{i(0.8)\pi}, 0.5e^{i(0.5)\pi})$, *and* $h(\eta_3) = (0.8e^{i(0.8)\pi}, 0.5e^{i(0.5)\pi})$. *Consider a C5-ROFS* τ_1 *of Y such that* $\tau_1 = \{(a_1, 0.8e^{i(0.8)\pi}, 0.6e^{i(0.6)\pi}), (a_2, 0.7e^{i(0.7)\pi}, 0.9e^{i(0.9)\pi}), (a_3, 0.8e^{i(0.8)\pi}, 0.5e^{i(0.5)\pi})\}$. *Note that,*

$$\begin{aligned} \eta_1^{(0.8e^{i(0.8)\pi}, 0.6e^{i(0.6)\pi})} &= \{a_1\},\ \eta_2^{(0.8e^{i(0.8)\pi}, 0.5e^{i(0.5)\pi})} = \{a_3\},\ \eta_3^{(0.8e^{i(0.8)\pi}, 0.5e^{i(0.5)\pi})} = \{a_3\}, \\ \tau_1^{(0.8e^{i(0.8)\pi}, 0.6e^{i(0.6)\pi})} &= \{a_1, a_3\},\ \tau_1^{(0.8e^{i(0.8)\pi}, 0.5e^{i(0.5)\pi})} = \{a_3\},\ \tau_1^{(0.8e^{i(0.8)\pi}, 0.5e^{i(0.5)\pi})} = \{a_3\}. \end{aligned}$$

Thus, we have $\eta_j^{h(\eta_j)} \cap \tau_1^{h(\eta_j)} \neq \emptyset$, *for all* $\eta_j \in \eta$. *Hence,* τ_1 *is a C5-ROFT of* $\mathcal{H}$. *Similarly,*

$$\begin{aligned} \tau_2 &= \{(a_1, 0.8e^{i(0.8)\pi}, 0.6e^{i(0.6)\pi}), (a_3, 0.8e^{i(0.8)\pi}, 0.5e^{i(0.5)\pi})\}, \\ \tau_3 &= \{(a_1, 0.8e^{i(0.8)\pi}, 0.6e^{i(0.6)\pi}), (a_3, 0.8e^{i(0.8)\pi}, 0.5e^{i(0.5)\pi}), (a_4, 0.6e^{i(0.6)\pi}, 0.8e^{i(0.8)\pi})\}, \\ \tau_4 &= \{(a_1, 0.8e^{i(0.8)\pi}, 0.6e^{i(0.6)\pi}), (a_3, 0.8e^{i(0.8)\pi}, 0.5e^{i(0.5)\pi}), (a_5, 0.7e^{i(0.7)\pi}, 0.5e^{i(0.5)\pi})\}, \end{aligned}$$

are C5-ROFTs of $\mathcal{H}$.

Definition 30. *A Cq-ROFHG* $\mathcal{H}_1 = (\mathcal{Q}_1, \eta_1)$ *is a partial Cq-ROFHG of* $\mathcal{H}_2 = (\mathcal{Q}_2, \eta_2)$ *if* $\eta_1 \subseteq \eta_2$, *denoted by* $\mathcal{H}_1 \subseteq \mathcal{H}_2$. *A Cq-ROFHG* $\mathcal{H}_1 = (\mathcal{Q}_1, \eta_1)$ *is ordered if the core set* $cor(\mathcal{H}) = \{\mathcal{H}^{(\mu_1 e^{i\theta_1}, \nu_1 e^{i\varphi_1})}, \mathcal{H}^{(\mu_2 e^{i\theta_2}, \nu_2 e^{i\varphi_2})}, \dots, \mathcal{H}^{(\mu_n e^{i\theta_n}, \nu_n e^{i\varphi_n})}\}$ *is ordered, i.e.,* $\mathcal{H}^{(\mu_1 e^{i\theta_1}, \nu_1 e^{i\varphi_1})} \subseteq \mathcal{H}^{(\mu_2 e^{i\theta_2}, \nu_2 e^{i\varphi_2})} \subseteq \dots \subseteq \mathcal{H}^{(\mu_n e^{i\theta_n}, \nu_n e^{i\varphi_n})}$. $\mathcal{H}$ *is simply ordered if* $\mathcal{H}$ *is ordered and* $\eta' \subset \eta^{(\mu_{l+1} e^{i\theta_{l+1}}, \nu_{l+1} e^{i\varphi_{l+1}})} \setminus \eta^{(\mu_l e^{i\theta_l}, \nu_l e^{i\varphi_l})} \Rightarrow \eta' \nsubseteq \mathcal{Q}^{(\mu_l e^{i\theta_l}, \nu_l e^{i\varphi_l})}$.

Definition 31. *A Cq-ROFS S on Y is elementary if S is single-valued on* $supp(S)$. *A Cq-ROFHG* $\mathcal{H} = (\mathcal{Q}, \eta)$ *is elementary if every* $Q_j \in \mathcal{Q}$ *and* η *are elementary.*

Proposition 1. *If* τ *is a Cq-ROFT of* $\mathcal{H} = (\mathcal{Q}, \eta)$, *then* $h(\tau) \geq h(\rho)$, *for all* $\rho \in \eta$. *Furthermore, if* τ *is minimal Cq-ROFT of* $\mathcal{H} = (\mathcal{Q}, \eta)$, *then* $h(\tau) = \max\{h(\rho) | \rho \in \eta\} = h(\mathcal{H})$.

Lemma 1. *Let* $\mathcal{H}_1 = (\mathcal{Q}_1, \eta_1)$ *be a partial Cq-ROFHG of* $\mathcal{H}_2 = (\mathcal{Q}_2, \eta_2)$. *If* τ_2 *is minimal Cq-ROFT of* $\mathcal{H}_2$, *then there is a minimal Cq-ROFT of* $\mathcal{H}_1$ *such that* $\tau_1 \subseteq \tau_2$.

Proof. Let S_1 be a Cq-ROFS on Y, which is defined as $S_1 = \tau_2 \cap (\cup_{Q_{1j} \in \mathcal{Q}_1} Q_{1j})$. Then, S_1 is a Cq-ROFT of $\mathcal{H}_1 = (\mathcal{Q}_1, \eta_1)$. Thus, there exists a minimal Cq-ROFT of $\mathcal{H}_1$ such that $\tau_1 \subseteq S_1 \subseteq \tau_2$. □

Lemma 2. *Let $\mathcal{H} = (\mathcal{Q}, \eta)$ be a Cq-ROFHG then $f_s(t_r(\mathcal{H})) \subseteq f_s(\mathcal{H})$.*

Proof. Let $f_s(\mathcal{H}) = \{(\mu_1 e^{i\theta_1}, \nu_1 e^{i\varphi_1}), (\mu_2 e^{i\theta_2}, \nu_2 e^{i\varphi_2}), \cdots, (\mu_n e^{i\theta_n}, \nu_n e^{i\varphi_n})\}$ and $\tau \in t_r(\mathcal{H})$. Suppose that for $u \in supp(\tau)$, $(T_\tau(u), F_\tau(u)) \in (\mu_{j+1}, \mu_j] \times (\nu_{j+1}, \nu_j]$, $\phi_\tau(u) \in (\theta_{j+1}, \theta_j]$, and $\psi_\tau(u) \in (\varphi_{j+1}, \varphi_j]$. Define a function λ by

$$T_\lambda(v)e^{i\phi} = \begin{cases} \mu_j e^{i\theta_j}, & \text{if } \; u = v, \\ T_\tau(u)e^{i\phi_\tau(u)}, & \text{otherwise.} \end{cases}, \quad F_\lambda(v)e^{i\psi} = \begin{cases} \mu_j e^{i\varphi_j}, & \text{if } \; u = v, \\ F_\tau(u)e^{i\psi_\tau(u)}, & \text{otherwise.} \end{cases}$$

From definition of λ, we have $\lambda^{(\mu_j e^{i\theta_j}, \nu_j e^{i\varphi_j})} = \tau^{(\mu_j e^{i\theta_j}, \nu_j e^{i\varphi_j})}$. Definition 28 implies that for every $t \in (\mu_{j+1} e^{i\theta_{j+1}}, \mu_j e^{\theta_j}] \times (\nu_{j+1} e^{i\varphi_{j+1}}, \nu_j e^{i\varphi_j}]$, $\mathcal{H}^t = \mathcal{H}^{(\mu_1 e^{i\theta_1}, \nu_1 e^{i\varphi_1})}$. Thus, $\lambda^{(\mu_j e^{i\theta_j}, \nu_j e^{i\varphi_j})}$ is a Cq-ROFT of $\mathcal{H}^t$. Since, τ is minimal Cq-ROFT and $\lambda^t = \tau^t$, for all $t \notin (\mu_{j+1} e^{i\theta_{j+1}}, \mu_j e^{\theta_j}] \times (\nu_{j+1} e^{i\varphi_{j+1}}, \nu_j e^{i\varphi_j}]$. This implies that λ is also a Cq-ROFT and $\lambda \leq \tau$ but the minimality of τ implies that $\lambda = \tau$. Hence, $\tau(u) = \lambda(u) = (\mu_j e^{i\theta_j}, \nu_j e^{i\varphi_j})$, which implies that for every Cq-ROFT $\tau \in t_r(\mathcal{H})$ and for each $u \in Y$, $\tau(u) \in f_s(\mathcal{H})$ and so we have $f_s(t_r(\mathcal{H})) \subseteq f_s(\mathcal{H})$. □

We now illustrate a recursive procedure to find $t_r(\mathcal{H})$ in Algorithm 1.

Algorithm 1: To find the family of minimal Cq-ROFTs $t_r(\mathcal{H})$.

Let $\mathcal{H} = (\mathcal{Q}, \eta)$ be a Cq-ROFHG having the fundamental sequence $f_s(\mathcal{H}) = \{(\mu_1 e^{i\theta_1}, \nu_1 e^{i\varphi_1}), (\mu_2 e^{i\theta_2}, \nu_2 e^{i\varphi_2}), \cdots, (\mu_n e^{i\theta_n}, \nu_n e^{i\varphi_n})\}$ and core set $cor(\mathcal{H}) = \{\mathcal{H}^{(\mu_1 e^{i\theta_1}, \nu_1 e^{i\varphi_1})}, \mathcal{H}^{(\mu_2 e^{i\theta_2}, \nu_2 e^{i\varphi_2})}, \cdots, \mathcal{H}^{(\mu_n e^{i\theta_n}, \nu_n e^{i\varphi_n})}\}$. The minimal transversal of $\mathcal{H} = (\mathcal{Q}, \eta)$ is determined as follows,

1. Determine a crisp minimal transversal t_1 of $\mathcal{H}^{(\mu_1 e^{i\theta_1}, \nu_1 e^{i\varphi_1})}$.
2. Determine a crisp minimal transversal t_2 of $\mathcal{H}^{(\mu_2 e^{i\theta_2}, \nu_2 e^{i\varphi_2})}$ satisfying the condition $t_1 \subseteq t_2$, i.e., obtain an hypergraph H_2 having the hyperedges $\eta^{(\mu_2 e^{i\theta_2}, \nu_2 e^{i\varphi_2})}$ and a loop at every vertex $u \in t_1$. Thus, we have $\eta(H_2) = \eta(\mu_2 e^{i\theta_2}, \nu_2 e^{i\varphi_2}) \cup \{\{u \in t_1\}\}$.
3. Let t_2 be the minimal transversal of H_2.
4. Obtain a sequence of minimal transversals $t_1 \subseteq t_2 \subseteq \cdots \subseteq t_j$ such that t_j is the minimal transversal of $\mathcal{H}^{(\mu_j e^{i\theta_j}, \nu_j e^{i\varphi_j})}$ satisfying the condition $t_{j-1} \subseteq t_j$.
5. Define an elementary Cq-ROFS S_j having the support t_j and $h(S_j) = (\mu_j e^{i\theta_j}, \nu_j e^{i\varphi_j}), 1 \leq j \leq n$.
6. Determine a minimal Cq-ROFT of $\mathcal{H}$ as $\tau = \bigcup_{j=1}^{n} \{S_j | 1 \leq j \leq n\}$.

Example 13. *Consider a C5-ROFHG $\mathcal{H} = (\mathcal{Q}, \eta)$ on $Y = \{v_1, v_2, v_3, v_4, v_5, v_6\}$ as shown in Figure 13. Let $(\mu_1 e^{i\theta_1}, \nu_1 e^{i\varphi_1}) = (0.9e^{i(0.9)\pi}, 0.7e^{i(0.7)\pi})$, $(\mu_2 e^{i\theta_2}, \nu_2 e^{i\varphi_2}) = (0.8e^{i(0.8)\pi}, 0.5e^{i(0.5)\pi})$, $(\mu_3 e^{i\theta_3}, \nu_3 e^{i\varphi_3}) = (0.6e^{i(0.6)\pi}, 0.4e^{i(0.4)\pi})$, and $(\mu_4 e^{i\theta_4}, \nu_4 e^{i\varphi_4}) = (0.3e^{i(0.3)\pi}, 0.2e^{i(0.2)\pi})$. Clearly, the sequence $\{(\mu_1 e^{i\theta_1}, \nu_1 e^{i\varphi_1}), (\mu_2 e^{i\theta_2}, \nu_2 e^{i\varphi_2}), (\mu_3 e^{i\theta_3}, \nu_3 e^{i\varphi_3}), (\mu_4 e^{i\theta_4}, \nu_4 e^{i\varphi_4})\}$ satisfies all the conditions of Definition 28. Hence, it is the fundamental sequence of $\mathcal{H}$.*

Note that, $t_1 = t_2 = \{v_4\}$ is the minimal transversal of $\mathcal{H}^{(\mu_1 e^{i\theta_1}, \nu_1 e^{i\varphi_1})}$ and $\mathcal{H}^{(\mu_2 e^{i\theta_2}, \nu_2 e^{i\varphi_2})}$, $t_3 = \{v_1\}$ is the minimal transversal of $\mathcal{H}^{(\mu_3 e^{i\theta_3}, \nu_3 e^{i\varphi_3})}$, and $t_4 = \{v_1, v_4\}$ is the minimal transversal of $\mathcal{H}^{(\mu_4 e^{i\theta_4}, \nu_4 e^{i\varphi_4})}$. Consider

$$\begin{aligned} S_1 &= \{(v_4, 0.9e^{i(0.9)\pi}, 0.7e^{i(0.7)\pi})\} = S_2, \\ S_3 &= \{(v_1, 0.8e^{i(0.8)\pi}, 0.5e^{i(0.5)\pi})\}, \\ S_4 &= \{(v_1, 0.8e^{i(0.8)\pi}, 0.5e^{i(0.5)\pi}), (v_4, 0.9e^{i(0.9)\pi}, 0.7e^{i(0.7)\pi})\}. \end{aligned}$$

Hence, $\bigcup_{j=1}^{4} = \{(v_1, 0.8e^{i(0.8)\pi}, 0.5e^{i(0.5)\pi}), (v_4, 0.9e^{i(0.9)\pi}, 0.7e^{i(0.7)\pi})\}$ *is a C5-ROFT of* $\mathcal{H}$.

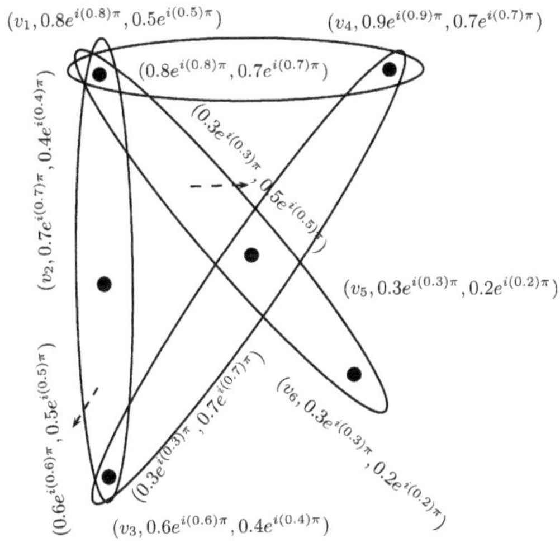

Figure 13. Complex five-rung orthopair fuzzy hypergraph.

Lemma 3. *Let* $\mathcal{H} = (\mathcal{Q}, \eta)$ *be a Cq-ROFHG with* $f_s(\mathcal{H}) = \{(\mu_1 e^{i\theta_1}, \nu_1 e^{i\varphi_1}), (\mu_2 e^{i\theta_2}, \nu_2 e^{i\varphi_2}), \cdots, (\mu_n e^{i\theta_n}, \nu_n e^{i\varphi_n})\}$. *If* τ *is a Cq-ROFT of* $\mathcal{H}$, *then* $h(\tau) \geq h(Q_j)$, *for every* $Q_j \in \mathcal{Q}$. *If* $\tau \in t_r(\mathcal{H})$ *then* $h(\tau) = \max\{h(Q_j) | Q_j \in \mathcal{Q}\} = (\mu_1 e^{i\theta_1}, \nu_1 e^{i\varphi_1})$.

Proof. Since τ is a Cq-ROFT of $\mathcal{H}$, implies that $\tau^{h(Q_j)} \cap Q_j^{h(Q_j)} \neq \emptyset$. Let $a \in supp(\tau)$, then $T_\tau(a) \geq T(h(Q_j))$, $F_\tau(a) \leq F(h(Q_j))$, $\phi_\tau(a) \geq \phi(h(Q_j))$, and $\psi_\tau(a) \leq \psi(h(Q_j))$. This shows that $h(\tau) \geq h(Q_j)$. If $\tau \in t_r(\mathcal{H})$, i.e., τ is minimal Cq-ROFT then $h(Q_j) = (\max T_{Q_j}(a) e^{i \max \phi_{Q_j}(a)}, \min F_{Q_j}(a) e^{i \min \psi_{Q_j}(a)}) = (\mu_1 e^{i\theta_1}, \nu_1 e^{i\varphi_1})$. Thus, we have $h(\tau) = \max\{h(Q_j) | Q_j \in \mathcal{Q}\} = (\mu_1 e^{i\theta_1}, \nu_1 e^{i\varphi_1})$. □

Lemma 4. *Let* β *be a Cq-ROFT of a Cq-ROFHG* $\mathcal{H}$. *Then, there exists* $\gamma \in t_r(\mathcal{H})$ *such that* $\gamma \leq \beta$.

Proof. Let $f_s(\mathcal{H}) = \{(\mu_1 e^{i\theta_1}, \nu_1 e^{i\varphi_1}), (\mu_2 e^{i\theta_2}, \nu_2 e^{i\varphi_2}), \cdots, (\mu_n e^{i\theta_n}, \nu_n e^{i\varphi_n})\}$. Suppose that $\lambda^{(\mu_k e^{i\theta_k}, \nu_k e^{i\varphi_k})}$ is a transversal of $\mathcal{H}^{(\mu_k e^{i\theta_k}, \nu_k e^{i\varphi_k})}$ and $\tau^{(\mu_k e^{i\theta_k}, \nu_k e^{i\varphi_k})} \in t_r(\mathcal{H}^{(\mu_k e^{i\theta_k}, \nu_k e^{i\varphi_k})})$, for $1 \leq k \leq n$ such that $\tau^{(\mu_k e^{i\theta_k}, \nu_k e^{i\varphi_k})} \subseteq \lambda^{(\mu_k e^{i\theta_k}, \nu_k e^{i\varphi_k})}$. Let β_k be an elementary Cq-ROFS having support λ_k and γ_k be an elementary Cq-ROFS having support τ_k, for $1 \leq k \leq n$. Then, Algorithm 1 implies that $\beta = \bigcup_{k=1}^{n} \beta_k$ is a Cq-ROFT of $\mathcal{H}$ and $\gamma = \bigcup_{k=1}^{n} \gamma_k$ is minimal Cq-ROFT of $\mathcal{H}$ such that $\gamma \leq \beta$. □

Theorem 3. *Let* $\mathcal{H}_1 = (\mathcal{Q}_1, \eta_1)$ *and* $\mathcal{H}_2 = (\mathcal{Q}_2, \eta_2)$ *be Cq-ROFHGs. Then,* $\mathcal{Q}_2 = t_r(\mathcal{H}_1) \Leftrightarrow \mathcal{H}_2$ *is simple,* $\mathcal{Q}_2 \subseteq \mathcal{Q}_1$, $h(\eta_k) = h(\mathcal{H}_1)$, *for every* $\rho_k \in \eta_2$, *and for every Cq-ROFS* $\xi \in \mathcal{P}(Y)$, *exactly one of the conditions must satisfy,*

- **(i)** $\rho \leq \xi$, *for some* $\rho \in \mathcal{Q}_2$ *or*
- **(ii)** *there is* $Q_j \in \mathcal{Q}_1$ *and* $(\mu e^{i\theta}, \nu e^{i\varphi})$, *where* $(\mu, \nu) \in [0, T_{h(Q_j)}] \times [0, F_{h(Q_j)}]$, $\theta \in [0, \phi_{h(Q_j)}]$, $\varphi \in [0, \psi_{h(Q_j)}]$ *such that* $Q_j^{(\mu e^{i\theta}, \nu e^{i\varphi})} \cap \xi^{(\mu e^{i\theta}, \nu e^{i\varphi})} = \emptyset$, *i.e.,* ξ *is not a Cq-ROFT of* $\mathcal{H}_1$.

Proof. Let $\mathcal{Q}_2 = t_r(\mathcal{H}_1)$. Since, the family of all minimal Cq-ROFTs form a simple Cq-ROFHG on $Y_1 \subseteq Y_2$. Lemma 3 implies that every edge of $t_r(\mathcal{H}_1)$ has height $(\mu_1 e^{i\theta_1}, \nu_1 e^{i\varphi_1}) = h(\mathcal{H}_1)$. Let ξ be an arbitrary Cq-ROFS.

Case(i) If ξ is a Cq-ROFT of $\mathcal{H}_1$), then Lemma 4 implies the existence of a minimal Cq-ROFT ρ such that $\rho \leq \xi$. Thus, the condition (i) holds and (ii) violates.

Case(ii) If ξ is not a Cq-ROFT of $\mathcal{H}_1$), then there is an edge $Q_j \in \mathcal{Q}_1$ such that $Q_j^{(\mu e^{i\theta}, \nu e^{i\varphi})} \cap \xi^{(\mu e^{i\theta}, \nu e^{i\varphi})} = \emptyset$. If condition (i) holds, $\rho \leq \xi$ implies that $Q_j^{(\mu e^{i\theta}, \nu e^{i\varphi})} \cap \rho^{(\mu e^{i\theta}, \nu e^{i\varphi})} = \emptyset$, which is the contradiction against the fact that ρ is Cq-ROFT. Hence, condition (i) does not hold and (ii) is satisfied.

Conversely, suppose that $\mathcal{Q}_2$ satisfies all properties as mentioned above and $\rho \in \mathcal{Q}_2$. Let $\rho = \xi$, then we obtain $\rho \leq \rho$ and conditions (ii) is not satisfied, so ρ is Cq-ROFT of $\mathcal{H}_1$. If t is minimal Cq-ROFT of $\mathcal{H}_1$ and $t \leq \rho$, t does not satisfy (ii), this implies the existence of $\rho_2 \in \mathcal{Q}_2$ such that $\rho_2 \leq t$, hence $\mathcal{Q}_2 \subseteq t_r(\mathcal{H}_1)$. Since, t is minimal Cq-ROF which implies that $\rho = t$, ρ and t were chosen arbitrarily therefore, we have $\mathcal{Q}_2 = t_r(\mathcal{H}_1)$. □

The construction of fundamental subsequence and subcore of Cq-ROFHG $\mathcal{H} = (\mathcal{Q}, \eta)$ is discussed in Algorithm 2.

Algorithm 2: Construction of fundamental subsequence and subcore.

Let $\mathcal{H} = (\mathcal{Q}, \eta)$ be a Cq-ROFHG and $\mathcal{H}_1 = (\mathcal{Q}_1, \eta_1)$ be a partial Cq-ROFHG of $\mathcal{H}$. The fundamental subsequence $f_{ss}(\mathcal{H})$ is constructed as follows:
Let $f_s(\mathcal{H}) = \{(\mu_1 e^{i\theta_1}, \nu_1 e^{i\varphi_1}), (\mu_2 e^{i\theta_2}, \nu_2 e^{i\varphi_2}), \cdots, (\mu_n e^{i\theta_n}, \nu_n e^{i\varphi_n})\}$ and
$cor(\mathcal{H}) = \{\mathcal{H}^{(\mu_1 e^{i\theta_1}, \nu_1 e^{i\varphi_1})}, \mathcal{H}^{(\mu_2 e^{i\theta_2}, \nu_2 e^{i\varphi_2})}, \cdots, \mathcal{H}^{(\mu_n e^{i\theta_n}, \nu_n e^{i\varphi_n})}\}$.

1. Construct $\widetilde{\mathcal{H}}^{(\mu_1 e^{i\theta_1}, \nu_1 e^{i\varphi_1})}$, a partial hypergraph of $\mathcal{H}^{(\mu_1 e^{i\theta_1}, \nu_1 e^{i\varphi_1})}$, by removing all hyperedges of $\mathcal{H}^{(\mu_1 e^{i\theta_1}, \nu_1 e^{i\varphi_1})}$, which contain properly any other hyperedge of $\mathcal{H}^{(\mu_1 e^{i\theta_1}, \nu_1 e^{i\varphi_1})}$.
2. In the same way, a partial hypergraph $\widetilde{\mathcal{H}}^{(\mu_2 e^{i\theta_2}, \nu_2 e^{i\varphi_2})}$ of $\mathcal{H}^{(\mu_2 e^{i\theta_2}, \nu_2 e^{i\varphi_2})}$ is constructed by removing all hyperedges of $\mathcal{H}^{(\mu_2 e^{i\theta_2}, \nu_2 e^{i\varphi_2})}$, which contain properly any other hyperedge of $\mathcal{H}^{(\mu_2 e^{i\theta_2}, \nu_2 e^{i\varphi_2})}$ or any other hyperedge of $\mathcal{H}^{(\mu_1 e^{i\theta_1}, \nu_1 e^{i\varphi_1})}$. $\widetilde{\mathcal{H}}^{(\mu_2 e^{i\theta_2}, \nu_2 e^{i\varphi_2})}$ is non-trivial iff there exists a Cq-ROFT $\tau \in t_r(\mathcal{H})$ and a vertex $u \in \mathcal{Q}^{(\mu_2 e^{i\theta_2}, \nu_2 e^{i\varphi_2})}$ such that $(T_\tau(u) e^{i\phi_\tau(u)}, F_\tau(u) e^{i\psi_\tau(u)}) = (\mu_2 e^{i\theta_2}, \nu_2 e^{i\varphi_2})$.
3. Continuing the same procedure, construct $\widetilde{\mathcal{H}}^{(\mu_k e^{i\theta_k}, \nu_k e^{i\varphi_k})}$, a partial hypergraph of $\mathcal{H}^{(\mu_k e^{i\theta_k}, \nu_k e^{i\varphi_k})}$, by removing all hyperedges of $\mathcal{H}^{(\mu_k e^{i\theta_k}, \nu_k e^{i\varphi_k})}$, which contain properly any other hyperedge of $\mathcal{H}^{(\mu_k e^{i\theta_k}, \nu_k e^{i\varphi_k})}$ or contain any other hyperedge of $\mathcal{H}^{(\mu_1 e^{i\theta_1}, \nu_1 e^{i\varphi_1})}, \mathcal{H}^{(\mu_2 e^{i\theta_2}, \nu_2 e^{i\varphi_2})}, \cdots, \mathcal{H}^{(\mu_{k-1} e^{i\theta_{k-1}}, \nu_{k-1} e^{i\varphi_{k-1}})}$. $\widetilde{\mathcal{H}}^{(\mu_k e^{i\theta_k}, \nu_k e^{i\varphi_k})}$ is non-trivial iff there exists a Cq-ROFT $\tau \in t_r(\mathcal{H})$ and an element $u \in \mathcal{Q}^{(\mu_k e^{i\theta_k}, \nu_k e^{i\varphi_k})}$ such that $(T_\tau(u) e^{i\phi_\tau(u)}, F_\tau(u) e^{i\psi_\tau(u)}) = (\mu_k e^{i\theta_k}, \nu_k e^{i\varphi_k})$.
4. Let $\{(\tilde{\mu}_1 e^{i\tilde{\theta}_1}, \tilde{\nu}_1 e^{i\tilde{\varphi}_1}), (\tilde{\mu}_2 e^{i\tilde{\theta}_2}, \tilde{\nu}_2 e^{i\tilde{\varphi}_2}), \cdots, (\tilde{\mu}_l e^{i\tilde{\theta}_l}, \tilde{\nu}_l e^{i\tilde{\varphi}_l})\}$ be the set of complex numbers such that the corresponding partial hypergraphs $\widetilde{\mathcal{H}}^{(\tilde{\mu}_1 e^{i\tilde{\theta}_1}, \tilde{\nu}_1 e^{i\tilde{\varphi}_1})}, \widetilde{\mathcal{H}}^{(\tilde{\mu}_2 e^{i\tilde{\theta}_2}, \tilde{\nu}_2 e^{i\tilde{\varphi}_2})}, \cdots, \widetilde{\mathcal{H}}^{(\tilde{\mu}_l e^{i\tilde{\theta}_l}, \tilde{\nu}_l e^{i\tilde{\varphi}_l})}$ are non-empty.
5. Then, $f_{ss}(\mathcal{H}) = \{(\tilde{\mu}_1 e^{i\tilde{\theta}_1}, \tilde{\nu}_1 e^{i\tilde{\varphi}_1}), (\tilde{\mu}_2 e^{i\tilde{\theta}_2}, \tilde{\nu}_2 e^{i\tilde{\varphi}_2}), \cdots, (\tilde{\mu}_l e^{i\tilde{\theta}_l}, \tilde{\nu}_l e^{i\tilde{\varphi}_l})\}$ and $\widetilde{cor}(\mathcal{H}) = \{\widetilde{\mathcal{H}}^{(\tilde{\mu}_1 e^{i\tilde{\theta}_1}, \tilde{\nu}_1 e^{i\tilde{\varphi}_1})}, \widetilde{\mathcal{H}}^{(\tilde{\mu}_2 e^{i\tilde{\theta}_2}, \tilde{\nu}_2 e^{i\tilde{\varphi}_2})}, \cdots, \widetilde{\mathcal{H}}^{(\tilde{\mu}_l e^{i\tilde{\theta}_l}, \tilde{\nu}_l e^{i\tilde{\varphi}_l})}\}$ are subsequence and subcore set of $\mathcal{H}$, respectively.

Definition 32. *Let $\mathcal{H} = (\mathcal{Q}, \eta)$ be a Cq-ROFHG having fundamental subsequence $f_{ss}(\mathcal{H})$ and subcore $\widetilde{cor}(\mathcal{H})$ of $\mathcal{H}$. The Cq-ROFT core of $\mathcal{H}$ is defined as an elementary Cq-ROFHG $\widehat{\mathcal{H}} = (\widehat{\mathcal{Q}}, \widehat{\eta})$ such that,*

(i) *$f_{ss}(\mathcal{H}) = f_{ss}(\widehat{\mathcal{H}})$, i.e., $f_{ss}(\mathcal{H})$ is also a fundamental subsequence of $\widehat{\mathcal{H}}$,*

(ii) *height of every* $\widehat{Q}_j \in \widehat{\mathcal{Q}}$ *is* $(\tilde{\mu}_j e^{i\tilde{\theta}_j}, \tilde{\nu}_j e^{i\tilde{\varphi}_j}) \in f_{ss}(\mathcal{H})$ *iff* $supp(\widehat{Q}_j)$ *is an hyperedge of* $\widehat{\mathcal{H}}^{(\tilde{\mu}_j e^{i\tilde{\theta}_j}, \tilde{\nu}_j e^{i\tilde{\varphi}_j})}$.

Theorem 4. *For every Cq-ROFHG, we have* $t_r(\mathcal{H}) = t_r(\widehat{\mathcal{H}})$.

Proof. Let $t \in t_r(\mathcal{H})$ and $\widehat{Q}_j \in \widehat{\mathcal{Q}}$. Definition 32 implies that $h(\widehat{Q}_j) = (\tilde{\mu}_j e^{i\tilde{\theta}_j}, \tilde{\nu}_j e^{i\tilde{\varphi}_j})$ and $\widehat{Q}_j^{(\tilde{\mu}_j e^{i\tilde{\theta}_j}, \tilde{\nu}_j e^{i\tilde{\varphi}_j})}$ is an hyperedge of $\widetilde{\mathcal{H}}^{(\tilde{\mu}_j e^{i\tilde{\theta}_j}, \tilde{\nu}_j e^{i\tilde{\varphi}_j})}$. Since $\widetilde{\mathcal{H}}^{(\tilde{\mu}_j e^{i\tilde{\theta}_j}, \tilde{\nu}_j e^{i\tilde{\varphi}_j})} \subseteq \mathcal{H}^{(\tilde{\mu}_j e^{i\tilde{\theta}_j}, \tilde{\nu}_j e^{i\tilde{\varphi}_j})}$ and $\tau^{(\mu_j e^{i\theta_j}, \nu_j e^{i\varphi_j})}$ is a transversal of $\mathcal{H}^{(\tilde{\mu}_j e^{i\tilde{\theta}_j}, \tilde{\nu}_j e^{i\tilde{\varphi}_j})}$ therefore $\widehat{Q}_j^{(\tilde{\mu}_j e^{i\tilde{\theta}_j}, \tilde{\nu}_j e^{i\tilde{\varphi}_j})} \cap \tau^{(\mu_j e^{i\theta_j}, \nu_j e^{i\varphi_j})} \neq \emptyset$. Thus, τ is a Cq-ROFT of $\widehat{\mathcal{H}}$.

Let $\widehat{\tau} \in t_r(\widehat{\mathcal{H}})$ and $Q_j \in \mathcal{Q}$. Definition 28 implies that $Q_j^{h(Q_j)} \in \mathcal{H}^{(\mu_j e^{i\theta_j}, \nu_j e^{i\varphi_j})}$, for $h(Q_j) \leq (\mu_j e^{i\theta_j}, \nu_j e^{i\varphi_j}) \in f_s(\mathcal{H})$. Definition of subcore $\widetilde{cor}(\mathcal{H})$ implies the existence of an hyperedge $\widehat{Q}_j^{(\mu_j e^{i\theta_j}, \nu_j e^{i\varphi_j})}$ of $\widetilde{\mathcal{H}}^{(\mu_j e^{i\theta_j}, \nu_j e^{i\varphi_j})}$ such that $\widehat{Q}_j^{(\mu_j e^{i\theta_j}, \nu_j e^{i\varphi_j})} \subseteq Q_j^{h(Q_j)}$ and $(\mu_k e^{i\theta_k}, \nu_k e^{i\varphi_k}) \geq (\mu_j e^{i\theta_j}, \nu_j e^{i\varphi_j})$ $\geq h(Q_j)$. For $\widehat{\tau} \in t_r(\widehat{\mathcal{H}})$, we have $u \in \widehat{Q}_j^{(\mu_j e^{i\theta_j}, \nu_j e^{i\varphi_j})} \cap \widehat{\tau}^{(\mu_j e^{i\theta_j}, \nu_j e^{i\varphi_j})} \subseteq \widehat{Q}_j^{h(Q_j)} \cap \widehat{\tau}^{(\mu_j e^{i\theta_j}, \nu_j e^{i\varphi_j})}$. Hence, $\widehat{\tau}$ is a Cq-ROFT of $\mathcal{H}$.

Let $\tau \in t_r(\mathcal{H}) \Rightarrow \tau$ is a Cq-ROFT of $\widehat{\mathcal{H}}$. This implies that there is $\widehat{\tau}$ such that $\widehat{\tau} \subseteq \tau$. But $\widehat{\tau}$ is a Cq-ROFT of $\mathcal{H}$ and $\tau \in t_r(\mathcal{H})$ implies that $\widehat{\tau} = \tau$. Thus, $t_r(\mathcal{H}) \subseteq t_r(\widehat{\mathcal{H}})$. Also $t_r(\widehat{\mathcal{H}}) \subseteq t_r(\mathcal{H})$ implies that $t_r(\mathcal{H}) = t_r(\widehat{\mathcal{H}})$. □

Although τ can be taken as a minimal transversal of $\mathcal{H}$, it is not necessary for $\tau^{(\mu e^{i\theta}, \nu e^{i\varphi})}$ to be the minimal transversal of $\mathcal{H}^{(\mu e^{i\theta}, \nu e^{i\varphi})}$, for all $\mu, \nu \in [0,1]$, and $\theta, \varphi \in [0, 2\pi]$. Furthermore, it is not necessary for the family of minimal Cq-ROFTs to form a hypergraph on Y. For those Cq-ROFTs that satisfy the above property, we have:

Definition 33. *A Cq-ROFT* τ *having the property that* $\tau^{(\mu e^{i\theta}, \nu e^{i\varphi})} \in t_r(\mathcal{H}^{(\mu e^{i\theta}, \nu e^{i\varphi})})$, *for all* $\mu, \nu \in [0,1]$, *and* $\theta, \varphi \in [0, 2\pi]$ *is called the locally minimal Cq-ROFT of* $\mathcal{H}$. *The collection of all locally minimal Cq-ROFTs of* $\mathcal{H}$ *is represented by* $t_r^*(\mathcal{H})$.

Note that, $t_r^*(\mathcal{H}) \subseteq t_r(\mathcal{H})$, *but the converse is not generally true.*

Example 14. *Consider a C6-ROFHG* $\mathcal{H} = (\mathcal{Q}, \eta)$ *as shown in Figure 14. The C6-ROFS*

$$\{(x_1, 0.6e^{i(0.6)\pi}, 0.4e^{i(0.4)\pi}), (x_5, 0.4e^{i(0.4)\pi}, 0.7e^{i(0.7)\pi}), (x_6, 0.4e^{i(0.4)\pi}, 0.7e^{i(0.7)\pi})\}$$

is a locally minimal C6-ROFT of $\mathcal{H}$.

Theorem 5. *Let* $\mathcal{H} = (\mathcal{Q}, \eta)$ *be an ordered Cq-ROFHG with* $f_s(\mathcal{H}) = \{(\mu_1 e^{i\theta_1}, \nu_1 e^{i\varphi_1}), (\mu_2 e^{i\theta_2}, \nu_2 e^{i\varphi_2}), \cdots, (\mu_n e^{i\theta_n}, \nu_n e^{i\varphi_n})\}$. *If* λ_k *is a minimal transversal of* $\mathcal{H}^{(\mu_k e^{i\theta_k}, \nu_k e^{i\varphi_k})}$, *then there exists* $\alpha \in t_r(\mathcal{H})$ *such that* $\alpha^{(\mu_k e^{i\theta_k}, \nu_k e^{i\varphi_k})} = \lambda_k$ *and* $\alpha^{(\mu_l e^{i\theta_l}, \nu_l e^{i\varphi_l})}$ *is a minimal transversal of* $\mathcal{H}^{(\mu_l e^{i\theta_l}, \nu_l e^{i\varphi_l})}$, *for all* $l < k$. *In particular, if* $\lambda_j \in t_r(\mathcal{H}^{(\mu_j e^{i\theta_j}, \nu_j e^{i\varphi_j})})$, *then there exists a locally minimal Cq-ROFT* $\alpha^{(\mu_j e^{i\theta_j}, \nu_j e^{i\varphi_j})} = \lambda_j$ *and* $t_r^*(\mathcal{H}) \neq \emptyset$.

Proof. Let $\lambda_k \in t_r(\mathcal{H}^{(\mu_k e^{i\theta_k}, \nu_k e^{i\varphi_k})})$. Since, $\mathcal{H} = (\mathcal{Q}, \eta)$ is an ordered Cq-ROFHG, therefore $\mathcal{H}^{(\mu_{k-1} e^{i\theta_{k-1}}, \nu_{k-1} e^{i\varphi_{k-1}})} \subseteq \mathcal{H}^{(\mu_k e^{i\theta_k}, \nu_k e^{i\varphi_k})}$. Also, there exists $\lambda_{k-1} \in t_r(\mathcal{H}^{(\mu_{k-1} e^{i\theta_{k-1}}, \nu_{k-1} e^{i\varphi_{k-1}})})$ such that $\lambda_{k-1} \subseteq \lambda_k$. Following this iterative procedure, we have a nested sequence $\lambda_1 \subseteq \lambda_2 \subseteq \cdots \subseteq \lambda_{k-1} \subseteq \lambda_k$ of minimal transversals, where every $\lambda_l \in t_r(\mathcal{H}^{(\mu_l e^{i\theta_l}, \nu_l e^{i\varphi_l})})$. Let α_l be an elementary Cq-ROFS having height $(\mu_l e^{i\theta_l}, \nu_l e^{i\varphi_l})$ and support α_l. Let us define $\alpha(x)$ such that $\alpha(x) = \{(\max T_{\alpha_l}(x) e^{i \max \phi_{\alpha_l}(x)}, \min F_{\alpha_l}(x) e^{i \min \psi_{\alpha_l}(x)}) | 1 \leq l \leq n\}$, that generates the required minimal Cq-ROFT of $\mathcal{H}$. If $k = n$, α is locally minimal Cq-ROFT of $\mathcal{H}$. Hence, $t_r^*(\mathcal{H}) \neq \emptyset$. □

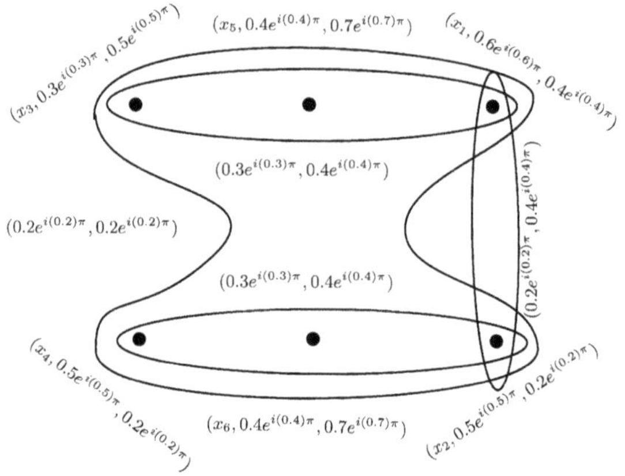

Figure 14. Complex six-rung orthopair fuzzy hypergraph.

Theorem 6. *Let $\mathcal{H} = (\mathcal{Q}, \eta)$ be a simply ordered Cq-ROFHG with $f_s(\mathcal{H}) = \{(\mu_1 e^{i\theta_1}, \nu_1 e^{i\varphi_1}), (\mu_2 e^{i\theta_2}, \nu_2 e^{i\varphi_2}), \cdots, (\mu_n e^{i\theta_n}, \nu_n e^{i\varphi_n})\}$. If $\lambda_k \in t_r(\mathcal{H}^{(\mu_k e^{i\theta_k}, \nu_k e^{i\varphi_k})})$, then there exists $\alpha \in t_r^*(\mathcal{H})$ such that $\alpha^{(\mu_k e^{i\theta_k}, \nu_k e^{i\varphi_k})} = \lambda_k$.*

Proof. Let $\lambda_k \in t_r(\mathcal{H}^{(\mu_k e^{i\theta_k}, \nu_k e^{i\varphi_k})})$ and $\mathcal{H} = (\mathcal{Q}, \eta)$ is a simply ordered Cq-ROFHG. Theorem 5 implies that a nested sequence $\lambda_1 \subseteq \lambda_2 \subseteq \cdots \subseteq \lambda_{k-1} \subseteq \lambda_k$ of minimal transversals can be constructed. Let Let α_l be an elementary Cq-ROFS having height $(\mu_l e^{i\theta_l}, \nu_l e^{i\varphi_l})$ and support α_l such that $\alpha(x) = \{(\max T_{\alpha_l}(x) e^{i \max \phi_{\alpha_l}(x)}, \min F_{\alpha_l}(x) e^{i \min \psi_{\alpha_l}(x)}) | 1 \leq l \leq n\}$ generates the locally minimal Cq-ROFT of $\mathcal{H}$ with $\alpha^{(\mu_k e^{i\theta_k}, \nu_k e^{i\varphi_k})} = \lambda_k$. □

6. Application

Most of the previous studies use crisp hypergraphs to analyze the co-authorship relation between two or more authors as a collaboration. In this section, we consider a Cq-ROFHG model of co-authorship network to represent the collaboration relations between authors having uncertainty and vagueness of periodic nature simultaneously. The next comparison law between Cq-ROFNs will be helpful in our application:

Definition 34. *Let $\mathcal{Q} = (Te^{i\phi}, Fe^{i\psi})$ be a Cq-ROFN. Then, the score function of $\mathcal{Q}$ is defined as,*

$$s(\mathcal{Q}) = (T^q - F^q) + \frac{1}{2^q \pi^q}(\phi^q - \psi^q).$$

The accuracy of $\mathcal{Q}$ is defined as,

$$a(\mathcal{Q}) = (T^q + F^q) + \frac{1}{2^q \pi^q}(\phi^q + \psi^q).$$

For two Cq-ROFNs $\mathcal{Q}_1$ and $\mathcal{Q}_2$,

1. *if $s(\mathcal{Q}_1) > s(\mathcal{Q}_2)$, then $\mathcal{Q}_1 \succ \mathcal{Q}_2$,*
2. *if $s(\mathcal{Q}_1) = s(\mathcal{Q}_2)$, then*
    - *if $a(\mathcal{Q}_1) > a(\mathcal{Q}_2)$, then $\mathcal{Q}_1 \succ \mathcal{Q}_2$,*
    - *if $a(\mathcal{Q}_1) = a(\mathcal{Q}_2)$, then $\mathcal{Q}_1 \sim \mathcal{Q}_2$.*

6.1. A C6-ROFHG Model of Research Collaboration Network

A collaboration network is a group of independent organizations or people that interact to complete a particular goal for achieving better collective results by means of the joint execution of a task. The entities of a collaborative network may be geographically distributed and heterogeneous in terms of their culture, goals, and operating environment but they collaborate to achieve compatible or common goals. For decades, science academies have been interested in research collaboration. The most common reasons for research collaboration are funding, more experts working on the same project imply the more chances for effectiveness, productivity, and innovativeness. Nowadays, most of the public research is based on the collaboration of different types of expertise from different disciples and different economic sectors. In this section, we study a research collaboration network model through C6-ROFHG. Consider a science academy that wants to select an author among a group of researchers that has the best collaborative skills. For this purpose, the following characteristics can be considered:

- Cooperative spirit
- Mutual respect
- Critical thinking
- Innovations
- Creativity
- Embrace diversity

We construct a C6-ROFHG $\mathcal{H} = (\mathcal{Q}, \eta)$ on $Y = \{A_1, A_2, A_3, A_4, A_5, A_6, A_7, A_8, A_9, A_{10}\}$. The universe Y represents the group of authors as the vertices of $\mathcal{H}$ and these authors are grouped through hyperedges if they have worked together on some projects. The truth-membership of each author represents the collaboration strength and falsity-membership describes the opposite behavior of the corresponding author. Suppose that a team of experts assigns that the collaboration power of A_1 is 60% and non-collaborative behavior is 50% after carefully observing the different attributes. The corresponding phase terms illustrate the specific period of time in which the collaborative behavior of an author varies. We model this data as $(A_1, 0.6e^{i(0.5)\pi}, 0.5e^{i(0.5)\pi})$. The C6-ROFHG $\mathcal{H} = (\mathcal{Q}, \eta)$ model of collaboration network is shown in Figure 15.

The membership degrees of hyperedges represent the collective degrees of collaboration and non-collaboration of the corresponding authors combined through an hyperedge. The adjacency matrix of this network is given in Tables 3–5.

Table 3. Adjacency matrix of collaboration network.

η	A_1	A_2	A_3	A_4
A_1	$(0,0)$	$(0.6e^{i(0.5)\pi}, 0.6e^{i(0.5)\pi})$	$(0.6e^{i(0.5)\pi}, 0.6e^{i(0.5)\pi})$	$(0.6e^{i(0.5)\pi}, 0.6e^{i(0.5)\pi})$
A_2	$(0.6e^{i(0.5)\pi}, 0.6e^{i(0.5)\pi})$	$(0,0)$	$(0.6e^{i(0.5)\pi}, 0.6e^{i(0.5)\pi})$	$(0,0)$
A_3	$(0.6e^{i(0.5)\pi}, 0.6e^{i(0.5)\pi})$	$(0.6e^{i(0.5)\pi}, 0.6e^{i(0.5)\pi})$	$(0,0)$	$(0,0)$
A_4	$(0.6e^{i(0.5)\pi}, 0.7e^{i(0.5)\pi})$	$(0,0)$	$(0,0)$	$(0,0)$
A_5	$(0,0)$	$(0,0)$	$(0,0)$	$(0,0)$
A_6	$(0,0)$	$(0,0)$	$(0,0)$	$(0,0)$
A_7	$(0,0)$	$(0,0)$	$(0,0)$	$(0,0)$
A_8	$(0,0)$	$(0,0)$	$(0.4e^{i(0.5)\pi}, 0.6e^{i(0.5)\pi})$	$(0,0)$
A_9	$(0,0)$	$(0,0)$	$(0,0)$	$(0,0)$
A_{10}	$(0,0)$	$(0,0)$	$(0,0)$	$(0,0)$

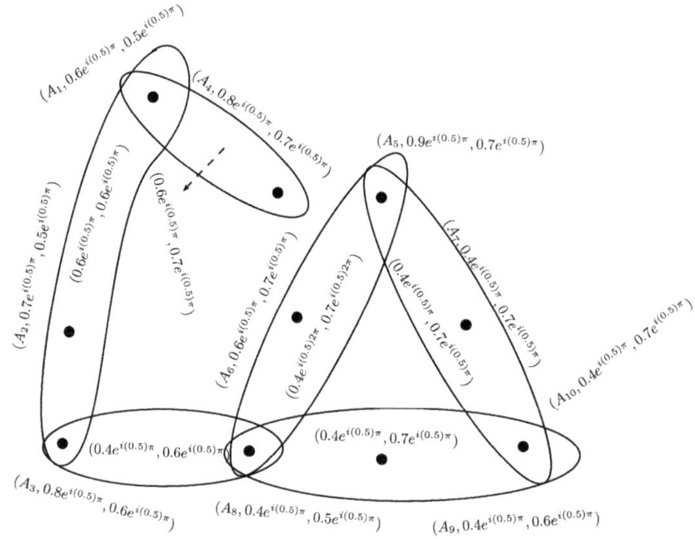

Figure 15. Complex six-rung orthopair fuzzy hypergraph model of collaboration network.

Table 4. Adjacency matrix of collaboration network.

η	A_5	A_6	A_7	A_8
A_1	$(0,0)$	$(0,0)$	$(0,0)$	$(0,0)$
A_2	$(0,0)$	$(0,0)$	$(0,0)$	$(0,0)$
A_3	$(0,0)$	$(0,0)$	$(0,0)$	$(0.4e^{i(0.5)\pi}, 0.6e^{i(0.5)\pi})$
A_4	$(0,0)$	$(0,0)$	$(0,0)$	$(0,0)$
A_5	$(0,0)$	$(0.4e^{i(0.5)\pi}, 0.7e^{i(0.5)\pi})$	$(0.6e^{i(0.5)\pi}, 0.7e^{i(0.5)\pi})$	$(0.4e^{i(0.5)\pi}, 0.7e^{i(0.5)\pi})$
A_6	$(0.4e^{i(0.5)\pi}, 0.7e^{i(0.5)\pi})$	$(0,0)$	$(0,0)$	$(0.4e^{i(0.5)\pi}, 0.7e^{i(0.5)\pi})$
A_7	$(0.4e^{i(0.5)\pi}, 0.7e^{i(0.5)\pi})$	$(0,0)$	$(0,0)$	$(0,0)$
A_8	$(0.4e^{i(0.5)\pi}, 0.7e^{i(0.5)\pi})$	$(0.4e^{i(0.5)\pi}, 0.7e^{i(0.5)\pi})$	$(0,0)$	$(0,0)$
A_9	$(0,0)$	$(0,0)$	$(0,0)$	$(0.4e^{i(0.5)\pi}, 0.7e^{i(0.5)\pi})$
A_{10}	$(0.4e^{i(0.5)\pi}, 0.7e^{i(0.5)\pi})$	$(0,0)$	$(0.4e^{i(0.5)\pi}, 0.7e^{i(0.5)\pi})$	$(0.4e^{i(0.5)\pi}, 0.7e^{i(0.5)\pi})$

Table 5. Adjacency matrix of collaboration network.

η	A_9	A_{10}
A_1	$(0,0)$	$(0,0)$
A_2	$(0,0)$	$(0,0)$
A_3	$(0,0)$	$(0,0)$
A_4	$(0,0)$	$(0,0)$
A_5	$(0,0)$	$(0.6e^{i(0.5)\pi}, 0.7e^{i(0.5)\pi})$
A_6	$(0,0)$	$(0,0)$
A_7	$(0,0)$	$(0.4e^{i(0.5)\pi}, 0.7e^{i(0.5)\pi})$
A_8	$(0.4e^{i(0.5)\pi}, 0.7e^{i(0.5)\pi})$	$(0.4e^{i(0.5)\pi}, 0.7e^{i(0.5)\pi})$
A_9	$(0,0)$	$(0.4e^{i(0.5)\pi}, 0.7e^{i(0.5)\pi})$
A_{10}	$(0.4e^{i(0.5)\pi}, 0.7e^{i(0.5)\pi})$	$(0,0)$

The score values and choice values of a C6-ROFHG $\mathcal{H} = (\mathcal{Q}, \eta)$ are calculated as follows,

$$s_{jk} = (T_{jk}^q + F_{jk}^q) + \frac{1}{2^q \pi^q}(\phi_{jk}^q + \psi_{jk}^q), \; c_j = \sum_k s_{jk} + (T_j^q + F_j^q) + \frac{1}{2^q \pi^q}(\phi_j^q + \psi_j^q),$$

respectively. These values are given in Table 6.

Table 6. Score and choice values.

s_{jk}	A_1	A_2	A_3	A_4	A_5	A_6	A_7	A_8	A_9	A_{10}	c_j
A_1	0	0.1245	0.1245	0.1245	0	0	0	0	0	0	0.88690
A_2	0.1245	0	0.1245	0	0	0	0	0	0	0	0.41377
A_3	0.1245	0.1245	0	0	0	0	0	0.0820	0	0	0.67105
A_4	0.1955	0	0	0	0	0	0	0	0	0	0.60654
A_5	0	0	0	0	0	0.1529	0.1955	0.1529	0	0.1955	1.37714
A_6	0	0	0	0	0.1529	0	0	0.1529	0	0	0.53480
A_7	0	0	0	0	0.1955	0	0	0	0	0.1529	0.50139
A_8	0	0	0.0820	0	0.1529	0.1529	0	0	0.1529	0.1529	0.74457
A_9	0	0	0	0	0	0	0	0.1529	0	0.1529	0.38780
A_{10}	0	0	0	0	0.1529	0	0.1529	0.1529	0.1529	0	0.76459

The choice values of Table 6 show that A_5 is the author having maximum strength of collaboration and good collective skills among all the authors. Similarly, the choice values of all authors represent the strength of their respective collaboration skills in a specific period of time. The method adopted in our model to select the author having best collaboration skills is given in Algorithm 3.

Algorithm 3: Selection of author having maximum collaboration skills.

1. Input the set of vertices (authors) $A_1, A_2, \cdots, A_j$.
2. Input the Cq-ROFS Q of vertices such that $Q(A_k) = (T_k e^{i\phi_k}, F_k e^{i\psi_k}), 1 \leq k \leq j$, $0 \leq T_k^q + F_k^q \leq 1, q \geq 1$. Here, $k = 1, 2, \cdots, j$ denotes the number of authors, $q \geq 1$ is the parameter, T and F characterize the truth and falsity membership degrees of corresponding authors.
3. Input the adjacency matrix $\eta = [(T_{kl} e^{i\phi_{kl}}, F_{kl} e^{i\psi_{kl}})]_{j \times j}$ of vertices.
4. **do** k from $1 \to j$
5. $c_k = 0$
6. **do** l from $1 \to j$
7. $s_{jk} = (T_{kl}^q + F_{kl}^q) + \frac{1}{2^q \pi^q}(\phi_{kl}^q + \psi_{kl}^q)$
8. $c_k = c_k + s_{jk}$
9. **end do**
10. $c_k = c_k + (T_k^q + F_k^q) + \frac{1}{2^q \pi^q}(\phi_k^q + \psi_k^q)$
11. **do**
12. Select a vertex of $\mathcal{H} = (\mathcal{Q}, \eta)$ having maximum choice value as the author possessing strong collaboration powers.

6.2. Comparative Analysis

The proposed Cq-ROF model is more flexible and compatible to the system when the given data ranges over complex subset with unit disk instead of real subset with $[0, 1]$. We illustrate the flexibility of our proposed model by taking an example. Consider an educational institute that wants to establish its minimum branches in a particular city in order to facilitate the maximum number of students according to some parameters such as transportation, suitable place, connectivity with the main branch, and expenditures. Suppose a team of three decision-makers selects the different places.

Let $Y = \{p_1, p_2, p_3\}$ be the set of places where the team is interested to establish the new branches. After carefully observing the different attributes, the first decision-makers assign the membership and non-membership degrees to support the place p_1 as 60% and 40%, respectively. The phase terms represent the period of time for which the place p_1 can attract maximum number of students. This information is modeled using a CIFS as $(p_1, 0.6e^{i(0.6)\pi}, 0.4e^{i(0.4)\pi})$. Note that, $0 \leq 0.6 + 0.4 \leq 1$ and $0 \leq (0.6)\pi + (0.4)\pi \leq \pi$. Similarly, he models the other places as, $(p_2, 0.7e^{i(0.7)\pi}, 0.2e^{i(0.2)\pi})$, $(p_3, 0.5e^{i(0.5)\pi}, 0.2e^{i(0.2)2\pi})$. We denote this CIF model as

$$I = \{(p_1, 0.6e^{i(0.6)\pi}, 0.4e^{i(0.4)\pi}), (p_2, 0.7e^{i(0.7)\pi}, 0.2e^{i(0.2)\pi}), (p_3, 0.5e^{i(0.5)\pi}, 0.2e^{i(0.2)\pi})\}.$$

All CIF grades are CPF as well as Cq-ROF grades. We find the score functions of the above values using the formulas $s(p_j) = (T - F) + \frac{1}{2\pi}(\phi - \psi)$, $s(p_j) = (T^2 - F^2) + \frac{1}{2^2\pi^2}(\phi^2 - \psi^2)$, and $s(p_j) = (T^3 - F^3) + \frac{1}{2^3\pi^3}(\phi^3 - \psi^3)$. The results corresponding to these three approaches are given in Table 7.

Table 7. Comparative analysis of CIF, CPF, and C3-ROF models.

Methods	Score Values	Ranking
CIF model	0.4 1.0 0.6	$p_2 > p_3 > p_1$
CPF model	0.4 0.9 0.42	$p_2 > p_3 > p_1$
C3-ROF model	0.104 0.67 0.234	$p_2 > p_3 > p_1$

Suppose that the second decision-maker assigns the membership values to these places as, $(p_1, 0.6e^{i(0.6)\pi}, 0.4e^{i(0.4)\pi})$, $(p_2, 0.7e^{i(0.7)\pi}, 0.2e^{i(0.2)\pi})$, $(p_3, 0.7e^{i(0.7)\pi}, 0.5e^{i(0.5)\pi})$. This information can not be modeled using CIFS as $0.7 + 0.5 = 1.2 > 1$. We model this information using a CPFS and the corresponding model is given as,

$$P = \{(p_1, 0.6e^{i(0.6)\pi}, 0.4e^{i(0.4)\pi}), (p_2, 0.7e^{i(0.7)\pi}, 0.2e^{i(0.2)\pi}), (p_3, 0.7e^{i(0.7)\pi}, 0.5e^{i(0.5)\pi})\}.$$

All CPF grades are also Cq-ROF grades. We find the score functions of the above values using the formulas $s(p_j) = (T^2 - F^2) + \frac{1}{2^2\pi^2}(\phi^2 - \psi^2)$ and $s(p_j) = (T^3 - F^3) + \frac{1}{2^3\pi^3}(\phi^3 - \psi^3)$. The results corresponding to these two approaches are given in Table 8.

Table 8. Comparative analysis of CPF, and C3-ROF models.

Methods	Score Values	Ranking
CPF model	0.4 0.9 0.48	$p_2 > p_3 > p_1$
C3-ROF model	0.104 0.67 0.436	$p_2 > p_3 > p_1$

We now suppose that the third decision-maker assigns the membership values to these places as, $(p_1, 0.6e^{i(0.6)\pi}, 0.4e^{i(0.4)\pi})$, $(p_2, 0.8e^{i(0.8)\pi}, 0.7e^{i(0.7)\pi})$, $(p_3, 0.7e^{i(0.7)\pi}, 0.5e^{i(0.5)\pi})$. This information can not be modeled using CIFS and CPFS as $0.7 + 0.8 = 1.5 > 1$, $0.7^2 + 0.8^2 = 1.13 > 1$. We model this information using a C3-ROFS and the corresponding model is given as,

$$Q = \{(p_1, 0.6e^{i(0.6)\pi}, 0.4e^{i(0.4)\pi}), (p_2, 0.8e^{i(0.8)\pi}, 0.7e^{i(0.7)\pi}), (p_3, 0.7e^{i(0.7)\pi}, 0.5e^{i(0.5)\pi})\}.$$

We find the score functions of the above values using the formula $s(p_j) = (T^3 - F^3) + \frac{1}{2^3\pi^3}(\phi^3 - \psi^3)$. The score values of C3-ROF information are given as,

$$s(p_1) = 0.304, \; s(p_2) = 0.438, \; s(p_3) = 0.436.$$

Note that p_2 is the best optimal choice to establish a new branch according to the given parameters. We see that every CIF grade is a CPF grade, as well as a Cq-ROF grade, however there are Cq-ROF grades that are not CIF nor CPF grades. This implies the generalization of Cq-ROF values. Thus the proposed Cq-ROF model provides more flexibility due to its most prominent feature that is the adjustment of the range of demonstration of given information by changing the value of parameter q, $q \geq 1$. The generalization of our proposed model can also be observed from the reduction of Cq-ROF model to CIF and CPF models for $q = 1$ and $q = 2$, respectively.

7. Conclusions and Future Directions

Fuzzy sets and intuitionistic fuzzy sets cannot handle imprecise, inconsistent, and incomplete information of periodic nature. They lack the capability to model two-dimensional phenomena. To vercome this difficulty, the concept of complex fuzzy sets was introduced by Ramot et al. [2]. Their phase term is the critical feature of the complex fuzzy set model. The potential of a complex fuzzy set for representing two-dimensional phenomena makes it superior when it comes to handle ambiguous and intuitive information, especially in time-periodic phenomena.

A Cq-ROF model is a generalized form of both the complex intuitionistic fuzzy and complex Pythagorean fuzzy models. Indeed, a Cq-ROF model reduces to a CIF model when $q = 1$, and it becomes a CPF model when $q = 2$. The Cq-ROF model provides a sufficiently wide space of permissible complex orthopairs.

Hypergraphs are mathematical tools for the representation and understanding of problems in a wide variety of scientific fields. In this article, we have applied the most fruitful concept of Cq-ROFSs to hypergraphs. We have defined the novel concepts of Cq-ROFSs, Cq-ROFGs, Cq-ROFHGs, level hypergraphs, and Cq-ROF transversals of Cq-ROFHGs. Further, we have proved that a C1-ROFHG is a CIFHG and a C2-ROFHG is a CPFHG. We have also designed algorithms to construct minimal transversals, fundamental subsequence and subcore of a Cq-ROFHG. Finally, we have illustrated a real-life application of Cq-ROFHGs in collaboration networks that enhances the motivation of this research article.

We aim to broaden our study in the future with the analysis of (1) Complex fuzzy directed hypergraphs, (2) Complex bipolar neutrosophic hypergraphs, (3) Fuzzy rough soft directed hypergraphs and (4) Fuzzy rough neutrosophic hypergraphs.

Author Contributions: investigation, A.L., M.A., A.N.A.-K. and J.C.R.A.; writing—original draft, A.L. and M.A.; writing—review and editing, A.N.A.-K. and J.C.R.A.

Funding: This research received no external funding.

Acknowledgments: This project was funded by the Deanship of Scientific Research (DSR), King Abdulaziz University, Jeddah, under grant No. (DF-121-130-1441). The authors, therefore, gratefully acknowledge DSR technical and financial support

Conflicts of Interest: The authors declare no conflict of interest.

References

1. Zadeh, L.A. Fuzzy sets. *Inf. Control* **1965**, *8*, 338–353. [CrossRef]
2. Ramot, D.; Milo, R.; Friedman, M.; Kandel, A. Complex fuzzy sets. *IEEE Trans. Fuzzy Syst.* **2002**, *10*, 171186. [CrossRef]
3. Yazdanbakhsh, O.; Dick, S. A systematic review of complex fuzzy sets and logic. *Fuzzy Sets. Syst.* **2018**, *338*, 1–22. [CrossRef]
4. Atanassov, K.T. Intuitionistic fuzzy sets. *Fuzzy Sets Syst.* **1983**, *20*, 87–96. [CrossRef]
5. Liu, X.; Kim, H.; Feng, F.; Alcantud, J.C.R. Centroid transformations of intuitionistic fuzzy values based on aggregation operators. *Mathematics* **2018**, *6*, 215. [CrossRef]
6. Feng, F.; Fujita, H.; Ali, M.I.; Yager, R.R.; Liu, X. Another view on generalized intuitionistic fuzzy soft sets and related multiattribute decision making methods. *IEEE Trans. Fuzzy Syst.* **2019**, *27*, 474–488. [CrossRef]
7. Shumaiza; Akram, M.; Al-Kenani, A.N.; Alcantud, J.C.R. Group decision-making based on the VIKOR method with trapezoidal bipolar fuzzy information. *Symmetry* **2019**, *11*, 1313. [CrossRef]

8. Akram, M.; Arshad, M. A novel trapezoidal bipolar fuzzy TOPSIS method for group decision-making. *Group Decis. Negot.* **2019**, *28*, 565–584. [CrossRef]
9. Alcantud, J.C.R.; Biondo, A.; Giarlotta, A. Fuzzy politics I: The genesis of parties. *Fuzzy Sets Syst.* **2018**, *349*, 71–98. [CrossRef]
10. Alcantud, J.C.R.; Giarlotta, A. Necessary and possible hesitant fuzzy sets: A novel model for group decision-making. *Inf. Fusion* **2019**, *46*, 63–76. [CrossRef]
11. Yager, R.R.; Abbasov, A.M. Pythagorean membership grades, complex numbers and dcision making. *Int. J. Intell. Syst.* **2013**, *28*, 436–452. [CrossRef]
12. Yager, R.R. Pythagorean membership grades in multi-criteria decision making. *IEEE Trans. Fuzzy Syst.* **2014**, *22*, 958–965. [CrossRef]
13. Yager, R.R. Generalized orthopair fuzzy sets. *IEEE Trans. Fuzzy Syst.* **2017**, *25*, 1222–1230. [CrossRef]
14. Liu, P.D.; Wang, P. Some q-rung orthopair fuzzy aggregation operators and their applications to multi-attribute decision making. *Int. J. Intell. Syst.* **2018**, *33*, 259–280. [CrossRef]
15. Alcantud, J.C.R.; Muñoz Torrecillas, M.J. Intertemporal choice of fuzzy soft sets. *Symmetry* **2018**, *10*, 371. [CrossRef]
16. Alcantud, J.C.R.; García-Sanz, M.D. Evaluations of infinite utility streams: Pareto efficient and egalitarian axiomatics. *Metroeconomica* **2013**, *64*, 432–447. [CrossRef]
17. Alcantud, J.C.R.; García-Sanz, M.D. Paretian evaluation of infinite utility streams: An egalitarian criterion. *Econ. Lett.* **2010**, *106*, 209–211. [CrossRef]
18. Wei, G.; Gao, H.; Wei, Y.; Some q-rung orthopair fuzzy heronian mean operators in multiple attribute decision making. *Int. J. Intell. Syst.* **2018**, *33*, 1426–1458. [CrossRef]
19. Bai, K.; Zhu, X.; Wang, J.; Zhang, R. Some partitioned maclaurin symmetric mean based on q-rung orthopair fuzzy information for dealing with multi-attribute group decision making. *Symmetry* **2018**, *10*, 383. [CrossRef]
20. Li, L.; Zhang, R.; Wang, J.; Shang, X.; Bai, K. A novel approach to multi-attribute group decision-making with q-rung picture linguistic information. *Symmetry* **2018**, *10*, 172. [CrossRef]
21. Alkouri, A.; Salleh, A. Complex intuitionistic fuzzy sets. *AIP Conf. Proc.* **2012**, *14*, 464–470.
22. Ullah, K.; Mahmood, T.; Ali, Z.; Jan, N. On some distance measures of complex Pythagorean fuzzy sets and their applications in pattern recognition. *Complex Intell. Syst.* **2019**, forthcoming. [CrossRef]
23. Rosenfeld, A. Fuzzy graphs. In *Fuzzy Sets and Their Applications*; Zadeh, L.A., Fu, K.S., Shimura, M., Eds.; Academic Press: New York, NY, USA, 1975; pp. 77–95.
24. Bhattacharya, P. Some remarks on fuzzy graphs. *Pattern Recognit. Lett.* **1987**, *6*, 297–302. [CrossRef]
25. Thirunavukarasu, P.; Suresh, R.; Viswanathan, K.K. Energy of a complex fuzzy graph. *Int. J. Math. Sci. Eng. Appl.* **2016**, *10*, 243–248.
26. Parvathi, R.; Karunambigai, M.G. Intuitionistic fuzzy graphs. In *Computational Intelligence, Theory and Applications*; Reusch, B., Ed.; Springer: Berlin/Heidelberg, Germany, 2006.
27. Akram, M.; Naz, S. Energy of Pythagorean fuzzy graphs with applications. *Mathematics* **2018**, *6*, 136. [CrossRef]
28. Akram, M.; Habib, A. q-Rung orthopair fuzzy competition graphs with application in the soil ecosystem. *Mathematics* **2018**, *7*, 91.
29. Akram, M.; Habib, A.; Koam, A.N. A novel description on edge-regular q-rung picture fuzzy graphs with application. *Symmetry* **2019**, *11*, 489. [CrossRef]
30. Yaqoob, N.; Gulistan, M.; Kadry, S.; Wahab, H. Complex intuitionistic fuzzy graphs with application in cellular network provider companies. *Mathematics* **2019**, *7*, 35. [CrossRef]
31. Yaqoob, N.; Akram, M. Complex neutrosophic graphs. *Bull. Comput. Appl. Math.* **2018**, *6*, 85–109.
32. Akram, M.; Naz, S. A novel decision-making approach under complex Pythagorean fuzzy environment. *Math. Comput. Appl.* **2019**, *24*, 73. [CrossRef]
33. Berge, C. *Graphs and Hypergraphs*; North-Holland Publishing Company: Amsterdam, The Netherlands, 1973.
34. Boulet, R.; Flávia Barros-Platiau, A.; Mazzega, P. Environmental and Trade Regimes: Comparison of Hypergraphs Modeling the Ratifications of UN Multilateral Treaties. In *Law, Public Policies and Complex Systems*; Boulet, R., Lajaunie, C., Mazzega, P., Eds.; Springer: Berlin/Heidelberg, Germany, 2019; Chapter 11.
35. Strzelecka, A.; Skworcow, P. Modelling and simulation of utility service provision for sustainable communities. *Int. J. Elect. Tel.* **2012**, *58*, 389–396. [CrossRef]

36. Han, Y.; Zhou, B.; Pei, J.; Jia, Y. Understanding importance of collaborations in co-authorship networks: A supportiveness analysis approach. In Proceedings of the 2009 SIAM International Conference on Data Mining, Sparks, NV, USA, 30 April–2 May 2009; pp. 1112–1123.
37. Zhang, Z.K.; Liu, C. A hypergraph model of social tagging networks. *J. Stat. Mech. Theory Exp.* **2010**, *10*, 10005. [CrossRef]
38. Ouvrard, X.; Goff, J.M.L.; Marchand-Maillet, S.; Networks of collaborations: Hypergraph modeling and visualisation. *arXiv* **2017**, arXiv:1707.00115.
39. Kaufmann, A. *Introduction a la Theorie des Sous-Ensemble Flous 1*; Masson: Paris, France, 1977.
40. Lee-kwang, H.; Lee, K.-M. Fuzzy hypergraph and fuzzy partition. *IEEE Trans. Syst. Man Cybern.* **1995**, *25*, 196–201. [CrossRef]
41. Mordeson, J.N.; Nair, P.S. *Fuzzy Graphs and Fuzzy Hypergraphs*, 2nd ed.; Physica Verlag: Heidelberg, Germany, 1998.
42. Goetschel, R.H.; Craine, W.L.; Voxman, W. Fuzzy transversals of fuzzy hypergraphs. *Fuzzy Sets. Syst.* **1996**, *84*, 235–254. [CrossRef]
43. Parvathi, R.; Thilagavathi, S.; Karunambigai, M.G. Intuitionistic fuzzy hypergraphs. *Cyber. Inf. Tech.* **2009**, *9*, 46–53.
44. Akram, M.; Dudek, W.A. Intuitionistic fuzzy hypergraphs with applications. *Inf. Sci.* **2013**, *218*, 182–193. [CrossRef]
45. Parvathi, R.; Akram, M.; Thilagavathi, S. Intuitionistic fuzzy shortest hyperpath in a network. *Inf. Process. Lett.* **2013**, *113*, 599–603.
46. Akram, M.; Luqman, A. Bipolar neutrosophic hypergraphs with applications. *J. Intell. Fuzzy Syst.* **2017**, *33*, 1699–1713. [CrossRef]
47. Akram, M.; Sarwar, M. Transversals of m-polar fuzzy hypergraphs with applications. *J. Intell. Fuzzy Syst.* **2017**, *33*, 351–364. [CrossRef]
48. Luqman, A.; Akram, M.; Al-Kenani, A.N. q-Rung orthopair fuzzy hypergraphs with applications. *Mathematics* **2019**, *7*, 260. [CrossRef]
49. Luqman, A.; Akram, M.; Koam, A.N. An m-polar fuzzy hypergraph model of granular computing. *Symmetry* **2019**, *11*, 483. [CrossRef]
50. Luqman, A.; Akram, M.; Koam, A.N. Granulation of hypernetwork models under the q-rung picture fuzzy environment. *Mathematics* **2019**, *7*, 496. [CrossRef]

symmetry

MDPI

Article

Proposal for the Identification of Information Technology Services in Public Organizations

Cristian Mera Macías [1,2,*] and Igor Aguilar Alonso [1,3]

1 School of Systems and Information Engineering, Universidad Nacional Mayor de San Marcos, Lima 15081, Peru; igor_aguilar@hotmail.com
2 Technical Area, Chone Extension, Universidad Laica Eloy Alfaro de Manabí, Manta 130802, Ecuador
3 Research Group: IT Governance and Management Platforms (IT-GOVMANPLA), Professional School of System Engineering, National Technological University of South Lima, Villa El Salvador 15834, Peru
* Correspondence: cristianmeramacias@hotmail.com

Received: 15 August 2019; Accepted: 1 October 2019; Published: 10 October 2019

Abstract: Handling complexity and symmetry in the identification of services for the management of information technology (IT) emerged as a serious challenge in recent times. One of the most important elements that must be defined in the management of information technology services is the construction and management of a service catalog. However, in order to create this catalog, it is necessary to correctly identify the services to be formed. So far, there are several proposals that serve to identify information technology services in public organizations. However, there are several inherent drawbacks to these processes, whereby many organizations are yet to adapt to the services. The main objective of this research is to present a proposal for the identification of information technology services and the construction of an information technology catalog. For this, the following methodology was applied: (a) a review of the literature, identifying the research that addressed the process of the identification of services; (b) a proposal based on automatic learning to identify information technology services in public organizations, adapting the catalog of services and taking as its main input the history of requests and incidents accredited by the department of information technologies in public organizations in the Republic of Ecuador. In conclusion, this work leads to satisfactory results for the identification of technology services used to construct its catalog.

Keywords: service identification; IT service; IT services catalog; IT services portfolio

1. Introduction

As the importance of information technology (IT) increases, the requirements placed on IT service (ITS) providers by ITS seekers are changing. In order to meet these requirements, many methods and approaches are being discussed and studied in science and in practice. Indeed, issues such as the orientation of ITS, the management of ITS (ITSM), and the industrialization of IT gained importance in recent years [1].

An ITS can be defined as a package of services provided by an IT system or the IT department to support business processes [2]. An ITS covers the development, customization, and operation of IT applications, as well as the IT infrastructure. In addition, an ITS must correspond to the needs of the clients and provide them with a benefit they can perceive [3]. This last criterion was reaffirmed by Simonova and Foltanova [4], who stated that the requirements of the users must be essential inputs for the development of services, as the initial identification of these requirements is essential.

In service engineering, the identification of services plays a fundamental role since this identification establishes the basis for subsequent processes [5]. Then, if the services are poorly identified, these subsequent processes are negatively influenced [6]. Although there are several techniques that identify each service [7], none of them are conventional or standard. Some techniques were not tested, while others are extremely difficult to execute, and some of them are not linked with the organization. According to Huego et al. [8], during the last few decades, several methods for the identification of services (SIM) were suggested. However, there is no consensus on the "best method" or a predominant approach to identify these services.

The identification of ITS is an inescapable process for the construction of an ITS catalog (ITSC). Indeed, Meister and Jetschni [9] affirmed that one of the first recommended steps toward service orientation is the implementation of an ITSC, even affirming that the ITSC is the cornerstone in the definition of IT business needs [10]. An ITSC can be defined as a subset of the ITS portfolio (ITSP) that includes the services offered to clients, whether internal or external. This requires that the services listed in the ITSC be standardized so that they can be offered to different clients [11]. Therefore, the ITSC is an ideal entry point when it comes to developing rich content and offering a functional capacity based on the portfolio strategy [12].

Frey et al. [13] affirmed that, in some projects, the process of identification is developed intuitively when it is based on the individual experiences of the people involved, without a defined or heuristic method [14]. For example, Kalia et al. [15] showed that the automation of services improves the efficiency of ITSM processes, despite the fact that these authors proposed an automatic method based on requests. This process allows users to specify their requests for sentence changes in natural language and recommends the most appropriate options within the ITSC. Therefore, when the management of services is developed, it is not used for the construction of the ITSC. Another investigation that works with natural language is the one carried out by Rosa et al. [16], which took a set of registered incidents to add services and categories to the ITSC, without this process being automatic. According to what was proposed, there is no proposal that takes as its input both requests for change and the registration of organization incidents for the identification of ITS or for the construction of the ITSC.

The main objective of this research is to present a proposal for the identification of ITS and the construction of an ITS catalog.

As a research methodology, three fundamental tasks were considered:

Task 1. A systematic review of the literature that allowed the identification of 30 studies related to the process of identifying ITS.

Task 2. A proposal for the identification of ITS and the construction of the ITSC, which takes as its main input the history of requests and incidents of the four organizations so that services can be identified through the use of machine learning, thus building the ITSC automatically.

Task 3. A case study applied to a public organization in the Republic of Ecuador, which was used to implement the proposal and obtain the necessary results to analyze its efficiency levels.

Through the tasks outlined above, we tried to find a technological solution that allows the identification of the ITS and the construction of an ITSC automatically for public organizations (since, according to the literature review carried out in this research, there were few SIMs tested for this type of organization) by applying machine learning, and using as the main input the change requests and the incidents provided by the IT users registered by the IT department or area of a public organization, so that these processes are more precise and based on the real needs of the organization's IT users.

This paper is structured as follows: Section 2 shows the conceptual framework, explaining the ITSP, the ITSC (including the process of identifying services and the management of the ITSC), and the management of IT demand. Section 3 shows the methodology that includes the guidelines used to review the literature, the framework of the proposal, and the considerations that serve to develop the case study. Section 4 details the results of the literature review. Firstly, the background of the investigation is exposed. Then, the list of selected articles inherent in the process of identification of services and construction of the ITSC is chronologically detailed. Finally, we provide an analysis of the studies found, which is the basis for the development of the proposal detailed in this document. Section 5 shows the structure of the proposal, explaining in detail the corresponding phases. Section 6 details the results of the application of the proposal in the case study, which in this case was a public institution in the province of Manabí, Republic of Ecuador. A discussion about the results obtained is also shown. Finally, we detail the conclusions.

2. Conceptual Framework

2.1. The Portfolio of ITS

The service portfolio is the complete set of services managed by a service provider and represents the commitments and investments of the service provider for all customers and market spaces. The portfolio also represents current contractual commitments, the development of new services, and the continuous service improvement plans initiated by the continuous improvement of the service [17]. The ITSP contains an ITSC and financial planning to execute the services offered by an organization. The service portfolio management (which includes the service catalog) and financial management (FM) of the ITS are included in the service strategy phase. Detailed information on the ITSC is also included in the ITSC management process (ITSCM) that is included in the service design phase [18].

The portfolio management process at a general level is defined as a dynamic decision-making process to evaluate, select, request adaptation or approval, or cancel the versions and variety of products and services [19]. The characteristics of the services contained in the portfolio may include the functionality of individual software components, as well as packages of software components, infrastructural elements, and additional services. Additional services are usually information services, consulting services, training services, problem-solving services, or update services [2]. One of the fundamental components of the ITSP is the ITSC, which is defined below.

2.2. The ITS Catalog

The ITSC (consisting of two words: service and catalog [20]) is a structure that contains the list of ITSs offered by IT departments to provide direct assistance to other departments of the organization [21]. Figure 1 shows part of a standard technical ITSC, which can be considered an ITS reference catalog (ITSRC). This catalog consists of categories at the top, and each category groups a list of the possible ITSs in a given organization. In other words, the catalog services are logically grouped according to activity, giving rise to a defined set of services that the IT department provides to a business [10].

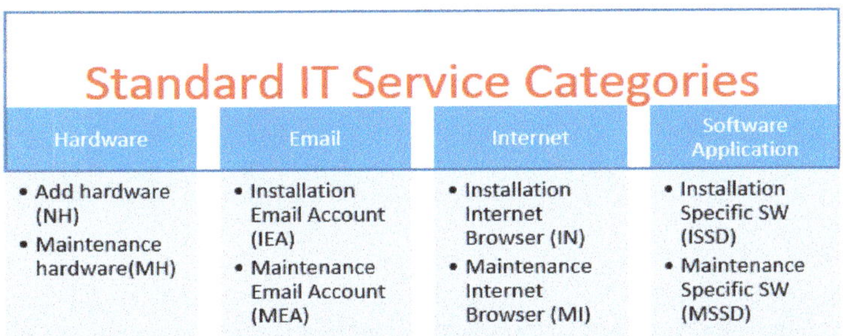

Figure 1. Part of a standard technical information technology service catalog (ITSC).

An ITSC presents an ITS that can be provided and supports customers/users. This influences the decisions that customers make regarding the IT help they require. The purpose of the ITSCM process is to ensure that this catalog is produced and maintained and that it contains accurate information on all operational services and those that are prepared to operate in an operational manner. Therefore, it is necessary to define the services and produce and maintain an accurate ITSC [18]. One of the most important processes for the construction of the ITSC is the identification of ITSs. This process is explained below.

2.2.1. The Process of Identifying ITSs

As mentioned above, in service engineering, the identification of services plays a fundamental role, since it establishes the basis for subsequent processes [5]. A study by Souza et al. [22] showed that the methods of analysis and the design of services aim to identify services and organize them into a manageable hierarchy of compound services to support the processes carried out by a company. These methods use different service identification techniques, which go from the top down (decomposition of the problem until the service level (SL) is reached), or downward (the composition of services in more general parts, to develop business processes) [23].

An important method for the identification of services was proposed by Lee et al. [24]. The methodology for the identification of services proposed contains four levels or stages: the level of business, the SL, the level of interaction, and the level of convergence.

- Business level: The objective of this level is to identify the final purpose of the system. At this level, it is assumed that the business requirements are given by an organization or a person who plays the role of business analyst.
- Service level: This level aims to identify the services that a system must provide an organization to meet the business requirements.
- Level of interaction: The level of interaction addresses the interactions between the system and its external entities, such as users and other systems. These interactions are necessary to achieve the services assigned to the system in the SL.
- Level of convergence: This level focuses on integrating the necessary technologies to provide the services that create a good "customer experience". By integrating several technologies, they can add additional requirements that must be addressed from the point of view of the joint creation of values.

The process of identifying services can be varied. Even natural language was already used as a basis to perform this task. Indeed, the process of identifying ITS can use both the requests (text) of users (as stated by Kalia et al. [15]) and the incidents (text) that may arise during the development of the ITSM (as stated by Rosa et al. [16]) as inputs. However, both inputs are yet to be used together to build an ITSC automatically.

2.2.2. The Management of the ITSC

Often, ITSCs are established within ITSM tools that include two views of the ITSC [21]: technical and business. For identification of the services that the IT area offers to its internal and external clients, a review of elements of the base infrastructure must be carried out. Based on subsequent revisions, an ITSC is built [25]. To start applying the ITSCM, organizations must begin with service identification, an activity that most organizations do not perform correctly. There are several types of information that should be included for all ITSs within an organization in the ITSC. This information includes service description, service type, policy, and service level agreement (SLA) [16]. Clients can use the ITSC to understand what service providers can do for them and to interact with the service provider to discuss those services. Users or individual consumers of a service can use the ITSC to understand the scope of the services available and to learn how to make service requests and/or report incidents associated with the services provided [17]. The ITSCM process is responsible for directing all catalog information and ensuring that the data are correct and up to date. Therefore, it is responsible for activities like defining, standardizing, refreshing, publishing, communicating, protecting, and ensuring the quality of an ITSC [21].

The activities for the identification of ITSs and the construction of the ITSC can be carried out automatically and precisely, including novel principles like text mining and machine learning, which offer improvements that could lead to the optimization of resources, which is precisely what is proposed in this paper.

2.3. IT Demand Management

Within an organization, whether public or private, it is necessary that the IT department properly manages the various requirements of their internal clients. This range of requests based on IT requirements is part of the "IT demand management". According to Legner and Löhe [26], organizations convert their IT demands into IT solutions through multiple steps as follows:

- Collection and detention;
- Evaluation, prioritization, and planning;
- Specification and realization;
- Deployment and operation.

For Aguilar et al. [27], the importance of demand management lies in the achievement of benefits for the company, and, to achieve them, it is necessary that they take into account the life cycle of business demand. Therefore, the demand for IT products and services comes from the needs of the different business processes of the clients. These processes can be in the form of ideas, new well-founded business opportunities, delivery dates, costs, and benefits [28], because the demand for IT is very broad. These processes are classified differently in different IT portfolios so they can be managed in an appropriate way. One of these portfolios is the ITSP, as explained above.

3. Methodology

3.1. Guidelines for the Systematic Review of Literature

For the development of the systematic review of the literature in this investigation, we followed the guidelines set forth by Kitchenham and Charters [29], who determined three phases that must be followed:

- Review planning: In this phase, the review process is planned. Here, the research questions, search chains, inclusion and exclusion criteria, consultation sources, and the review protocol must be considered.
- Carrying out the review: In this phase, the process is developed, following the guidelines outlined in the planning phase. Primary studies are selected here in a methodical manner.
- Results of the review: In this phase, the results and analysis of the studies are shown.

3.1.1. Planning the Review

In order to understand the various investigations carried out to identify the ITSs and construct the ITSC, the following sources were searched: IEEE Xplore, ACM Digital Library, SpringerLink, AIS Electronic Library, and Science Direct.

To carry out the literature review, the following research question was posed:

What research was done to identify services in the conformation of the ITSC?

Likewise, the terms for the search string were defined as follows: (IT OR TECHNOLOGY) AND SERVICE AND (CATALOG OR IDENTIFICATION).

Likewise, the inclusion and exclusion criteria detailed in Table 1 were established.

Table 1. Inclusion and exclusion criteria. ITSC—information technology service catalog.

Inclusion Criteria	Exclusion Criteria
Articles related to research questions	Studies in a language other than English
Studies from 2008 to 2018	Studies that do not meet the inclusion criteria
Studies related to service identification and ITSC	
Articles that are in journals or congresses	
Complete studies	

3.1.2. Carrying out the Review

To develop the review, a discrimination process was followed according to the criteria established in Table 1 and applying the terms of the search chain set forth above. The protocol that was used to perform the review is shown in Figure 2.

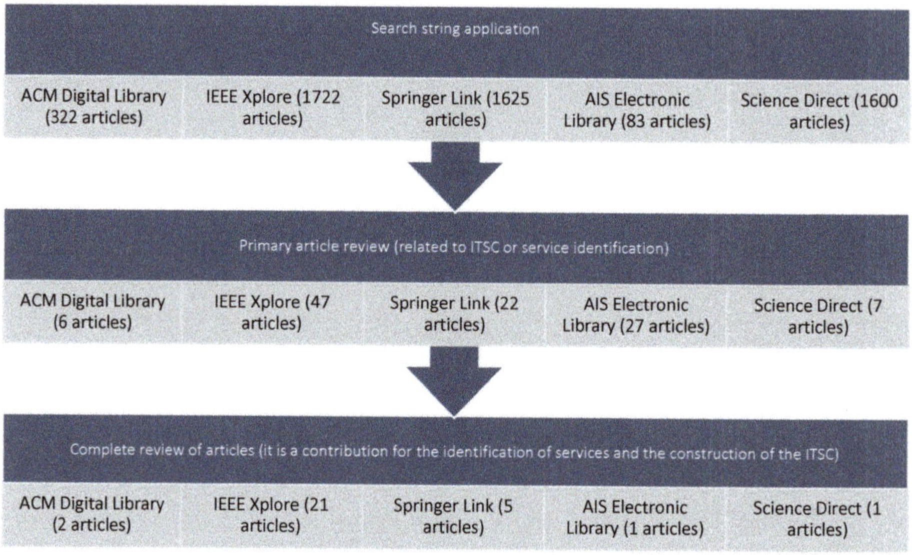

Figure 2. Protocol for the review and selection of articles.

3.1.3. Results of the Review

Initially, 5352 articles were obtained via this systematic search process, of which 109 were relevant. Ultimately, 30 studies were selected, as shown in Table 2.

Table 2. Potentially eligible studies, relevant studies, and selected studies.

Reference Source	Potentially Eligible Studies	Relevant Studies	Selected Studies	Percentage
ACM Digital Library	322	6	2	7%
IEEE Xplore	1722	47	21	70%
Springer Link	1625	22	5	17%
AIS Electronic Library	83	27	1	3%
Science Direct	1600	7	1	3%
Total	5352	109	30	100%

Most of the studies found corresponded to IEEE Xplore (with 70% of the total), and the sources with the fewest studies were the AIS Electronic Library and Science Direct with 3%, as shown in Table 2. A detailed explanation of the 30 selected articles can be found in Sections 4.2 and 4.3.

3.2. Framework for the Construction of the Proposal

This proposal was constructed based on the framework shown in Figure 3.

As shown in Figure 3, phase 1 covers the construction of the solution, where a model is initially defined to learn from a knowledge base. This learning is done using the requests (or incidents) and its cataloging of the departments or IT areas of organizations that have the ITSC implemented. Phase 2 can be executed N times in institutions that wish to build an ITSC.

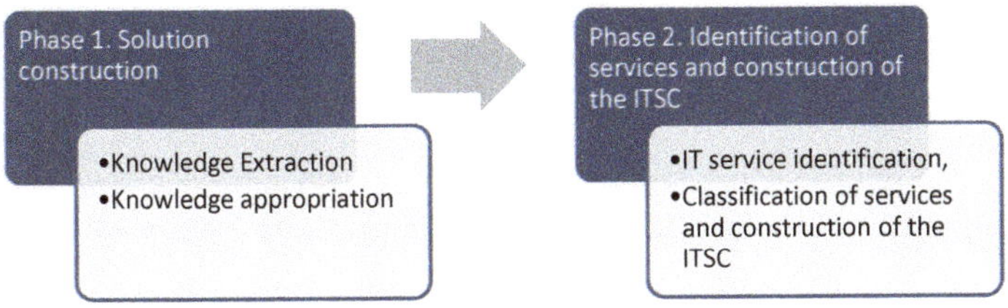

Figure 3. Framework for the construction of the proposal.

3.3. Considerations for the Development of the Case Study

Phase 1, corresponding to the "construction of the solution", was created with a knowledge base of four public institutions, in this case, corresponding to decentralized municipal autonomous governments, from which 1699 requests (requirements or incidents) were recovered with their respective catalogs.

To develop the case study (that is, to apply phase 2 of the proposal), a public institution was chosen from the province of Manabí, Republic of Ecuador. Specifically, we chose a municipal decentralized autonomous government with an IT department and no ITSC. The IT department has a chief or area coordinator and six IT workers who attend to the requests of IT users that provide ITSs to the department in which the proposal was applied. Therefore, a registry of requests available for the department to identify their services and build an ITSC was included. To evaluate this proposal, we used the quality factors to evaluate artifacts for the construction of an ITSC, as proposed by Moody et al. [30] (see Appendix A), since it was used in several investigations, such as those carried out by Mendes et al. [6], Rosa et al. [16], and Gama et al. [31].

4. Review of the Literature

4.1. Background

When talking about ITSM, it is necessary to mention the fundamental role that the ITS Management Forum (ITSMF) has. The ITSMF is a global, independent, and non-profit organization dedicated to continuous improvement of ITSM [32]. According to Clacy and Jennings [32], the ITSMF was established in the United Kingdom in 1991, and the founding chapter of the United Kingdom made significant contributions over the years in establishing the ITSMF as an international organization, as well as supporting the development of the library of infrastructure and information technologies (ITIL) and the associated schemes for qualification and certification. These authors state that this forum continues to be one of the most mature chapters, although it now manages several initiatives to continue developing and sharing these activities throughout the world. The identification, classification, and deregistration of ITSs are very important tasks for the ITSM—as important as the management of IT demand, as well as the financial planning, which represents the cost for the provision of these services.

Since 1998, Niessin and van Vliet [33] worked on a mature model of ITS capabilities that originated from the idea of developing a framework for quality improvement that was oriented to help service organizations become more efficient. One of the most important parts of the work was the ITSC, which requires catalogs to have experience with SLA and services. In 2001, Walker [34] included a section dedicated to the maintenance of the ITSC in his book, "IT problem management", where he addressed the process to add services and to remove services. However, no evidence was found to corroborate the effectiveness of the processes proposed. Then, in 2002, Sullivan, Edmond, and Hofstede [35] worked on a description of the general nature of these services based on a review of the literature, where they defined the ITSC as the list of services categorized according to classification schemes.

In 2004, Sallé [36] reviewed the literature up to that point and noted the importance of the design, development, operation, and delivery of services as fundamental aspects of the management of services, in reference to several frameworks, such as ITIL, British Standards (BS) 15,000, the Hewlett-Packard (HP)

ITS Management Reference Model, the Microsoft Operations Framework (MOF), and IBM's Systems Management Solution Life Cycle. The ITSC is one of the most valuable elements of a comprehensive approach in the provision of services and, as such, should receive due care and attention from its construction.

4.2. Research Conducted to Identify ITS

According to the systematic review of the literature that was carried out in this research, since 2008, 30 articles were identified to be related to the process of identifying ITSs. These articles are detailed in Table 3.

Table 3. Research conducted on the identification of information technology services (ITS). SOA—service-oriented architecture; DEMO—design and engineering methodology for organizations.

Year	Title
2008	Decomposition of IT service processes and alternative service identification using ontologies
2008	A method of service identification for product line
2009	Design rules for user-oriented IT service descriptions
2009	Toward an operationalization of governance and strategy for service identification and design
2009	Rule-based service modeling
2009	Service identification in SOA governance literature review and implications for a new method
2009	Information security pre-evaluation model for u-IT services
2010	Implementing the service catalog management
2010	Research on service identification methods based on SOA
2010	Dynamic life-cycle management of IT services in corporate information systems
2011	An approach for service identification using value co-creation and IT convergence
2011	A survey of service identification strategies
2011	Conceptualization of hybrid service models: an open model approach
2011	Towards a method for service design
2012	Supplier portfolio management for IT services considering diversification effects
2012	A method for identifying IT services using incidents
2012	Using DEMO to identify IT services
2012	Creating composite IT services in the global enterprise
2012	A conceptual framework of service innovation and its implications for future research
2013	IT services reference catalog
2013	From service design to innovation through services: emergence of a methodological and systemic framework
2013	Building an IT service catalog in a small company as the main input for the IT financial manager
2014	A method to identify services using master data and artifact-centric modeling approach
2014	A decision model for optimizing the service portfolio in SOA governance
2014	Process-oriented dependency modeling for service identification
2015	Capability-based service identification in service-oriented legacy modernization
2017	Implementation of quality principles for IT service requirements analysis
2017	An approach to align business and IT perspectives during the SOA services identification
2017	Cataloger—catalog recommendation service for IT change request
2018	Review of proposals for the construction and management of the catalog of information technology services

According to Table 3, the first work identified in this review is the research carried out by Bartsch et al. [37] in 2008, who proposed an approach to the identification and decomposition of hierarchical services based on ontologies to support service providers in the management of operational service processes through the characterization and exploitation of the processes of elementary services. In that same year, Kang et al. [38] proposed an SIM for a product line with appropriate granularity, which used ontology to avoid ambiguous inconsistencies, reaching the conclusion that the proposed method provides adequate granularity for the service and improves reuse in a service-oriented product line. Next, Brocke et al. [39] proposed a way to describe ITS that follows the paradigm of "dominant service logic". If a service provider manages to understand the needs of their client and can create a solution for him or her, the description reflects this and, therefore, offers an advantage over the competitors for organizations that provide ITSs for other entities. In the same way, Borner et al. [40] contributed to governance and strategy of management issues with a focus on individual services, focusing especially on the identification and design of services that belong to the initial phases of the life cycle of the service architecture.

In 2009, Gebhart and Abeck [41] proposed a set of rules to systematically derive a service inventory model from a customized business model that captures the most relevant requirements. An investigation conducted by Boerner and Goeken [42] proposed a process-oriented SIM. This approach incorporated a business point of view, strategic and economic aspects, and technical feasibility. Heo et al. [43] decided to present a methodology for the prior evaluation of information security for a u-ITS that analyzes information security threats and vulnerabilities for a u-ITS provider that develops a service and provides the methods and procedures for preparing countermeasures to support the identification of potential risks and how to address them.

In 2010, an investigation carried out by Zimin and Kulakov [44] formulated an approach for the organizational regulation of the ITS in external and internal conditions that change rapidly. This study also has a relationship with the ITSC and the ITSP. Mendes and Silva [45] affirmed that the ITSC is integrated with processes like the SL, FM, request management, and demand management. This research proposed some solutions that tried to mitigate the risks of implementing an ITSC without neglecting the identification of ITSs. Tian et al. [46] conducted an investigation that analyzed how the implementation of a service-oriented architecture (SOA) materializes in the achievements of several applicable IT functions in the form of the encapsulation of services and the interconnection and interoperation of services in the form of flexible coupling. For this, the authors analyzed the structural model in the SOA, in combination with the development code of the unified rational process (RUP), and showed the three SIMs most commonly used in service oriented analysis and design (SOAD) procedures. However, future research should include analyses of different services at all levels of abstraction.

Continuing with the work related to the identification of services, Lee et al. [24] (2011) proposed an SIM based on the modeling of scenarios and the joint creation of values. However, as part of future work, they proposed to refine the approach and apply it to other domains. A prototype environment was also developed to help identify the services and perform an empirical validation of the approach to ensure that unexpected changes in the companies can be treated with care. Another work on the identification of services was developed by Cai et al. [47], who proposed a complete understanding of the service's identification with the service engineering process and the SOA adoption objectives, illustrated the different meanings, positions, and activities of service identification in existing jobs for top-down and bottom-up approaches, presented the common high-value activities in several SIMs, and presented details for these activities.

However, these authors stated that it remains to be determined how high-value activities could be integrated to form an appropriate SIM for a given environment. Research conducted by Utz et al. [48] presented a hybrid modeling approach for services. This research was based on meta-model concepts that resulted in six axioms for combining perspective and modeling aspects in an open environment. However, the additional work needed is related to the identification and formalization of the conceptualization process at a level of detail where the compilation can build on the definition of the identified model goal and provide mechanisms for the reuse of concepts, references, and remarks, as well as their translation and transformation. In that same year (2011), Levina et al. [49] proposed a service design approach that combines several existing methods and approaches. The objective was to develop a method for "service design" in research and industry. The authors also stated that verifiable propositions should be defined and applied in a subsequent case study. Therefore, to be able to refer to an approach as a "method", that approach must provide the definition and description of the construct, the revision of the principles of form and function, and verifiable propositions.

In 2012, Schwarz et al. [50] proposed a research portfolio that classifies the research questions that need to be addressed. This portfolio was developed based on a conceptual framework for service innovation research. The main results show the specific challenges of research in improving the knowledge base of service innovation. Then, Probst and Buhl [51] analyzed how the design of SOA processes is based on selection decisions among the ITSs offered by different providers. They also developed a procedural model for value-based management that considers the dependencies between selection decisions. However, the authors stated that this approach needs to be further developed in the future. Also, in 2012, a study by Rosa et al. [16] stated that the main objective of the ITSM is to guarantee the quality of the ITS and that ITIL is the most commonly adopted best practice framework for implementing ITSM within organizations. They also proposed an SIM through incidents based on an ITSRC. This document's contribution was to help organizations provide quality ITS. However, despite the good results of this work, it remains necessary to continue putting the proposed artifact into practice. The goal of future research should be based on applying the artifact to other organizations to confirm its applicability, ease, and efficiency in achieving a more accurate and better quality ITSRC.

Also in 2012, Mendes et al. [6] analyzed how one of the most important elements of the ITSM is the ITSC, which is described in a formal document listing the available services provided by IT organizations. Likewise, a method based on the design and engineering methodology for organizations (DEMO) was proposed to identify the ITS. However, in future work, the authors intend to apply this proposal to more complex situations, in which the service provider has a broader catalog of services. Likewise, McCarthy et al. [12] described the process used in the IBM company to create and distribute service packages. Finally, the authors stated that an IT organization or service provider can have hundreds of individual services in their catalog, which could be optimized through simple packages or integrated composite services.

In 2013, Gama et al. [31] stated that, until then, there were difficulties in implementing an ITSC. Due to the complexity of an ITSC and other aspects, including the identification of services, the authors proposed an ITSRC to resolve the absence of a foundation for an ITSC. Similarly, Bugeaud et al. [52] presented a dedicated framework based on four basic components throughout the service design process: knowledge, software tools, communities, and places. This research focused mainly on the first two components. In the same year, Arcilla et al. [18] focused on providing useful information for companies interested in defining their own ITSC from a standard ITSC. The purpose of this research was to create a standard ITSC to help microenterprises and small businesses define their own ITSCs as one of the main inputs for their IT FM processes.

In 2014, Huergo et al. [8] proposed an SIM that uses master data and logical data models as inputs. The proposed method also uses an artifact-centered modeling technique to detail the life cycle of the master data and the business rules it contains. However, this method must still be tested in a real business scenario to identify problems and opportunities for improvement. In that same year, the authors Kim et al. [53] analyzed how the governance of SOA requires an adequate process through which the services described by the service model become candidates to enter the portfolio of services. In addition, the authors presented a decision model to evaluate the services according to metrics, where a comparison of the relative value of each service with its development or maintenance cost should be used to establish priorities. This work sought to complement existing governance standards, such as the SOA governance framework, thereby creating a reasonable service development priority. As in 2014, in an investigation carried out by Rong et al. [5], a SIM was proposed from the perspective of the business process. In addition, the authors stated that the proposed framework is being created but still requires some improvements. The application of this method challenges researchers to concentrate their work on designing a special service identification tool and validate it in other domains in the future.

In 2015, Frey et al. [13] introduced a capacity-based service pattern, which helps overcome the challenges associated with identifying the correct services in the analysis phase of large IT modernization programs. This pattern was discovered and derived from practical experience. However, additional research is needed to clarify how capacity-based services can be leveraged to define the architecture, evaluate legacy applications, and determine the roadmap for a program.

Recently, in 2017, Simonova and Foltanova [4] conducted research that focused on the application of business process tools and principles to identify and model the requirements of individual ITSs, with the aim of increasing the quality of information services as part of the quality of business performance. The answers to this research serve as a basis for further discussion and analysis of requirements.

Likewise, in 2017, Souza et al. [54] proposed an approach that focused on business value modeling and used model-based techniques to generate the information required by current methods for identifying software services, thereby aligning business and software perspectives. The results show that this proposal is a promising approach for the alignment of business and IT perspectives during SOA service identification activities. Kalia et al. [15] stated that the automation of services improves the efficiency of ITSM processes. Therefore, the authors proposed a tool called Cataloger, which is a recommendation system that allows humans to specify their requests for change in natural language sentences and generates the most appropriate recommendations. However, this approach has several limitations. Firstly, the dataset that was used is not balanced for all actions. Therefore, grouping-based approaches must be used to minimize the labeling effort and obtain more labels. Secondly, the datasets that were created for specific actions to identify parameters are not large. Thirdly, in the feedback approach, a heuristic approach to decision-making is proposed. In the future, the authors plan to improve this heuristic approach to improve the results.

Recently, in 2018, Mera and Aguilar [55] carried out a systematic review of the literature on existing proposals for the construction and management of the ITSC. In this study, 14 proposals were identified to include methods, framework, approaches, and models, showing that 43% of these proposals were not applied to real environments (that is, they were not checked in public or private organizations), leaving the field open to check these proposals or propose more feasible methods to be applied and evaluated in real environments.

4.3. Analysis of the Studies Found in the Review

This sub-section shows a comparison of the studies found with their respective analyses. The compared attributes are detailed in Table 4.

According to Table 4, three types of studies were found. There were 11 articles that detailed literature reviews. There were also 17 articles that corresponded to proposals without verification or application. Finally, there were proposals that were verified to relate them to 10 other articles (33.33% of the studies found). These results show the approximate systematic review of the literature shown by Mera and Aguilar [55]. The percentage relates to the studies tested or applied in a real environment, since most of the articles found appear to correspond to literature reviews and proposals without verification.

Regarding the general activities for the construction of the ITSC covered by the proposals found, the most studied activity was to "define the ITS to be provided", which 22 of the studies developed. This activity is directly related to the ITS identification process. However, of this total, only eight items were tested in real environments. Therefore, for the identification of ITS, there are only a few proven proposals that carried out this process.

Another general activity for the construction of the ITSC that stands out and relates to the identification of ITS is the "initial collection of information", which seven of the studies developed. However, of that total, only one study was carried out, reaffirming the scarce application of the proposals in real environments.

A relevant aspect that should be highlighted is that only two of the studies that correspond to proposals with results were tested in public organizations, and none applied automatic mechanisms (machine learning) for the identification of ITSs or the construction of an ITSC. Therefore, the possibility is open to develop proposals for the identification of ITSs and the construction of ITSCs in public organizations, which provides sufficient information to develop the proposal detailed in this paper.

Table 4. Comparison of the studies found in the review. SLA—service level agreement.

References	Type of Study			General Activities for the Construction of the ITSC Covered by the Proposals Found										
	Revision of the Literature	Proposal without Results	Proposal with Results	Initial Collection of Information	Draft Proposal	Development and Continuous Maintenance	Defining the ITSs to Be Provided	Defining the ITS Reference Points	Developing and Documenting SLA	Starting up the Technical Architecture	Refining Catalog Offers	Removing Redundant Services	Publishing the Catalog	Tested in Public Organizations
[37]		X		X			X							
[38]		X	X				X							
[39]		X		X			X	X	X					
[40]	X	X					X							
[41]			X	X	X									
[42]	X	X												
[43]		X					X							
[45]			X				X	X	X				X	
[46]	X													
[44]	X	X		X			X				X			
[24]	X	X		X			X							
[47]	X													
[48]	X	X					X							
[49]	X	X					X							
[51]	X		X											
[16]			X				X			X	X		X	X
[6]			X				X							
[12]		X				X	X				X	X	X	
[50]	X	X		X		X	X							
[31]			X				X				X		X	X
[52]		X					X							
[18]			X				X				X		X	
[8]			X											
[53]			X											
[5]		X	X				X							
[13]		X												
[4]		X					X							
[54]		X					X							
[15]			X				X							
[55]	X			X		X	X			X	X	X	X	

5. Proposal

The proposed solution is based on the structure of the framework shown in Figure 3, which has two phases that are explained below.

5.1. Phase 1—Construction of the Solution

Taking into account the study conducted by Mera and Aguilar [56], Figure 4 shows the scheme currently used by the few organizations that implemented an ITSC to manage requests from the various departments of the organization to the IT department or area.

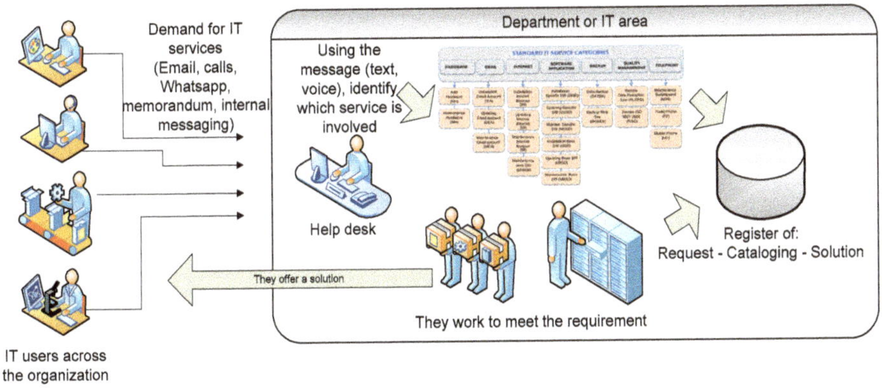

Figure 4. Process of IT demand management.

According to the literature reviewed, the ITSC, in addition to containing properly categorized services, must also provide an access method to request such services. However, the field study on public organizations conducted by Mera and Aguilar [56] determined that IT users do not mostly use ITSC to choose the service they need (in this study, only 20% of users claimed to use a catalog to request a service). Instead, these users prefer to perform requests through e-mail, calls, WhatsApp messages, memoranda, and internal messaging, in that order. Thus, the "message" is firstly processed by the person in charge of the help desk. Then, one must locate or verify that message within the ITSC of the IT service to which this request belongs. This information is registered in a database together with the solution offered or delivered to the IT user.

With this background, phase 1 of the proposal is proposed. This phase consists of constructing a solution, where, from the "knowledge" generated by organizations that have a record of requests for requirements or IT incidents within their respective catalogs, this knowledge can be used to build a solution tailored to the real needs of public organizations, including the identification of ITSs and the construction of an ITSC.

5.1.1. Extraction of Knowledge

In this part, we extract the knowledge of organizations (in this case, municipal autonomous governments) that implemented an ITSC that registers requests, catalogs, and solutions, to empty them into their own structure. This process is detailed in Figure 5.

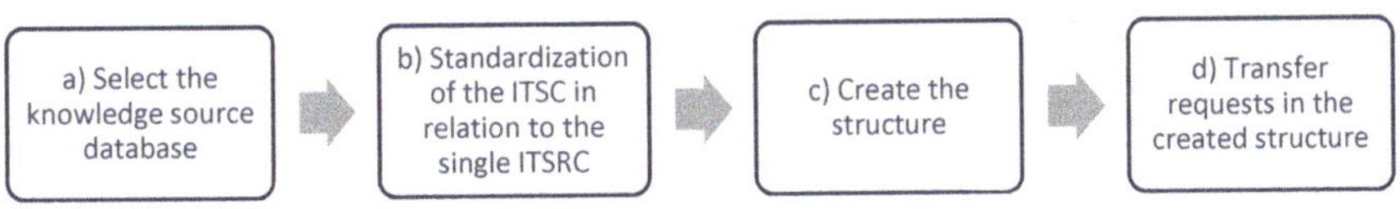

Figure 5. Process for the extraction of knowledge.

(a) Select the knowledge source database

Each selected database corresponds to an organization that fully implemented the ITSC and necessarily registers requests (requirements and incidents), catalogs, and solutions (optional) that were registered based on the management of the demand for ITS in recent years, with the aim of obtaining a list of categories, services, and requests (messages) *M*. Therefore, in this study, four institutions that meet these parameters were selected (located via the field study conducted by Mera et al. [57]), whose request databases were named *BD1*, *BD2*, *BD3*, and *BD4*.

(b) Standardization of the ITSC in relation to the single ITSRC

Using the unique ITSRC obtained through the process detailed in Appendix B, the names of the categories and services of the ITSC for the four organizations chosen as the knowledge base were normalized, so that the services and categories had uniform names or denominations. In this way, the conformation of the resulting structure detailed in the next section could be created correctly. The procedure to normalize the ITSC of each organization or entity was the same as that used to create the unique ITSRC. Firstly, the catalog that was taken as the primary reference was a unique ITSRC. Moreover, all the requests or messages were placed according to the services and categories, thereby yielding the ITSC of each organization after the comparison process. A new service was also included, as detailed below.

Service, security cameras, in **Category**, security management

This service arose from the comparison made with the ITSC of the organizations that had this service registered in their ITSC.

(c) Create the structure

The next step was the creation of the structure. To build a structure called "knowledge" *C*, from the *BD* database of each organization, the first level corresponded to the *Cat* categories, which were the IT categories established in the ITSC used in these institutions. The second level corresponded to the *S* services identified in these institutions, which were categorized in the ITSC. In this way, we obtained the following:

Organization 1: *BD*1

$$
\begin{aligned}
&BD1 = \{Cat1, Cat2, \ldots\ldots Catn\},\\
&\quad Cat1 \subset BD1 = \{S1,\ S2, \ldots\ldots Sx\},\\
&\qquad S1 \subset Cat1 \subset BD1 = \{M1,\ M2, \ldots\ldots Mn\},\\
&\qquad S2 \subset Cat1 \subset BD1 = \{M1,\ M2, \ldots\ldots Mn\},\\
&\qquad Sx \subset Cat1 \subset BD1 = \{M1,\ M2, \ldots\ldots Mn\},\\
&\quad Cat2 \subset BD1 = \{S4,\ S5, \ldots\ldots Sy\},\\
&\qquad S4 \subset Cat2 \subset BD1 = \{M1,\ M2, \ldots\ldots Mn\},\\
&\qquad S5 \subset Cat2 \subset BD1 = \{M1,\ M2, \ldots\ldots Mn\},\\
&\qquad Sy \subset Cat2 \subset BD1 = \{M1,\ M2, \ldots\ldots Mn\},\\
&\quad Catn \subset BD1 = \{S7,\ S8, \ldots\ldots Sz\},\\
&\qquad S7 \subset Catn \subset BD1 = \{M1,\ M2, \ldots\ldots Mn\},\\
&\qquad S8 \subset Catn \subset BD1 = \{M1,\ M2, \ldots\ldots Mn\},\\
&\qquad Sz \subset Catn \subset BD1 = \{M1,\ M2, \ldots\ldots Mn\},
\end{aligned}
$$

Organization 2: $BD2$

$$
\begin{aligned}
&BD2 = \{Cat1, Cat2, \ldots\ldots Catn\},\\
&\quad Cat1 \subset BD2 = \{S1,\ S2, \ldots\ldots Sx\},\\
&\qquad S1 \subset Cat1 \subset BD2 = \{M1,\ M2, \ldots\ldots Mn\},\\
&\qquad S2 \subset Cat1 \subset BD2 = \{M1,\ M2, \ldots\ldots Mn\},\\
&\qquad Sx \subset Cat1 \subset BD2 = \{M1,\ M2, \ldots\ldots Mn\},\\
&\quad Cat2 \subset BD2 = \{S4,\ S5, \ldots\ldots Sy\},\\
&\qquad S4 \subset Cat2 \subset BD2 = \{M1,\ M2, \ldots\ldots Mn\},\\
&\qquad S5 \subset Cat2 \subset BD2 = \{M1,\ M2, \ldots\ldots Mn\},\\
&\qquad Sy \subset Cat2 \subset BD2 = \{M1,\ M2, \ldots\ldots Mn\},\\
&\quad Catn \subset BD2 = \{S7,\ S8, \ldots\ldots Sz\},\\
&\qquad S7 \subset Catn \subset BD2 = \{M1,\ M2, \ldots\ldots Mn\},\\
&\qquad S8 \subset Catn \subset BD2 = \{M1,\ M2, \ldots\ldots Mn\},\\
&\qquad Sz \subset Catn \subset BD2 = \{M1,\ M2, \ldots\ldots Mn\},
\end{aligned}
$$

Organization 3: $BD3$

$$
\begin{aligned}
&BD3 = \{Cat1, Cat2, \ldots\ldots Catn\},\\
&\quad Cat1 \subset BD3 = \{S1,\ S2, \ldots\ldots Sx\},\\
&\qquad S1 \subset Cat1 \subset BD3 = \{M1,\ M2, \ldots\ldots Mn\},\\
&\qquad S2 \subset Cat1 \subset BD3 = \{M1,\ M2, \ldots\ldots Mn\},\\
&\qquad Sx \subset Cat1 \subset BD3 = \{M1,\ M2, \ldots\ldots Mn\},\\
&\quad Cat2 \subset BD3 = \{S4,\ S5, \ldots\ldots Sy\},\\
&\qquad S4 \subset Cat2 \subset BD3 = \{M1,\ M2, \ldots\ldots Mn\},\\
&\qquad S5 \subset Cat2 \subset BD3 = \{M1,\ M2, \ldots\ldots Mn\},\\
&\qquad Sy \subset Cat2 \subset BD3 = \{M1,\ M2, \ldots\ldots Mn\},\\
&\quad Catn \subset BD3 = \{S7,\ S8, \ldots\ldots Sz\},\\
&\qquad S7 \subset Catn \subset BD3 = \{M1,\ M2, \ldots\ldots Mn\},\\
&\qquad S8 \subset Catn \subset BD3 = \{M1,\ M2, \ldots\ldots Mn\},\\
&\qquad Sz \subset Catn \subset BD3 = \{M1,\ M2, \ldots\ldots Mn\},
\end{aligned}
$$

Organization 4: $BD4$

$$
\begin{aligned}
&BD4 = \{Cat1, Cat2, \ldots\ldots Catn\},\\
&\quad Cat1 \subset BD4 = \{S1,\ S2, \ldots\ldots Sx\},\\
&\qquad S1 \subset Cat1 \subset BD4 = \{M1,\ M2, \ldots\ldots Mn\},\\
&\qquad S2 \subset Cat1 \subset BD4 = \{M1,\ M2, \ldots\ldots Mn\},\\
&\qquad Sx \subset Cat1 \subset BD4 = \{M1,\ M2, \ldots\ldots Mn\},\\
&\quad Cat2 \subset BD4 = \{S4,\ S5, \ldots\ldots Sy\},\\
&\qquad S4 \subset Cat2 \subset BD4 = \{M1,\ M2, \ldots\ldots Mn\},\\
&\qquad S5 \subset Cat2 \subset BD4 = \{M1,\ M2, \ldots\ldots Mn\},\\
&\qquad Sy \subset Cat2 \subset BD4 = \{M1,\ M2, \ldots\ldots Mn\},\\
&\quad Catn \subset BD4 = \{S7,\ S8, \ldots\ldots Sz\},\\
&\qquad S7 \subset Catn \subset BD4 = \{M1,\ M2, \ldots\ldots Mn\},\\
&\qquad S8 \subset Catn \subset BD4 = \{M1,\ M2, \ldots\ldots Mn\},\\
&\qquad Sz \subset Catn \subset BD4 = \{M1,\ M2, \ldots\ldots Mn\},
\end{aligned}
$$

where, to obtain the desired structure in relation to the *Cat* categories, the following operation was carried out:

$$C = BD1 \cup BD2 \cup BD3 \cup BD4.$$

In relation to the ***S*** services that go into each ***Cat*** category, the following operations were carried out:

$$Cat1 \subset C = Cat1 \subset BD1 \cup Cat1 \subset BD2 \cup Cat1 \subset BD3 \cup Cat1 \subset BD4;$$

$$Cat2 \subset C = Cat2 \subset BD1 \cup Cat2 \subset BD2 \cup Cat2 \subset BD3 \cup Cat2 \subset BD4;$$

$$Catn \subset C = Catn \subset BD1 \cup Catn \subset BD2 \cup Catn \subset BD3 \cup Catn \subset BD4.$$

As a result, at the level of the directory structure, we obtained a result similar to the one shown in Figure 6.

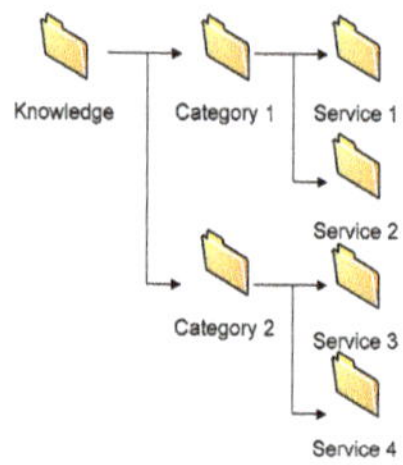

Figure 6. Knowledge structure C and directories.

(d) Transfer requests in the created structure

Once the directory structure was created, each request (message) was distributed in the service that corresponds to it in the structure created. What was expressed up to now is summarized in the following processes:

$$S1 \subset Cat1 \subset C = S1 \subset Cat1 \subset BD1 \cup S1 \subset Cat1 \subset BD2 \cup S1 \subset Cat1 \subset BD3 \cup S1 \subset Cat1 \subset BD4;$$

$$S2 \subset Cat1 \subset C = S2 \subset Cat1 \subset BD1 \cup S2 \subset Cat1 \subset BD2 \cup S2 \subset Cat1 \subset BD3 \cup S2 \subset Cat1 \subset BD4;$$

$$S3 \subset Cat2 \subset C = S1 \subset Cat2 \subset BD1 \cup S1 \subset Cat2 \subset BD2 \cup S1 \subset Cat2 \subset BD3 \cup S1 \subset Cat2 \subset BD4;$$

$$S4 \subset Cat2 \subset C = S2 \subset Cat2 \subset BD1 \cup S2 \subset Cat2 \subset BD2 \cup S2 \subset Cat2 \subset BD3 \cup S2 \subset Cat2 \subset BD4.$$

Explained graphically, the entire process was as shown in Figure 7:

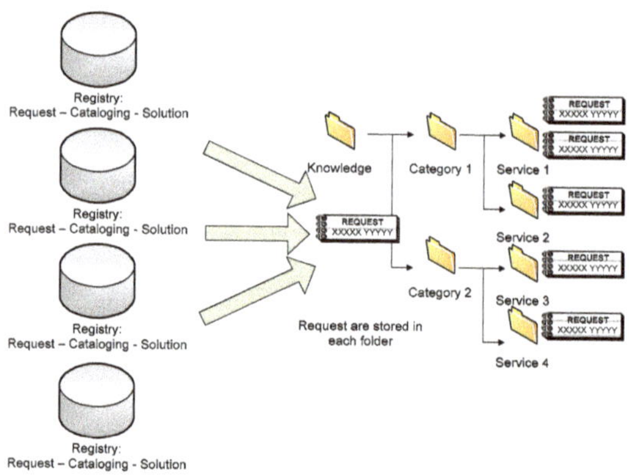

Figure 7. Knowledge extraction process.

Once the procedures explained above were applied, the resulting catalog corresponding to the "knowledge" structure was as shown in Figure 8.

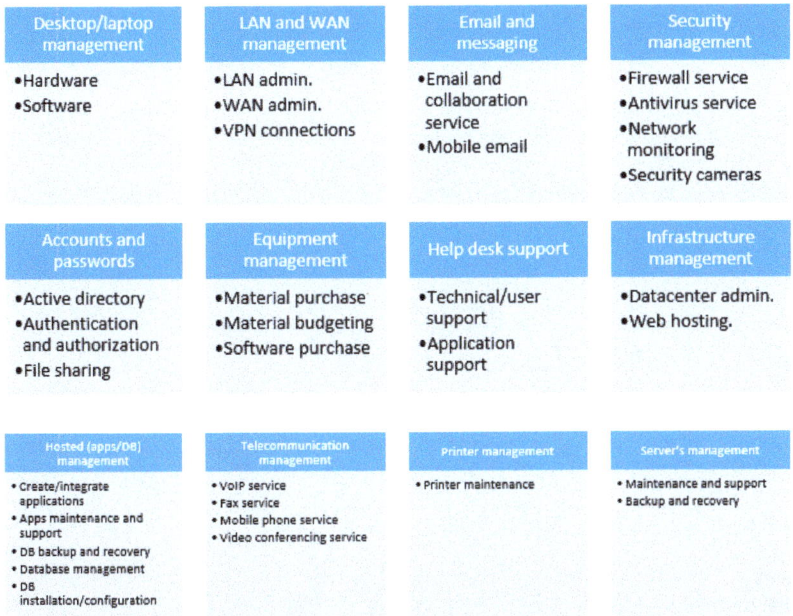

Figure 8. An ITSC that corresponds to the knowledge structure.

5.1.2. Appropriation of Knowledge

In this phase, a process based on text mining is proposed. This process was founded on the requests categorized in the "knowledge" structure that allowed a model based on Bayesian networks to "learn" where each request is located (that is, to which service and to what category each request corresponds). For the construction of this process, the Rapid Miner tool was used. This tool is generally structured as shown in Figure 9.

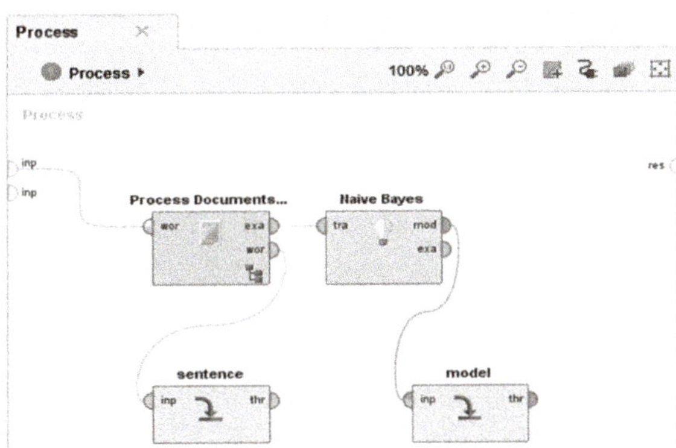

Figure 9. Rapid Miner-based learning process.

Below, each of the operators involved in knowledge learning is described.

(a) Process Documents from Files Operator

With this operator, requests are collected from the "knowledge" structure. Within this "process documents from files" structure, the operations proposed in Figure 10 are specified.

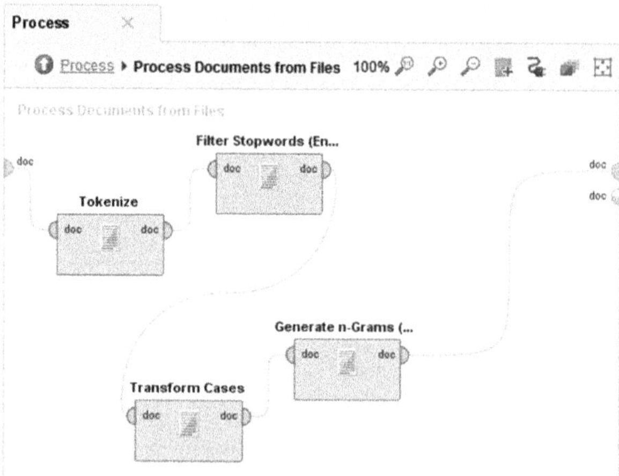

Figure 10. Internal process of the "process documents from files" operator.

The internal operations performed in this operator are the following:

- **Tokenize.** This operator divides the text of the document into a sequence of tokens that are used to construct the word vector.
- **Filter Stop words.** This operator filters the words considered "stop words", i.e., words like in, the, if, for, where, etc.
- **Transform Cases.** This operator transforms the characters to lowercase.
- **Generate n-Grams.** With this operator, "phrases" of up to three words are generated.

(b) Operator Naïve Bayes

This is the model that is used to "learn" the word structure of the different requests of each service in the categories specified in the "knowledge" structure. This learning is done through a simple probabilistic classification model. This operator can establish the characteristics to build classifications based on few data.

(c) Storage Operators (Sentence and Model)

Storage operators are used to store the characteristics and models already learned from the requests found in the knowledge structure. With this last process, the solution is ready to be used in the identification of ITSs and, therefore, for the construction of an ITSC in a number of organizations.

5.2. *Phase 2—Identification of Services and the Construction of an ITSC*

Once the solution was finished, it was ready to be executed in a number of organizations. To accomplish this task, it was necessary to follow the processes detailed below.

5.2.1. Identification of ITS

A fundamental task in the construction of an ITSC is the identification and subsequent classification of services, since a service is the most fundamental component of an ITSC. Therefore, its correct identification is crucial when initiating an ITSCM. In fact, identification is the first activity. In order to execute the identification of ITSs, the processes detailed in Figure 11 were used.

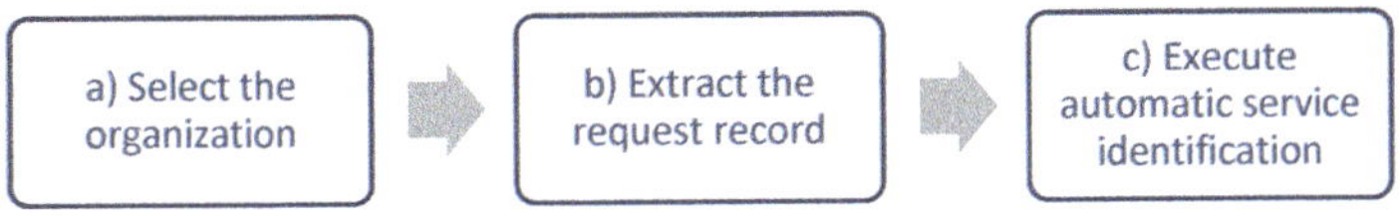

Figure 11. General processes for the identification of services.

(a) Select the organization

The first thing to do in this phase was to select the organization, which must meet the following requirements:

- It must have a department or IT area.
- It must have a record of service requests or IT incidents (for at least the last year, preferably).

(b) Extract the request record

After selecting the organization, it was necessary to extract the record of requests, remembering that the M requests could come from different FR files, such as Excel sheets, emails, internal messaging, etc. The idea was to extract the requests from all the data sources and centralize them in a BDS database, as follows:

According to the request files:

$$FR1 = \{M1, M2, \ldots .Mx\},$$

$$FR2 = \{M4, M5, \ldots .My\},$$

$$FR3 = \{M6, M7, \ldots .Mz\}.$$

The operation to be performed was as follows:

$$BDS = FR1 \cup FR2 \cup FR3 \ldots,\ FRn,$$

which graphically translates to the scheme shown in Figure 12.

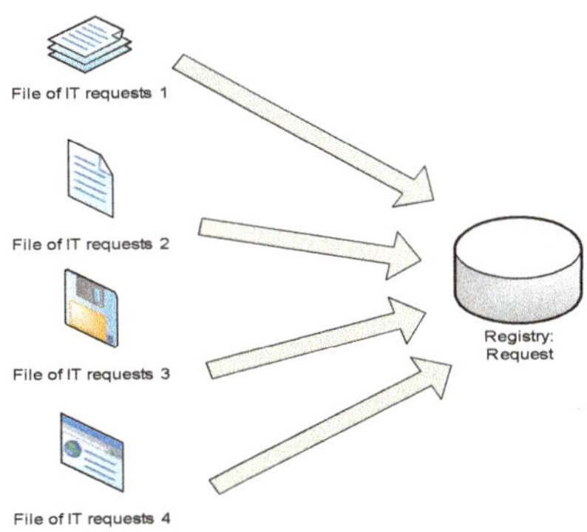

Figure 12. Graphical scheme of the collection of requests.

(c) Execute automatic service identification

Once the requests were ready in a BDS central database and in the same format, a process was applied to identify the services based on the requests made in the "construction of the solution" phase using the Rapid Miner tool to process each request. This process was structured as shown in Figure 13.

- **Storage operators (models and sentences).** The first step is to establish that the service identification activity is carried out based on the characteristics and the models that were "learned". For this purpose, the operators specify that these inputs leave the previously prepared repositories.
- **Process Document from File Operator.** With this operator, requests are collected from ***BDS***. Within this "process documents from files", the solutions for phase 1 are specified via the same operations used in Section 5.1.2 ("appropriation of knowledge").
- **Apply model operator.** With this operator, the service identification process is performed based on the requests that the model "learned" in the "construction of the solution" phase from the "knowledge" database C. The process then verifies each request stored in the *BDS* to review its characteristics to determine which ITS corresponds to that service. This ITS is then registered in the "list of found ITSs". The process of identifying such services is summarized in the algorithm shown in Figure 14.

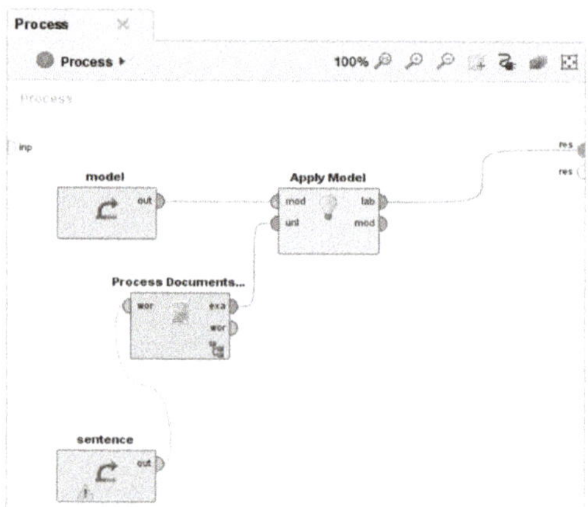

Figure 13. Text mining process for the automatic identification of services.

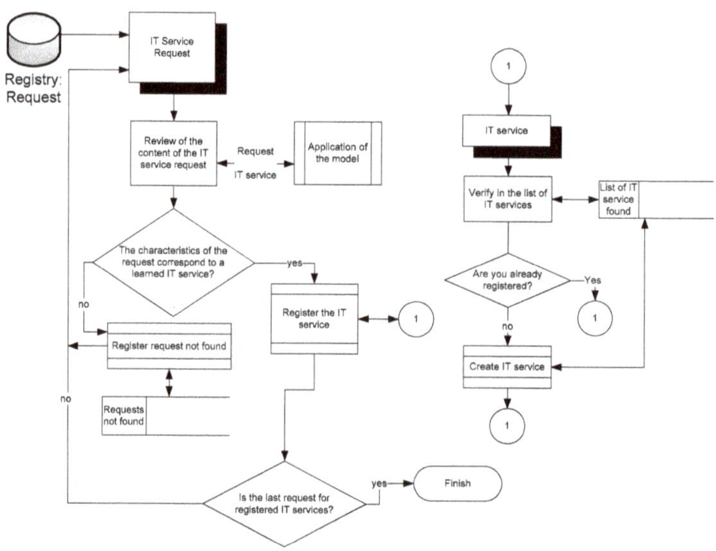

Figure 14. Algorithm for the automatic identification of services.

In this classification process, there is the possibility that some of the requests do not correspond to any service that was learned. In this case, the IT worker(s) must discern which services correspond to the request or discard the request because its content is inconsistent and, thus, possibly a wrong request registered with the *BDS*.

5.2.2. Classification of Services and Generation of the ITSC

Once the ITSs were identified, it was necessary to classify them. For this task, it was necessary to compare the list of services found with the ITSRC, thereby registering each service with its corresponding category. The algorithm for this process is represented in Figure 15.

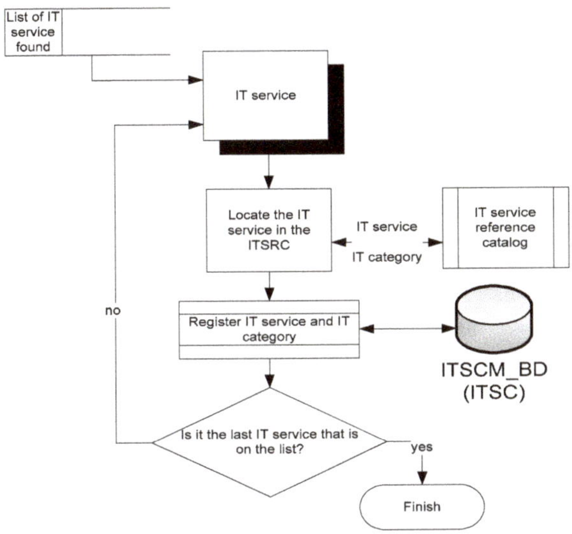

Figure 15. Algorithm for the classification of services and generation of the ITSC.

With the application of the previous algorithm, a result similar to that shown in Table 5 was generated.

Table 5. Result of the classification of services at the table level.

Category	Service
Category 1	*Service 1*
Category 1	*Service 2*
Category 2	*Service 3*
Category 2	*Service 4*

6. Case Study

To develop the "identification of services and construction of the ITSC" (that is, phase 2 of the proposal), a municipal autonomous decentralized government of the province of Manabí, Republic of Ecuador, was chosen. This government met the following conditions: (a) it had a department or IT area with a coordinator or head and several IT workers (six); (b) it did not have an ITSC implemented; and (c) it had a record of requests (requirements or incidents).

From this institution, 415 requests expressed in natural language were obtained. These requests served as a basis to identify ITSs and generate an appropriate ITSC. From the 415 requests (requirements or incidents), 23 ITSs were ultimately identified: LAN admin, WAN admin, authentication and authorization, file sharing, material purchase, software purchase, backup and recovery, create/integrate applications, active directory, hardware, printer maintenance, maintenance and support, apps maintenance and support, network monitoring, antivirus service, firewall service, video conferencing, VoIP service, software, application support, technical/user support, security cameras, email, and collaboration services. These ITSs corresponded to 11 categories. As a result of the classification process, the ITSC of the institution in question was formed as shown in Figure 16.

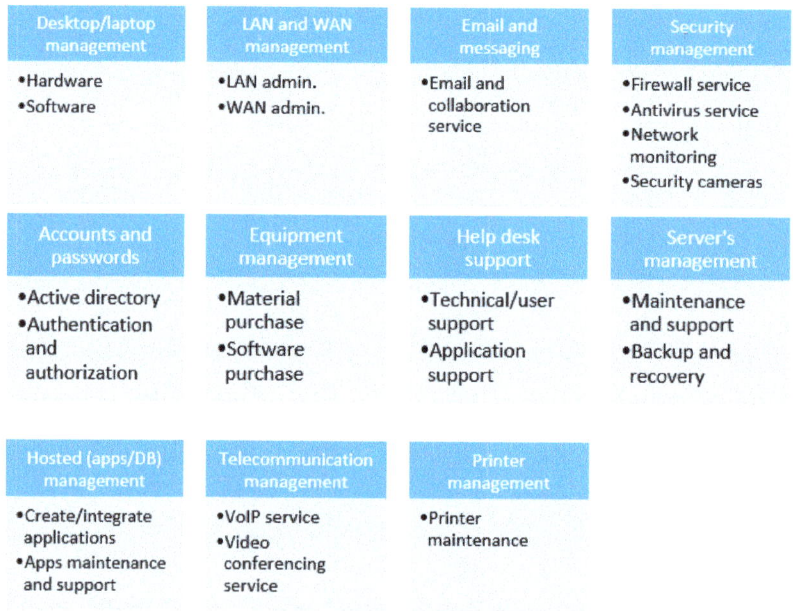

Figure 16. An ITSC corresponding to the knowledge structure.

Once the ITSC of the institution was generated, a survey was applied to the head and the six workers of the IT department. This survey was elaborated based on the quality factors of Moody et al. [30] to evaluate the artifacts of this type. These factors (F) were as follows: (F1) completeness, (F2) integrity, (F3) flexibility, (F4) understandability, (F5) correctness, (F6) simplicity, (F7) integration, and (F8) implementation, where each factor corresponded to a survey question applied. These questions are shown in Appendix A.

6.1. Results

According to Figure 17, the results were very satisfactory, since, in most of the factors, the respondents answered affirmatively. Next, each of the factors was analyzed.

To qualify the assessment of the respondents' responses, a Likert scale was used with the following values for each questioned factor: 1 (poor), 2 (fair), 3 (average), 4 (good), 5 (excellent). To verify the consistency of the survey, Cronbach's alpha coefficient was used to quantify the level of reliability of the measurement scale. The result obtained was 0.845, which means that our instrument is reliable. Below, the results obtained for each of the factors analyzed in our proposal are shown.

In relation to F1 (completeness), the respondents were asked if the device is sufficiently complete for the identification of the ITS required by the organization. Ultimately, 57.1% of the respondents said that the ITS is excellent in this aspect, and 42.9% said that the ITS is good (that is, that it is sufficiently complete).

Regarding F2 (integrity), the respondents were asked if the device has an integrated structure for the identification of services and the construction of the ITSC. A total of 71.4% of respondents said that the structure is excellent, while 28.6% said that it was good (i.e., sufficiently intact).

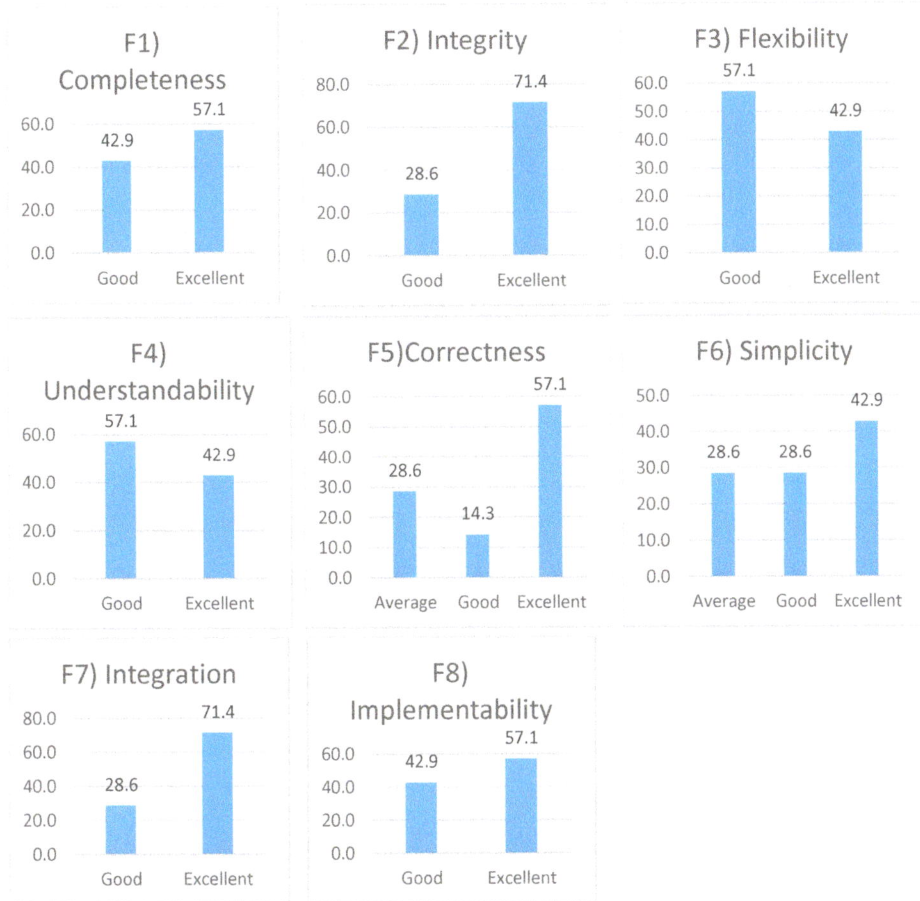

Figure 17. Survey results statistics.

For F3 (flexibility), the question to the respondents was the following: Is the device flexible enough to adapt to the diverse requirements of the organization in relation to the identification of services and the construction of the ITSC? To this question, 42.9% of the respondents stated that the artifact is excellent, while 57.1% of the respondents said that it was good (i.e., that the proposed artifact is very flexible).

In relation to F4 (understandability), it was asked if the device is easy to understand for those in charge and for IT workers. To this, 42.9% said that the device is excellent, while 57.1% said that it is good. This shows that the proposed artifact is very easy to understand.

Regarding F5 (correctness), respondents were asked if the structure of the artifact is clear due to the notation used. A total of 57.1% of respondents said it is excellent, while 14.3% said it is good, and 28.6% said that it is average. This shows that the structure of the artifact is clear.

For F6 (simplicity), it was asked if the artifact is simple in its ability to identify the services and construction of the ITSC. A total of 42.9% of the respondents said the artifact is excellent, while 28.6% said it is good; 28.6% also said that it is average, which shows that the simplicity levels are high.

For F7 (integration), respondents were asked if the ITSC built after the application of the device reflects the ITS required by the institution. A total of 71.4% said that the tool is excellent, while 28.6% said it is good, which shows that the device integrates very well with the organization's IT aspects.

Finally, for F8 (implementation), respondents were asked if it is feasible to implement the device in organizations that wish to identify their ITSs and build their ITSC. A total of 57.1% of respondents said that it is excellent, while 42.9% stated that it is good, meaning that the tool has high levels of implementation in municipal organizations.

Thus, the assessment provided by the respondents is highly positive. To this, we add that the number of services identified corresponded 100% to the services that should have been identified based on the historical data of the organization for which the identification process was carried out for the ITS and construction of the ITSC.

6.2. Discussion

The proposal detailed in this document allows identification of the ITSs and the construction of the ITSC. This proposal was developed considering several important aspects obtained from other SIM and research works dedicated to the construction of the ITSC (mainly in relation to the identification process of ITS). As described above, in phase 1 of the proposal, machine learning was used to generate a model that carried out its "learning" process based on 1699 applications (IT requirements and incidents) cataloged in departments or areas of IT from four public organizations (municipal autonomous decentralized government of Ecuador). Therefore, the possibility remains that more public organizations can be used as a reference to enrich the initial "knowledge" and, thus, have a greater capacity for automatic identification of the ITS.

Although identification of the ITS is a process that can be carried out through different SIMs, with the proposed method, it was possible to use requests and IT incidents registered by the coordinator and IT workers of a public organization as the input (case study, the municipal autonomous decentralized government of Ecuador) to identify their ITSs and to build their ITSC. This approach achieved very satisfactory results, as evidenced by the survey applied to the coordinator and the IT area workers. The success achieved was subject to the fact that the learning carried out in phase 1 of the proposal was based on the applications cataloged by four public organizations of the same type as the organization chosen to develop the case study, since the nature of the requests was similar, and they belonged to the same context at the level of daily operations carried out by IT users.

When technological solutions, such as the one described in this paper, are proposed, they are expected to have the approval of the personnel working in the area involved, since these personnel are the people who will use this tool. In this case, one of the most relevant aspects is the necessarily high level of understanding of the staff needed to use the proposal to identify ITSs and construct an ITSC. This was evidenced in the responses of respondents regarding factors of understandability, correctness, and simplicity. These results should be considered since there are studies (such as that conducted by Mera et al. [57]) that showed how many public institutions do not have an ITSC. One of the factors highlighted in that study is the "complexity of applying existing standards" to build the ITSC.

Another important result that should be highlighted is the high levels of efficiency and effectiveness of the proposal. For efficiency, we highlighted the high levels obtained for the factors of completeness, integrity, flexibility, integration, and implementation, which evaluated the proposal itself. At the level of structure, the levels of adaptability, the degree of integration of the results with the needs of those involved, and the levels of implementation of the proposal are important. On the other hand, when talking about effectiveness, it is necessary to highlight that, according to the historical data of requests that were analyzed with those involved, the results obtained for the amount of identified ITS were very satisfactory, based on all possible services that could result after applying the proposal for that data.

An important contribution of this proposal is the inclusion of machine learning for the identification of ITSs and the automatic construction of the ITSC, since this technique was previously used for the identification of ITSs, but only at the level of request management when the ITSC was already in operation [15] and not for the construction of the ITSC (in addition to the registration of IT requests and also to the registration of IT incidents for these tasks).

7. Conclusions

Once the present research was concluded, the conclusions below were reached.

As science and technology advances, it is necessary to have efficient SIMs built with a consistent knowledge base, which allows workers and IT coordinators to have appropriate tools to carry out their activities related to ITSM, taking advantage of technologies like machine learning to give added value to these activities. The ITS identification process is a fundamental aspect in building an ITSC; therefore, it is very important to perform ITS identification correctly, since the ITSC is a fundamental component of the ITSP, which in turn is the basis for deploying a correct ITSM.

According to the systematic review of the literature, 5352 articles were identified, and 30 articles were selected for our study. It can be seen that there are several SIMs that were used, none of which became a standard for the identification of ITSs, nor is there a standard for the construction of the ITSC. Of the SIMs identified, there were proposals that used natural language as an input to identify ITS. However, none of the investigations conducted were able to integrate the change requests and the incident register to build an ITSC.

A proposal was submitted for the identification of ITS and construction of the ITSC, which consists of two phases. The first phase corresponds to a learning stage, from which knowledge of the four public institutions that have an ITSC generated a model that uses Bayesian networks as a classification system. The second phase corresponds to the implementation stage of the ITSC, which can be replicated in a number of organizations. This process can be applied to identify ITSs and build the ITSC in several institutions, based on their history of IT requests and incidents (text).

The proposal presented was applied to a public institution, specifically, a municipal autonomous decentralized government of Ecuador, whose history of requests and incidents was used to identify ITSs and build the ITSC. After creating the ITSC, it was possible to show that the proposal is highly effective, since, through the application of a survey, the coordinator and the IT workers were able to assess the quality factors of the proposal presented for the creation of the ITSC.

Making a comparison between our proposal and the SIMs detailed in Table 4, three aspects were highlighted. Firstly, the proposal present in this document is a "proposal with results", which were obtained through its application in a real environment, adding to the limited list of proposals with results detailed in Table 4. Secondly, this proposal complies with four of the general activities for the construction of the ITSC: the initial collection of information, the identification of ITSs (ITS definition), and the refinement and the publication of the ITSC (which is generated from the classification of the ITSs). What makes the difference, compared to the SIMs detailed in Table 4, is the application of machine learning with the characteristics described in Section 5, which include the requests and incidents of registered ITSs, taking them as input to build the ITSC, unlike the SIMs identified in Table 4. Thirdly, another important aspect is that the proposal was applied in a public organization, adding to only two SIMs in Table 4 that were tested in this type of organization, contributing to the generation of important data for this type of proposal, which must continue to be tested in real environments.

In future work, it will be necessary to replicate the proposals in other public organizations in order to obtain the necessary feedback to improve them. It will also be necessary to integrate them with other ITSCM activities at the ITSC architecture level, such as ITSC feedback, and the elimination or withdrawal of services, as well as other activities.

Author Contributions: C.M.M. participated in the development of all research work, which includes conceptualization, methodology, literature review, development and proof of the proposal; and during the drafting and correction of the final document. I.A.A. participated directly in all the research work, through his advice and supervision in its development.

Funding: This research received no external funding.

Acknowledgments: Recognition goes to the Research Group of Artificial Intelligence (AI) of the Faculty of Systems Engineering and Informatics at the National University of San Marcos for the facilities provided, as well as the anonymous reviewers for their constructive comments and suggestions. We also thank the Universidad Laica Eloy Alfaro de Manabi, Chone Extension for its financial contributions and the human resources that participated in this research.

Conflicts of Interest: The authors declare no conflict of interest.

Appendix A

Table A1. Quality factors and questions to evaluate the artifacts for ITSC construction.

Factor	Question
(F1) Completeness	is the device sufficiently complete to identify the IT services required by the organization?
(F2) Integrity	Does the artifact have an integrated structure for the identification of services and construction of the ITSC?
(F3) Flexibility	Is the device flexible enough to adapt to different requirements of the organization in relation to the identification of the services and construction of the ITSC?
(F4) Understandability	Is the device easy to understand for managers and IT workers?
(F5) Correctness	Is the structure of the artifact clear due to the notation used?
(F6) Simplicity	Is the artifact simple in its application for the identification of services and the construction of the ITSC?
(F7) Integration	Does the ITSC built after the application of the device reflect the IT services required by the institution?
(F8) Implementation	Is it feasible to implement the device in organizations that wish to identify their IT services and build their ITSC?

Appendix B

To carry out the process of obtaining the ITSRC, it was necessary to start from several previously proposed catalogs; the catalogs considered for obtaining the single ITSRC are shown below.

Table A2. ITSCs found in the review.

Number	Title of the Research Work Year	Year
1	A method for identifying IT services using incidents	2012
2	IT services reference catalog	2013
3	Building an IT service catalog in a small company as the main input for the IT financial manager	2013
4	Service catalog implementation model	2016

After showing each catalog, it is necessary to specify that catalog 2 is the result of an improvement made by the researchers who are the authors of said work. We needed catalogs 2 and 4 to have the same structure; therefore, the comparison was done using catalogs 2 and 3.

With these two catalogs, where each catalog C has several Cat categories,

$$C = \{Cat1,\ Cat2, \ldots .Catn\}.$$

Each category has one or more services, that is,

$$Cat1 \subset C = \{S1,\ S2, \ldots \ldots Sn\},$$

$$Cat2 \subset C = \{S3,\ S4, \ldots \ldots Sy\},$$

$$Catn \subset C = \{S5,\ S6, \ldots \ldots Sz\}.$$

A comparison was made between the catalogs to obtain a unique ITSRC, according to the relationship between the definitions of the categories and the services provided by the researchers who developed those catalogs, considering the following criteria:

(a) Catalog 2 was taken as the first reference, because it contains a greater number of ITSs.
(b) If a category from catalog 2 did not appear in catalog 3, the category of catalog 2 was included in the single ITSRC with all of its services.
(c) If a category from catalog 3 did not appear in catalog 2, the category of catalog 3 was included in the single ITSRC.
(d) If there was a relationship between a service from one catalog and several from another catalog, this service was included because it is of greater importance, together with its category (also, the category that has more services prevails) in the single ITSRC, and the other related services were absorbed.
(e) If there was a unique relationship between a service from catalog 2 and one from catalog 3, the service of greater hierarchy was maintained and its category was the one of greater importance, according to the previous precepts.
(f) If a service from catalog 2 was not included in catalog 3, the service of catalog 2 was included in the single ITSRC together with its category, if it did not violate the precepts of previous categories.
(g) If a service of catalog 3 was not included in catalog 2, the service of catalog 3 was included in the single ITSRC together with its category, if it did not violate the precepts of previous categories.

The results of the comparison between the services and the categories of each ITSC are shown below in the relationship table.

Catálogo 2 \ Catálogo 3	Categoria	Hardware	Hardware	Email	Email	Email	Internet	Internet	Internet	Internet	Software application	Software application	Software application	Software application	Software application	Software application	Backup	Backup	Quality manag	Quality manag	Telephony	Telephony	Telephony
Categoria	**Servicio**	Add hardware	Maintenance hardware	Installation Email Account	Updating Email account	Maintenance Email account	Installation internet browser	Updating internet browser	Maintenance internet browser	Maintenance Web site	Installation specific software	Updating specific software	Maintenance specific software	Installation Base software	Updating Base software	Maintenance Base software	Data Backup	Backup web site	Review data protection law	Review ISO 9001:2008	Maintenance switchboard	Fixed phone	Mobile phone
Desktop/laptop management	Hardware	X	X																				
Desktop/laptop management	Software													X	X	X							
Desktop/laptop management	Material request																						
LAN and WAN management	LAN admin.																						
LAN and WAN management	WAN admin.																						
LAN and WAN management	VPN conections																						
Email and messaging	Email and collaboration service			X	X	X																	
Email and messaging	Mobile email			X	X	X																	
Security management	Firewall service																						
Security management	Antivirus service																						
Security management	Network monitoring																						
Accounts and passwords	Active Directory																						
Accounts and passwords	Authentication and authorisation																						
Accounts and passwords	File Sharing																						
Equipment management	Loan of material																						
Equipment management	Material purchase																						
Equipment management	Material budgeting																						
Equipment management	Software purchase																						
Help desk support	Technical/user support																						
Help desk support	Application support						X	X	X		X	X	X										
Infraestructure management	Datacenter admin.																						
Infraestructure management	Web hosting																						
Infraestructure management	Housing																						
Hosted (apps/DB) management	Create/integrate applications																						
Hosted (apps/DB) management	Apps maintenance and support									X													
Hosted (apps/DB) management	DB backup and recovery																X	X					
Hosted (apps/DB) management	Database management																						
Hosted (apps/DB) management	DB installation/configuration																						
Telecommunication management	VoIP service																					X	
Telecommunication management	Fax service																					X	
Telecommunication management	Mobile phone service																						X
Telecommunication management	Video conferencing service																						
Printer management	External support to printer																						
Printer management	Printer maintenance		X																				
Servers management	Maintenance and support		X																				
Servers management	Backup and recovery																X						

Figure A1. Comparison of services and ITSC categories.

Based on the above relationship chart, the resulting unique ITSRC consisted of the following categories (Cat.) and services:

(Cat. 2) Desktop/laptop management;

- Hardware ((Cat. 3) add hardware, maintenance of hardware);
- Software ((Cat. 3) installation of base software, updating base software, maintenance of base software);
- Material Request

(Cat. 2) LAN and WAN management;

- LAN admin;
- WAN admin;
- VPN connections

(Cat. 2) Email and messaging;

- Email and collaboration service ((Cat. 3) installation of email account, updating email account, maintenance email account);
- Mobile email ((Cat. 3) installation of email account, updating email account, of maintenance email account);

(Cat. 2) Security management;

- Firewall service;
- Antivirus service;
- Network monitoring;

(Cat. 2) Accounts and passwords;

- Active directory;
- Authentication and authorization;
- File sharing;

(Cat. 2) Equipment management;

- Loan of material;
- Material purchase;
- Material budgeting;
- Software purchase;

(Cat. 2) Help desk support;

- Technical/user support;
- Application support ((Cat. 3) installation of internet browser, updating internet browser, maintenance of internet browser, installation of specific software, updating specific software, maintenance of specific software);

(Cat. 2) Infrastructure management;

- Datacenter admin;
- Web hosting;
- Housing;

(Cat. 2) Hosted (apps/DB) management;

- Create/integrate applications;
- Apps maintenance and support ((Cat. 3) maintenance of web site);
- DB backup and recovery ((Cat. 3) data backup, backup web site);

- Database management;
- DB installation/configuration;

(Cat. 2) Telecommunication management;

- VoIP service ((Cat. 3) fixed phone);
- Fax service ((Cat. 3) fixed phone);
- Mobile phone service ((Cat. 3) mobile phone);
- Video conferencing service;
- (Cat. 3) Maintenance switchboard;

(Cat. 2) Printer management;

- External support to printer;
- Printer maintenance ((Cat. 3) maintenance of hardware);

(Cat. 2) Server management;

- Maintenance and support;
- Backup and recovery ((Cat. 3) data backup);

(Cat. 3) Quality management;

- Review data protection law;
- Review ISO 9001:2008.

References

1. Kozlova, E.; Hasenkamp, U.; Kopanakis, E. Use of IT Best Practices for Non-IT Services. In Proceedings of the 2012 Annual SRII Global Conference, San Jose, CA, USA, 24–27 July 2012; pp. 725–734.
2. Braun, C.; Winter, R. Integration of IT Service Management into Enterprise Architecture. In Proceedings of the Proceedings of the 2007 ACM Symposium on Applied Computing, Seoul, Korea, 11–15 March 2007; ACM: New York, NY, USA, 2007; pp. 1215–1219.
3. Zarnekow, R.; Brenner, W.; Pilgram, U. *Integrated Information Management: Applying Successful Industrial Concepts in IT*; Business Engineering; Springer: Berlin/Heidelberg, Germany, 2006; ISBN 978-3-540-32306-8.
4. Simonova, S.; Foltanova, N. Implementation of quality principles for IT service requirements analyse. In Proceedings of the 2017 International Conference on Information and Digital Technologies (IDT), Zilina, Slovakia, 5–7 July 2017; pp. 365–372.
5. Rong, W.; Li, T.; Ouyang, Y.; Li, C.; Xiong, Z. Process Oriented Dependency Modelling for Service Identification. In Proceedings of the Service Science and Knowledge Innovation, Shanghai, China, 23–24 May 2014; Liu, K., Gulliver, S.R., Li, W., Yu, C., Eds.; Springer: Berlin/Heidelberg, Germany, 2014; pp. 166–175.
6. Mendes, C.; Ferreira, J.; da Silva, M.M. Using DEMO to Identify IT Services. In Proceedings of the 2012 Eighth International Conference on the Quality of Information and Communications Technology, Lisbon, Portugal, 3–6 September 2012; pp. 166–171.
7. Hubbers, J.-W.; Ligthart, A.; Terlouw, L. Ten Ways to Identify Services. Available online: https://searchmicroservices.techtarget.com/tip/Ten-ways-to-identify-services (accessed on 6 November 2018).
8. Huergo, R.S.; Pires, P.F.; Delicato, F.C. A Method to Identify Services Using Master Data and Artifact-centric Modeling Approach. In Proceedings of the 29th Annual ACM Symposium on Applied Computing, Gyeongju, Korea, 24–28 March 2014; pp. 1225–1230.
9. Meister, V.G.; Jetschni, J. Towards a semantic information system for IT services. In Proceedings of the 2015 IEEE Seventh International Conference on Intelligent Computing and Information Systems (ICICIS), Cairo, Egypt, 12–14 December 2015; pp. 58–65.
10. Magdalena, A.; Cerrada, J.A.; Calvo-Manzano, J.A. A practical approach for implemeting the service catalogue in micro, small and medium enterprises. In Proceedings of the 7th Iberian Conference on Information Systems and Technologies (CISTI 2012), Madrid, Spain, 20–23 June 2012; pp. 1–8.

11. Teubner, A.; Remfert, C. IT Service Management Revisited—Insights from Seven Years of Action Research. In Proceedings of the CONF-IRM 2012, Vienna, Austria, 21–23 May 2012.
12. McCarthy, M.A.; Herger, L.M. Creating Composite IT Services in the Global Enterprise. In Proceedings of the 2012 IEEE Ninth International Conference on Services Computing, Honolulu, HI, USA, 24–29 June 2012; pp. 687–691.
13. Frey, F.J.; Hentrich, C.; Zdun, U. Capability-based Service Identification in Service-oriented Legacy Modernization. In Proceedings of the 18th European Conference on Pattern Languages of Program, Irsee, Germany, 10–14 July 2013; pp. 10:1–10:12.
14. Börner, R.; Goeken, M.; Rabhi, F. *SOA Development and Service Identification: A Case Study on Method Use, Context and Success Factors*; Working Paper Series; Frankfurt School of Finance & Management: Frankfurt, Germany, 2012.
15. Kalia, A.K.; Xiao, J.; Bulut, M.F.; Vukovic, M.; Anerousis, N. Cataloger: Catalog Recommendation Service for IT Change Requests. In Proceedings of the Service-Oriented Computing, Malaga, Spain, 13–16 November 2017; Springer: Cham, Switzerland, 2017; pp. 545–560.
16. Do Mar Rosa, M.; Gama, N.; da Silva, M.M. A Method for Identifying IT Services Using Incidents. In Proceedings of the 2012 Eighth International Conference on the Quality of Information and Communications Technology, Lisbon, Portugal, 3–6 September 2012; pp. 172–177.
17. Lloyd, V.; Rudd, C. *ITIL Versión 3 Service Design*; APM Group: High Wycombe, UK, 2007.
18. Arcilla, M.; Calvo-Manzano, J.A.; San Feliu, T. Building an IT service catalog in a small company as the main input for the IT financial management. *Comput. Stand. Interfaces* **2013**, *36*, 42–53. [CrossRef]
19. Garbi, G.; Loureiro, G. Business-Product-Service Portfolio Management. Available online: https://www.researchgate.net/publication/279138118_Business-Product-Service_Portfolio_Management (accessed on 25 May 2019).
20. Sembiring, M.; Surendro, K. Service catalogue implementation model. In Proceedings of the 2016 4th International Conference on Information and Communication Technology (ICoICT), Bandung, Indonesia, 25–27 May 2016; pp. 1–6.
21. Nord, F.; Dörbecker, R.; Böhmann, T. Structure, Content and Use of IT Service Catalogs—Empirical Analysis and Development of a Maturity Model. In Proceedings of the 2016 49th Hawaii International Conference on System Sciences (HICSS), Koloa, HI, USA, 5–8 January 2016; pp. 1642–1651.
22. Saura, J.R.; Palos-Sanchez, P.; Reyes-Menendez, A. Marketing a través de aplicaciones móviles de turismo (m-tourism): Un estudio exploratorio. *Int. J. World Tour.* **2017**, *4*, 45–56. [CrossRef]
23. Josuttis, N. *Soa in Practice: The Art of Distributed System Design*; O'Reilly Media, Inc.: Newton, MA, USA, 2007; ISBN 978-0-596-52955-0.
24. Lee, J.; Sugumaran, V.; Park, S.; Sansi, D. An Approach for Service Identification Using Value Co-creation and IT Convergence. In Proceedings of the 2011 First ACIS/JNU International Conference on Computers, Networks, Systems and Industrial Engineering, Jeju Island, Korea, 23–25 May 2011; pp. 441–446.
25. Cisneros, C.; Alberto, C.; del Castillo, S.; Fernanda, C. *Diseño de un modelo de procesos para construir el portafolio de servicios de tics en la Corporación Financiera Nacional*; Escuela Politécnica Nacional: Quito, Ecuador, 2014; Available online: http://bibdigital.epn.edu.ec/handle/15000/8518 (accessed on 9 October 2019).
26. Legner, C.; Löhe, J. Improving the Realization of IT Demands: A Design Theory for End-to-End Demand Management. In Proceedings of the ICIS 2012, Orlando, FL, USA, 16–19 December 2012.
27. Alonso, I.A.; Verdún, J.C.; Caro, E.T. The Importance of IT Strategic Demand Management in Achieving the Objectives of the Strategic Business Planning. In Proceedings of the 2008 International Conference on Computer Science and Software Engineering, Wuhan, China, 12–14 December 2008; Volume 2, pp. 235–238.
28. Aguilar Alonso, I.; Carrillo Verdún, J.; Tovar Caro, E. Description of the structure of the IT demand management process framework. *Int. J. Inf. Manag.* **2017**, *37*, 1461–1473. [CrossRef]
29. Kitchenham, B.; Charters, S. *Guidelines for Performing Systematic Literature Reviews in Software Engineering*; Version 2.3; Keele University, Staffs and University of Durham: Durham, UK, 2007.
30. Moody, D.L.; Sindre, G.; Brasethvik, T.; Solvberg, A. Evaluating the quality of information models: Empirical testing of a conceptual model quality framework. In Proceedings of the 25th International Conference on Software Engineering, Portland, OR, USA, 3–10 May 2003; pp. 295–305.

31. Gama, N.; do Mar Rosa, M.; da Silva, M.M. IT Services Reference Catalog. In Proceedings of the 2013 IFIP/IEEE International Symposium on Integrated Network Management (IM 2013), Ghent, Belgium, 27–31 May 2013; pp. 764–767.
32. Clacy, B.; Jennings, B. Service Management: Driving the Future of IT. *Computer* **2007**, *40*, 98–100. [CrossRef]
33. Niessin, F.; Van Vliet, H. Towards Mature IT Services. Available online: https://onlinelibrary.wiley.com/doi/abs/10.1002/%28SICI%291099-1670%28199806%294%3A2%3C55%3A%3AAID-SPIP97%3E3.0.CO%3B2-T (accessed on 17 March 2018).
34. Walker, G. *IT Problem Management*; Prentice Hall PTR: Upper Saddle River, NJ, USA, 2001; ISBN 0-13-030770-X.
35. Sullivan, J.; Edmond, D.; Hofstede, A. Service Description: A survey of the general nature of services. *Distrib. Parallel Databases J.* **2002**, *12*, 117–133. [CrossRef]
36. Sallé, M. *IT Service Management and IT Governance: Review, Comparative Analysis and their Impact on Utility Computing*; Hewlett-Packard Company: Palo Alto, CA, USA, 2004.
37. Bartsch, C.; Shwartz, L.; Ward, C.; Grabarnik, G.; Buco, M.J. Decomposition of IT service processes and alternative service identification using ontologies. In Proceedings of the NOMS 2008—2008 IEEE Network Operations and Management Symposium, Salvador, Bahia, Brazil, 7–11 April 2008; pp. 714–717.
38. Kang, D.; Song, C.; Baik, D. A Method of Service Identification for Product Line. In Proceedings of the 2008 Third International Conference on Convergence and Hybrid Information Technology, Busan, Korea, 11–13 November 2008; Volume 2, pp. 1040–1045.
39. Brocke, H.; Hau, T.; Vogedes, A.; Schindlholzer, B.; Uebernickel, F.; Brenner, W. Design Rules for User-Oriented IT Service Descriptions. In Proceedings of the 2009 42nd Hawaii International Conference on System Sciences, Waikoloa, HI, USA, 5–8 January 2009; IEEE: Waikoloa, HI, USA, 2009; pp. 1–10.
40. Borner, R.; Looso, S.; Goeken, M. Towards an operationalisation of governance and strategy for service identification and design. In Proceedings of the 2009 13th Enterprise Distributed Object Computing Conference Workshops, Auckland, New Zealand, 1–4 September 2009; pp. 180–188.
41. Gebhart, M.; Abeck, S. Rule-Based Service Modeling. In Proceedings of the 2009 Fourth International Conference on Software Engineering Advances, Porto, Portugal, 20–25 September 2009; pp. 271–276.
42. Boerner, R.; Goeken, M. Service identification in SOA Governance literature review and implications for a new method. In Proceedings of the 2009 3rd IEEE International Conference on Digital Ecosystems and Technologies, Istanbul, France, 1–3 June2009; pp. 588–593.
43. Heo, J.; Kim, H.; Lee, W.; Won, Y. Information security pre-evaluation model for U-IT services. In Proceedings of the 2009 First International Conference on Networked Digital Technologies, Ostrava, Czech Republic, 28–31 July 2009; pp. 500–503.
44. Zimin, V.V.; Kulakov, S.M. Dynamic lifecycle management of IT services in corporate information systems. *Steel Transl.* **2010**, *40*, 539–548. [CrossRef]
45. Mendes, C.; da Silva, M.M. Implementing the Service Catalogue Management. In Proceedings of the 2010 Seventh International Conference on the Quality of Information and Communications Technology, Porto, Portugal, 29 September–2 October 2010; pp. 159–164.
46. Tian, Y.; Su, Y.; Zhuang, X. Research on service identification methods based on SOA. In Proceedings of the 2010 3rd International Conference on Advanced Computer Theory and Engineering (ICACTE), Chengdu, China, 20–22 August 2010; Volume 6, pp. V6-27–V6-31.
47. Cai, S.; Liu, Y.; Wang, X. A Survey of Service Identification Strategies. In Proceedings of the 2011 IEEE Asia-Pacific Services Computing Conference, Jeju Island, Korea, 12–15 December 2011; pp. 464–470.
48. Utz, W.; Woitsch, R.; Karagiannis, D. Conceptualisation of Hybrid Service Models: An Open Models Approach. In Proceedings of the 2011 IEEE 35th Annual Computer Software and Applications Conference Workshops, Munich, Germany, 18–22 July 2011; pp. 494–499.
49. Levina, O.; Nguyen Thanh, T.; Holschke, O.; Rake-Revelant, J. Towards a Method for Service Design. In Proceedings of the Engineering Methods in the Service-Oriented Context, Paris, France, 20–22 April 2011; Ralyté, J., Mirbel, I., Deneckère, R., Eds.; Springer: Berlin/Heidelberg, Germany, 2011; pp. 91–96.
50. Schwarz, S.; Durst, C.; Bodendorf, F. A Conceptual Framework of Service Innovation and Its Implications for Future Research. In Proceedings of the 2012 Annual SRII Global Conference, San Jose, CA, USA, 24–27 July 2012; pp. 172–182.
51. Probst, F.; Buhl, H. Supplier Portfolio Management for IT Services Considering Diversification Effects. *Bus. Inf. Syst. Eng.* **2012**, *4*, 71–83. [CrossRef]

52. Bugeaud, F.; Pietyra, P.; Liger, V. From Service Design to Innovation through Services: Emergence of a Methodological and Systemic Framework. In Proceedings of the Collaborative Systems for Reindustrialization, Dresden, Germany, 30 September–2 October 2013; Camarinha-Matos, L.M., Scherer, R.J., Eds.; Springer: Berlin/Heidelberg, Germany, 2013; pp. 431–438.
53. Kim, Y.; Choi, J.; Shin, Y. A decision model for optimizing the service portfolio in SOA governance. In Proceedings of the 2014 4th World Congress on Information and Communication Technologies (WICT 2014), Bandar Hilir, Malaysia, 8–11 December 2014; pp. 57–62.
54. Souza, E.; Moreira, A.; De Faveri, C. An approach to align business and IT perspectives during the SOA services identification. In Proceedings of the 2017 17th International Conference on Computational Science and Its Applications (ICCSA), Trieste, Italy, 3–6 July 2017; pp. 1–7.
55. Mera, C.; Aguilar, I. Review of Proposals for the Construction and Management of the Catalog of Information Technology Services. *IEEE Access* **2018**, *6*, 45335–45346. [CrossRef]
56. Mera, C.; Aguilar, I. Field Study of the Management of the IT Services Catalog in Public Organizations in the Manabí Province, Ecuador. In Proceedings of the 31st International Business Information Management Association Conference (IBIMA), Milán, Italia, 25–26 April 2018; pp. 4450–4465.
57. Mera, C.; Aguilar, I.; Vera, D. Evaluation of the Management of the Information Technology Services Catalog in Public Organizations in the Province of ManabÍ, Ecuador. In Proceedings of the 2018 10th International Conference on Information Management and Engineering, Manchester, UK, 22–24 September 2018; ACM: New York, NY, USA, 2018; pp. 193–199.

symmetry

MDPI

Article

Directional Thermodynamic Formalism

Mourad Ben Slimane [1,*], Moez Ben Abid [2], Ines Ben Omrane [3] and Borhen Halouani [1]

1 Department of Mathematics, College of Science, King Saud University, P.O. Box 2455, Riyadh 11451, Saudi Arabia; halouani@ksu.edu.sa
2 Ecole Supérieure des Sciences et de la Technologie de Hammam Sousse, Université de Sousse, Sousse 4011, Tunisia; moezbenabid@yahoo.fr
3 Department of Mathematics, Faculty of Science, Imam Mohammad Ibn Saud Islamic University (IMSIU), P.O. Box 90950, Riyadh 11623, Saudi Arabia; imbenomrane@imamu.edu.sa
* Correspondence: mbenslimane@ksu.edu.sa

Received: 5 April 2019; Accepted: 4 June 2019; Published: 21 June 2019

Abstract: The usual thermodynamic formalism is uniform in all directions and, therefore, it is not adapted to study multi-dimensional functions with various directional behaviors. It is based on a scaling function characterized in terms of isotropic Sobolev or Besov-type norms. The purpose of the present paper was twofold. Firstly, we proved wavelet criteria for a natural extended directional scaling function expressed in terms of directional Sobolev or Besov spaces. Secondly, we performed the directional multifractal formalism, i.e., we computed or estimated directional Hölder spectra, either directly or via some Legendre transforms on either directional scaling function or anisotropic scaling functions. We obtained general upper bounds for directional Hölder spectra. We also showed optimal results for two large classes of examples of deterministic and random anisotropic self-similar tools for possible modeling turbulence (or cascades) and textures in images: Sierpinski cascade functions and fractional Brownian sheets.

Keywords: directional hölder regularity; anisotropic hölder regularity; directional scaling function; anisotropic scaling function; directional multifractal formalism; wavelet bases; sierpinski cascade functions; fractional brownian sheets

MSC: 26A15; 26A16; 26B35; 26B05; 46E35; 46E99; 42C40

---

1. Introduction

Multifractal models were originally proposed to describe the intermittent behavior of fully-developed turbulence [1,2] and also chaotic features in dynamical systems [3,4]. If T is the time and $v(T)$ is a the stream-wise component of the velocity of a turbulent flow at a given point, Kolmogorov [5] expected a power law behavior:

$$\int_{\mathbb{R}} |v(T+t)-v(T)|^p \, dT \sim |t|^{\eta_L(p)} \qquad \text{for small } |t|, \tag{1}$$

with $\eta_L(p) = p/3$. Kolmogorov and Oboukhov and [6,7] have refined this prediction into a quadratic behavior.

Various experimental results and other models have confirm the nonlinear behavior. Mandelbrot [2,8,9] has introduced multiplicative cascades for the dissipation of energy in turbulent flows and thus has associated fractals to measures (or functions). Frisch and Parisi [10] conjectured that $\eta_L(p)$ describes the statistical repartition of the pointwise Lipschitz regularities. The Lipschitz spectrum of v is the map which associates to each $0 < H < 1$ the Hausdorff dimension $d(H)$ of the set of times T where v has a given pointwise Lipschitz regularity $h_v(T) = H$ in the sense that:

$$|v(T+t) - v(T)| \sim |t|^H \qquad \text{for small } |t| \,. \tag{2}$$

By heuristic arguments, the thermodynamic formalism [10] states that $d(H)$ is given by the following Legendre transform of $\eta_L(p)$:

$$d(H) = \inf_p (Hp - \eta_L(p) + 1) \,. \tag{3}$$

A similar formalism for measures has been derived; if μ is a probability measure on $\mathbb{R}^d$, the notion of pointwise Lipschitz regularity is replaced by the local regularity of μ defined as:

$$h_\mu(x) = \liminf_{r\to 0} \frac{\log \mu(B(x,r))}{\log r} \,, \tag{4}$$

(see [3,4,11–20]....). Note that if $d = 1$ and $h_\mu(T) \in [0,1)$, then $h_f(T) = h_\mu(T)$ for $f(x) = \mu((-\infty, x])$.

In [21], Daubechies and Lagarias proved the validity of the thermodynamic formalism (3) for some refinement functions used in the construction of orthonormal wavelet bases in one dimension.

An alternative formulation of the Lipschitz scaling function $\eta_L(p)$ in terms of continuous wavelet transform $C_{a,b} = \frac{1}{a^d}\int_{\mathbb{R}^d} f(x)\psi(\frac{x-b}{a})\, dx$ ($a > 0$ and $b \in \mathbb{R}^d$) was proposed by Arneodo, Bacry and Muzy in $\mathbb{R}^d$ with $d = 1$ (see [22–24]). The corresponding thermodynamic formalism was proved in [22] for the primitive of a multinomial measure or a C^1 perturbation of such a measure.

Jaffard [25] extended these formulae in $\mathbb{R}^d$ and showed the link between them via the function space interpretation of the scaling function in terms of (isotropic) Sobolev or Besov-type norms. Such spaces for smoothness index higher than 1 are also characterized by finite order differences. The corresponding scaling function $\eta(p)$ (given in (14)) is expected to give information for the Hölder spectrum $d(H)$ for any value of H. The scaling function $\eta(p)$ is also characterized by either isotropic wavelet bases (i.e., decompositions on tensor products of one-dimensional wavelets with the same dilation factor 2^j at scale j in all coordinate axes) or continuous wavelet transform (in fact, $\eta(p) = \liminf_{a\to 0} \frac{\log \int |C_{a,b}|^p\, db}{\log a}$).

The scope of the mathematical validity of the thermodynamic formalism,

$$d(H) = \inf_p (Hp - \eta(p) + d), \tag{5}$$

has become an important issue. The general rule for a good multifractal formalism is to get optimal upper bounds for the spectra. Optimality is obtained for examples that saturate the upper bound, i.e., the upper bound becomes equality.

If the range P of p's over which one computes the Legendre transform is chosen appropriately:

$$P = \{p : \quad \eta(p) > d\} \,, \tag{6}$$

then (5) yields an upper bound for all functions [25].

The optimality has been either studied or proved under self-similarity assumptions [21,22,25–31], or for a class of particular random processes [32], or for specific functions [33,34], or even generically in either Baire's categories [35] or prevalence sense [36,37].

Alternatively, Kestener and Arneodo (see [38] and references therein) proposed, also for $d = 2$ and $d = 3$, different vectoriel wavelet transform formulas. They applied them to turbulent velocity and vorticity 3D numerical data.

Unfortunately, for $d \geq 2$, the scaling functions $\eta_L(p)$ and $\eta(p)$ are uniform in all directions, and, therefore, are not adapted to study images and multi-dimensional signals with various directional behaviors. These behaviors are important for detection of edges, efficient image compression, texture classification, etc., (see, for instance, [39–51] and the references therein). They also appear in partial differential equations, pseudo-differential operators theory, approximation theory, etc. (for example, see [52] or [53] and references therein). In that case, signals present anisotropies quantified through regularity characteristics and features that strongly differ when measured in different directions [17,19,46,47,49,54–56]. Classes of functions that satisfy different scaling properties according the coordinate axes have been studied in [29,42,44,46,49–51,53–55,57–72]. For an a priori prescribed anisotropy $\alpha = (\alpha_1, \cdots, \alpha_d)$ with $0 < \alpha_1, \cdots, \alpha_d$ and $\sum_{i=1}^{d} \alpha_i = d$, such signals can be expanded in Triebel bases (i.e., tensor products of one-dimensional wavelets that allow dilations factors about $2^{j\alpha_1}, \cdots, 2^{j\alpha_d}$ in coordinate axes [69,70]). Alternatively, we can also use anisotropic continuous wavelet transform $\frac{1}{a^d}\int_{\mathbb{R}^d} f(x)\psi(\frac{x_1-b_1}{a^{\alpha_1}}, \cdots, \frac{x_d-b_d}{a^{\alpha_d}})\, dx$ (see [45,48]). Signals where no a priori anisotropy is prescribed, can be expanded in DeVore, Konyagin, and Temlyakov hyperbolic wavelet bases [73], i.e., tensor products of one-dimensional wavelets that allow different dilations factors $2^{j_1}, \cdots, 2^{j_d}$ in coordinate axes (see also [39,74–76]. Hyperbolic wavelet bases contain all possible anisotropies.

Both Triebel and hyperbolic bases characterize anisotropic Besov spaces [39,70].

Let $d \geq 2$ be a positive integer. Let e be a fixed vector in the unit sphere S^{d-1}. We will focus on the following global and local directional behaviors:

- A natural directional Lipschitz scaling function $\eta_L(p, e)$ of v in direction e can be given by:

$$\int_{\mathbb{R}^d} |v(y+te) - v(y)|^p \, dy \sim |t|^{\eta_L(p,e)} \qquad \text{for small } |t| \, . \tag{7}$$

  It can be extended to a directional scaling function $\eta(p, e)$ that involves any finite order differences in direction e (see Definition 1). It can be also restricted to a bounded domain (see Definition 3).
- A natural directional pointwise Lipschitz regularity $h_L(y, e)$ of v at y in direction e can be given by:

$$|v(y+te) - v(y)| \sim |t|^{h_L(y,e)} \qquad \text{for small } |t| \, . \tag{8}$$

  It can be extended to a directional pointwise Hölder regularity $h(y, e)$ (see Definition 5).

In this paper, we want to understand how singularities given by directional pointwise Hölder regularities fluctuate from point to point for a fixed direction e. These singularities may share a given value on a fractal set. One wishes to compute the Hausdorff dimension of this set. One also wishes to derive this size from global quantities extracted from the signal, given by either a directional scaling function or anisotropic scaling functions. Firstly, we prove wavelet criteria for the directional scaling function $\eta(p, e)$ (see Sections 1–3). Secondly, we perform directional multifractal formalism (see Sections 4–6), i.e., we compute or estimate directional Hölder spectra either directly or via some Legendre transforms on either directional scaling function or anisotropies scaling functions.
Two types of results will be performed:

- We will obtain general upper bounds for the directional Hölder spectra.
- We will show optimal results for two large classes of examples of deterministic and random anisotropic self-similar tools for possible modeling turbulence (or cascades) and textures in images (see [50]): Sierpinski cascade functions introduced by the first author in [29] and fractional Brownian sheets introduced by both Kamont in [67] and Pesquet-Popesu and Lévy-Véhel in [77], and revisited by Ayache, Léger, and Pontier [78] for extra properties.

Note that the heuristic classical arguments from which the thermodynamic formalism for pointwise Lipschitz regularity was derived (see, for example, [25], pp. 947–948) cannot be applied to the directional pointwise Lipschitz regularity; near a point y such that $h(y) = H$, we have

$|v(y+t) - v(y)| \sim |t|^H$ in a cube of radius $|t|$. There are about $|t|^{-d(H)}$ such cubes, so that the total contribution to $\int_{\mathbb{R}^d} |v(y+t) - v(y)|^p \, dy$ is $|t|^{Hp+d-d(H)}$. It follows that $\eta_L(p) = \inf_H (Hp - d(H) + d)$. If $d(H)$ is concave, it is recovered by an inverse Legendre transform formula which yields $d(H) = \inf_p (Hp - \eta_L(p) + d)$. In the context of directional pointwise Lipschitz regularity, we want to calculate the contribution of critical directional pointwise Lipschitz regularity of order H in direction e to the integral $\int_{\mathbb{R}^d} |v(y+te) - v(y)|^p \, dy$: near a point y such that $h(y,e) = H$, we have $|v(y+te) - v(y)| \sim |t|^H$ in a small **segment** of length $|t|$. So, we cannot pursue the above heuristic arguments.

To overcome this inconvenience, we will use a criterion of directional pointwise Hölder regularity $h(y,e)$ in terms of rapid decay of highly oriented multi-scaled wavelet coefficients. Actually, $h(y,e)$ is related to anisotropic pointwise Hölder regularities $h_\alpha(y)$, for all anisotropies α (see [43,44]). The anisotropic pointwise regularity $h_\alpha(y)$ was already introduced in [29] and characterized with either anisotropic continuous wavelets or in Triebel bases [43,44].

Alternatively, we will use a characterization of directional pointwise Lipschitz regularity obtained in [58] directly (i.e., without passing through anisotropies) in terms of decay conditions for the coefficients of the expansion in the hyperbolic Schauder basis.

Note that partial characterizations for the directional pointwise Hölder regularity have been obtained by Sampo and Sumetkijakan [51,79,80] (respective to Jaffard [48]), when using parabolic basis, i.e., curvelets and Hart-Smith transform (respective to the anisotropic Gabor-wavelet transform).

Note that other directional behaviors have been studied. Donoho [81], Guo, and Labate [82] have used wedgelets and shearlets for the detection of discontinuities along smooth edges. Candes, Donoho [83], and Mallat [84] have used wavelet bases elongated in particular directions (ridgelets and bandelets) to deal with singularities along lines, along hyperplanes, etc. Fell, Führ, and Voigtlaender [85] characterized the wavefront set in terms of rapid continuous wavelet decay, for a large variety of dilation groups. For the shearlet groups single wavelets suffice, whereas similitude and diagonal groups need suitable families of wavelets. Recently, by using the harmonic wavelet, Sun, Leng, and Cattani [86] constructed a new multilevel system in the Fourier domain, which has the circular shape. This new system is more suitable for the distribution of general images in the Fourier domain.

In the next section, we first give the definition of the directional scaling function $\eta(p,e)$ in terms of directional Sobolev or Besov spaces (expressed by finite order differences) to which v belongs. We then make the connection between directional scaling function and anisotropic scaling functions analyzed in anisotropic function spaces (see Theorem 1).

In Section 3, using the characterization of the anisotropic scaling function in Triebel wavelet bases [69,70], we deduce a criterion of directional scaling function in these bases (see Theorem 2). Then, using the characterization of the anisotropic scaling function in hyperbolic wavelet bases [39], we deduce a criterion of directional scaling function in these bases (see Theorem 3). Finally, using a result of Kamont [87], we deduce a criterion of directional Lipschitz scaling function in hyperbolic Schauder bases without passing through anisotropies (see Theorem 4).

In Section 4, we recall the connection between both directional and anisotropic Hölder regularities (see Proposition 9). We first deduce a general upper bound for the directional Hölder spectrum by anisotropic Hölder spectra (i.e., Hausdorff dimension of anisotropic Hölder sets) (see Theorem 5). Note that in [58], we characterized directional pointwise Lipschitz regularity in terms of decay conditions for the coefficients of the expansion of f in the hyperbolic basis of tensor products of Schauder functions, but we do not yet succeed to deduce a general upper bound for the directional Lipschitz spectrum. We will instead recall a result of [59] in which we adapted the notion of Hausdorff dimension to the anisotropy, we used a criterion of [43] for anisotropic pointwise regularity in terms of conditions on Triebel wavelet coefficients and deduced a general upper bound for the adapted anisotropic Hölder spectrum by means a Legendre transform of the anisotropic scaling function.

Both this upper bound and Theorem 5 yield a general upper bound for the directional spectrum (see Theorem 6).

In Section 5, we apply Theorem 4 for fractional Brownian sheets to show that unlike the Lipschitz scaling function $\eta_L(p)$ and the Lipschitz spectum $d(H)$ which are uniform in all directions, the directional scaling function $\eta_L(p,e)$ and the directional spectrum $d(H,e)$ are tools that detect directional behaviors (see Theorems 7 and 8). We also prove that if the corresponding appropriate range P of p's over which one will computes the Legendre transform is:

$$P = \{p \geq 1 \; : \; \eta_L(p) > 1\} = \{p \geq 1 \; : \; \eta_L(p,e_i) > 1 \;\; \forall\, i \in \{1,\cdots,d\}\} \tag{9}$$

then $\inf_{p\in P}(Hp - \eta_L(p,e) + 1)$ provides a common directional Lipschitz scaling based directional thermodynamic formalism for these examples (see Theorem 8).

In Section 6, we apply Theorem 4 for Sierpinski cascade functions to show that the directional scaling function $\eta_L(p,e)$ and the directional spectrum $d(H,e)$ are tools that detect directional behaviors. We also show that contrary to $\eta_L(p,e)$, the directional spectrum $d(H,e)$ depends on the geometric disposition of the chosen contractions for each cascade function. We also provide non common directional Lipschitz scaling based directional thermodynamic formalisms for these examples (see Theorems 9 and 11). These formalisms depend on the geometric disposition of contractions for each cascade function. Nevertheless, all obtained formalisms share the same corresponding appropriate range P of p's over which one will compute the Legendre transform given in (9). Moreover, we show the optimality of Theorem 6 for Sierpinski cascade functions corresponding to a large class of geometric disposition of contractions (see Theorem 12). Finally, we modify the notion of the Hausdorff dimension to provide a new common directional Lipschitz scaling based directional thermodynamic formalism for all Sierpinski cascade functions (see Theorem 13).

Finally Section 7 motivates the anisotropic cascade model on the physics side.

2. Directional Scaling Function and Its Connection with Anisotropic Scaling Functions

2.1. Directional Scaling Function

For the definitions of Besov spaces stated in this section, we refer the reader to [69]. Let $d \geq 2$ be a positive integer. Let e be a fixed vector in the unit sphere S^{d-1}. For $t \in \mathbb{R}$ and $y \in \mathbb{R}^d$, define the difference $\Delta_{t,e}f$ in direction e by the standard formula:

$$\Delta_{t,e}f(y) = f(y+te) - f(y)\,. \tag{10}$$

Define the iterated differences in direction e inductively by

$$\Delta^1_{t,e}f = \Delta_{t,e}f \quad \text{and} \quad \Delta^{n+1}_{t,e}f = \Delta^1_{t,e}(\Delta^n_{t,e}f)\,.$$

Definition 1. *Let $0 < s < M$ and $M \in \mathbb{N}$, $1 \leq p < \infty$ and $f \in L^p(\mathbb{R}^d)$.*
We say that $f \in B^s_p(\mathbb{R}^d,e)$ if there exists $C > 0$ such that:

$$\forall\, 0 < t \leq 1 \qquad \int_{\mathbb{R}^d} |\Delta^M_{t,e}f(y)|^p\,dy \leq C|t|^{sp}\,. \tag{11}$$

Define the directional scaling function $\eta(p,e)$ of f in direction e by:

$$\eta(p,e) = \sup\{sp \;\; : \;\; f \in B^s_p(\mathbb{R}^d,e)\}\,. \tag{12}$$

We say that f belongs to the usual isotropic Besov space $B^s_{p,\infty}(\mathbb{R}^d)$ (we will ignore ∞ and write $B^s_p(\mathbb{R}^d)$) if there exists $C > 0$ such that:

$$\forall\, e \in S^{d-1} \ \ \forall\, 0 < t \leq 1 \qquad \int_{\mathbb{R}^d} |\Delta^M_{t,e} f(y)|^p \, dy \leq C|t|^{sp} \,. \tag{13}$$

Define the scaling function $\eta(p)$ of f by:

$$\eta(p) = \sup\{sp \ : \ f \in B^s_p(\mathbb{R}^d)\} \,. \tag{14}$$

Anisotropic Besov spaces were introduced for the study of semi-elliptic pseudo-differential operators whose symbols have different degrees of smoothness along different directions (see [52]; see also [53] and references therein, for a recent use of such spaces for optimal regularity results for the heat equation).

Definition 2. *Let $1 \leq p < \infty$. Denote by $\mathcal{D}$ the set $\{1, \cdots, d\}$. For $i \in \mathcal{D}$, let $M_i \in \mathbb{N}$, $0 < s_i < M_i$ and $e_i = (\delta_{1,i}, \cdots, \delta_{d,i})$ denotes the i-th coordinate vector in $\mathbb{R}^d$. The so-called classical anisotropic Besov space $B_p^{(s_1,\cdots,s_d)}(\mathbb{R}^d)$ is defined as*

$$B_p^{(s_1,\cdots,s_d)}(\mathbb{R}^d) = \bigcap_{i \in \mathcal{D}} B_p^{s_i}(\mathbb{R}^d, e_i) \,.$$

Remark 1. *When $s_1, \cdots, s_d = s$, the space $B_p^{(s_1,\cdots,s_d)}(\mathbb{R}^d)$ coincides with $B^s_p(\mathbb{R}^d)$.*

We will be interested in the characterization of $\eta(p,e)$ in terms of decay conditions for a structure function of the coefficients of the expansion of f in either Triebel anisotropic wavelet bases [69,70] (see Section 3), or hyperbolic wavelet bases [68,73–76] (see Section 4). **Without any loss of generality, we can assume that $e = e_1$, because we can take coordinates on an orthonormal basis $\mathcal{B}$ of $\mathbb{R}^d$ that starts with** e. Using the partial ordering property,

$$B_p^{(s_1,\cdots,s_d)}(\mathbb{R}^d) \subset B_p^{(s'_1,\cdots,s'_d)}(\mathbb{R}^d) \qquad \forall\, s'_i \leq s_i \ \ \forall\, i \in \mathcal{D} \,, \tag{15}$$

we introduce the following substitute for $\eta(p,e)$,

$$\breve{\eta}(p,e) = \ \sup\left\{ s_1 p : \exists 0 < \varepsilon \leq s_1 \quad f \in B_p^{(s_1,\varepsilon,\cdots,\varepsilon)}(\mathbb{R}^d) \right\} \,. \tag{16}$$

We obtain the following result.

Proposition 1. 1. *If $\eta(p,e) = 0$ then $\breve{\eta}(p,e) = 0$.*
2. *We have always:*

$$\breve{\eta}(p,e) \leq \eta(p,e) \,. \tag{17}$$

3. *If $\eta(p) > 0$ then $\breve{\eta}(p,e) = \eta(p,e)$.*

Proof. Both first and second points follow directly from Definition 2.

Assume that $\eta(p) > 0$, then $f \in B_p^{(\delta,\cdots,\delta)}(\mathbb{R}^d)$ for $0 < p\delta < \eta(p)$. Clearly, $f \in B^\delta_p(\mathbb{R}^d, e_i)$ for all $i \in \mathcal{D}$. Let $p\beta_1 < \eta(p,e)$. Since $f \in B_p^{\beta_1}(\mathbb{R}^d, e)$, then Definition 2 and the partial ordering property (15) yield the third point. □

2.2. Connection Between the Directional Scaling Function and Anisotropic Scaling Functions

Space $B_p^{(s_1,\cdots,s_d)}(\mathbb{R}^d)$ is related to the anisotropic Besov space $B_{p,\infty}^{\hat{s},(\alpha_1,\ldots,\alpha_d)}(\mathbb{R}^d)$ introduced in [60,61] using an anisotropic Littlewood Paley analysis. We will drop ∞. Let $\alpha = (\alpha_1,\ldots,\alpha_d) \in \mathbb{R}^d$ be an anisotropy, i.e.,

$$\alpha_1 > 0, \cdots, \alpha_d > 0 \;\text{ and }\; \sum_{i\in\mathcal{D}} \alpha_i = d\,. \tag{18}$$

For $r > 0$ and $x = (x_1,\ldots,x_d) \in \mathbb{R}^d$, we define the anisotropic map:

$$r^\alpha x = (r^{\alpha_1}x_1,\ldots,r^{\alpha_d}x_d)\,. \tag{19}$$

Define the mean smoothness $\hat{s}$ of $(s_1,\cdots,s_d)$ and the anisotropic indices by:

$$\frac{1}{\hat{s}} = \frac{1}{d}\sum_{i\in\mathcal{D}}\frac{1}{s_i} \quad\text{and}\quad \alpha_i = \frac{\hat{s}}{s_i}\,. \tag{20}$$

Then,

$$B_p^{(s_1,\cdots,s_d)}(\mathbb{R}^d) = B_p^{\hat{s},\alpha}(\mathbb{R}^d)\,. \tag{21}$$

If $s_1,\cdots,s_d = s$ then $\hat{s} = s$ and α is the isotropy $(1,\ldots,1)$. Thus,

$$B_p^{(s,\cdots,s)}(\mathbb{R}^d) = B_p^s(\mathbb{R}^d) = B_p^{s,(1,\ldots,1)}(\mathbb{R}^d)\,. \tag{22}$$

For a fixed anisotropy α, define the anisotropic scaling function by:

$$\eta_\alpha(p) = \sup\left\{\tau p \;:\; f \in B_p^{\tau,\alpha}(\mathbb{R}^d)\right\}\,. \tag{23}$$

Clearly, using relation (21), the substitute $\breve{\eta}(p,e)$ for $\eta(p,e)$ given in (16) satisfies the following result.

Theorem 1. *Let $\mathcal{B}$ denotes any orthonormal basis of $\mathbb{R}^d$ starting with e. Let Ω be the set of all anisotropies α satisfying (18) and $\alpha_2 = \cdots = \alpha_d$. Then:*

$$\breve{\eta}(p,e) = \sup_{\alpha\in\Omega}\left(\frac{\eta_\alpha(p)}{\alpha_1}\right). \tag{24}$$

If $\eta(p) > 0$ then:

$$\eta(p,e) = \sup_{\alpha\in\Omega}\left(\frac{\eta_\alpha(p)}{\alpha_1}\right). \tag{25}$$

3. Criteria of Directional Scaling Function

3.1. Criterion of Directional Scaling Function in Triebel Wavelet Bases

We will use Theorem 1 to characterize the directional scaling function $\eta(p,e)$ in Triebel anisotropic wavelet bases [69,70].

Triebel anisotropic wavelets characterize anisotropic Besov spaces; if ψ_{-1} and ψ_1 are the Lemarié-Meyer [88,89] (respective to Daubechies [90]) father and mother wavelets in the Schwartz class (respective to compactly supported and finitely smooth with a large enough smoothness), such that all moments (respective to a large enough finite number of moments) of ψ_1 vanish, $\int_{\mathbb{R}}\psi_{-1}(x)dx = 1$ and the collection $\left(\psi_{-1}(x-k)\right)_{k\in\mathbb{Z}}$ and $\left(2^{j/2}\psi_1(2^jx-k)\right)_{j\in\mathbb{N}_0,k\in\mathbb{Z}}$ is an orthonormal basis of $L^2(\mathbb{R})$. Let α

an anisotropy as in (18). For $j \in \mathbb{N}_0$, let $I_{j,\alpha}$ be the set of pairs $(G, \boldsymbol{l})$ where $G = (G_1, \ldots, G_d) \in \{-1,1\}^d$ such that at least one component G_i is -1 and $\boldsymbol{l} = (l_1, \ldots, l_d) \in \mathbb{N}_0^d$ where:

$$l_i = [j\alpha_i] \text{ if } G_i = -1, \tag{26}$$

$$[j\alpha_i] \le l_i < [(j+1)\alpha_i] \text{ if } G_i = 1 \text{ and } [(j+1)\alpha_i] > [j\alpha_i], \tag{27}$$

and

$$l_i = [j\alpha_i] \text{ if } G_i = 1 \text{ and } [(j+1)\alpha_i] = [j\alpha_i]. \tag{28}$$

(Clearly, in the isotropic setting $\alpha = (1, \ldots, 1)$ and $\boldsymbol{l} = (j, \ldots, j)$).

The cardinality of $I_{j,\alpha}$ is bounded independently of j, more precisely:

$$1 \le \sharp I_{j,\alpha} \le (2^d - 1) \prod_{i \in \mathcal{D}} (2 + \alpha_i). \tag{29}$$

The following proposition is given in [69,70].

Proposition 2. *For all $x = (x_1, \ldots, x_d) \in \mathbb{R}^d$ and all $k = (k_1, \ldots, k_d) \in \mathbb{Z}^d$, we put:*

$$\Psi_{-1,\cdots,-1,k}(x) := \prod_{i \in \mathcal{D}} \psi_{-1}(x_i - k_i), \tag{30}$$

and:

$$\Psi^{(G,\boldsymbol{l})}_{j,k,\alpha}(x) = \prod_{i \in \mathcal{D}} \psi_{G_i}(2^{l_i} x_i - k_i) . \tag{31}$$

Set $|\boldsymbol{l}| := \sum_{i \in \mathcal{D}} l_i$. The collection of the union of $(\Psi_{-1,\cdots,-1,k})$ for $k \in \mathbb{Z}^d$ and $(2^{|\boldsymbol{l}|/2}\, \Psi^{(G,\boldsymbol{l})}_{j,k,\alpha})$ for $j \in \mathbb{N}_0$, $(G, \boldsymbol{l}) \in I_{j,\alpha}$ and $k \in \mathbb{Z}^d$, is then an orthonormal basis of $L^2(\mathbb{R}^d)$. Thus any function $f \in L^2(\mathbb{R}^d)$ can be written as:

$$f(x) = \sum_{k \in \mathbb{Z}^d} c_{-1,\cdots,-1,k}\, \Psi_{-1,\cdots,-1,k}(x), \tag{32}$$

$$+ \sum_{j \in \mathbb{N}_0} \sum_{k \in \mathbb{Z}^d} \sum_{(G,\boldsymbol{l}) \in I_{j,\alpha}} c^{(G,\boldsymbol{l})}_{j,k,\alpha} \Psi^{(G,\boldsymbol{l})}_{j,k,\alpha}(x), \tag{33}$$

with:

$$c_{-1,\cdots,-1,k} = \int_{\mathbb{R}^d} f(x) \Psi_{-1,\cdots,-1,k}(x)\, dx , \tag{34}$$

and:

$$c^{(G,\boldsymbol{l})}_{j,k,\alpha} = 2^{|\boldsymbol{l}|} \int_{\mathbb{R}^d} f(x) \Psi^{(G,\boldsymbol{l})}_{j,k,\alpha}(x)\, dx . \tag{35}$$

The following result was obtained in [69,70].

Proposition 3. *Let $p > 0$ and $s \in \mathbb{R}$. Then, $f \in B^{s,\alpha}_p(\mathbb{R}^d)$ if and only if:*

$$\sup_{j \in \mathbb{N}_0} 2^{(sp-d)j} \sum_{k \in \mathbb{Z}^d} \sum_{(G,\boldsymbol{l}) \in I_{j,\alpha}} \left| c^{(G,\boldsymbol{l})}_{j,k,\alpha} \right|^p < \infty . \tag{36}$$

Using (25), we deduce the following theorem.

Theorem 2. *The anisotropic Besov exponent is given by:*

$$\eta_\alpha(p) = \liminf_{j\to\infty} \frac{\log\left(2^{-dj}\sum_{k\in\mathbb{Z}^d}\sum_{(G,I)\in I_{j,\alpha}} |c_{j,k,\alpha}^{(G,I)}|^p\right)}{\log(2^{-j})}. \tag{37}$$

Let $\mathcal{B}$ and Ω be as in Theorem 1. If $\eta(p) > 0$ then:

$$\eta(p,e) = \sup_{\alpha\in\Omega}\left(\liminf_{j\to\infty} \frac{\log\left(2^{-dj}\sum_{k\in\mathbb{Z}^d}\sum_{(G,I)\in I_{j,\alpha}} |c_{j,k,\alpha}^{(G,I)}|^p\right)}{\log(2^{-j\alpha_1})}\right). \tag{38}$$

3.2. Criterion of Directional Scaling Function in Hyperbolic Wavelet Bases

We will use Theorem 1 to characterize the directional scaling function $\eta(p,e)$ in hyperbolic wavelet bases [68,73–76]. As in [39], without any loss of generality take $d = 2$. For $x = (x_1, x_2) \in \mathbb{R}^2$ and $k = (k_1, k_2) \in \mathbb{Z}^2$, we put $\Psi_{-1,-1,k}(x)$ as in (30) for $d = 2$,

$$\Psi_{j_1,j_2,k}(x) = \psi_1(2^{j_1}x_1 - k_1)\psi_1(2^{j_2}x_2 - k_2),$$

$$\Psi_{j_1,-1,k}(x) = \psi_1(2^{j_1}x_1 - k_1)\psi_{-1}(x_2 - k_2),$$

$$\Psi_{-1,j_2,k}(x) = \psi_{-1}(x_1 - k_1)\psi_1(2^{j_2}x_2 - k_2).$$

The collection of the union of $\{\Psi_{-1,-1,k} : k \in \mathbb{Z}^2\}$, $\{2^{(j_1+j_2)/2}\,\Psi_{j_1,j_2,k} : (j_1,j_2) \in \mathbb{N}_0^2, k \in \mathbb{Z}^2\}$, $\{2^{j_1/2}\,\Psi_{j_1,-1,k} : j_1 \in \mathbb{N}_0, k \in \mathbb{Z}^2\}$, and $\{2^{j_2/2}\,\Psi_{-1,j_2,k} : j_2 \in \mathbb{N}_0, k \in \mathbb{Z}^2\}$, is then an orthonormal basis of $L^2(\mathbb{R}^2)$. Thus any function $f \in L^2(\mathbb{R}^2)$ can be written as:

$$f(x) = \sum_{k\in\mathbb{Z}^2} c_{-1,-1,k}\,\Psi_{-1,-1,k}(x) + \sum_{(j_1,j_2)\in\mathbb{N}_0^2}\sum_{k\in\mathbb{Z}^2} c_{j_1,j_2,k}\Psi_{j_1,j_2,k}(x), \tag{39}$$

$$+ \sum_{j_1\in\mathbb{N}_0}\sum_{k\in\mathbb{Z}^2} c_{j_1,-1,k}\Psi_{j_1,-1,k}(x) + \sum_{j_2\in\mathbb{N}_0}\sum_{k\in\mathbb{Z}^2} c_{-1,j_2,k}\Psi_{-1,j_2,k}(x), \tag{40}$$

with $c_{-1,-1,k}$ as in (34) for $d = 2$,

$$c_{j_1,j_2,k} = 2^{(j_1+j_2)}\int_{\mathbb{R}^2} f(x)\Psi_{j_1,j_2,k}(x)\,dx\,, \tag{41}$$

$$c_{j_1,-1,k} = 2^{j_1}\int_{\mathbb{R}^2} f(x)\Psi_{j_1,-1,k}(x)\,dx, \tag{42}$$

and:

$$c_{-1,j_2,k} = 2^{j_2}\int_{\mathbb{R}^2} f(x)\Psi_{-1,j_2,k}(x)\,dx\,. \tag{43}$$

Let (α_1, α_2) an anisotropy as in (18) (with $d = 2$). For $j \in \mathbb{N}_0$, set:

$$\Gamma_j(\alpha_1,\alpha_2) = \Gamma_j^{h,l}(\alpha_1,\alpha_2) \cup \Gamma_j^{l,h}(\alpha_1,\alpha_2) \cup \Gamma_j^{h,h}(\alpha_1,\alpha_2), \tag{44}$$

with

$$\Gamma_j^{h,h}(\alpha_1,\alpha_2) = \prod_{i=1}^{2}\{[(j-1)\alpha_i] - 1, \cdots, [j\alpha_i] + 1\}, \tag{45}$$

$$\Gamma_j^{h,l}(\alpha_1,\alpha_2) = \{[(j-1)\alpha_1] - 1, \cdots, [j\alpha_1] + 1\} \times \{0, \cdots, [(j-1)\alpha_2] - 1\}, \tag{46}$$

and

$$\Gamma_j^{l,h}(\alpha_1,\alpha_2) = \{0,\cdots,[(j-1)\alpha_1]-1\} \times \{[(j-1)\alpha_2]-1,\cdots,[j\alpha_2]+1\}\,. \tag{47}$$

The following result was obtained in [39].

Proposition 4. *Let $p>0$ and $s\in\mathbb{R}$. Then, $f\in B_p^{s,(\alpha_1,\alpha_2)}(\mathbb{R}^2)$ if and only if:*

$$\sup_{j\in\mathbb{N}_0}\ \sup_{(j_1,j_2)\in\Gamma_j(\alpha_1,\alpha_2)} 2^{spj-(j_1+j_2)}\sum_{k\in\mathbb{Z}^2}\left|c_{j_1,j_2,k}\right|^p < \infty\,. \tag{48}$$

Using (25), we deduce the following theorem.

Theorem 3. *The anisotropic Besov exponent is given by:*

$$\eta_{(\alpha_1,\alpha_2)}(p) = \liminf_{j\to\infty\,,\,(j_1,j_2)\in\Gamma_j(\alpha_1,\alpha_2)} \frac{\log\left(2^{-(j_1+j_2)}\sum_{k\in\mathbb{Z}^2}\left|c_{j_1,j_2,k}\right|^p\right)}{\log(2^{-j})}\,. \tag{49}$$

Let $\mathcal{B}$ and Ω be as in Theorem 1. If $\eta(p)>0$ then:

$$\eta(p,e) = \sup_{(\alpha_1,\alpha_2)\in\Omega}\left(\liminf_{j\to\infty\,,\,(j_1,j_2)\in\Gamma_j(\alpha_1,\alpha_2)} \frac{\log\left(2^{-(j_1+j_2)}\sum_{k\in\mathbb{Z}^2}\left|c_{j_1,j_2,k}\right|^p\right)}{\log(2^{-\alpha_1 j})}\right)\,. \tag{50}$$

3.3. Criterion of Directional Lipschitz Scaling Function of f on Hyperbolic Schauder Bases

We will characterize the directional Lipschitz scaling function restricted to a bounded domain on hyperbolic Schauder functions without passing through anisotropies as done previously. Without any loss of generality, we will work on the unit cube I^d. For this purpose we have to adapt the difference $\Delta_{t,e}f$ in direction e by the standard formula

$$\Delta_{t,e}f(y) = \begin{cases} f(y+te)-f(y) & \text{if } y+te\in I^d \\ 0 & \text{if } y+te\notin I^d\,. \end{cases} \tag{51}$$

Definition 3. *Let $0<s<1$, $1\le p<\infty$ and $f\in L^p(I^d)$. We say that $f\in Lip_p^s(I^d,e)$ if there exists $C>0$ such that:*

$$\forall\, 0<t\le 1 \qquad \int_{I^d}|\Delta_{t,e}f(y)|^p\,dy \le C|t|^{sp}\,. \tag{52}$$

Define the directional Lipschitz scaling function of f in direction e by:

$$\eta_L(p,e) = p\ \sup\{0<s<1\ :\ f\in Lip_p^s(I^d,e)\}\,. \tag{53}$$

We say that $f\in Lip_p^s(I^d)$ if there exists $C>0$ such that:

$$\forall\, e\in S^{d-1}\ \forall\, 0<t\le 1 \qquad \int_{I^d}|\Delta_{t,e}f(y)|^p\,dy \le C|t|^{sp}\,. \tag{54}$$

Define the Lipschitz scaling function of f by:

$$\eta_L(p) = p\ \sup\{0<s<1\ :\ f\in Lip_p^s(I^d)\}\,. \tag{55}$$

Remark 2. *Fix $1 \leq p < \infty$. As in Bonami and Estrade [55], using triangular inequality, we can prove that if there exists e_0 such that $0 < \eta_L(p,e_0) < p$ then the map $e \mapsto \eta_L(p,e)$ takes at most d different values. Moreover, it is constant except, perhaps, on the intersection of unit sphere S^{d-1} with a subspace of dimension at most $d-1$ where it may take larger values. Therefore,*

$$\eta_L(p) = \min_{e \in S^{d-1}} \eta_L(p,e) = \min_{i \in \mathcal{D}} \eta_L(p,e_i) . \tag{56}$$

Since we will use a Kamont result [67], we follow the same notations. Let e_i and $\mathcal{D}$ be as in Definition 2. Write $\Delta_{t,i}f$ instead of $\Delta_{t,e_i}f$. Denote by $\mathbf{0}$ and $\mathbf{1}$, respectively, the vectors $(0,\cdots,0)$ and $(1,\cdots,1)$ in $\mathbb{R}^d$. Let $\mathbf{a} = (a_1,\cdots,a_d)$ and $\mathbf{b} = (b_1,\cdots,b_d)$ be two vectors of $\mathbb{R}^d$. Put $|\mathbf{a}| = |a_1| + \cdots + |a_d|$. If $A \subset \mathcal{D}$, put $\mathbf{a}(A) = (\tilde{a}_1,\cdots,\tilde{a}_d)$ where $\tilde{a}_i = a_i$ if $i \in A$, and $\tilde{a}_i = 0$ if $i \notin A$. Write $\mathbf{a} \leq \mathbf{b}$ if $a_i \leq b_i$ for all $i \in \mathcal{D}$, and $\mathbf{a} < \mathbf{b}$ if $a_i < b_i$ for all $i \in \mathcal{D}$. Finally, write $\mathbf{a}^{\mathbf{b}} = \prod_{i \in \mathcal{D}} a_i^{b_i}$.
For $\mathbf{h} = (h_1,\cdots,h_d) \in \mathbb{R}^d$ and $A = \{i_1,\cdots,i_k\} \subset \mathcal{D}$, set:

$$\Delta_{\mathbf{h},A}f = \Delta_{h_{i_1},i_1} \circ \cdots \circ \Delta_{h_{i_k},i_k} f . \tag{57}$$

Clearly,

$$\Delta_{h_i,i} \circ \Delta_{h_j,j} f = \Delta_{h_j,j} \circ \Delta_{h_i,i} f . \tag{58}$$

For $f \in L^p(I^d)$, define:

$$\omega_{p,A}(f,\mathbf{t}) = \sup_{\mathbf{0}<\mathbf{h}\leq\mathbf{t}} ||\Delta_{\mathbf{h},A}f||_p \qquad \text{for} \quad \mathbf{t} \in \mathbb{R}^d, \ \ \mathbf{0} < \mathbf{t} \leq \mathbf{1} . \tag{59}$$

Remark 3. *Clearly, $f \in Lip_p^{s_i}(I^d,e_i)$ is equivalent to $\omega_{p,\{i\}}(f,\mathbf{t}) = O(t_i^{s_i})$.*

Let $\mathbf{0} < (s_1,\cdots,s_d) < \mathbf{1}$. For $\mathbf{t} = (t_1,\cdots,t_d)$, define:

$$\omega^{(s_1,\cdots,s_d)}(\mathbf{t}) = \prod_{i \in \mathcal{D}} t_i^{s_i} \tag{60}$$

and

$$\omega^{(s_1,\cdots,s_d),\frac{1}{2}}(\mathbf{t}) = (\prod_{i \in \mathcal{D}} t_i^{s_i}) \ (1 - \sum_{i \in \mathcal{D}} \log(t_i))^{1/2} . \tag{61}$$

For a function g given on I^d, $A \subset \mathcal{D}$, and $\mathbf{t} \in I^d$, put:

$$g(\mathbf{t};A) = g(\mathbf{t}(A) + \mathbf{1}(\mathcal{D} \setminus A)) . \tag{62}$$

Set:

$$\mathcal{D}^* = \{A \subset \mathcal{D} \ : \ A \neq \emptyset\} . \tag{63}$$

In [67], Kamont considered the following spaces described in terms of moduli of smoothness in the L^p-norm:

$$Lip_p^{(s_1,\cdots,s_d)}(I^d) = \{f \in L^p(I^d) : \forall A \in \mathcal{D}^* \ \ \omega_{p,A}(f,\mathbf{t}) = O(\omega^{(s_1,\cdots,s_d)}(\mathbf{t};A))\} , \tag{64}$$

$$Lip_p^{(s_1,\cdots,s_d),\frac{1}{2}}(I^d) = \{f \in L^p(I^d) : \forall A \in \mathcal{D}^* \ \ \omega_{p,A}(f,\mathbf{t}) = O(\omega^{(s_1,\cdots,s_d),\frac{1}{2}}(\mathbf{t};A))\}, \tag{65}$$

and

$$lip_p^{(s_1,\cdots,s_d),\frac{1}{2}}(I^d) = \{f \in Lip_p^{(s_1,\cdots,s_d),\frac{1}{2}}(I^d) : \forall A \in \mathcal{D}^* \ \ \omega_{p,A}(f,\mathbf{t}) = o(\omega^{(s_1,\cdots,s_d),\frac{1}{2}}(\mathbf{t};A))\}, \tag{66}$$

where $O(\mathbf{t}(A))$ and $o(\mathbf{t}(A))$ refer to $\min(t_i : i \in A) \to 0$.

The following embeddings hold.

Proposition 5. *1.*

$$Lip_p^{(s_1,\cdots,s_d)}(I^d) \subset \bigcap_{i\in\mathcal{D}} Lip_p^{s_i}(I^d,e_i)\,. \tag{67}$$

2.

$$\bigcap_{i\in\mathcal{D}} Lip_p^{s_i}(I^d,e_i) \subset Lip_p^{(\theta_1 s_1,\cdots,\theta_d s_d)}(I^d) \qquad \forall\, \mathbf{0}<\theta=(\theta_1,\cdots,\theta_d) \text{ with } |\theta|\le 1\,. \tag{68}$$

Proof of Proposition 5:

1. If $f\in Lip_p^{(s_1,\cdots,s_d)}(I^d)$, then $\omega_{p,\{i\}}(f,\mathbf{t})=O(t_i^{s_i})$ for all $i\in\mathcal{D}$. The result follows from Remark 3.
2. Conversely, assume that $f\in Lip_p^{s_i}(I^d,e_i)$ for all $i\in\mathcal{D}$. Let $A\subset\mathcal{D}$ be non-empty. Write $A=\{i_1,\cdots,i_k\}$, we have $\Delta_{h,A}f=\Delta_{h_{i_1},i_1}g$ where $g=\Delta_{h_{i_2},i_2}\circ\cdots\circ\Delta_{h_{i_k},i_k}f$. Since $f\in Lip_p^{s_i}(I^d,e_i)$ and g is a linear combination of translated copies of f, then $\omega_{p,A}(f,\mathbf{t})=O(t_{i_1}^{s_{i_1}})$. Similarly, using property (58), we have $\omega_{p,A}(f,\mathbf{t})=O(t_{i_l}^{s_{i_l}})$ for all $2\le l\le k$. On the other hand, since $f\in L^p(I^d)$ then $\omega_A(f,\mathbf{t})=O(1)$ for all $k+1\le l\le d$. Therefore, (68) holds. □

The following embeddings hold too.

Proposition 6. *1. We have $Lip_p^{(s_1,\cdots,s_d)}(I^d)\subset Lip_p^{(s_1,\cdots,s_d),\frac{1}{2}}(I^d)$ and $lip_p^{(s_1,\cdots,s_d)}(I^d)\subset lip_p^{(s_1,\cdots,s_d),\frac{1}{2}}(I^d)$.*
2. If $(s'_1,\cdots,s'_d)<(s_1,\cdots,s_d)$ then $Lip_p^{(s_1,\cdots,s_d)}(I^d)\subset lip_p^{(s'_1,\cdots,s'_d)}(I^d)$.
3. If $(s'_1,\cdots,s'_d)<(s_1,\cdots,s_d)$ then $Lip_p^{(s_1,\cdots,s_d),\frac{1}{2}}(I^d)\subset Lip_p^{(s'_1,\cdots,s'_d)}(I^d)$.

Proof of Proposition 6:

1. The first point is a consequence of $\omega^{(s_1,\cdots,s_d)}(\mathbf{t})\le\omega^{(s_1,\cdots,s_d),\frac{1}{2}}(\mathbf{t})$.
2. Let $f\in Lip_p^{(s_1,\cdots,s_d)}(I^d)$ and $(s'_1,\cdots,s'_d)<(s_1,\cdots,s_d)$. We know from (69) that $f\in Lip_p^{(s'_1,\cdots,s'_d)}(I^d)$. Let $A\in\mathcal{D}^*$. Since $(s'_1,\cdots,s'_d)<(s_1,\cdots,s_d)$ then

$$\frac{\omega_A(f,\mathbf{t})}{\omega^{(s'_1,\cdots,s'_d)}(\mathbf{t};A)}\le C\omega^{(s_1,\cdots,s_d)-(s'_1,\cdots,s'_d)}(\mathbf{t};A)=o(\mathbf{t}(A))\,.$$

   Hence $f\in lip_p^{(s'_1,\cdots,s'_d)}(I^d)$.
3. Let $f\in Lip_p^{(s_1,\cdots,s_d),\frac{1}{2}}(I^d)$ and $(s'_1,\cdots,s'_d)<(s_1,\cdots,s_d)$. Let $A\in\mathcal{D}^*$. Since $(s'_1,\cdots,s'_d)<(s_1,\cdots,s_d)$ and $t\log t=o(1)$ when t goes to 0 then

$$\frac{\omega_A(f,\mathbf{t})}{\omega^{(s'_1,\cdots,s'_d)}(\mathbf{t};A)}\le C\omega^{(s_1,\cdots,s_d)-(s'_1,\cdots,s'_d)}(\mathbf{t};A)(1-\sum_{i\in A}\log(t_i))^{1/2}=o(\mathbf{t}(A))\,.$$

   It follows that $f\in Lip_p^{(s'_1,\cdots,s'_d)}(I^d)$. □

We will characterize $\eta_L(p,e)$ in terms of decay conditions for the coefficients of the expansion of f in the basis of tensor products of Schauder functions. **Without any loss of generality, the orthonormal basis $\mathcal{B}$ (on which coordinates are considered) can start with** e.
Using the partial ordering property

$$Lip_p^{(s_1,\cdots,s_d)}(I^d)\subset Lip_p^{(s'_1,\cdots,s'_d)}(I^d) \qquad \forall\,(s'_1,\cdots,s'_d)\le(s_1,\cdots,s_d)\,, \tag{69}$$

we introduce the following definition as a substitute for $\eta_L(p,e)$

$$\widetilde{\eta_L}(p,e)=p\ \sup\left\{s_1\in(0,1):\ \exists 0<\varepsilon<1\quad f\in Lip_p^{(s_1,\varepsilon,\cdots,\varepsilon)}(I^d)\right\}\,. \tag{70}$$

We will show the following proposition.

Proposition 7. *1. If $\eta_L(p,e) = 0$ then $\widetilde{\eta_L}(p,e) = 0$.*
2. We have always

$$\widetilde{\eta_L}(p,e) \leq \eta_L(p,e)\ . \tag{71}$$

3. If $\eta_L(p) > 0$ then $\widetilde{\eta_L}(p,e) = \eta_L(p,e)$.

Proof of Proposition 7:

Both first and second results are consequences of the first part of Proposition 5.

Assume that $\eta_L(p) > 0$, then $f \in Lip_p^{\delta}(I^d)$ for $0 < p\delta < \eta_L(p)$. Clearly, $f \in Lip_p^{\delta}(I^d, e_i)$ for all $i \in \mathcal{D}$ and $\eta_L(p,e) \geq p\delta$. Let $ps_1 < \eta_L(p,e)$. Since $f \in Lip_p^{s_1}(I^d, e)$, then the second result in Proposition 5 implies that $f \in Lip_p^{((1-(d-1)\theta)s_1, \theta\delta, \cdots, \theta\delta)}(I^d)$ for all $0 < \theta \leq \dfrac{1}{d-1}$. Letting θ tends to 0, we obtain $\widetilde{\eta_L}(p,e) \geq \eta_L(p,e)$. □

In [67,87], Kamont characterized the space $Lip_p^{(s_1, \cdots, s_d)}(I^d)$ in terms of decay conditions for the coefficients of the expansion of f in the basis of tensor products of Schauder functions.

Let $\{\phi_k, k \geq 0\}$ be the family of Schauder functions on I, normed in L^∞, i.e., $\phi_0 = 1$, $\phi_1(t) = t$, and for $k \geq 2$, $k = 2^j + n$ with $j \geq 0$ and $1 \leq n \leq 2^j$, $\phi_k(t) = \phi(2^{j+1}t - 2n + 1)$ (with support $[(n-1)2^{-j}, n2^{-j}]$), where $\phi(t) = \max(0, 1 - |t|)$ (the so-called Schauder function).

In several dimensions, we consider the family $\{\Phi_{\mathbf{k}}, \mathbf{k} \geq 0\}$ of tensor products of Schauder functions, i.e., $\Phi_{\mathbf{k}}(x) = \phi_{k_1}(x_1) \cdots \phi_{k_d}(x_d)$ for $\mathbf{k} = (k_1, \cdots, k_d)$.

For $j \in M = \{-2, -1, 0, 1, 2, \cdots\}$, let

$$\tilde{N}_{-2} = \{0\},\ \tilde{N}_{-1} = \{1\},\ \text{and}\ \tilde{N}_j = \{2^j + n : n = 1, \cdots, 2^j\} \ \text{ for } j \geq 0, \tag{72}$$

and for a vector $\mathbf{j} = (j_1, \cdots, j_d)$ we put

$$\tilde{N}_{\mathbf{j}} = \tilde{N}_{j_1} \times \cdots \tilde{N}_{j_d}\ . \tag{73}$$

Let for $f \in C(I^d)$, $i \in \mathcal{D}$, $x \in I^d$ and $k \geq 0$

$$c_{i,0}(f)(x) = f(x - x_i e_i)\ ,\ c_{i,1}(f)(x) = f(x + (1 - x_i)e_i) - f(x - x_i e_i)\ , \tag{74}$$

and for $k = 2^j + n \in \tilde{N}_j$ with $j \geq 0$

$$c_{i,k}(f)(x) = f(x + (\frac{2n-1}{2^{j+1}} - x_i)e_i) - \frac{1}{2}(f(x + (\frac{n-1}{2^j} - x_i)e_i) + f(x + (\frac{n}{2^j} - x_i)e_i)). \tag{75}$$

For $\mathbf{k} = (k_1, \cdots, k_d)$ we put

$$C_{\mathbf{k}}(f) = c_{1,k_1} \circ \cdots \circ c_{d,k_d}(f)\ . \tag{76}$$

Then for any $f \in C(I^d)$ we have

$$f = \sum_{\mathbf{j} \in M^d} \sum_{\mathbf{k} \in \tilde{N}_{\mathbf{j}}} C_{\mathbf{k}}(f)\Phi_{\mathbf{k}}\ . \tag{77}$$

In $\sum_{\mathbf{j} \in M^d}$ we assume the following order: for $\mathbf{j} = (j_l, ..., j_d)$ and $\mathbf{j}' = (j'_l, ..., j'_d)$, if $\max(j_1, ..., j_d) < \max(j'_1, ..., j'_d)$, then $\mathbf{j}$ precedes $\mathbf{j}'$.

For f given by (77) we put

$$\tau_{\mathbf{j},p}(f) = 2^{-|\mathbf{j}|/p} \left(\sum_{\mathbf{k} \in \tilde{N}_{\mathbf{j}}} |C_{\mathbf{k}}(f)|^p \right)^{1/p}\ . \tag{78}$$

The following wavelet characterization of spaces $Lip_p^{(s_1,\cdots,s_d)}(I^d)$ is due to Kamont [67].

Proposition 8. *Let:*

$$\mathbf{t_j} = (2^{-\max(j_1,0)}, \cdots, 2^{-\max(j_d,0)}) . \tag{79}$$

Then, for $(1/p, \cdots, 1/p) < (s_1, \cdots, s_d) < \mathbf{1}$,

$$f \in Lip_p^{(s_1,\cdots,s_d)}(I^d) \quad \Leftrightarrow \quad \tau_{\mathbf{j},p}(f) = O(\omega^{(s_1,\cdots,s_d)}(\mathbf{t_j})) \text{ as } |\mathbf{j}| \to \infty . \tag{80}$$

Thanks to Proposition 7, the last result leads to the following characterization.

Theorem 4. *Assume that* $\eta_L(p) > 0$. *If:*

$$\forall\, i \in \mathcal{D} \qquad \liminf_{|\mathbf{j}|\to\infty} \frac{\log \tau_{\mathbf{j},p}(f)}{\log(2^{-\max(j_i,0)})} > 1/p, \tag{81}$$

then:

$$\widetilde{\eta_L}(p,e_i) = \eta_L(p,e_i) = p\ \min\left(1, \liminf_{|\mathbf{j}|\to\infty} \frac{\log \tau_{\mathbf{j},p}(f)}{\log(2^{-\max(j_i,0)})}\right) .$$

Remark 4. *We will see that assumption (81) yields the appropriate range P given in (9) (thanks to (56)) for the directional thermodynamic formalisms that we will find in Section 7 (respective to Section 8) for fractional Brownian sheets (respective to Sierpinski cascade functions).*

4. General Upper Bound for the Directional Hölder Spectrum

Let us first recall the notions of Hölder regularity, directional Hölder regularity and anisotropic Hölder regularity.

Definition 4. *Let* $h > 0$ *be non integer,* $y \in \mathbb{R}^d$ *and* $f : \mathbb{R}^d \to \mathbb{C}$. *We say that* $f \in C^h(y)$ *if there exists* $C > 0$ *and a polynomial* P_y *of degree at most the integer part* $[h]$ *of* h *such that in a neighborhood of* y *we have*

$$|f(x) - P_y(x)| \leq C|x-y|^h . \tag{82}$$

The Hölder exponent (or regularity) of f *at* y *is*

$$h(y) = \sup\left\{h : \quad f \in C^h(y)\right\}.$$

We say that $f \in C^h(\mathbb{R}^d)$ *if* $f \in L^\infty(\mathbb{R}^d)$ *and if (82) holds for any* x *and* y *in* $\mathbb{R}^d$ *with uniform constant* C.

The Hölder (upper-Hölder) spectrum of f is the map which associates to each H the Hausdorff dimension $d(H)$ (respective to $D(H)$) of the set of points y where $h(y) = H$ (respective to $h(y) \leq H$).

Definition 5. *Let* $h > 0$ *be non integer,* $y \in \mathbb{R}^d$ *and* $f : \mathbb{R}^d \to \mathbb{C}$. *Let* $e \in S^{d-1}$. *We say that* $f \in C^h(y,e)$ *if there exists* $\delta > 0$, $C > 0$ *and a polynomial* P_y *of degree at most the integer part* $[h]$ *of* h *such that for all* $t \in (-\delta, \delta)$

$$|f(y+te) - P_y(y+te)| \leq C|t|^h . \tag{83}$$

The directional Hölder exponent (or regularity) of f *at* y *in direction* e *is*

$$h(y,e) = \sup\left\{h : \quad f \in C^h(y,e)\right\}.$$

We say that $f \in C^h(\mathbb{R}^d, e)$ if $f \in L^\infty(\mathbb{R}^d)$ and if (83) holds for any $y \in \mathbb{R}^d$ and $t \in \mathbb{R}$ with uniform constant C.

The directional Hölder (upper-Hölder) spectrum of f in direction e is the map which associates to each H the Hausdorff dimension $d(H,e)$ (respective to $D(H,e)$) of the set of points y where $h(y,e) = H$ (respective to. $h(y,e) \leq H$).

In [44], we found a connection between both the notion of directional Hölder regularity and the anisotropic version of Definition 4 (see [29,43]).

Let $\alpha \in \mathbb{R}^d$ be an anisotropy as in (18). For $x = (x_1, \dots, x_d) \in \mathbb{R}^d$, we set

$$|x|_\alpha = \max(|x_1|^{1/\alpha_1}, \dots, |x_d|^{1/\alpha_d}). \tag{84}$$

The corresponding α-ball $R_\alpha(x,r) := \left\{y \in \mathbb{R}^d : \ |x-y|_\alpha < r\right\}$ of α-radius r centered on x is a rectangle with sides parallel to the axes of coordinates, centered at x and with side-length $2r^{\alpha_i}$ in the x_i-direction. If $P = \sum_{(i_1,\dots,i_d)\in\mathbb{N}_0^d} a_{(i_1,\dots,i_d)} x_1^{i_1} \cdots x_d^{i_d}$ is a polynomial, define its α-homogeneous degree by

$$d_\alpha^o P = \max\left\{\sum_{l\in\mathcal{D}} \alpha_l \, i_l \ : \ a_{(i_1,\dots,i_d)} \neq 0\right\}.$$

Definition 6. *Let $h > 0$ and $y \in \mathbb{R}^d$. A function $f : \mathbb{R}^d \to \mathbb{C}$ belongs to $C_\alpha^h(y)$ if there exist $C > 0$ and a polynomial P of α-homogeneous degree smaller than h such that in a neighborhood of y*

$$|f(x) - P_y(x)| \leq C|x-y|_\alpha^h . \tag{85}$$

The α-Hölder exponent of f at y is defined by:

$$h_\alpha(y) = \sup\left\{h : \quad f \in C_\alpha^h(y)\right\}. \tag{86}$$

In [44], we found a connection between both directional and anisotropic pointwise Hölder exponents of f.

Proposition 9. *Let $\mathcal{B}$ and Ω be as in Theorem 1. Then*

$$h(x,e) = \sup_{\alpha\in\Omega}\left(\frac{h_\alpha(x)}{\alpha_1}\right). \tag{87}$$

Proposition 9 yields the following general upper bound for the directional Hölder spectrum.

Theorem 5. *Let $\mathcal{B}$ and Ω be as in Theorem 1. Then*

$$D(H,e) \leq \inf_{\alpha\in\Omega} dim\left\{x \in \mathbb{R}^d \ : \ h_\alpha(x) \leq \alpha_1 H\right\}. \tag{88}$$

Remark 5. *In [58], we characterized directional pointwise Lipschitz regularity in terms of decay conditions for the coefficients of the expansion of f in the hyperbolic basis of tensor products of Schauder functions (see Section 8.2). Nevertheless, we do not yet deduce a general upper bound for the directional Lipschitz spectrum.*

Let us now show how to use Theorem 5 in order to obtain a general upper bound for the directional spectrum. In [29,43], we adapted the notion of Hausdorff dimension to the anisotropy α; if $E \subset \mathbb{R}^d$, we define its α-diameter to be $|E|_\alpha := \sup_{x,y\in E} |x-y|_\alpha$. By replacing in the definition of Hausdorff measure, the usual notion of diameter by the α-diameter, we easily check (see [91]) that we get the following notion of anisotropic dimension.

Definition 7. *Let $E \subset \mathbb{R}^d$, $\varepsilon > 0$ and R_ε the set of all coverings $R = (E_n)_{n\in\mathbb{N}}$ of E by sets E_n of α-diameter $|E_n|_\alpha$ at most ε. Let*

$$M^\delta_{\varepsilon,\alpha}(E) \;=\; \inf_{R\in R_\varepsilon} \sum_{n\in\mathbb{N}} |E_n|^\delta_\alpha \, .$$

The δ-dimensional α-Hausdorff measure of E is

$$M^\delta_\alpha(E) \;=\; \limsup_{\varepsilon\to 0} M^\delta_{\varepsilon,\alpha}(E).$$

The α-Hausdorff dimension of E is

$$dim_\alpha(E) \;=\; \inf\left\{\delta \;:\; M^\delta_\alpha(E) = 0\right\} \;=\; \sup\left\{\delta \;:\; M^\delta_\alpha(E) = \infty\right\}.$$

Note that we get the same value of $dim_\alpha(E)$ if we use coverings $R = (E_n)_{n\in\mathbb{N}}$ of E by rectangles E_n with sides parallel to the axes of coordinates and with side-length $2\varepsilon^{\alpha_i}$ in the x_i-direction.

In the isotropic case, $|.|_{(1,\ldots,1)}$ is equivalent to the Euclidean norm on $\mathbb{R}^d$ and $dim_{(1,\ldots,1)}(E)$ coincides with $dimE$. But if $\alpha \neq (1,\ldots,1)$, then $dim_\alpha(E)$ doesn't necessarily coincide with $dimE$. Actually, if

$$\alpha_{min} = \min_{i\in\mathcal{D}} \alpha_i \quad \text{and} \quad \alpha_{max} = \max_{i\in\mathcal{D}} \alpha_i \, , \tag{89}$$

then there exists $C \geq 1$ such that

$$\forall x \in \mathbb{R}^d \quad \frac{1}{C}\min\left\{|x|^{1/\alpha_{min}}, |x|^{1/\alpha_{max}}\right\} \leq |x|_\alpha \leq C\max\left\{|x|^{1/\alpha_{min}}, |x|^{1/\alpha_{max}}\right\} \, . \tag{90}$$

and

$$\alpha_{min}\, dim(E) \leq dim_\alpha(E) \leq \alpha_{max}\, dim(E) \, . \tag{91}$$

Definition 8. *The α-spectrum is:*

$$d_\alpha(H) = dim_\alpha \left\{x \in \mathbb{R}^d \;:\; h_\alpha(x) = H\right\} \, . \tag{92}$$

The α-upper-spectrum is:

$$D_\alpha(H) = dim_\alpha \left\{x \in \mathbb{R}^d \;:\; h_\alpha(x) \leq H\right\} \, . \tag{93}$$

The following upper bound was proved in [59].

Proposition 10. *If f is uniform Hölder on $\mathbb{R}^d$ in the sense that $f \in C^\varepsilon(\mathbb{R}^d)$ for $\varepsilon > 0$, then:*

$$D_\alpha(H) \leq \inf_{p\geq p_\alpha} (Hp - \eta_\alpha(p) + d), \tag{94}$$

where η_α is the anisotropic scaling function of f given in (23) and p_α satisfies

$$\eta_\alpha(p_\alpha) = d \, . \tag{95}$$

Consequently, we obtain the following result.

Theorem 6. *Let $\mathcal{B}$ and Ω be as in Theorem 1. If f is uniform Hölder on $\mathbb{R}^d$, then*

$$D(H,e) \leq \inf_{\alpha\in\Omega} \frac{1}{\alpha_{min}} \inf_{p\geq p_\alpha} (\alpha_1 Hp - \eta_\alpha(p) + d) \, . \tag{96}$$

Proof. By (91)

$$D(H,e) \leq \inf_{\alpha \in \Omega} \frac{1}{\alpha_{min}} D_\alpha(\alpha_1 H)\,.$$

Thus (94) yields (96). □

Remark 6. *Let us come back to the optimality of (91). For that, we will consider a general anisotropic Sierpinski carpet; let s and t be two integers with $s \leq t$. We divide the unit square $\mathfrak{R} = [0,1]^2$ into a uniform grid of rectangles of height $1/t$ and width $1/s$. Choose $A \subset \{0,1,\ldots,s-1\} \times \{0,1,\ldots,t-1\}$. For $\omega = (u,v) \in A$, the contraction $S_\omega(x_1,x_2) = \left(\frac{x_1}{s} + \frac{u}{s}, \frac{x_2}{t} + \frac{v}{t}\right)$ maps the unit square $\mathfrak{R}$ into the rectangle:*

$$\mathfrak{R}_\omega = [\frac{u}{s}, \frac{u+1}{s}] \times [\frac{v}{t}, \frac{v+1}{t}]. \tag{97}$$

The (general) Sierpinski carpet K (see [17,19,29]) and references therein) is the unique non-empty compact set (see [16]) satisfying

$$K = \bigcup_{\omega \in A} S_\omega(K). \tag{98}$$

It is given by:

$$\begin{aligned} K &= \{x \in \mathfrak{R} \;:\; (S_{\omega_1} \circ \cdots \circ S_{\omega_n})^{-1}(x) \in \bigcup_{\omega \in A} \mathfrak{R}_\omega \quad \forall \omega = (\omega_1, \ldots, \omega_n) \in A^n\} \\ &= \bigcap_{n \in \mathbb{N}} (\bigcup_{\omega \in A^n} \mathfrak{R}_\omega) \end{aligned}$$

where

$$\mathfrak{R}_\omega = (S_{\omega_1} \circ \cdots \circ S_{\omega_n})(\mathfrak{R}) \quad for \quad \omega = (\omega_1, \ldots, \omega_n)\,.$$

Let $\sigma = \dfrac{\log s}{\log t}$ and

$$(\alpha_1, \alpha_2) = (\frac{2\sigma}{1+\sigma}, \frac{2}{1+\sigma}) = (\frac{2\log s}{\log(st)}, \frac{2\log t}{\log s})\,. \tag{99}$$

By arguments similar to those of [15] pages 118–119, we can prove that $dim_{(\alpha_1,\alpha_2)}(K) = 2\dfrac{\log a}{\log(st)}$ where a is the cardinality of A, whereas $dim(K) = \dfrac{\log(\sum_{i=1}^{s} N_i^\sigma)}{\log s}$ where N_i is the number of selected rectangles in A from the i-th column of the grid (see [15] page 129).

If $s \leq t$ then $(\alpha_{min}, \alpha_{max}) = (\alpha_1, \alpha_2)$. By restricting ourselves to the two cases below, we will show that the optimality of (91) may depend on the geometric arrangement of the chosen ω's in A. Actually, the left-right side of (91) is optimal in case1 and non optimal in case2 if $s < t$.

1. *Case1: assume that each column of the grid contains at most one $\mathfrak{R}_\omega$, $\omega \in A$. Then $dimK = \dfrac{\log a}{\log s}$. Therefore, $\alpha_1 dimK = dim_{(\alpha_1,\alpha_2)}K$.*
2. *Case2: assume that there is only one column containing all the $\mathfrak{R}_\omega, \omega \in A$. Then $dimK = \dfrac{\sigma \log a}{\log s}$. Therefore, $\dfrac{\alpha_2}{2} dimK = dim_{(\alpha_1,\alpha_2)}K$.*

5. Fractional Brownian Sheets

We will apply Theorem 4 for fractional Brownian sheets to show that unlike the Lipschitz scaling function $\eta_L(p)$ and the Lipschitz spectum $d(H)$ which are uniform in all directions, the directional scaling function $\eta_L(p,e)$ and the directional spectrum $d(H,e)$ are tools that detect directional behaviors. We also provide a directional thermodynamic formalism valid for all fractional Brownian sheets.

Actually, we will prove that if the corresponding appropriate range P of p's over which one will computes the Legendre transform is given by (9), then $\inf_{p\in P}(Hp - \eta_L(p,e) + 1)$ provides a common directional Lipschitz scaling based directional thermodynamic formalism.

5.1. Computation of the Directional Scaling Function

The fractional Brownian sheet $\{B^{(H_1,\cdots,H_d)}(y) \ : \ y = (y_1,\cdots,y_d) \in \mathbb{R}^d\}$ was introduced by Kamont in [67], then redefined by Ayache, Léger, and Pontier in [78] through its harmonizable representation, for any $(H_1,\cdots,H_d) \in (0,1)^d$

$$B^{(H_1,\cdots,H_d)}(y) = \int_{\mathbb{R}^d} \prod_{i\in\mathcal{D}} (e^{iy_i\xi_i} - 1)|\xi_i|^{-H_i-\frac{1}{2}} \ d\widehat{W}_{(\xi_1,\cdots,\xi_d)}, \tag{100}$$

where $\widehat{W}_{(\xi_1,\cdots,\xi_d)}$ is the Fourier transform of a Brownian measure $W_{(\xi_1,\cdots,\xi_d)}$ on $\mathbb{R}^d$.

Fractional Brownian Sheet has stationary rectangular increments and satisfies the following anisotropic scaling relation

$$\{B^{(H_1,\cdots,H_d)}(a_1y_1,\cdots,a_dy_d)\}_{y\in\mathbb{R}^d} = \{(\prod_{i\in\mathcal{D}} a_i^{H_i}) \ B^{(H_1,\cdots,H_d)}(y)\}_{y\in\mathbb{R}^d} \quad \text{(same law)}\,. \tag{101}$$

In [67], Kamont proved that, if $\frac{1}{p} < H_i < 1$ for all $i \in \mathcal{D}$, then with Probability 1, the restrictions $B_{I^d}^{(H_1,\cdots,H_d)}$ of realizations of $B^{(H_1,\cdots,H_d)}$ to I^d satisfy

$$B_{I^d}^{(H_1,\cdots,H_d)} \in Lip_p^{(H_1,\cdots,H_d),\frac{1}{2}}(I^d), \tag{102}$$

and

$$B_{I^d}^{(H_1,\cdots,H_d)} \notin lip_p^{(H_1,\cdots,H_d),\frac{1}{2}}(I^d)\,. \tag{103}$$

Put

$$H_{min} = min(H_1,\cdots,H_d)\,. \tag{104}$$

We will prove the following result.

Theorem 7. *If* $\frac{1}{p} < H_i < 1$ *for all* $i \in \mathcal{D}$*, then with Probability 1,* $B_{I^d}^{(H_1,\cdots,H_d)}$ *satisfy*

$$\forall\, i \in \mathcal{D} \qquad \eta_L(p,e_i) = pH_i, \tag{105}$$

and

$$\eta_L(p) = pH_{\min}\,. \tag{106}$$

Proof. Using the third point in Proposition 6, relation (102) implies that, with Probability 1

$$B_{I^d}^{(H_1,\cdots,H_d)} \in Lip_p^{(s_1,\cdots,s_d)}(I^d) \qquad \forall (s_1,\cdots,s_d) < (H_1,\cdots,H_d)\,. \tag{107}$$

Using the second point in Proposition 6, relation (103) implies that, with Probability 1

$$B_{I^d}^{(H_1,\cdots,H_d)} \notin Lip_p^{(s_1,\cdots,s_d)}(I^d) \qquad \forall (s_1,\cdots,s_d) > (H_1,\cdots,H_d)\,. \tag{108}$$

Thanks to the first point in Proposition 5, relation (107) yields the lower bound in (106).

The optimality of this lower bound cannot be deduced from (108). Nevertheless, the coefficients of $B_{I^d}^{(H_1,\cdots,H_d)}$ in the tensor product Schauder basis were obtained in [67]; in fact

$$B_{I^d}^{(H_1,\cdots,H_d)} = \sum_{\mathbf{j}\in M^d} \sum_{\mathbf{k}\in\tilde{N}_{\mathbf{j}}} C_{\mathbf{k}}\Phi_{\mathbf{k}}, \tag{109}$$

where $(C_{\mathbf{k}})_{\mathbf{k}\geq\mathbf{0}}$ is a Gaussian sequence, with $EC_{\mathbf{k}} = 0$, and the variance given by the formula

$$E|C_{\mathbf{k}}|^2 = \prod_{i\in\mathcal{D}} a_{k_i} \tag{110}$$

where

$$a_0 = 0,\ a_1 = 1 \text{ and } a_{k_i} = (2^{-2H_i} - 2^{-2})2^{-2j_iH_i} \ \text{ for } \ k_i \in \tilde{N}_{j_i}\ j_i \geq 0\,. \tag{111}$$

The above optimality follows immediately from Theorem 4 and arguments similar to those in [92] (p. 236), since both (102), the third result in Proposition 6 and the first result in Proposition 5 imply that $\eta_L(p) \geq pH_{min} > 0$. The latest lower bound for $\eta_L(p)$ turned out to be equality thanks to (56). □

Remark 7. *If the unit cube I^d is replaced by any arbitrary cube $Q \subset \mathbb{R}^d$ then the same arguments applied to the dilated and shifted field $\{\rho^{|(H_1,\cdots,H_d)|}B^{(H_1,\cdots,H_d)}(\rho^{-1}t - c) : t \in I^d\}$, $(\rho > 0, c \in \mathbb{R}^d)$ give the same result as in Theorem 7.*

5.2. Lipschitz and Directional Spectra and Thermodynamic Formalisms

We will now compute both Lipschitz spectrum $d(H)$ and directional Lipschitz spectra $d(H,e)$ for $B_{I^d}^{(H_1,\cdots,H_d)}$. Let us first recall these notions.

Definition 9. *Let f be a continuous function on I^d (we write $f \in C(I^d)$). Let $y \in I^d$. Let $0 < H < 1$. We say that $f \in C^H(y)$ if there exists $C > 0$, such that:*

$$|f(y+t) - f(y)| \leq C|t|^H \quad \forall\, y+t \in I^d\,. \tag{112}$$

The pointwise Lipschitz regularity of f at y is:

$$h(y) = \sup\{0 < H < 1: \quad f \in C^H(y)\}\,. \tag{113}$$

Define the Lipschitz spectrum (respective to upper Lipschitz spectrum) of f as the function $d(H)$ (respective to $D(H)$) given by the Hausdorff dimension of the set of points y where $h(y) = H$ (respective to $h(y) \leq H$).
We say that $f \in C^H(I^d)$, if there exists $C > 0$ such that

$$|f(y+t) - f(y)| \leq C|t|^H \quad \forall\, (y, y+t) \in (I^d)^2\,. \tag{114}$$

Definition 10. *Let $0 < H < 1$ and $f \in C(I^d)$. Let $e \in S^{d-1}$. Let $y \in I^d$.*
We say that $f \in C^H(y,e)$ if there exists $C > 0$ such that

$$|f(y+te) - f(y)| \leq C|t|^H \quad \forall\, y+te \in I^d\,. \tag{115}$$

The directional pointwise Lipschitz regularity of f at y in direction e is

$$h(y,e) = \sup\{0 < H < 1: \quad f \in C^H(y,e)\}\,. \tag{116}$$

Define the directional Lipschitz spectrum (respective to directional upper Lipschitz spectrum) of f in direction e as the function $d(H,e)$ (respective to $D(H,e)$) given by the Hausdorff dimension of the set of points y where $h(y,e) = H$ (respective to $h(y,e) \leq H$).

The following theorem provides a common directional thermodynamic formalism for all fractional Brownian sheets.

Theorem 8. *With probability 1, both Lipschitz and upper Lipschitz spectra of the restrictions $B_{I^d}^{(H_1,\cdots,H_d)}$ of realizations of $B^{(H_1,\cdots,H_d)}$ are trivial and satisfy the following thermodynamic formalism*

$$d(H) = \begin{cases} -\infty & \text{if } H \neq H_{min} \\ d & \text{if } H = H_{min} \end{cases} = \inf_{p>1/H_{min}} (Hp - \eta_L(p) + d) \quad \forall\, H \leq H_{min}\,, \tag{117}$$

$$D(H) = \begin{cases} -\infty & \text{if } H < H_{min} \\ d & \text{if } H \geq H_{min} \end{cases} = \inf_{p>1/H_{min}} (Hp - \eta_L(p) + d) \quad \forall\, H \leq 1\,. \tag{118}$$

With probability 1, both directional Lipschitz and directional upper Lipschitz spectra of the restrictions $B_{I^d}^{(H_1,\cdots,H_d)}$ of realizations of $B^{(H_1,\cdots,H_d)}$ are trivial and satisfy the following directional thermodynamic formalism

$$d(H,e_i) = \begin{cases} -\infty & \text{if } H \neq H_i \\ d & \text{if } H = H_i \end{cases} = \inf_{p>1/H_{min}} (Hp - \eta_L(p,e_i) + d) \quad \forall\, H \leq H_i, \tag{119}$$

and

$$D(H,e_i) = \begin{cases} -\infty & \text{if } H < H_i \\ d & \text{if } H \geq H_i \end{cases} = \inf_{p>1/H_{min}} (Hp - \eta_L(p,e_i) + d) \quad \forall\, H \leq 1\,. \tag{120}$$

Moreover, Remark 4 holds.

Proof. In [67], we have

$$B_{I^d}^{(H_1,\cdots,H_d)} \in C^{H_{min}}(I^d)\,. \tag{121}$$

Let $y = (y_1, \cdots, y_d) \in I^d$. The unidimensional process $X(x_1) = B_{I^d}^{(H_1,\cdots,H_d)}(x_1, y_2, \cdots, y_d)$ is Gaussian, self-similar, with stationary increments and has H_1 as Hurst index. From the uniqueness of the fractional Brownian motion with Hurst index H_1, we deduce that $h(y,e_1) = H_1$. Similarly, we get $h(y,e_i) = H_i$ for all $i \in \mathcal{D}$. Using (121), we deduce that $h(y) = H_{min}$. The rest of the proof is straightforward. □

6. Sierpinski Cascade Functions

We will apply Theorem 4 for Sierpinski cascade functions to show that unlike the Lipschitz scaling function $\eta_L(p)$ and the Lipschitz spectum $d(H)$ which are uniform in all directions, the directional scaling function $\eta_L(p,e)$ and the directional spectrum $d(H,e)$ are tools that detect directional behaviors. We also show that contrary to $\eta_L(p,e)$, the directional spectrum $d(H,e)$ depends on the geometric disposition of the chosen contractions for each cascade function. We also provide non common directional Lipschitz scaling based directional thermodynamic formalisms for these examples. These formalisms depend on the geometric disposition of contractions for each cascade function. Nevertheless, all obtained formalisms share the same corresponding appropriate range P of p's over which one will compute the Legendre transform given in (9). Moreover, we show the optimality of Theorem 6 for Sierpinski cascade functions corresponding to a large class of geometric disposition of contractions corresponding to case 1 described in Remark 6. Finally, we modify the notion of the Hausdorff dimension to provide a new common directional Lipschitz scaling based directional thermodynamic formalism for all Sierpinski cascade functions.

Without any loss of generality, we take $d = 2$. A Sierpinski cascade function is a self-similar function adapted to the subdivision A used for the construction of Sierpinski carpet K given in (98). It is written as the superposition of similar anisotropic structures at different scales, reminiscent of some possible modelization of turbulence or cascade models. In [29], we proved that some Sierpinski cascade functions do not satisfy the thermodynamic formalism (5). Put $g(x) = \Lambda(x_1)\Lambda(x_2)$ with $\Lambda(t) = min(t, 1-t)$ if $t \in [0,1]$ and 0 else. Clearly, $\Lambda(t) = \frac{1}{2}\Phi_2(t)$.

The Sierpinski cascade function adapted to the subdivision A satisfies

$$\forall\, x \in \mathfrak{R} \qquad F(x) = \sum_{\omega \in A} \lambda_\omega F(S_\omega^{-1}(x)) \;+\; g(x). \tag{122}$$

Define

$$|\lambda|_{max} = \max_{\omega \in A} |\lambda_\omega| \;,\; |\lambda|_{min} = \min_{\omega \in A} |\lambda_\omega| ,\; H_{min} = -\frac{\log|\lambda|_{max}}{\log t} \text{ and } H_{max} = -\frac{\log|\lambda|_{min}}{\log t} \;.$$

The following result was obtained in [29].

Proposition 11. *Suppose that* $\sum_{\omega \in A} |\lambda_\omega| < st$, *then the series:*

$$F(x) = g(x) + \sum_{n=1}^{\infty} \sum_{(\omega_1,\dots,\omega_n) \in A^n} \lambda_{\omega_1} \cdots \lambda_{\omega_n}\, g\left(S_{\omega_n}^{-1} \cdots S_{\omega_1}^{-1}(x)\right). \tag{123}$$

is a unique solution in $L^1(\mathfrak{R})$ *for Equation (122).*

If, furthermore, $\frac{1}{t} < |\lambda|_{max} < 1$, *then* $F \in C^{H_{min}}(\mathfrak{R})$ *with* $0 < H_{min} < 1$.

Clearly, if $\omega_l = (u_l, v_l)$ then

$$g\left(S_{\omega_n}^{-1} \cdots S_{\omega_1}^{-1}(x)\right) = \Lambda(s^n x_1 - s^{n-1}u_1 - \cdots - s u_{n-1} - u_n)\; \Lambda(t^n x_2 - t^{n-1}v_1 - \cdots - t v_{n-1} - v_n)\;.$$

In [29], we proved that unlike the spectrum $d(H)$, the Lipschitz scaling function $\eta_L(p)$ (given in (55)) does not depend on the geometrical arrangement of the chosen $\mathfrak{R}_\omega$, $\omega \in A$, and, so, the multifractal formalism $d(H) = \inf_p(Hp - \eta_L(p) + 2)$ may fail.

6.1. Computation of the Directional Lipschitz Scaling Function

Using Theorem 4 and Remark 2, we obtain the following result which shows that, unlike the Lipschitz scaling function $\eta_L(p)$ which is uniform in all directions, the directional scaling function $\eta_L(p, e)$ is a tool to detect directional behaviors.

Theorem 9. *Let* S *and* T *be two positive integers. Assume that* $s = 2^S$ *and* $t = 2^T$ *and* $s \le t$. *Assume that* $\frac{1}{t} < |\lambda|_{max} < 1$.

Let $1 \le p < \infty$. *Set* $\sigma = S/T$ *and* $\tau(p) = -\frac{\log_2(\sum_{\omega \in A} |\lambda_\omega|^p)}{S}$. *Let* F *be the Sierpinski cascade function that corresponds to* A.

- *Suppose that* $s < t$ *(i.e.,* $\sigma < 1$*).*

- *We have* $\frac{t}{s^{p-1}} < \sum_{\omega\in A} |\lambda_\omega|^p < s$ *is equivalent to* $(1 < \sigma+1+\sigma\tau(p) < p$ *and* $1 < 1+\frac{1}{\sigma}+\tau(p) < p)$. *In that case,*

$$\eta_L(p,e_1) = 1+\frac{1}{\sigma}+\tau(p)\ ,\ \eta_L(p,e_2) = \sigma+1+\sigma\tau(p)\ .$$

and

$$\forall\, e \neq \pm e_1 \qquad \eta_L(p,e) = \sigma+1+\sigma\tau(p) = \eta_L(p)\ .$$

- *We have* $\frac{s}{t^{p-1}} < \sum_{\omega\in A} |\lambda_\omega|^p < s$ *and* $\sum_{\omega\in A} |\lambda_\omega|^p \leq \frac{t}{s^{p-1}}$ *is equivalent to* $(1 < \sigma+1+\sigma\tau(p) < p$ *and* $1+\frac{1}{\sigma}+\tau(p) \geq p)$. *In that case*

$$\eta_L(p,e_1) = p\ ,\ \eta_L(p,e_2) = \sigma+1+\sigma\tau(p)$$

and

$$\forall\, e \neq \pm e_1 \qquad \eta_L(p,e) = \sigma+1+\sigma\tau(p) = \eta_L(p)\ .$$

- *In the case* $\sum_{\omega\in A} |\lambda_\omega|^p \leq \frac{s}{t^{p-1}}$, *we have* $\eta_L(p,e_1) = \eta_L(p,e_2) = p$.

- *Suppose that* $s = t$ *(i.e.,* $\sigma = 1$*).*
  - *We have* $\frac{s}{s^{p-1}} < \sum_{\omega\in A} |\lambda_\omega|^p < s$ *is equivalent to* $1 < 2+\tau(p) < p$. *In that case,*

$$\eta_L(p,e_1) = 2+\tau(p)\ \text{ and }\ \eta_L(p,e_2) = 2+\tau(p)\ .$$

  - *In the case where* $\sum_{\omega\in A} |\lambda_\omega|^p \leq \frac{s}{s^{p-1}}$, *we have* $2+\tau(p) \geq p$, *therefore,*

$$\eta_L(p,e_1) = \eta_L(p,e_2) = p\ .$$

Proof. Of course, if $\frac{1}{t} < |\lambda|_{max} < 1$, then by Proposition 11 $F \in C^{H_{min}}(\mathfrak{R})$ with $0 < H_{min} < 1$ and, so, $\eta_L(p) > 0$.

For $\mathbf{j} = (j_1, j_2) = (nS, nT)$, we have

$$\tau_{\mathbf{j},p}(f) = 2^{-n(S+T)/p}\left(\sum_{\omega\in A} |\lambda_\omega|^p\right)^{n/p}\ .$$

It follows that

$$\frac{\log \tau_{\mathbf{j},p}(f)}{\log(2^{-j_1})} = \frac{1}{p}\left(1+\frac{T}{S}-\frac{1}{S}\log_2(\sum_{\omega\in A}|\lambda_\omega|^p)\right) = \frac{1}{p}\left(1+\frac{1}{\sigma}+\tau(p)\right).$$

and

$$\frac{\log \tau_{\mathbf{j},p}(f)}{\log(2^{-j_2})} = \sigma\ \frac{\log \tau_{\mathbf{j},p}(f)}{\log(2^{-j_1})} = \frac{1}{p}\left(\sigma+1+\sigma\tau(p)\right)\ .$$

In order to apply Theorem 4, we need that $\liminf_{|\mathbf{j}|\to\infty} \frac{\log \tau_{\mathbf{j},p}(f)}{\log(2^{-j_i})} > 1/p$ for every $i = 1,2$. Clearly,

$$\liminf_{|\mathbf{j}|\to\infty} \frac{\log \tau_{\mathbf{j},p}(f)}{\log(2^{-j_2})} > 1/p \Leftrightarrow \sigma+1+\sigma\tau(p) > 1 \Leftrightarrow \tau(p) > -1 \Leftrightarrow \sum_{\omega\in A}|\lambda_\omega|^p < s\ ,$$

$$\liminf_{|\mathbf{j}|\to\infty} \frac{\log \tau_{\mathbf{j},p}(f)}{\log(2^{-j_2})} < 1 \Leftrightarrow \sigma + 1 + \sigma\tau(p) < p \Leftrightarrow \frac{s}{t^{p-1}} < \sum_{\omega\in A} |\lambda_\omega|^p ,$$

$$\liminf_{|\mathbf{j}|\to\infty} \frac{\log \tau_{\mathbf{j},p}(f)}{\log(2^{-j_1})} > 1/p \Leftrightarrow 1 + \frac{1}{\sigma} + \tau(p) > 1 \Leftrightarrow \tau(p) > -1/\sigma \Leftrightarrow \sum_{\omega\in A} |\lambda_\omega|^p < t$$

and

$$\liminf_{|\mathbf{j}|\to\infty} \frac{\log \tau_{\mathbf{j},p}(f)}{\log(2^{-j_1})} < 1 \Leftrightarrow 1 + \frac{1}{\sigma} + \tau(p) < p \Leftrightarrow \frac{t}{s^{p-1}} < \sum_{\omega\in A} |\lambda_\omega|^p .$$

Clearly, since $s \le t$ then $\frac{s}{t^{p-1}} \le \frac{t}{s^{p-1}}$.

- Suppose that $s < t$ (i.e., $\sigma < 1$).
  - If $\frac{t}{s^{p-1}} < \sum_{\omega\in A} |\lambda_\omega|^p < s$ then $p > 1/\sigma$, $\eta_L(p,e_1) = 1 + \frac{1}{\sigma} + \tau(p)$ and $\eta_L(p,e_2) = \sigma + 1 + \sigma\tau(p)$. So, using Remark 2, we deduce that $\eta_L(p,e) = \sigma + 1 + \sigma\tau(p) = \eta_L(p)$ for all $e \neq \pm e_1$.
  - If $\frac{s}{t^{p-1}} < \sum_{\omega\in A} |\lambda_\omega|^p < s$ and $\sum_{\omega\in A} |\lambda_\omega|^p \le \frac{t}{s^{p-1}}$ then $\eta_L(p,e_1) = p$ and $\eta_L(p,e_2) = \sigma + 1 + \sigma\tau(p)$. So, using Remark 2, we deduce that $\eta_L(p,e) = \sigma + 1 + \sigma\tau(p) = \eta_L(p)$ for all $e \neq \pm e_1$. Note that $s \le \frac{t}{s^{p-1}}$ iff $p \le 1/\sigma$.
  - If $\sum_{\omega\in A} |\lambda_\omega|^p \le \frac{s}{t^{p-1}}$ then $\eta_L(p,e_1) = \eta_L(p,e_2) = p$.
- Suppose that $s = t$.
  - We have $\frac{s}{s^{p-1}} < \sum_{\omega\in A} |\lambda_\omega|^p < s$ is equivalent to $1 < 2 + \tau(p) < p$. In that case

    $$\eta_L(p,e_1) = 2 + \tau(p) \quad \text{and} \quad \eta_L(p,e_2) = 2 + \tau(p) .$$

  - In the case where $\sum_{\omega\in A} |\lambda_\omega|^p \le \frac{s}{s^{p-1}}$, we have $2 + \tau(p) \ge p$, therefore,

    $$\eta_L(p,e_1) = \eta_L(p,e_2) = p .$$

□

Remark 8. *Theorem 9 improves previous results in [29] without any assumptions on the choice of $\omega \in A$ and the positivity of the corresponding λ_ω. Recall that in [29], we were interested in the computation of $\eta_L(p)$ by the increments method.*

6.2. Directional Pointwise Lipschitz Regularity

We will now compute the pointwise directional Lipschitz regularity of the Sierpinski function. In [29], we were interested in the computation of the pointwise Lipschitz regularity. Let us recall the obtained results. Consider the "separated open set condition":

$$\forall\, (\omega,\omega') \in A^2 \quad \omega \neq \omega' \Rightarrow \mathfrak{R}_\omega \cap \mathfrak{R}_{\omega'} = \emptyset . \tag{124}$$

Recall that K is the Sierpinski carpet (98).

$$\forall\, x \notin K \qquad h(x) = 1 . \tag{125}$$

Define for $x \in K$, $\omega = \omega(x) = (\omega_1, \omega_2, \ldots, \omega_n, \cdots) \in A^{\mathbb{N}}$ by $\omega_l = (u_l, v_l) \in A$ with $x = \left(\sum_{l=1}^{\infty} \frac{u_l}{s^l}, \sum_{l=1}^{\infty} \frac{v_l}{t^l}\right)$.

Denote by

$$\omega(n,x) = (\omega_1, \ldots, \omega_n) \ , \ \lambda_{\omega(n,x)} = \lambda_{\omega_1} \cdots \lambda_{\omega_n}$$

$$a_t(x) = \liminf_{n \mapsto \infty} \frac{\log |\lambda_{\omega(n,x)}|}{\log t^{-n}} \quad \text{and} \quad a_s(x) = \liminf_{n \mapsto \infty} \frac{\log |\lambda_{\omega(n,x)}|}{\log s^{-n}} \ .$$

In Proposition 3 in [29], using increments method for the Sierpinski cascade function, we proved that $h(x) \geq a_t(x)$ under assumptions (124), $a_t(x) < 1$ and

$$\forall \omega \in A \, , \quad \mathfrak{R}_\omega \subset [1/s, 1-1/s] \times [1/t, 1-1/t] \ . \tag{126}$$

This yields

$$h(x, e_2) \geq a_t(x) \ . \tag{127}$$

Similar arguments allow us to obtain

$$h(x, e_1) \geq a_s(x) \quad \text{if} \quad a_s(x) < 1 \ . \tag{128}$$

In Proposition 4 in [29], using increments method, we proved that $h(x) \leq a_t(x)$ under assumptions (124), $a_t(x) \leq 1$

$$0 < \lambda_\omega < 1 \quad \forall \, \omega \in A \, , \tag{129}$$

and either:

$$\forall \omega \in A \, , \quad \mathfrak{R}_\omega \subset [1/s, 1-1/s] \times [1/t, 1/2], \tag{130}$$

or

$$\forall \omega \in A \, , \quad \mathfrak{R}_\omega \subset [1/s, 1-1/s] \times [1/2, 1-1/t] \ . \tag{131}$$

Actually, we proved that $h(x, e_2) \leq a_t(x)$. We deduce that:

$$h(x) = h(x, e_2) = a_t(x) \ . \tag{132}$$

We will improve result (132) and obtain a similar result for $h(x, e_1)$ without adding assumption (130) nor (131). For that we will use Theorem 4 obtained in [58], in which we characterized directional pointwise Lipschitz regularity in terms of decay conditions for the coefficients $C_{\mathbf{k}}(f)$ (given in (76)) of the expansion of f in the basis of tensor products of Schauder functions. Let us recall this result for $d = 2$; if $k_i \in \tilde{N}_{j_i}$, with $k_i \geq 2$ then ϕ_{k_i} has support $[(n_i - 1)2^{-j_i}, n_i 2^{-j_i}]$. It follows that for $j_i \in M$ with $j_i \geq 0$ and $x \in I^2$, there exists a unique value of $k_i(x_i)$ for which $x_i \in [(n_i - 1)2^{-j_i}, n_i 2^{-j_i})$. We keep the notation $k_i(x))$ even if $j_i \in \{-2, -1\}$.

Proposition 12. *Let $x \in I^2$ and $f \in C(I^2)$. Set*

$$\rho(x, e_1) = \liminf_{j_1 \to \infty} \inf_{k_1 \in \tilde{N}_{j_1}} \inf_{j_2 \in M} \frac{\log \left(|C_{(k_1, k_2(x_2))}(f)| \phi_{k_2(x_2)}(x_2) \right)}{\log \left(2^{-j_1} + |n_1 2^{-j_1} - x_1| \right)}$$

and

$$\rho(x, e_2) = \liminf_{j_2 \to \infty} \inf_{k_2 \in \tilde{N}_{j_2}} \inf_{j_1 \in M} \frac{\log \left(|C_{(k_1(x_1), k_2)}(f)| \phi_{k_1(x_1)}(x_1) \right)}{\log \left(2^{-j_2} + |n_2 2^{-j_2} - x_2| \right)}$$

Assume that f is uniformly Lipschitz regular on I^2 in direction e_1 in the sense that there exists $\delta > 0$ and $C > 0$, such that $|f(x + te_1) - f(x)| \leq C|t|^\delta$ for all x and $x + te_1 \in I^2$.

If:

$$\forall \, \mathbf{k} \quad C_{(k_1, k_2(x_2))}(f) \geq 0, \tag{133}$$

then:

$$h(x,e_1) = \min(1, \rho(x,e_1)) \ . \quad (134)$$

Assume that f is uniformly Lipschitz regular on I^2 in direction e_2.

If:

$$\forall\, \mathbf{k} \quad C_{(k_1(x_1),k_2)}(f) \geq 0, \quad (135)$$

then:

$$h(x,e_2) = \min(1, \rho(x,e_2)) \ . \quad (136)$$

Theorem 10. *Assume $\lambda_{max} > 1/t$, (124), (126), (129) and (142). Then, for the Sierpinski cascade function:*

$$h(x,e_1) = \begin{cases} 1 & \text{if } x \notin K \\ a_s(x) & \text{if } x \in K \end{cases} \quad (137)$$

and

$$h(x,e_2) = h(x) = \begin{cases} 1 & \text{if } x \notin K \\ a_t(x) & \text{if } x \in K\,. \end{cases} \quad (138)$$

Proof. Results for $x \notin K$ follow from (125).

Assumption (129) yields both (133) and (135). On the other hand, by assumption $\lambda_{max} > 1/t$, Proposition 11 implies that $F \in C^{H_{min}}(\mathfrak{R})$ with $0 < H_{min} < 1$. This implies that F is uniformly Lipschitz regular on I^2 in any direction. By Proposition 12, for $x \in K$ and $i = 1,2$:

$$h(x,e_i) = \min(1, \rho(x,e_i)) \ . \quad (139)$$

Of course,

$$\rho(x,e_1) \leq \liminf_{j_1 \to \infty} \inf_{k_1(x_1) \in \tilde{N}_{j_1}} \inf_{j_2 \in M} \frac{\log\left(C_{(k_1(x_1),k_2(x_2))}(f)\phi_{k_2(x_2)}(x_2)\right)}{\log\left(2^{-j_1} + |n_1 2^{-j_1} - x_1|\right)} \ .$$

Assumption (126) yields:

$$\rho(x,e_1) \leq a_s(x) \quad (140)$$

because from the definition of $(k_1(x_1), k_2(x_2))$, we have $|n_1 2^{-j_1} - x_1| \leq 2^{-j_1}$ and if $x_2 = \sum_{l=1}^{\infty} \frac{v_l}{t^l}$ then thanks to assumption (126):

$$\phi_{k_2(x_2)}(x_2) = 2\Lambda(t^n x_2 - t^{n-1}v_1 - \cdots - t v_{n-1} - v_n) = 2\Lambda(\sum_{l=1}^{\infty} \frac{v_{n+l}}{t^l}) \geq \frac{1}{t} \ .$$

Similarly, assumption (126) yields:

$$\rho(x,e_2) \leq a_t(x) \ . \quad (141)$$

Properties (139), (140), and (141) make equalities in results (127) and (128) (under assumptions (124) and (126)). □

6.3. Directional Pointwise Lipschitz Spectrum and Directional Thermodynamic Formalisms

We will now compute the directional Lipschitz spectrum of the Sierpinski function and provide directional thermodynamic formalisms. We will see that, unlike the directional Lipschitz scaling function $\eta_L(p,e)$, the directional spectrum $d(H,e)$ (and, so, the directional thermodynamic formalisms) may depend on the geometric arrangement of the chosen R_ω. Actually, in [29], we proved a similar property for the Lipschitz scaling function $\eta_L(p)$ and the Lipschitz spectrum $d(H)$. Nevertheless, we will show that unlike the Lipschitz spectrum $d(H)$ which is uniform in all directions, the directional Lipschitz spectrum $d(H,e)$ may depend on e and consequently is a tool to detect directional behaviors.

Assume that if column u of the grid contains points of K, then the two adjacent columns do not, i.e.,

$$\text{if} \quad \omega = (u, v) \in A \quad \text{then} \quad (u \pm 1, v) \notin A \, . \tag{142}$$

The following theorem holds. It provides directional thermodynamic formalisms valid for the Sierpinski function.

Theorem 11. *Assume $\lambda_{max} > 1/t$, (124), (126), (129), and (142). Let F be the corresponding Sierpinski cascade function.*

The set P given in (9) is:

$$P = \{p \geq 1 \; : \; \frac{t}{s^{p-1}} < \sum_{\omega \in A} |\lambda_\omega|^p < s\} \, . \tag{143}$$

Put:

$$\tau'(P) = \{\tau'(p) \; : \; p \in P\} \text{ and } \sigma\tau'(P) = \{\sigma\tau'(p) \; : \; p \in P\} \, . \tag{144}$$

1. *Case 1: Assume that each column of the grid contains at most one $\mathfrak{R}_\omega$, $\omega \in A$. Then:*

$$d(H, e_2) = d(H) = \begin{cases} -\infty & \text{if } H < H_{min} \\ \inf_q (q\sigma^{-1}H - \tau(q)) & \text{if } H \in [H_{min}, min(1, H_{max})] \end{cases} \tag{145}$$

and

$$d(H, e_1) = d(H\sigma, e_2) = \begin{cases} -\infty & \text{if } H\sigma < H_{min} \\ \inf_q (qH - \tau(q)) & \text{if } H\sigma \in [H_{min}, min(1, H_{max})] \, . \end{cases} \tag{146}$$

The following directional thermodynamic formalisms hold:

$$\forall H \in \tau'(P) \cap (-\infty, \frac{1}{\sigma}\min(1, H_{max})] \quad d(H, e_1) = \inf_{p \in P} (pH - \eta_L(p, e_1) + 1 + \frac{1}{\sigma}) \tag{147}$$

and

$$\forall H \in \sigma\tau'(P) \cap (-\infty, \min(1, H_{max})] \quad d(H, e_2) = \frac{1}{\sigma} \inf_{p \in P} (pH - \eta_L(p, e_2) + 1 + \sigma) \, . \tag{148}$$

2. *Case 2: Assume that there is only one column containing all the $\mathfrak{R}_\omega$, $\omega \in A$. Then:*

$$d(H, e_2) = d(H) = \begin{cases} -\infty & \text{if } H < H_{min} \\ \inf_q (qH - \sigma\tau(q)) & \text{if } H \in [H_{min}, min(1, H_{max})] \end{cases} \tag{149}$$

and

$$d(H, e_1) = d(H\sigma, e_2) = \begin{cases} -\infty & \text{if } H\sigma < H_{min} \\ \sigma \inf_q (qH - \tau(q)) & \text{if } H\sigma \in [H_{min}, min(1, H_{max})] \, . \end{cases} \tag{150}$$

The following directional thermodynamic formalisms hold:

$$\forall H \in \tau'(P) \cap (-\infty, \frac{1}{\sigma}\min(1, H_{max})] \quad d(H, e_1) = \sigma \inf_{p \in P} (pH - \eta_L(p, e_1) + 1 + \frac{1}{\sigma}) \tag{151}$$

and

$$\forall H \in \sigma\tau'(P) \cap (-\infty, \min(1, H_{max})] \quad d(H, e_2) = \inf_{p \in P} (pH - \eta_L(p, e_2) + 1 + \sigma) \, . \tag{152}$$

Proof. Relation (143) is direct consequence of Theorem 9.

Since $s \leq t$ then (132) yields $d(H) = d(H, e_2)$, therefore, results of Proposition 5 in [29] (respective to Proposition 6 in [29]) remain valid for $d(H)$ replaced by $d(H, e_2)$ with the same conditions except for (130) or (131) which can be replaced by the weaker condition (126) (see the previous section). Moreover, results (146) and (150) follow from the fact that $a_s(t) = \frac{1}{\sigma} a_t(x)$. The above thermodynamic formalisms follow directly from Theorem 9. □

Remark 9. *When $s = t$, the above thermodynamic formalisms coincide with the classical formalism.*

6.4. Optimality of Theorem 6 in Case 1

We will prove that Theorem 6 is optimal in case1, in the sense that the upper bound (96) becomes equality.

Theorem 12. *Let $\mathcal{B}$ and Ω be as in Theorem 1. Assume $\lambda_{max} > 1/t$, (124), (126), (129) and (142). Let F be the corresponding Sierpinski cascade function.*

Assume that each column of the grid contains at most one $\mathfrak{R}_\omega$. Then:

$$\forall H \leq 1 \quad D(H, e) = \inf_{\alpha \in \Omega} \frac{1}{\alpha_{min}} \inf_{p \geq p_\alpha} (\alpha_1 H p - \eta_\alpha(p) + d) \,. \tag{153}$$

Proof. If (α_1, α_2) is given by (99), then $\alpha_{min} = \alpha_1$, $p_{(\alpha_1, \alpha_2)}$ given in (95) satisfies $\tau(p_{(\alpha_1, \alpha_2)}) = 0$, and

$$\forall\, i \in \{1, 2\} \quad \eta_L(p, e_i) = \frac{\eta_{(\alpha_1, \alpha_2)}(p)}{\alpha_i} \,.$$

Relations (147) and (148) (in case 1), respectively, can be rewritten as:

$$\forall H \in \tau'(P) \cap (-\infty, \frac{1}{\sigma} \min(1, H_{max})] \quad d(H, e_1) = \frac{1}{\alpha_{min}} \inf_{p \in P} (\alpha_1 H p - \eta_{(\alpha_1, \alpha_2)}(p) + 2)$$

and

$$\forall H \in \sigma\tau'(P) \cap (-\infty, \min(1, H_{max})] \quad d(H, e_2) = \frac{1}{\alpha_{min}} \inf_{p \in P} (\alpha_2 H p - \eta_{(\alpha_1, \alpha_2)}(p) + 2) \,.$$

Since:

$$\{p \geq p_{(\alpha_1, \alpha_2)}\} \subset P,$$

then the previous upper bounds become equalities and (153) holds.

6.5. Directional Thermodynamic Formalisms Independent on the Choice of A

We will modify the notion of the Hausdorff dimension to provide a new directional thermodynamic formalism independent on the choice of A. For any $n \geq 1$, $\mathcal{T}_n := \{\mathfrak{R}_\omega \,; \omega \in A^n\}$ is a partition of K. Let $\mathcal{T} = \bigcup_{n \geq 1} \mathcal{T}_n$. Define $dim_{\mathcal{T}}$ in a similar way to the Hausdorff dimension but by considering only coverings by elements of $\mathcal{T}$. Note that such 'restriction' to the elements of dynamics was done by many authors (see [11–14,20,30]). Of course, the diameter $|\mathfrak{R}_\omega|$ for $\omega \in A^n$ can be replaced by s^{-n} (because it is equivalent to s^{-n}). Define the modified directional Lipschitz $\mathcal{T}$ spectrum (respective to directional upper Lipschitz $\mathcal{T}$ spectrum) of f in direction e as the function $d_{\mathcal{T}}(H, e)$ (respective to $D_{\mathcal{T}}(H, e)$) given by the $dim_{\mathcal{T}}$ of the set of points y where $h(y, e) = H$ (respective to $h(y, e) \leq H$).

Theorem 13. *Assume $\lambda_{max} > 1/t$, (124), (126), (129) and (142). Let F be the corresponding Sierpinski cascade function. Let P as in (143). Then the following directional thermodynamic formalisms hold:*

$$\forall\, H \in \tau'(P) \cap (-\infty, \min(1, H_{max})] \qquad d_{\mathcal{T}}(H, e_1) = \inf_{p \in P}(pH - \eta_L(p, e_1) + 1 + \frac{1}{\sigma}) \tag{154}$$

and

$$\forall\, H \in \sigma\tau'(P) \cap (-\infty, \min(1, H_{max})] \qquad d_{\mathcal{T}}(H, e_2) = \frac{1}{\sigma} \inf_{p \in P}(pH - \eta_L(p, e_2) + 1 + \sigma)\,. \tag{155}$$

Proof. Set $P_\omega(q) = \lambda_\omega^q s^{\tau(q)}$ and let μ_q be a probability measure on K such that:

$$\forall (\omega_1, \ldots, \omega_n) \in A^n \qquad \mu_q(\mathfrak{R}_{\omega_1,\ldots,\omega_n}) = P_{\omega_1}(q) \ldots P_{\omega_n}(q).$$

Since:

$$\mu_q(\mathfrak{R}_{\omega_1,\ldots,\omega_n,\omega'_1,\ldots,\omega'_m}) = \mu_q(\mathfrak{R}_{\omega_1,\ldots,\omega_n})\mu_q(\mathfrak{R}_{\omega'_1,\ldots,\omega'_m})\,,$$

then, as in [25,30], we can concentrate a Gibbs measure ν_p on $E_F^{\varphi'(p)}$, i.e.,

$$\forall\, \omega \qquad \nu_p(\mathfrak{R}_\omega) \simeq (\mu(\mathfrak{R}_\omega))^p |\mathfrak{R}_\omega|^{-\varphi(p)}$$

where φ is defined as in [13], and, thus, we obtain:

$$d_{\mathcal{T}}(H, e_1) = \inf_q (Hq - \tau(q))$$

and

$$d_{\mathcal{T}}(H, e_2) = \inf_q (\frac{H}{\sigma} q - \tau(q)) = \frac{1}{\sigma} \inf_q (Hq - \sigma\tau(q))\,.$$

Therefore, (154) and (155) hold. □

Remark 10. *The new formalism shows also that if $D(H,e)$ in Theorem 6 is replaced by $D_{\mathcal{T}}(H,e)$ then we get optimality independently on the choice of A.*

7. Motivation of the Anisotropic Cascade Model on the Physics Side

In all realistic flows in turbulence, there always exists some anisotropy at all scales (for example, see [93] and references therein); the statistical properties of the velocity field are effected by the geometry of the boundaries or the driving mechanism, which are never rotationally invariant [94]. For example, all geophysical flows are subject to the rotation of the globe, which introduces anisotropy via the Coriolis forces [95]. There is also a whole literature on anisotropy in turbulence created by vortex stretching (for example, see [96] and references therein). It has been a grand challenge in the mathematical fluid mechanics community to try to explain/quantify the process of anisotropic dissipation in turbulent flows directly from the mathematical model the 3D Navier–Stokes equations NSE. In [97], Constantin derived a singular integral representation of the stretching factor in the evolution of the vorticity magnitude featuring a geometric kernel that is depleted by local coherence of the vorticity direction. In [98], Ran showed that there are dynamical systems that are much simpler than the NSE but that can still have turbulent states and for which many concepts developed in the theory of dynamical systems can be successfully applied.

Clearly, cascade models introduced to model turbulence (the cascade picture of turbulent flows takes its origin from Richardson in 1922) should be able to take into account anisotropy. If we want to be able to make model selection with, we must use an anisotropic multifractal formalism as the equality in (94).

Author Contributions: All authors contributed to this article.

Funding: Mourad Ben Slimane and Borhen Halouani extend their appreciation to the Deanship of Scientific Research at King Saud University for funding this work through research group No (RG-1435-063).

Acknowledgments: Mourad Ben Slimane would like to thank Marianne Clausel, Stéphane Jaffard and Béatrice Vedel for stimulating discussions. We are very grateful to the referees for their comments and detailed suggestions which helped us to improve the manuscript.

Conflicts of Interest: The authors declare no conflict of interest.

References

1. Benzi, R.; Paladin, G.; Parisi, G.; Vulpani, A. On the multifractal nature of turbulence and chotic systems. *J. Phys. A* **1984**, *17*, 3521–3531. [CrossRef]
2. Mandelbrot, B. Intermittent turbulence in selfsimilar cascades: Divergence of high moments and dimension of the carrier. *J. Fluid Mech.* **1974**, *62*, 331–358. [CrossRef]
3. Eckmann, B.; Ruelle, D. Ergodic theory of chaos and strange attractors. *Rev. Mod. Phys.* **1985**, *57*, 617–656. [CrossRef]
4. Halsey, T.-C.; Jensen, M.-H.; Kadaroff, L.-P.; Procaccia, I.; Shraiman, B.-I. Fractal measures and their singularities: The charaterization of strange sets. *Phys. Rev. A* **1986**, *33*, 1141–1151. [CrossRef]
5. Kolmogorov, A.N. Energy dissipation in locally isotropic turbulence. *Dokl. Akad. Nauk.* **1941**, *32*, 16–18. [CrossRef]
6. Kolmogorov, A.N. A refinement of previous hypotheses concerning the local structure of turbulence in a viscous incompressible fluid at high Reynolds number. *J. Fluid Mech.* **1962**, *12*, 8285. [CrossRef]
7. Oboukhov, A.M. Some specific features of atmospheric tubulence. *J. Fluid Mech.* **1962**, *12*, 7781. [CrossRef]
8. Mandelbrot, B. *Les Objets Fractals: Forme, Hasard et Dimension*; Flammarian: Paris, France, 1975.
9. Mandelbrot, B. *The Fractal Geometry of Nature*; W. H. Freeman: New York, NY, USA, 1982.
10. Frisch, U.; Parisi, G. Fully developped turbulence and intermittency. In *Proceedings of the International Summer School in Physics*; Fermi, E., Ed.; North-Holland: Amsterdam, The Netherlands, 1985; pp. 84–88.
11. Ben Nasr, F. Multifractal analysis of measures. *C. R. Acad. Sci. Paris Sér. I Math.* **1994**, *319*, 807–810.
12. Ben Nasr, F.; Bhouri, I.; Heurtaux, Y. A necessary condition and sufficientn condition for a valid multifractal formalism. *Math. Adv. Math.* **2002**, *165*, 264–284. [CrossRef]
13. Brown, G.; Michon, G.; Peyrière, J. On the multifractal analysis of measures. *J. Statist. Phys.* **1992**, *66*, 775–790. [CrossRef]
14. Collet, P.; Lebowitz, J.; Porzio, A. The dimension specrum of some dynamical systems. *J. Statist. Phys.* **1987**, *47*, 609–644. [CrossRef]
15. Falconer, K.-J. *Fractal Geometry: Mathematical Foundations and Applications*; John Wiley and Sons: Toronto, ON, Canada, 1990.
16. Hutchinson, J. Fractals and self-similarity. *Indiana Univ. Math. J.* **1981**, *30*, 713–747. [CrossRef]
17. King, J. The singularity spectrum for general Sierpinski carpets. *Adv. Math.* **1995**, *116*, 1–11. [CrossRef]
18. Olsen, L. A multifractal formalism. *Adv. Math.* **1995**, *116*, 92–195. [CrossRef]
19. Olsen, L. Self-affine multifractal Sierpinski sponges in R^d. *Pac. J. Math.* **1998**, *183*, 143–199. [CrossRef]
20. Rand, D.A. The singularity spectrum $f(\alpha)$ for cookie-cutters. *Ergodic Theory Dyn. Syst.* **1989**, *9*, 527–541. [CrossRef]
21. Daubechies, I.; Lagarias, J.-C. On the thermodynamic formalism for functions. *Rev. Math. Phys.* **1994**, *6*, 1033–1070. [CrossRef]
22. Arneodo, A.; Bacry, E.; Muzy, J.-F. Singularity spectrum of fractal signals from wavelet analysis: Exact results. *J. Statist. Phys.* **1993**, *70*, 635–674.
23. Arneodo, A.; Bacry, E.; Muzy, J.-F. The thermodynamics of fractals revisited with wavelets. *Physica A* **1995**, *213*, 232–275. [CrossRef]
24. Muzy, J.-F.; Arneodo, A.; Bacry, E. A multifractal formalism revisited with wavelets. *Internat. J. Bifur. Chaos Appl. Sci. Engrg.* **1994**, *4*, 245. [CrossRef]
25. Jaffard, S. Multifractal formalism for functions. Part 1: Results valid for all functions and Part 2: Selfsimilar functions. *SIAM J. Math. Anal.* **1997**, *28*, 944–998. [CrossRef]
26. Ben Abid, M.; Seuret, S. Hölder regularity of μ-similar functions. *Const. Approx.* **2010**, *31*, 69–93. [CrossRef]

27. Ben Slimane, M. Etude du Formalisme Multifractal pour les Fonctions. Ph.D. Thesis, Ecole Nationale des Ponts et Chaussées, Paris, France, 1996.
28. Ben Slimane, M. Formalisme Multifractal pour quelques généralisations des fonctions autosimilaires. *C. R. Acad. Sci. Paris Sér. I Math.* **1997**, *324*, 981–986. [CrossRef]
29. Ben Slimane, M. Multifractal formalism and anisotropic selfsimilar functions. *Math. Proc. Camb. Philos. Soc.* **1998**, *124*, 329–363. [CrossRef]
30. Ben Slimane, M. Multifractal formalism for selfsimilar functions under the action of nonlinear dynamical systems. *Constr. Approx.* **1994**, *15*, 209–240. [CrossRef]
31. Ben Slimane, M. Multifractal formalism for selfsimilar functions expanded in singular basis. *Appl. Comput. Harmon. Anal.* **2001**, *11*, 387–419. [CrossRef]
32. Jaffard, S. The multifractal nature of the Lévy processes. *Probab. Theory Related Fields* **1999**, *114*, 207–227. [CrossRef]
33. Ben Slimane, M. Some functional equations revisited: The multifractal properties. *Integral Transf. Spec. Funct.* **2003**, *14*, 333–348. [CrossRef]
34. Jaffard, S. The spectrum of singularities of Riemann's function. *Rev. Math. Iberoam.* **1996**, *12*, 441–460. [CrossRef]
35. Jaffard, S. On the Frisch-Parisi conjecture. *J. Math. Pures Appl.* **2000**, *79*, 525–552. [CrossRef]
36. Fraysse, A. Generic validity of the multifractal formalism. *SIAM J. Math. Anal. Soc. Ind. Appl. Math.* **2007**, *39*, 593–607. [CrossRef]
37. Fraysse, A.; Jaffard, S. How smooth is almost every function in Sobolev space? *Rev. Math. Iberoam.* **2006**, *22*, 663–682. [CrossRef]
38. Kestener, P.; Arneodo, A. Generalizing the wavelet-based multifractal formalism to vector-valued random fields: Application to turbulent velocity and vorticity 3D numerical data. *Phys. Rev. Lett.* **2004**, *93*, 044501. [CrossRef] [PubMed]
39. Abry, P.; Clausel, M.; Jaffard, S.; Roux, S.G.; Vedel, B. Hyperbolic wavelet transform: An efficient tool for multifractal analysis of anisotropic textures. *Rev. Math. Iberoam.* **2015**, *31*, 313–348. [CrossRef]
40. Abry, P.; Roux, S.-G.; Wendt, H.; Messier, P.; Klein, A.-G.; Tremblay, N.; Borgnat, P.; Jaffard, S.; Vedel, B.; Coddington, J.; et al. Multiscale Anisotropic Texture Analysis and Classification of Photographic Prints: Art scholarship meets image processing algorithms. *IEEE Signal Proc. Mag.* **2015**, *32*, 18–27. [CrossRef]
41. Arneodo, A.; Audit, B.; Decoster, N.; Muzy, J.-F.; Vaillant, C. Wavelet-based multifractal formalism: Applications to DNA sequences, satellite images of the cloud structure and stock market data. In *The Science of Disasters*; Bunde, A., Kropp, J., Schellnhuber, H.J., Eds.; Springer: Berlin/Heidelberg, Germany, 2002; pp. 27–102.
42. Aubry, J.-M.; Maman, D.; Seuret, S. Local behavior of traces of Besov functions: Prevalent results. *J. Funct. Anal.* **2013**, *264*, 631–660. [CrossRef]
43. Ben Braiek, H.; Ben Slimane, M. Directional regularity criteria. *C. R. Acad. Sci. Paris Sér. I Math.* **2011**, *349*, 385–389. [CrossRef]
44. Ben Slimane, M.; Ben Braiek, H. Directional and anisotropic regularity and irregularity criteria in Triebel wavelet bases. *J. Fourier Anal. Appl.* **2012**, *18*, 893–914. [CrossRef]
45. Ben Slimane, M. Wavelet characterizations of multi-directional regularity. *Fractals* **2012**, *20*, 245–256. [CrossRef]
46. Clausel, M.; Vedel, B. Explicit constructions of operator scaling Gaussian fields. *Fractals* **2011**, *19*, 101–111. [CrossRef]
47. Davies, S.; Hall, P. Fractal analysis of surface roughness by using spatial data. *J. R. Stat. Soc. Ser. B Stat. Methodol.* **1999**, *61*, 3–37. [CrossRef]
48. Jaffard, S. Pointwise and directional regularity of nonharmonic Fourier series. *Appl. Comput. Harmon. Anal.* **2010**, *28*, 251–266. [CrossRef]
49. Ponson, L.; Bonamy, D.; Auradou, H.; Mourot, G.; Morel, S.; Bouchaud, E.; Guillot, C.; Hulin, J.P. Anisotropic self-affine properties of experimental fracture surfaces. *Int. J. Fracture* **2006**, *140*, 27–37. [CrossRef]
50. Roux, S.-G.; Clausel, M.; Vedel, B.; Jaffard, S.; Abry, P. Self-Similar Anisotropic Texture Analysis: The Hyperbolic Wavelet Transform Contribution. *IEEE Trans. Image Proc.* **2013**, *22*, 4353–4363. [CrossRef] [PubMed]

51. Sampo, J.; Sumetkijakan, S. Estimations of Hölder Regularities and Direction of Singularity by Hart Smith and Curvelet Transforms. *J. Fourier Anal. Appl.* **2009**, *15*, 58–79. [CrossRef]
52. Triebel, H. *Interpolation Theory, Function Spaces, Differential Operators;* North-Holland: Amsterdam, The Netherlands, 1978.
53. Aimar, H.; Gomez, I. Parabolic Besov regularity for the heat equation. *Constr. Approx.* **2012**, *36*, 145–159. [CrossRef]
54. Biermé, H.; Meerschaert, M.M.; Scheffler, H.-P. Operator scaling stable random fields. *Stoch. Proc. Appl.* **2007**, *117*, 312–332. [CrossRef]
55. Bonami, A.; Estrade, A. Anisotropic analysis of some Gaussian models. *J. Fourier Anal. Appl.* **2003**, *9*, 215–236. [CrossRef]
56. Khalil, A.; Joncas, G.; Nekka, F.; Kestener, P.; Arneodo, A. Morphological Analysis of H I Features. II. Wavelet-based multifractal formalism. *Astrophys. J. Suppl. Ser.* **2006**, *165*, 512–550. [CrossRef]
57. Ben Mabrouk, A. An adapted group dilation anisotropic multifractal formalism for functions. *J. Nonlinear Math. Phys.* **2008**, *15*, 1–23. [CrossRef]
58. Ben Slimane, M.; Ben Abid, M.; Ben Omrane, I.; Halouani, B. Criteria of pointwise and uniform directional Lipschitz regularities on tensor products of Schauder functions. *J. Math. Anal. Appl.* **2018**, *460*, 496–515. [CrossRef]
59. Ben Slimane, M.; Ben Braiek, H. Baire generic results for the anisotropic multifractal formalism. *Rev. Mater. Complut.* **2016**, *29*, 127–167. [CrossRef]
60. Bownik, M. Atomic and molecular decomposition of anisotropic Besov spaces. *Math. Z.* **2005**, *250*, 539–571. [CrossRef]
61. Bownik, M.; Ho, K.-P. Atomic and molecular decomposition of anisotropic Triebel- Lizorkin spaces. *Trans. Am. Math. Soc.* **2005**, *385*, 1469–1510.
62. Farkas, W. Atomic and subatomic decompositions in anisotropic function spaces. *Math. Nachr.* **2000**, *209*, 83–113. [CrossRef]
63. Führ, H. Vanishing moment conditions for wavelet atoms in higher dimensions. *Adv. Comput. Math.* **2016**, *42*, 127–153. [CrossRef]
64. Garrigós, G.; Tabacco, A. Wavelet decompositions of anisotropic Besov spaces. *Math. Nachr.* **2002**, *239*, 80–102. [CrossRef]
65. Garrigós, G.; Hochmuth, R.; Tabacco, A. Wavelet characterizations for anisotropic Besov spaces with $0 < p < 1$. *Proc. Edinb. Math. Soc.* **2004**, *47*, 573–595.
66. Hochmuth, R. Wavelet characterizations for anisotropic Besov spaces. *Appl. Comput. Harmon. Anal.* **2002**, *12*, 179–208. [CrossRef]
67. Kamont, A. On the fractional anisotropic Wiener field. *Probab. Math. Statist.* **1996**, *16*, 85–98.
68. Rosiene, C.-P.; Nguyen, T.-Q. Tensor-product wavelet vs. Mallat decomposition: A comparative analysis. In Proceedings of the 1999 IEEE International Symposium on Circuits and Systems VLSI (Cat. No.99CH36349), Orlando, FL, USA, 30 May–2 June 1999; Volume 3, p. 431434.
69. Triebel, H. *Theory of Function Spaces III*; Monographs in Mathematics, 78; Birkhäuser: Basel, Switzerland, 2006.
70. Triebel, H. Wavelet Bases in Anisotropic Function Spaces. *Funct. Space Differ. Oper. Nonlinear Anal.* **2004**, 370–387.
71. Berkolako, M.Z.; Novikov, I.-Y. Wavelet bases in spaces of differentiable functions of anisotropic smoothness. (Russian). *Dokl. Akad. Nauk.* **1992**, *324*, 615–618.
72. Berkolako, M.Z.; Novikov, I.-Y. Unconditional bases in spaces of functions of anisotropic smoothness. (Russian). *Trudy Mat. Inst. Steklov. Issled. Teor. Differ. Funktsii Mnogikh Peremen. Prilozh.* **1993**, *204*, 35–51.
73. DeVore, R.-A.; Konyagin, S.-V.; Temlyakov, V.-N. Hyperbolic wavelet approximation. *Constr. Approx.* **1998**, *14*, 1–26. [CrossRef]
74. Westerink, P.-H. Subband Coding of Images. Ph.D. Thesis, Delft University of Technology, Delft, The Netherlands, 1989.
75. Yu, T.-P.; Stoschek, A.; Donoho, D.-L. Translation and direction invariant denoising of 2D and 3D images: Experience and algorithms. *Proc. SPIE* **1996**, *2825*, 608619.
76. Zavadsky, V. Image approximation by rectangular wavelet transform. *J. Math. Imaging Vis.* **2007**, *27*, 129–138. [CrossRef]

77. Pesquet-Popesu, B.; Lévy-Véhel, J.; Stochastic Fractal Models for Image Processing. *IEEE Signal Proces. Mag.* **2002**, *19*, 48–62. [CrossRef]
78. Ayache, A.; Léger, S.; Pontier, M. Drap brownien fractionnaire. *Potential Anal.* **2002**, *17*, 31–43. [CrossRef]
79. Lakhonchai, P.; Sampo, J.; Sumetkijakan, S. Shearlet transforms and Hölder regularities. *Int. J. Wavelets Multiresolut. Inform. Proc.* **2010**, *8*, 743–771. [CrossRef]
80. Nualtong, K.; Sumetkijakan, S. Analysis of Hölder regularities by wavelet-like transforms with parabolic scaling. *Thai J. Math.* **2005**, *3*, 275–283.
81. Donoho, D. Wedgelets: Nearly minimax estimation of edges. *Ann. Stat.* **1999**, *27*, 353–382. [CrossRef]
82. Guo, K.; Labate, D. Analysis and detection of surface discontinuities using the 3D continuous shearlet transform. *Appl. Comp. Harm. Anal.* **2011**, *30*, 231–242. [CrossRef]
83. Candès, E.; Donoho, D. Ridgelets: A key to higher-dimensional intermittency? *Philos. Trans. R. Soc. Lond. Ser. A Math. Phys. Eng. Sci.* **1999**, *357*, 2495–2509. [CrossRef]
84. Mallat, S. Challenges for the 21st century. In *Applied Mathematics Meets Signal Processing, In Proceedings of the the International Conference on Fundamental Sciences: Mathematics and Theoretical Physics (ICFS 2000), Singapore, 13–17 March 2000;* World Scientific: Singapore, 2001; pp. 138–161.
85. Fell, J.; Führ, H.; Voigtlaender, F. Resolution of the wavefront set using general continuous wavelet transforms. *J. Fourier Anal. Appl.* **2016**, *22*, 997–1058. [CrossRef]
86. Sun, G.; Leng, J.; Cattani, C. A framework for circular multilevel systems in the frequency domain. *Symmetry* **2018**, *10*, 101. [CrossRef]
87. Kamont, A. Isomorphism of some anisotropic Besov and sequence spaces. *Studia Math.* **1994**, *110*, 169–189. [CrossRef]
88. Lemarié, P.-G.; Meyer, Y. Ondelettes et bases hilbertiennes. *Rev. Mat. Iberoam.* **1986**, *1*, 1–8. [CrossRef]
89. Meyer, Y. *Ondelettes et Opérateurs*; Hermann: Paris, France, 1990.
90. Daubechies, I. Orthonormal bases of compactly supported wavelets. *Commun. Pure Appl. Math.* **1988**, *41*, 909–996. [CrossRef]
91. Rogers, C.A. *Hausdorff Measures*; Cambridge University Press: Cambridge, UK, 1970.
92. Jaffard, S.; Lashermes, B.; Abry, P. Wavelet Leaders in Multifractal Analysis. In *Wavelet Analysis and Applications*; Tao Q., Vai, M.I., Xu Y., Eds.; Applied and Numerical Harmonic Analysis; Birkhaüser Verlag: Basel, Switzerland, 2006; pp. 219–264.
93. Biferalea, L.; Procaccia, I. Anisotropy in turbulent flows and in turbulent transport. *Phys. Rep.* **2005**, *414*, 43–164. [CrossRef]
94. Hinze, J.O. *Turbulence*; McGraw-Hill: New York, NY, USA, 1975.
95. Greenspan, H.P. *The Theory of Rotating Fluids*; Cambridge University Press: Cambridge, UK, 1968.
96. Grujic, Z. Vortex stretching and anisotropic diffusion in the 3D Navier-Stokes equations. *Contemp. Math.* **2016**, *666*, 240–251.
97. Constantin, P. Geometric statistics in turbulence. *SIAM Rev.* **1994**, *36*, 73–98. [CrossRef]
98. Ran, Z. Statistical Theory of Isotropic Turbulence. Part IV: Multiscales and Cascade. Available online: https://arxiv.org/pdf/1012.5151(accessed on 23 December 2010).

symmetry

MDPI

Article

Reversible Data Hiding Scheme Using Adaptive Block Truncation Coding Based on an Edge-Based Quantization Approach

Chia-Chen Lin [1,*], Ching-Chun Chang [2] and Zhi-Ming Wang [3]

1 Department of Computer Science and Information Management, Providence University, Taichung 43301, Taiwan
2 Department of Computer Science, University of Warwick, Coventry CV4 7AL, UK; ching-chun.chang@warwick.ac.uk
3 Department of Information Engineering and Computer Science, Feng Chia University, Taichung 40724, Taiwan; rock410186@gmail.com
* Correspondence: mhlin3@pu.edu.tw; Tel.: +886-426-328-001 (ext. 18108 or 11100); Fax: +886-426-324-045

Received: 8 May 2019; Accepted: 1 June 2019; Published: 5 June 2019

Abstract: In this paper, we provide a novel reversible data hiding method using adaptive block truncation coding based on an edge-based quantization (ABTC-EQ) approach. We exploit the characteristic not being used in ABTC-EQ. To accomplish this, we first utilized a Canny edge detector to obtain an edge image and classify each block in a cover image into two versions, edge-block and non-edge-block. Subsequently, k-means clustering was used to obtain three quantization levels and derive the corresponding bit map while the current processing block was the case of an edge-block. Then Zero-Point Fixed Histogram Shifting (ZPF-HS) was applied to embed the secret information into compressed code. The experimental results show that our method provides a high embedding capacity for each test image and performance is better than other methods.

Keywords: BTC; edge-based quantization; reversible data hiding; histogram shifting technique

---

1. Introduction

Due to the continuing advance of networks in recent years, it has become increasingly convenient and necessary for users to transmit messages to each other through the Internet. This, however, also creates many security problems, including the opportunity for a malicious attacker to destroy the transmitted information or tamper with data due to the openness of the Internet. To address these issues, researchers have explored different approaches, such as conventional cryptographic algorithms and information hiding methods. The former transforms the encrypted message into a meaningless format, but may leave clues for attackers. In contrast, the latter un-perceptively embeds the protected message into cover media. In terms of avoiding attacker attention, the information hiding approach outperforms conventional cryptographic algorithms.

Over the past decade, a variety of information hiding schemes have been proposed [1–19]. These information hiding schemes can be divided into two categories based on the subject that is embedded into a cover media. One is used for secret message transmission [1–4,6,7,10–19] and the other is used for claim of ownership [5,8,9] which is also called watermark scheme. The cover media used to carry a secret message can be image, text, audio or video. Currently, images are the primary media used to conceal secret messages because they can be easily found from the Internet. To embed a secret message into a cover image, there are three alternatives, including: spatial domain [1–4], frequency domain [5–8] and compression domain [9–20]. Spatial domain-based information hiding schemes conceal a secret message into a cover image by simply modifying pixel values of the cover image. A representative example is Least Significant Bit (LSB) substitution [1]. Frequency domain-based information hiding schemes need to transform a cover image into the frequency domain by using discrete wavelet transform (DWT) [21], discrete cosine transform (DCT) [22], etc. The frequency coefficients are then modified to carry a secret message. For compression domain-based information hiding, a secret message is embedded into the compression codes of a cover image and the compression codes are generated by any kind of compression algorithm, such as VQ [23], SMVQ [24], block truncation coding (BTC) [20] or JPEG. Among the above three types of information hiding schemes, frequency domain methods offer relatively higher protection compared to the others. Based on the reversibility feature of the proposed information hiding schemes, information hiding schemes can be further classified into those that are irreversible [1,2,5,6] and reversible [3,4,7–12,14–20,25–33]. The former can only extract information that is embedded in the media. Decoders still cannot completely restore the original cover image even after the hidden message has been extracted.

For example, in 2004, Chen et al. provided an irreversible scheme that embeds the secret data into a cover image by exploiting the Least Significant Bit (LSB) [1]. Decoders can determine the secret bit according to the LSB value of each pixel. However, decoders cannot recover each pixel back to the original, because this method directly changes the LSB value without recording any information regarding the replaced bits. However, irreversible information hiding schemes are not suitable for concealing a secret message into a cover image that requires exact restoration after data extraction, such as in military or medical applications.

In 1997, Barton [27] first proposed a reversible data hiding method. In this approach, the bits to be overlaid were compressed in advance and added to the bitstring. After that, the bitstring carrying hidden compressed bits was embedded into data block in the cover image. In 2002, Celik et al. [28] presented a method called generalized least significant bit, G-LSB for short, where they utilized a variant of an arithmetic compression algorithm (CALIC) [29] to encode a message and hide the resulting interval number along with extra information that was exploited to recover the cover image. In 2003, Tian [30] proposed a novel reversible information hiding method called difference expansion (DE) by embedding the secret message into the difference values between each pixel pair in a cover image. In 2004, Alattat [31] improved Tian's method by exploiting the difference in expansion of vectors instead of two adjacent pixels to enhance embedding capacity. In 2006, Ni et al. proposed a reversible scheme that hides secret data using histogram shifting [3]. They calculated the frequency of each pixel in the cover image and found zero and peak points to embed the secret data based on the histogram modification. When the receiver extracts the secret message from the cover image, the modified pixel can be recovered back to the original pixel value according to the modified method.

In 2009, Tai et al. [4] designed an efficient extension of the histogram modification technique by constructing a histogram of a cover image based on the differences between pixel values of each pixel pair to enhance the hiding capacity of Ni et al.'s scheme. In 2011, Li et al. [32] proposed a novel reversible watermarking scheme by exploiting prediction-error expansion (PEE), adaptive hiding and pixel selection. Their scheme concentrated on highly relevant regions and pixels of the cover image, and it obtains a high embedding capacity with less distortion. In 2012, in order to provide good visual quality and higher embedding capacity, Chang et al. [33] proposed a reversible data hiding scheme that determines whether a pixel is embeddable or not by calculating the absolute difference of its neighboring pixels. In Chang et al.'s scheme, once the derived absolute difference is larger than the predetermined threshold, the corresponding pixel remains unchanged to maintain a high image quality. However, these methods described above are mainly designed for the spatial domain rather than the compression domain. In general applications, images needed to be compressed before they are transmitted over the Internet because the size of raw images can be large. Since image compression is very popular, it is necessary to design reversible data hiding techniques for the compression domain.

Over the last few years, many hiding schemes designed for the compression domain have been proposed to reduce the transmission size of multimedia files during transmission and to increase the number of alternatives for cover media. Among these methods, many hiding schemes have been proposed based on block truncation coding (BTC) [14–19], which has been the most efficient and fastest compression method. In 2008, Chang et al. presented an information hiding scheme based on BTC [14]. They applied a genetic algorithm to substitute the original three bitmaps by finding an approximate optimal common bit map. Subsequently, the common bit map and block quantization levels for each block are used to hide the secret information. Side matching and quantization level orders are utilized to make the method reversible. In 2011, Li et al. proposed a reversible data hiding scheme based on BTC [15]. In their scheme, they utilized two quantization levels to generate a histogram. Histogram shifting and bitplane flipping are used to hide the secret data into a compressed code stream to improve the hiding capacity and to retain acceptable image quality. For example, if the secret bit is 1 then the high value and low value will be swapped with each other in the compression code, etc. In 2013, Sun et al. presented a novel BTC-based reversible hiding scheme by adopting a joint neighbor coding technique to embed the secret data into quantization levels [16]. In 2015, Lin et al. also proposed a reversible information hiding method based on BTC. In their scheme, they embed the secret information into the bit map of each image block [19]. However, their method only utilized the concept of BTC, and they did not compress the image so that the stego-image is not the BTC codestream. Although many BTC-based reversible data hiding schemes have been proposed, we found that these schemes are limited by a blocking effect problem. As such, in this paper, we try to propose a BTC-based reversible data hiding scheme without a blocking effect problem. To solve the blocking effect problem while offering a reversibility feature, we utilized Zero-Point Fixed Histogram Shifting (ZPF-HS) to embed the secret information and adaptive block truncation coding based on edge-based quantization (ABTC-EQ) to improve image quality and obtain a high embedding capacity.

The reminder of this paper is divided into five sections. Section 2 introduces the ABTC-EQ method, which forms the basis of our proposed reversible data hiding scheme. Section 3 briefly describes our proposed reversible data hiding scheme. Section 4 presents experiments to prove the performance of the proposed scheme. Finally, conclusions are given in Section 5.

2. Related Work

2.1. Histogram Shifting Technique (HS)

In 2006, Ni et al. presented an information hiding method based on the histogram shifting technique (HS) [3]. HS is a simple and efficient reversible data hiding method. In their scheme, they calculated the frequency of each pixel value in a cover image and generated an image histogram. Some pixel values from the histogram are selected and modified to embed the secret data. The modified pixel values can be recovered when the secret information is extracted, such that reversible data hiding is achieved. Their scheme is described as follows:

Step 1. Input an $H \times W$ sized cover image I.
Step 2. Compute the frequency of each pixel value and construct an image histogram. *Peak* and *zero* are the values of peak point and zero point, respectively.
Step 3. Shift the pixel values according to a pair for *peak* and *zero*. If *peak* > *zero*, the histogram ranging from $zero + 1$ to $peak - 1$ will be shifted to the left side by decreasing 1. Otherwise, the histogram ranging from $peak + 1$ to $zero - 1$ will be shifted to the right side by adding 1.

$$I'_{row,\ col} = \begin{cases} I_{row,\ col} + 1, \text{ if } peak + 1 \le I_{row,\ col} \le zero - 1 \text{ and } peak < zero \\ I_{row,\ col} - 1, \text{ if } zero + 1 \le I_{row,\ col} \le peak - 1 \text{ and } peak > zero \end{cases} \quad (1)$$

where $I_{row,\ col}$ and $I'_{row,\ col}$ are the pixel values at the locations $(row,\ col)$ of cover image I and modified cover image I', respectively.

Step 4. Embed the secret information into the modified cover image I'. If the secret bit S is "1" and the pixel value is equal to *peak*, it will be increased or decreased by 1. Otherwise, its value remains unchanged.

$$I''_{row,\ col} = \begin{cases} I'_{row,\ col} + 1, \text{ if } I'_{row,\ col} = peak \text{ and } peak < zero,\ \mathrm{S} = 1 \\ I'_{row,\ col} - 1, \text{ if } I'_{row,\ col} = peak \text{ and } peak > zero,\ \mathrm{S} = 1 \\ I'_{row,\ col} \quad ,\text{ if } I'_{row,\ col} = peak \text{ and } peak < zero,\ \mathrm{S} = 0 \\ I'_{row,\ col} \quad ,\text{ if } I'_{row,\ col} = peak \text{ and } peak > zero,\ \mathrm{S} = 0 \end{cases} \quad (2)$$

Step 5. Repeat **Step 4** until all $I'_{row,\ col}$ are processed.
Step 6. Output stego-cover image I''.

2.2. ABTC-EQ

In 2015, Mathews et al. [23] proposed a novel adaptive block truncation coding technique called ABTC-EQ. It is introduced in detail in this section to offer a better understanding of our proposed method. The cover image is compressed according to the result presented in the edge image that is derived by Canny edge detection [21]. Next, a quantization approach is processed based on the edge information of each block. If a block is determined as non-edge-block, it proceeds with bi-clustering. In contrast, an edge-block proceeds with tri-clustering. All steps are described as follows:

Step 1. Input cover image I sized as $H \times W$ pixels and divide it into $k \times k$ non-overlapping blocks b_i's, where $i = 0,\ 1,\ \ldots,\ \frac{H}{k} \times \frac{W}{k} - 1$ and $k = 4,\ 8, \ldots 32$.

$$B = \begin{bmatrix} b_0 & \cdots & b_{\frac{H}{k}-1} \\ \vdots & \ddots & \vdots \\ . & \cdots & b_{\frac{H}{k}\times\frac{W}{k}-1} \end{bmatrix}$$

Step 2. Utilize Canny edge detection to obtain the edge map of the whole cover image denoted as *emp*.

Canny edge detection is an optimal algorithm including three steps to detect edge information from the given cover image. The first step is to reduce the noise by using Gaussian filter. Next, find the gray levels and apply a non-maximum suppression technique to thin the edge. Then, utilize double thresholds and connectivity analysis to indicate the edge map *emp* for the given cover image I.

Step 3. Divide the *emp* into $k \times k$ non-overlapping edge-blocks e_i's.

$$emp = \begin{bmatrix} e_0 & \cdots & e_{\frac{H}{k}-1} \\ \vdots & \ddots & \vdots \\ . & \cdots & e_{\frac{H}{k}\times\frac{W}{k}-1} \end{bmatrix}$$

Step 4. Perform block classification based on edge-blocks generated by **Step 3**.

If there is only one edge value, it is 1 in edge-block e_i and the rest of the values are 0, and block b_i can be determined as an edge-block with three quantization levels and goes to **Step 5**. Otherwise, it belongs to the non-edge-block with two quantization levels and goes to **Step 6**.

Step 5. Employ k-means clustering [22] to partition the pixels in the current block b_i into three clusters, C_0, C_1 and C_2, respectively.

$$C_f = \begin{cases} C_0 = \left\{x_0^0,\ x_1^0,\ \ldots,\ x_r^0\right\} \\ C_1 = \left\{x_0^1,\ x_1^1,\ \ldots,\ x_r^1\right\} \\ C_2 = \left\{x_0^2,\ x_1^2,\ \ldots,\ x_r^2\right\} \end{cases}$$

Then calculate the mean values of each cluster using Equation (3), and these three mean values will serve as three quantization levels.

$$\mu_f = \frac{1}{m_f} \sum\nolimits_{r=0}^{m_f-1} x_r^f, \tag{3}$$

where $f = 0,\ 1$ or 2, $0 \le r \le k \times k - 1$, m_f is the member of each cluster and x_r^f's mean the members in each cluster.

The bp_n^i in BMP^i will be defined according to Equation (4).

$$BMP^i = \begin{bmatrix} bp_0^i & \cdots & bp_{k-1}^i \\ \vdots & \ddots & \vdots \\ \cdot & \cdots & bp_{k\times k-1}^i \end{bmatrix}, \text{ where } bp_n^i = \begin{cases} 00\ ,\ \text{if } x_r^f \epsilon\ C_0 \\ 01\ \text{ if } x_r^f \epsilon\ C_1 \\ 10\ \text{ if } x_r^f \epsilon\ C_2 \end{cases}, \tag{4}$$

where BMP^i is the bit map of b_i, bp_n^i is the value in BMP^i and $n = 0,\ 1,\ \ldots,\ k \times k - 1$.

Step 6. Find the maximum (*max*) and minimum (*min*) values of gray levels in block b_i. Then, compute the average value avg of block b_i.

Calculate the value of threshold T using Equation (5).

$$T = \frac{max + min + avg}{3}. \tag{5}$$

Construct the BMP^i by using Equation (6) and calculate the two quantization levels h^i and l^i by using Equations (7) and (8).

$$BMP^i = \begin{bmatrix} bp_0^i & \cdots & bp_{k-1}^i \\ \vdots & \ddots & \vdots \\ \cdot & \cdots & bp_{k\times k-1}^i \end{bmatrix}, \text{ where } bp_n^i = \begin{cases} 1\ ,\ \text{if } p_n^i > T \\ 0\ ,\ \text{if } p_n^i \le T \end{cases} \tag{6}$$

$$h^i = \frac{1}{num_0} \sum_{r=0}^{num_0-1} p_n^i,\ \text{if } p_n^i > T \tag{7}$$

$$l^i = \frac{1}{num_1} \sum_{r=0}^{num_1-1} p_n^i,\ \text{if } p_n^i \le T \tag{8}$$

Here p_n^i is the pixel value in block b_i, num_0 is the number of pixels that are greater than T, num_1 means the numbers that are smaller than or equal to T, h^i is the high value in b_i and l^i is the low value.

Step 7. Repeat **Step 4** to **Step 6** until all block b_i's are processed and then obtain ABTC-EQ compressed codes.

Figure 1a,b show the encoding flowcharts of BTC [13] and ABTC-EQ [23], respectively. To simplify our example shown in Figure 1, a single block b_i sized 4×4 pixels using BTC and ABTC-EQ, respectively, is demonstrated. We used Equation (9) to calculate the Mean Square Error (*MSE*) of BTC and ABTC-EQ, whose values were 698 and 55, respectively. Obviously, ABTC-EQ has good performance when a block is in the complexity area.

$$MSE = \frac{1}{H \times W} \sum_{row=0}^{H-1} \sum_{col=0}^{W-1} \left(I'_{row,col} - I_{row,col}\right)^2 \tag{9}$$

where $I'_{row,col}$ and $I_{row,col}$ are the values of the decompressed pixel and the original pixel values.

Original block b_i

108	115	152	187
107	130	178	193
111	147	190	195
121	167	199	190

(a) The flowchart of BTC

Step 1. Calculate the mean value μ and standard deviation σ of block b_i.
$\mu = 155$,
and $\sigma = 8.62183$.

Step 2. Compute $high^i$ and low^i

$$high^i = \mu + \sigma\sqrt{\frac{k\times k-q}{q}},$$

$$low^i = \mu - \sigma\sqrt{\frac{q}{k\times k-q}},$$

where q is the number of pixels greater than μ and the values of $high^i$ and low^i are 164 and 146, respectively.

$\{high^i, low^i\} \longrightarrow high^i\|low^i\|$.

Step 3. Construct BMP^i using Eq. (4).

$BMP^i =$

0	0	0	1
0	0	1	1
0	0	1	1
0	1	1	1

$\{high^i, low^i, BMP^i\} \longrightarrow high^i\|low^i\|BMP^i$.

Step 4. Output code stream:
Encoding format: $high^i\|low^i\|BMP^i$
CS: 10100100||10010010||
0001001100110111.

(b) The flowchart of ABTC-EQ

Step 1. Utilize Canny edge detector and obtain the edge image *emp*.

$emp =$

0	1	0	0
0	1	0	0
0	1	0	0
1	0	0	0

Step 2. Determine the case of block b_i.
b_i is edge block.

$\{\text{indicator}\} \longrightarrow 1\|$

Step 3. Partition the p_n^i into three cluster using k-means clustering and calculate the mean values of μ_0, μ_1 and μ_2 using Eq. (1).

$\{00\}C_0 = \{187, 178, 193, 190, 195, 199, 190\}$
$\{01\}C_1 = \{152, 147, 167\}$
$\{10\}C_2 = \{108, 115, 107, 130, 111, 121\}$
$\mu_0 = 190$, $\mu_1 = 155$ and $\mu_2 = 115$

$\{\text{indicator}, \mu_0, \mu_1, \mu_2\} \longrightarrow 1\|\mu_0\|\mu_1\|\mu_2$.

Step 4. Construct BMP^i using Eq. (2).

$BMP^i =$

10	10	01	00
10	10	00	00
10	01	00	00
10	01	00	00

$\{\text{indicator}, \mu_0, \mu_1, \mu_2, BMP^i\} \longrightarrow$

Step 5. Output code stream:
Encoding format: $1\|\mu_0\|\mu_1\|\mu_2\|BMP^i$.
CS: 1||10111110||10011011||01110011||
10100100101000001001000010010000.

Figure 1. Compression flowcharts of block truncation coding (BTC) and adaptive block truncation coding based on edge-based quantization (ABTC-EQ algorithms): (**a**) BTC encoding and (**b**) ABTC-EQ encoding.

3. Proposed Scheme

This section presents the proposed scheme. In our method, we utilized ABTC-EQ to compress the cover image because its reconstructed image quality is relatively good compared to other BTC variant techniques. Next, ZPF-HS was used to embed the secret information into an ABTC-EQ compressed code stream. To further enlarge the hiding capacity of our proposed method, we also embed the secret data into quantization levels. As background for our proposed scheme, Section 3.1 reviews the zero-point fixed histogram shifting (ZPF-HS) that will be used for data embedding in our approach. Our proposed scheme contains two phases: a data embedding phase and the data extraction and recovery phase, which are demonstrated in Sections 3.2 and 3.3, respectively.

3.1. Zero-Point Fixed Histogram Shifting (ZPF-HS)

The histogram shifting technique [3], called HS for short, is a simple and efficient hiding method, and has been widely adopted in various reversible data hiding schemes. In this section, the features of HS are explored and then expanded to support a zero-point fixed scenario as zero-point fixed histogram shifting, called ZPF-HS for short. Finally, ZPF-HS is adopted in our proposed scheme.

In our proposed method, there are only three histogram bins that need to addressed if the compressed blocks are determined as edge-blocks and the corresponding bit map is the source for our ZPF-HS. Figure 2 shows examples of three possible cases of the bit map for an edge-block. In ABTC-EQ, bits 11 are not being used, as shown in Figure 2. In our scheme, zero point (*zero*) is always set as 11 and the peak point (*peak*) is defined as the bit values in the bit map which has a large population.

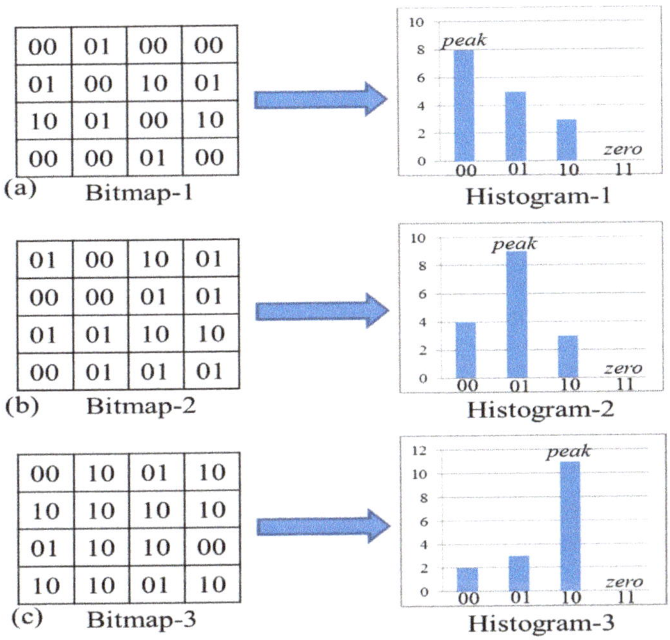

Figure 2. (a–c) are the bit maps and histograms of each case.

Take Figure 2a for example: there are 8 bit values "00", 5 bit values "01" and 3 bit values "10" in bit map-1. Therefore, peak point is defined as "00". We exploit the first case shown in Figure 2a as an example to explain in detail our proposed ZFP-HS in Figure 3. Figure 3a shows the original bit map and its corresponding histogram, Figure 3b presents the secret data and Figure 3c is the result of the modified bit map and its corresponding histogram after embedding. In this example, *peak* is defined as "00" and *zero* is defined as "11", then according to Equation (10) with a zig-zag scan, the secret data can be embedded into the original bit map and the modified bit map is shown in Figure 3c.

$$peak = \begin{cases} peak\,, & \text{if secret bit is 0} \\ 11\,, & \text{if secret bit is 1} \end{cases} \tag{10}$$

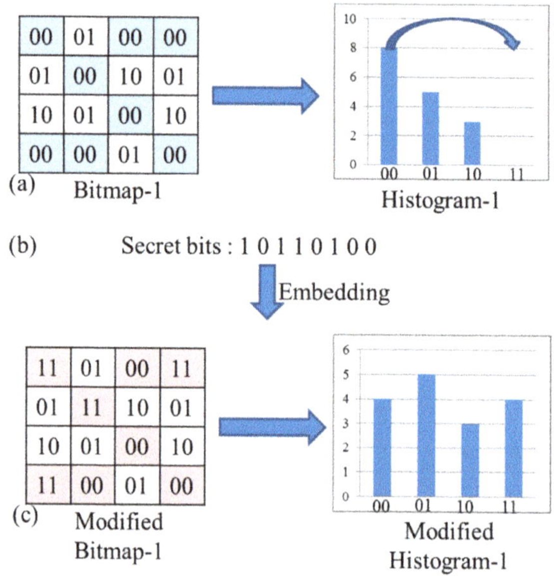

Figure 3. Example of operations in our ZPF-HS. (**a**) Original bit map and histogram, (**b**) secret bits and (**c**) modified bit map and histogram.

3.2. Data Embedding Phase

In our proposed data embedding phase, the embedding operations and encoding phase of ABTC-EQ are merged seamlessly. Blocks are identified as non-edge-block and edge-block after Canny edge detection. Therefore, two block types are identified and two cases of data hiding operations need to be explored in our embedding phases as shown in Figure 4. For an edge-block case, both quantization levels and a bit map are used for data hiding. By contrast, only quantization levels are used for data embedding in a non-edge-block.

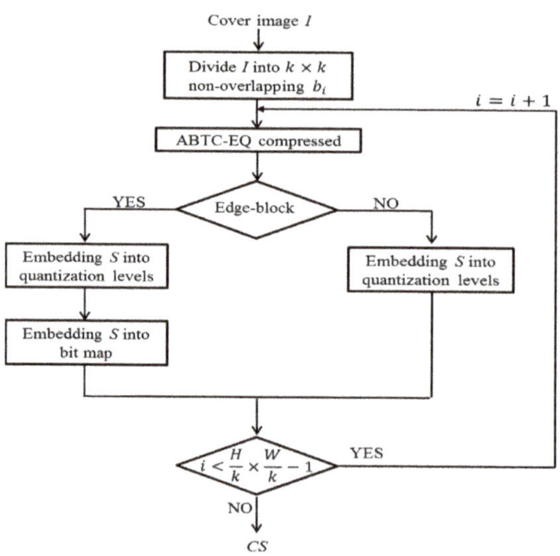

Figure 4. The flowchart of data embedding phase.

In our data embedding phase, the input cover image is sized as $H \times W$ pixels. Each block b_i is sized $k \times k$ pixels, where $i = 0, 1, \ldots, \frac{H}{k} \times \frac{W}{k} - 1$. Note that the ABTC-EQ procedure is also included as shown in Figure 4. Secret information S is a bitstream in binary form, and s_l is the value of a secret bit in S, where $s_l = 0$ or 1 and $l = 0, 1, 2, \ldots, N$. N is the number of maximum capacity of cover image I. And S is embedded into the ABTC-EQ compressed code stream of cover image I.

Input: Cover image I and secret information S.
Output: Code stream CS.
Step 1. Divide I into $k \times k$ non-overlapping blocks b_i's.
Step 2. Utilize ABTC-EQ to compress the current processing block b_i.
Step 3. Determine block b_i to be edge-block or non-edge-block. If block b_i is an edge-block, then go to **Step 4**. Otherwise, go to **Step 8**.
Step 4. Insert one bit to serve as the indicator and set it as 1. Then, use Equation (3) to compute the mean values μ_0, μ_1 and μ_2 of three clusters C_0, C_1 and C_2, respectively. Finally, cluster C_{y_1}w, which has a large population will be encoded as $1||\mu_{y_1}||$, where $||$ represents the concatenation operation and $y_1 = 0, 1$ or 2.
Step 5. Read the next s_l from S, if $s_l = 0$, and the remaining clusters will be encoded as $1||\mu_{y_1}||\max\{\mu_{f-\{y_1\}}\}||\min\{\mu_{f-\{y_1\}}\}$, where y_2 and $y_3 \in \{0, 1, 2\}$. Otherwise, encode by $1||\mu_{y_1}||\min\{\mu_{f-\{y_1\}}\}||\max\{\mu_{f-\{y_1\}}\}$.
Step 6. Embed the next s_l from S into the BMP^i and obtain a modified BMP^i by using Equation (10).
Step 7. Output $1||\mu_{y_1}\min\{\mu_{f-\{y_1\}}\}||\max\{\mu_{f-\{y_1\}}\}||$modified BMP^i to be part of CS.
Step 8. Insert one bit as the indicator and set it as 0. Then, use Equations (7) and (8) to compute two quantization levels h_i and l_i.
Step 9. Determine the next s_l, if the next $s_l = 0$, indicator, h_i and l_i will be encoded by $0||h_i||l_i$. Otherwise, it will be encoded by $0||l_i||h_i$.
Step 10. Output the indicator, that is the sequence according to the corresponding embedding order of two quantization levels, and the original bit map BMP^i to be part of CS.
Step 11. Repeat **Step 2** to **Step 10** until all blocks b_i's are processed.
Step 12. Obtain output code stream CS.

We obtain the modified code stream CS, which concealed the S after all the steps are completed. An example of our proposed data embedding phase is shown in Figure 5 to explain each step in detail. Figure 5a shows an example of a 4×4 sized block b_i. Figure 5b presents the histogram of three clusters corresponding bp_n^i in block b_i. Figure 5c,d present the original BMP^i and the modified BMP^i, respectively. Figure 5e provides the code stream of a modified BMP^i. Figure 5f is the sequence of the indicator, three quantization levels and modified BMP^i. Figure 5f presents the binary form of Figure 5g. In Figure 5, all pixels in block b_i will be partitioned into three clusters exploiting k-means clustering. Then, compute the mean values μ_f of three clusters using Equation (3). Because C_0 has the largest population, the C_0 corresponding to bp_n^i is *peak*. The indicator and three quantization levels μ_f's will be encoded by $1||\mu_0||\mu_2||\mu_1$ while the next s_l is 1. In the next step, construct BMP^i, embed the next s_l into BMP^i using Equation (10) and obtain the modified BMP^i. Finally, we obtain the modified code stream CS.

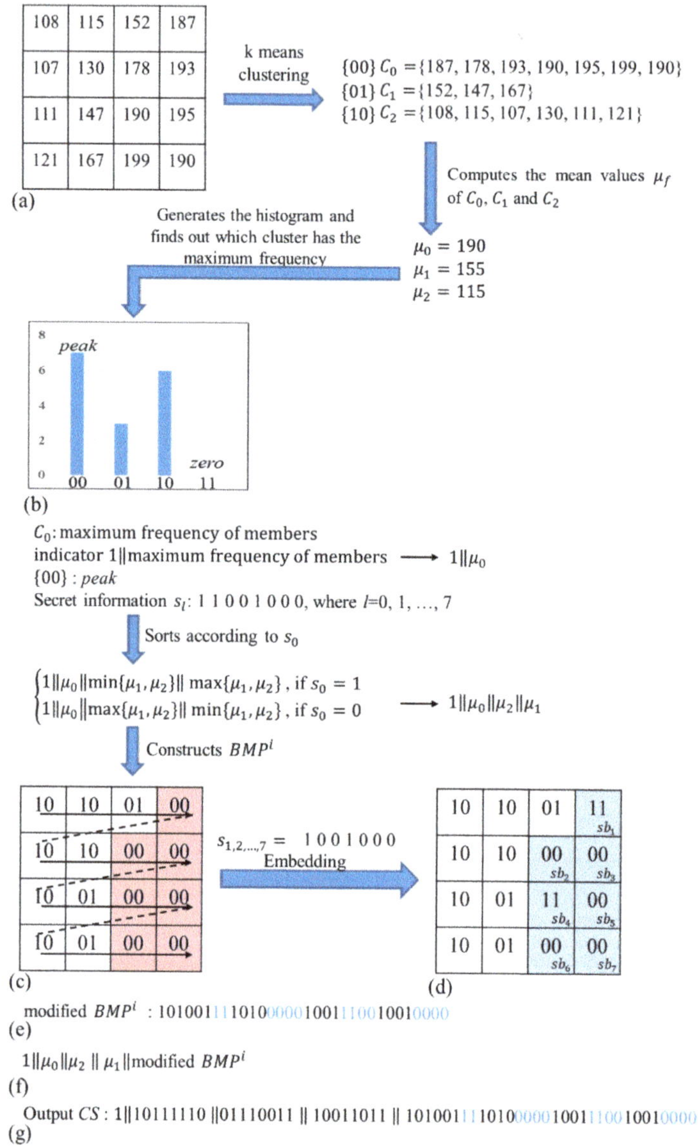

Figure 5. (**a**) Block b_i sized 4×4, (**b**) histogram of block b_i, (**c**) original BMP^i, (**d**) modified BMP^i, (**e**) the code stream of modified BMP^i, (**f**) structure of code stream CS and (**g**) output code stream CS.

3.3. Extraction and Recovery Phase

In this section, hidden secret information S is extracted from code stream CS. Because one indicator has been added during our data embedding phase, a decoder can be guided by the indicator to conduct the extraction operation. If the indicator is 1, block b_i will be judged as an edge-block. Three quantization levels will be extracted and among three quantization levels of bp_n^i will serve as the *peak*. In other words, our proposed scheme does not need extra information to record the value of *peak* to recover the BMP^i, as the histogram shifting technique is adopted in our scheme. Flowchart for extraction and recovery phase is shown in Figure 6.

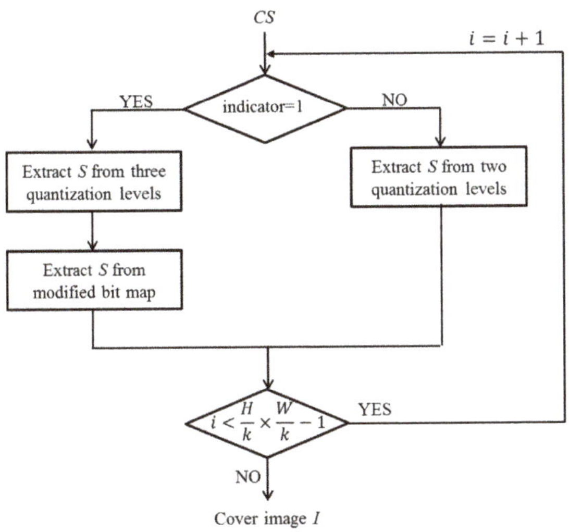

Figure 6. Flowchart for extraction and recovery phase.

Input: Code stream CS.
Output: Cover image I and secret information S.
Step 1. Read the 1-bit indicator in the CS and determine the value of the indicator, if the indicator value is 1, then go to **Step 2**. Otherwise, go to **Step 8**.
Step 2. Read the next 56 bits, then obtain the bit stream of three quantization levels μ_0', μ_1' and μ_2', and the modified BMP^i. Its sequence is $1||\mu_0'||\mu_1'||\mu_2'||$modified BMP^i.
Step 3. Determine the maximum of μ_1' and μ_2'. If $\mu_1' > \mu_2'$, the hidden $s_l = 0$. Otherwise, the hidden $s_l = 1$.
Step 4. Construct the modified BMP^i and sort μ_0', μ_1' and μ_2' in descending order. The value of *peak* is μ_0''s corresponding bp_n^i.
Step 5. Extract the next s_l from the modified BMP^i. If $bp_n^i = peak$, the hidden $s_l = 0$. And if $bp_n^i = 11$, the hidden $s_l = 1$.
Step 6. Modify *zero* back to *peak* where $zero = 11$.
Step 7. Decompress block b_i according to each bp_n^i's corresponding quantization level.
Step 8. Read the next 32 bits, then obtain the bit stream of two quantization levels μ_0' and μ_1', and the original BMP^i. Its sequence is $0||\mu_0'||\mu_1'||$original BMP^i.
Step 9. Determine the maximum of μ_0' and μ_1'. If $\mu_0' > \mu_1'$, the hidden $s_l = 0$. Otherwise, the hidden $s_l = 1$.
Step 10. Sort μ_0' and μ_1' in descending order and decompress block b_i according to each bp_n^i's corresponding quantization level.
Step 11. Repeat **Steps 1** to **10** until all bits in CS are read and proceeded.
Step 12. Obtain secret information S and decompressed cover image I.

After all steps are completed, decompressed cover image I and secret information S are obtained. We also provide an example to further clarify the extraction and recovery phases, which is shown in Figure 7. Figure 7a shows the CS in binary form, Figure 7b presents the sequence of indicator, three quantization levels and modified BMP^i, Figure 7c shows the modified BMP^i, Figure 7d presents the original BMP^i and Figure 7e provides the extracted S. In Step 1, three quantization levels μ_0', μ_1' and μ_2'

are converted into decimal values. Because $\mu'_1 = 115$ is less than $\mu'_2 = 155$, hidden s_0 is judged as 1. In Step 2, $\mu'_0 = 190$, $\mu'_1 = 115$ and $\mu'_2 = 155$ are sorted in descending order, and μ'_0 corresponding to bp^i_n is *peak*, so the bp^i_n of *peak* is determined as 00. As the next step, the modified BMP^i is constructed and $s_{1,\,2,\,\ldots,\,7}$ are extracted from a modified BMP^i. If $bp^i_n = peak$, the hidden $s_l = 0$. If $bp^i_n = 11$, the hidden $s_l = 1$. After extracting all S from the modified BMP^i, change all bp^i_n values of 11 into *peak*. Finally, we can obtain the original BMP^i as shown in Figure 7d.

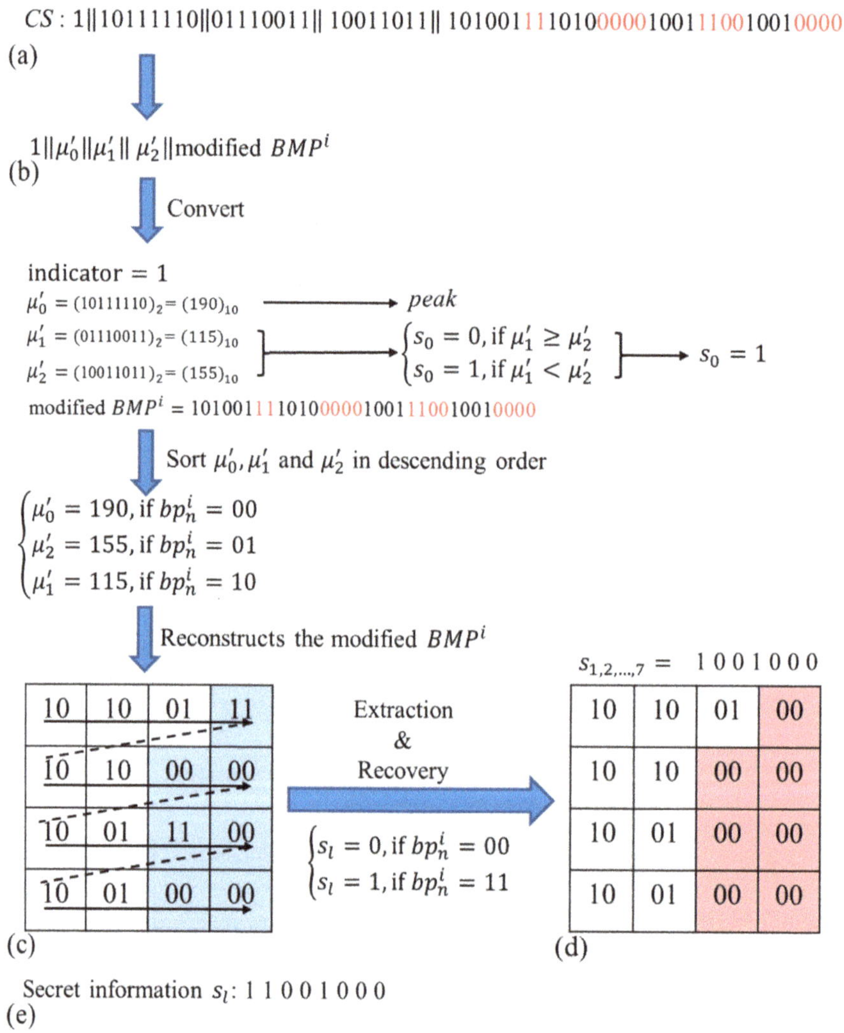

Figure 7. (**a**) Code stream CS, (**b**) example of output format, (**c**) modified BMP^i, (**d**) recovery BMP^i and (**e**) secret information S.

4. Experimental Results

We describe some experimental results in this section to demonstrate hiding capacity, output code stream size and the compression ratio in our proposed method. The eleven 512 × 512 test grayscale cover images as shown in Figure 8 were used for our experiments. The results of their edge images based on Canny edge detection are shown in Figure 9.

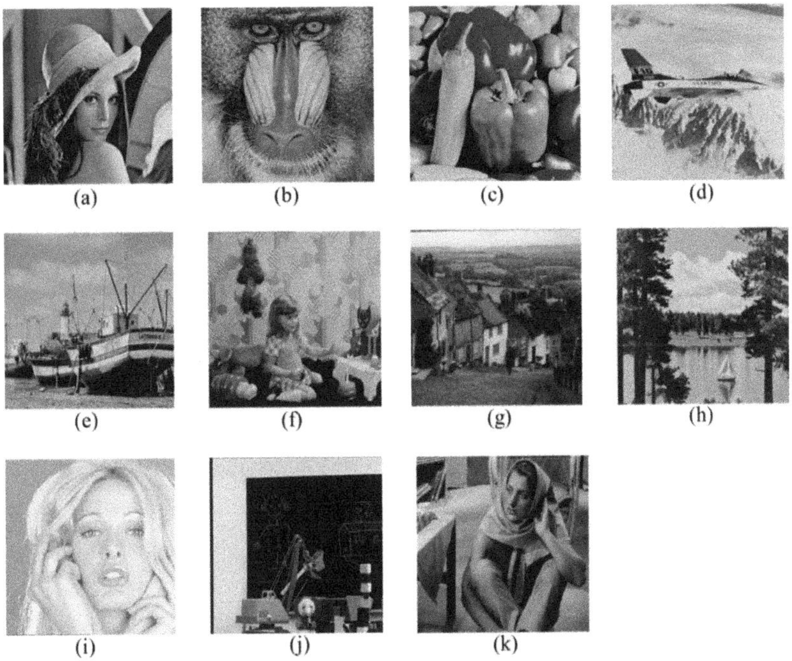

Figure 8. Test images: (**a**) Lena, (**b**) Baboon, (**c**) Peppers, (**d**) F-16, (**e**) Fishing Boat, (**f**) Girl, (**g**) Gold hill, (**h**) Sailboat, (**i**) Tiffany, (**j**) Toys and (**k**) Barbara.

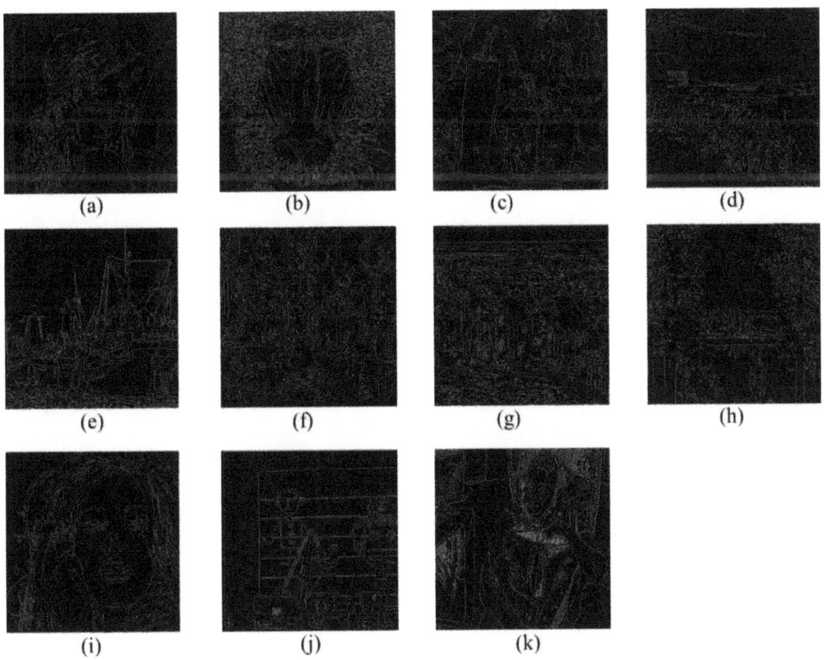

Figure 9. Canny edge detection images of test images: (**a**) Lena, (**b**) Baboon, (**c**) Peppers, (**d**) F-16, (**e**) Fishing Boat, (**f**) Girl, (**g**) Gold hill, (**h**) Sailboat, (**i**) Tiffany, (**j**) Toys, and (**k**) Barbara.

To illustrate the performance of our proposed method, the results of our scheme with two different block sizes, 4×4 pixels and 8×8 pixels, are shown in Tables 1 and 2, respectively. In Tables 1 and 2, we present embedding capacity (number of bits), the size of *CS* (number of bits), compressed ratio (*CR*) (%) and peak signal-to-noise-ratio (*PSNR*) (dB) of ABTC-EQ and BTC in two different block sizes, 4×4 pixels and 8×8 pixels, respectively. Obviously, compressing the image to exploit ABTC-EQ can obtain an overall better image quality than BTC, as seen in Tables 1 and 2 by exploiting Equation (12). Because our scheme embeds the secret data into the compression code stream, a decompressed image cannot be directly obtained from the *CS* that carries the hidden secret data. As for *PSNR* (db), it denotes the decompressed image of the recovery *CS*. The *CR* of conventional BTC is 0.25 using Equation (11). The size of output *CS* (number of bits) and *PSNR* (dB) using ABTC-EQ in the case of an 8×8 block size for b_i is similar to the result of the BTC of the 4×4 block size for b_i. In our scheme, we utilize the characteristic of ABTC-EQ to apply our proposed ZPF-HS to embed the secret data, and we see that the size of *CS* before and after hiding are the same in our method. Despite the size, our *CS* (number of bits) is very large because of the cost of bits, while b_i is the edge-block. But the problem of a blocking effect can be better solved with our method than with other compression methods. The average hiding capacity (number of bits) and *PSNR* (dB) in our experiment are 74,138 (number of bits) and 36.327 (number of bits), respectively. Additionally, the PSNR (dB) means the resulting image after extracting the secret information in Tables 1 and 2.

$$CR = \frac{CS}{H \times W \times n} \tag{11}$$

$$PSNR = 10 \times \log_{10}\left(\frac{255^2}{MSE}\right). \tag{12}$$

Table 1. Performance of our proposed method in 4×4 block sizes for each block b_i.

Image with Block Size 4×4	Capacity (Number of Bits)	*CS* (Number of Bits)	*CR* (%)	ABTC-EQ *PSNR* (dB)	BTC *PSNR* (dB)
Lena	63,342	675,384	0.3220	37.115	33.659
Baboon	102,999	796,152	0.3796	31.255	27.752
Peppers	63,603	670,944	0.3199	37.486	34.151
F-16	66,094	675,240	0.3220	37.405	33.359
Fishing boat	70,813	695,712	0.3317	35.944	32.000
Girl	88,051	739,800	0.3528	38.157	34.706
Gold hill	89,745	751,176	0.3582	37.075	33.659
Sailboat	69,283	689,736	0.3289	34.653	31.139
Tiffany	72,395	691,536	0.3298	40.153	36.991
Toys	56,115	650,696	0.3103	37.666	33.216
Barbara	73,083	710,496	0.3388	32.688	29.868
Average	74,138	704,261	0.3358	36.327	32.773

From Table 1, we can see that the average capacity is around 74,000 bits and the *CR* is about 0.3358% when the block size is 4×4 pixels.

Table 2. Performance of our proposed method in 8×8 block sizes for each block b_i.

Image with Block Size 8×8	Capacity (Number of Bits)	*CS* (Number of Bits)	*CR* (%)	ABTC-EQ *PSNR* (dB)	BTC *PSNR* (dB)
Lena	76,671	480,096	0.2289	33.892	30.273
Baboon	110,340	565,488	0.2696	29.025	25.843
Peppers	83,774	486,792	0.2321	33.995	30.273
F-16	80,009	474,264	0.2261	34.276	30.204
Fishing boat	84,791	489,960	0.2336	32.821	29.042
Girl	108,217	544,104	0.2594	34.934	31.055

Table 2. *Cont.*

Image with Block Size 8 × 8	Capacity (Number of Bits)	*CS* (Number of Bits)	*CR* (%)	ABTC-EQ *PSNR* (dB)	BTC *PSNR* (dB)
Gold hill	114,533	559,440	0.2668	33.859	30.723
Sailboat	87,997	495,432	0.2362	31.887	28.129
Tiffany	90,900	497,088	0.2370	37.239	33.979
Toys	70,889	455,112	0.2170	34.198	30.069
Barbara	87,264	510,048	0.2432	30.917	27.832
Average	90,489	505,257	0.2409	33.368	29.766

In comparison, we can see that the average capacity is up to 90,000 bits and the *CR* is about 0.2409% when the block size is changed to 8 × 8 pixels as shown in Table 2. Certainly, the average image quality will be slightly decreased to 33.368 dB, but it is significantly higher than the average PSNR offered by conventional BTC.

To demonstrate the performance results for our proposed scheme, the proposed method in this experiment was compared to previous schemes, i.e., Chang et al. [14], Li et al. [15], Sun et al. [16] and Lin et al. [19] in terms of embedding capacity (number of bits) and embedding efficiency (*EF*) (%), the results of which are shown in Table 3. These four existing schemes are selected and compared with our proposed scheme because they are reversible data hiding schemes and they are either designed for BTC or AMBTC. Moreover, their hiding strategies are embedding secrets into bitmap and two quantization levels, which are the same as ours. Here, *EF* was used to evaluate embedding efficiency, which is defined as follows:

$$EF = \frac{\text{Capacity}}{||CS||}, \tag{13}$$

where $||CS||$ is the size of the output *CS* and Capacity is the embedding capacity of each test image.

Table 3. Embedding capacity (number of bits) and *EF* (%) for the proposed scheme and four previous schemes.

Schemes	Parameters	Lena	F-16	Sailboat	Girl	Toys	Barbara
Chang et al. [14]	Capacity	31,011	30,518	28,766	30,962	27,870	30,151
	CS	524,288	524,288	524,288	524,288	524,288	524,288
	EF	0.0591	0.0582	0.0549	0.0591	0.0532	0.0575
Li et al. [15]	Capacity	16,789	17,659	17,082	16,990	17,761	16,755
	CS	524,288	524,288	524,288	524,288	524,288	524,288
	EF	0.032	0.0337	0.0326	0.0324	0.0339	0.032
Sun et al. [16]	Capacity	64,008	64,008	64,008	64,008	64,008	64,008
	CS	524,288	524,288	524,288	524,288	524,288	524,288
	EF	0.1221	0.1221	0.1221	0.1221	0.1221	0.1221
Lin et al. [19]	Capacity	262,112	261,984	262,096	262,128	262,112	262,128
	CS	2,097,152	2,097,152	2,097,152	2,097,152	2,097,152	2,097,152
	EF	0.125	0.1249	0.125	0.125	0.125	0.125
Our scheme	Capacity	76,671	80,009	84,791	108,217	70,889	87,264
	CS	480,096	474,264	495,432	544,104	455,112	510,048
	EF	0.1597	0.1687	0.1731	0.1989	0.1558	0.1711

In this experiment, the size of all test images were 512 × 512 pixels and the block size was set as 8 × 8 pixels. In this experiment, our embedding capacity was better than three previous schemes [14,16,17]. While Lin et al.'s scheme provides good hiding capacity performance, their scheme extracts the secret data from the 512 × 512 resulting images instead of extracting the secret information from the output *CS* (number of bits). Therefore, the size of each *CS* (number of bits) in Lin et al.'s scheme is 512 × 512 × 8. The size of our *CS* (number of bits) remains unchanged even after embedding the secret information. In our scheme, the sizes of *CS*'s for, "Lena," "F-16," "Sailboat," "Girl," "Toys" and "Barbara" are 480,096 (number of bits); 474,264 (number of bits); 495,432 (number of bits); 544,104 (number of bits); 455,112 (number of bits) and 510,048 (number of bits), respectively, and are shown in Table 3. For the purpose of having a better comparison with the previous four methods, we utilize *EF* (%) to analyze the performance of our scheme and compare to other schemes using Equation (13). Our proposed scheme obtained a higher *EF* than the previous four methods. Moreover, the *EF* offered by Lin et al.'s scheme is lower than ours because their results are presented as images rather than from the code stream.

5. Conclusions

This paper presented a novel reversible data hiding method using block truncation coding based on an edge-based quantization approach. By applying two embedding levels and our proposed ZPF-HS to hide the secret information, it was possible to have a high capacity, high *PSNR* and high *EF* despite the generation of a large *CS* size. In addition, we utilized n bits after the indicator to record the *peak* while blocks are edge-block to ensure that our method exactly restores the original cover image. The experimental results show that our proposed method is indeed suitable for hiding large volumes of information in multimedia. However, it still remains that one value 11 of bitmap cannot be used in the ABTC-EQ compressed method. Our future work will concentrate on how to utilize this value that is not being used in ABTC-EQ to enhance image quality and how to exploit this feature to embed more secret information into compressed code. Moreover, two possible approaches, i.e., CNN and hyperchaos, will be explored and applied when we try to study the above two objectives.

Author Contributions: Conceptualization and funding acquisition, C.-C.L.; software and writing-original draft preparation, Z.-M.W.; validation, C.-C.C.

Funding: This research was funded by Ministry of Science and Technology grant number 105-2410-H-126-005-MY3.

Conflicts of Interest: The authors declare no conflict of interest.

References

1. Chan, C.K.; Cheng, L.M. Hiding data in images by simple LSB sub-stitution. *Pattern Recognit.* **2004**, *37*, 469–474. [CrossRef]
2. Zhang, X.P.; Wang, S.Z. Efficient steganographic embedding by exploiting modification direction. *IEEE Commun. Lett.* **2006**, *10*, 781–783. [CrossRef]
3. Ni, Z.; Shi, Y.Q.; Ansari, N.; Su, W. Reversible data hiding. *IEEE Trans. Circuits Syst. Video Technol.* **2006**, *16*, 354–362.
4. Tai, W.L.; Yeh, C.M.; Chang, C.C. Reversible data hiding based on histogram modification of pixel differences. *IEEE Trans. Circuits Syst. Video Technol.* **2009**, *19*, 906–910.
5. Zhang, D.X.; Pan, Z.E.; Li, H.H. A contour-based semi-fragile image watermarking algorithm in DWT domain. In Proceedings of the 2nd International Workshop on Education Technology and Computer Science (ETCS), Wuhan, China, 6–7 March 2010; Volume 3, pp. 228–231.
6. Wu, X.; Sun, W. Robust copyright protection scheme for digital images using overlapping DCT and SVD. *Appl. Soft Comput.* **2013**, *13*, 1170–1182. [CrossRef]
7. Chan, Y.K.; Chen, W.T.; Yu, S.S.; Ho, Y.A.; Tsai, C.S.; Chu, Y.P. A HDWT-based reversible data hiding method. *J. Syst. Softw.* **2009**, *82*, 411–421. [CrossRef]
8. Thabit, R.; Khoo, B.E. Robust reversible watermarking scheme using slantlet transform matrix. *J. Syst. Softw.* **2014**, *88*, 74–86. [CrossRef]
9. Zhang, X.P.; Wang, S.Z.; Qian, Z.X.; Feng, G. Reversible fragile watermarking for locating tampered blocks in JPEG images. *Signal Process.* **2010**, *90*, 3026–3036. [CrossRef]
10. Wang, K.; Lu, Z.M.; Hu, Y.J. A high capacity lossless data hiding scheme for JPEG images. *J. Syst. Softw.* **2013**, *86*, 1965–1975. [CrossRef]
11. Chang, C.C.; Kieu, T.D.; Wu, W.C. A lossless data embedding technique by joint neighboring coding. *Pattern Recognit.* **2009**, *42*, 1597–1603. [CrossRef]

12. Lee, J.D.; Chiou, Y.H.; Guo, J.M. Reversible data hiding based on histogram modification of SMVQ indices. *IEEE Trans. Inf. Forensics Secur.* **2010**, *5*, 638–648. [CrossRef]
13. Chang, C.C.; Lin, C.Y.; Fan, F.H. Lossless data hiding for color images based on block truncation coding. *Pattern Recognit. Lett.* **2008**, *41*, 2347–2357. [CrossRef]
14. Chang, C.C.; Lin, C.Y.; Fan, Y.H. Reversible steganography for BTC-compressed images. *Fundam. Inform.* **2011**, *109*, 121–134.
15. Li, C.H.; Lu, Z.M.; Su, Y.X. Reversible data hiding for BTC-compressed images based on bitmplane flipping and histogram shifting of mean tables. *Inf. Technol. J.* **2011**, *10*, 1421–1426.
16. Sun, W.; Lu, Z.M.; Wen, Y.C. High performance reversible data hiding for block truncation coding compressed images. *Signal Image Video Process.* **2013**, *7*, 297–306. [CrossRef]
17. Lo, C.C.; Hu, Y.C.; Chen, W.L.; Wu, C.M. Reversible data hiding scheme for BTC-compressed images based on histogram shifting. *Int. J. Secur. Appl.* **2014**, *8*, 301–314. [CrossRef]
18. Chang, I.C.; Hu, Y.C.; Chen, W.L.; Lo, C.C. High capacity reversible data hiding scheme based on residual histogram shifting for block truncation coding. *Signal Process.* **2015**, *108*, 376–388. [CrossRef]
19. Lin, C.C.; Liu, X.L.; Tai, W.L.; Yuan, S.M. A novel reversible data hiding scheme based on AMBTC compression technique. *Multimed. Tools Appl.* **2015**, *74*, 3823–3842. [CrossRef]
20. Delp, E.J.; Mitchell, O.R. Image compression using block truncation coding. *IEEE Trans. Commun.* **1979**, *27*, 1335–1342. [CrossRef]
21. Canny, J. A computational approach to edge detection. *IEEE Trans. Pattern Anal. Mach. Intell.* **1986**, *8*, 679–698. [CrossRef]
22. Kanungo, T.; Mount, D.M.; Netanyahu, N.S.; Piatko, C.D.; Silverman, R.; Wu, A.Y. An efficient k-means clustering algorithm: Analysis and implementation. *IEEE Trans. Pattern Anal. Mach. Intell.* **2002**, *24*, 881–892. [CrossRef]
23. Gray, R.M. Vector quantization. *IEEE Assp Mag.* **1984**, *1*, 4–29. [CrossRef]
24. Kim, T. Side match and overlap match vector quantizers for images. *IEEE Trans. Image Process.* **1992**, *1*, 170–185. [CrossRef] [PubMed]
25. Mathews, J.; Nair, M.S. Adaptive block truncation coding technique using edge-based quantization approach. *Comput. Electr. Eng.* **2015**, *43*, 169–179. [CrossRef]
26. Goljan, M.; Fridrich, J.; Du, R. Distortion-free data embedding for images. In Proceedings of the 4th International Workshop on Information Hiding, London, UK, 25–27 April 2001; pp. 27–41.
27. Barton, J.M. Method and Apparatus for Embedding Authentication Information within Digital Data. U.S. Patent US5646997A, 8 July 1997.
28. Celik, M.U.; Sharma, G.; Tekalp, A.M.; Saber, E. Reversible data hiding. In Proceedings of the IEEE International Conference on Image Processing, Rochester, NY, USA, 22–25 September 2002; Volume 2, pp. 157–160.
29. Wu, X. Lossless compression of continuous-tone images via context selection, quantization, and modeling. *IEEE Trans. Image Process.* **1997**, *6*, 656–664. [PubMed]
30. Tian, J. Reversible data embedding using a difference expansion. *IEEE Trans. Circuits Syst. Video Technol.* **2003**, *13*, 890–896. [CrossRef]
31. Alattar, A.M. Reversible watermark using the difference expansion of a generalized integer transform. *IEEE Trans. Image Process.* **2004**, *13*, 1147–1156. [CrossRef] [PubMed]
32. Li, X.L.; Yang, B.; Zeng, T.Y. Efficient reversible watermarking based on adaptive prediction-error expansion and pixel selection. *IEEE Trans. Image Process.* **2011**, *20*, 3524–3533.
33. Chang, C.C.; Huang, Y.H.; Tsai, H.Y.; Qin, C. Prediction-based reversible data hiding using the difference of neighboring pixels. *Int. J. Electron. Commun.* **2012**, *66*, 758–766. [CrossRef]

symmetry

MDPI

Article

On a SIR Model in a Patchy Environment Under Constant and Feedback Decentralized Controls with Asymmetric Parameterizations

Manuel De la Sen [1,*], Asier Ibeas [2], Santiago Alonso-Quesada [1] and Raul Nistal [1]

1 Institute of Research and Development of Processes IIDP, University of the Basque Country, Campus of Leioa, Barrio Sarriena, 48940 Leioa, Bizkaia, Spain; santiago.alonso@ehu.eus (S.A.-Q.); raul.nistal@gmail.com (R.N.)

2 Department of Telecommunications and Systems Engineering, Universitat Autònoma de Barcelona, UAB, 08193 Barcelona, Spain; Asier.Ibeas@uab.cat

* Correspondence: manuel.delasen@ehu.eus

Received: 15 February 2019; Accepted: 14 March 2019; Published: 22 March 2019

Abstract: This paper presents a formal description and analysis of an SIR (involving susceptible-infectious-recovered subpopulations) epidemic model in a patchy environment with vaccination controls being constant and proportional to the susceptible subpopulations. The patchy environment is due to the fact that there is a partial interchange of all the subpopulations considered in the model between the various patches what is modelled through the so-called travel matrices. It is assumed that the vaccination controls are administered at each community health centre of a particular patch while either the total information or a partial information of the total subpopulations, including the interchanging ones, is shared by all the set of health centres of the whole environment under study. In the case that not all the information of the subpopulations distributions at other patches are known by the health centre of each particular patch, the feedback vaccination rule would have a decentralized nature. The paper investigates the existence, allocation (depending on the vaccination control gains) and uniqueness of the disease-free equilibrium point as well as the existence of at least a stable endemic equilibrium point. Such a point coincides with the disease-free equilibrium point if the reproduction number is unity. The stability and instability of the disease-free equilibrium point are ensured under the values of the disease reproduction number guaranteeing, respectively, the un-attainability (the reproduction number being less than unity) and stability (the reproduction number being more than unity) of the endemic equilibrium point. The whole set of the potential endemic equilibrium points is characterized and a particular case is also described related to its uniqueness in the case when the patchy model reduces to a unique patch. Vaccination control laws including feedback are proposed which can take into account shared information between the various patches. It is not assumed that there are in the most general case, symmetry-type constrains on the population fluxes between the various patches or in the associated control gains parameterizations.

Keywords: epidemic model; irreducible matrix; Metzler matrix; disease transition and transmission matrices; decentralized control; disease-free and endemic equilibrium points; Moore–Penrose pseudoinverse; next generation matrix; patchy environment; vaccination controls

1. Introduction

Usually, populations mutually interact through migrations and immigrations to and from other environments. Therefore, the study of more general epidemic models based on interacting subsystems, patches or frame-worked in patchy environments is of a major interest. See, for instance [1–8], and references therein. Then, the implementation of decentralized treatment or vaccination strategies in health centres [9] is of interest, so as to increase their efficiency, by taking into account not only the fixed population assigned to them but also the available information about the fluctuant population associated with migration and punctual travelling. It can be pointed out that the topic of Decentralized Control is very important in a variety of complex problems where control decisions have to be locally taken for the integrated subsystems due to a lack of full information on the coupling dynamics from and to the remaining coupled subsystems taking part of the whole dynamic systems [10–12], the first one concerning with decentralized control while the two last ones are concerned with positivity. In [13], some useful numerical tools are given concerning the non-singularity of perturbed matrices which are used in this paper. Background literature on dynamic systems, including its role on epidemic modelling, is given in [14–19]. In this context, typical situations which need relevant attention when dealing with epidemic models, thinking of their usefulness in their practical implementation in health centers are:

(a) The implementation of mixed constant and feedback controls with eventual alternative controller parameterizations and supervisory switching actions between them according to optimization trade-off criteria on the vaccine costs, or their availability, and the infection evolution through time [15,20]. The supervisory scheme chooses online the best appropriate controller parameterization that minimizes the loss function. These considerations could be also of potential applicability interests in the cases of quarantine evaluation on certain parts of the population [17], or occurring transfers from infectious to susceptible individuals [21].
(b) The need for a development of adequate strategies for online either commissioning data [22], or intervention strategies [23], or even the programming of useful strategies for vaccine procurement in due time towards its application to the population [24].
(c) The design of control strategies to fight against the epidemic spreading on multiplex networks which are subject to nonlinear mutual interaction [25], or in cases when the vaccination [16,26–28] is imperfect so that certain amounts of vaccinated susceptible subpopulation are not, in fact, removed from the susceptible subpopulation and transferred to the recovered one.

It can be pointed out that patch models have also been used for description of diseases spreading in the real world. In particular, these kind of models have been used to simulate and predict the spatial spreading of infectious diseases. For instance, it is concluded in [29] that the analysis the disease dynamics by considering the effective distances leads to understand complex contagion mechanisms in multiscale networks. The performed analysis showed that network and flux information are sufficient to predict the dynamics and the arrival times. Finally, it was pointed out that the study could be extended to other contagion phenomena, such as activated bio invasion or the spread of rumors. On the other hand, an operational forecast system was developed and verified in [30] that can successfully predict the spatial transmission of influenza in the United States at the state and county levels. On the other hand, we point out that there are other epidemic problems which involve couplings of dynamics between different compartments and subsystems like, for instance, when there are combined diseases and/or the influence of vectors in their propagation. See, for instance [31]. The designed system included processes of surveillance data from multiple locations, forecast accuracy for onset week, peak week, and peak intensity. This paper is focused on the study of the disease-free and endemic equilibrium points as well as the global stability in a patchy environment with multiple patches when there are travelling populations coming into and leaving the various patches. Vaccination strategies are proposed so that each health centre at a particular patch can have and use some certain crossed shared complete or partial information from the remaining patches. It is not assumed, in the most general case, that there are symmetry-type constraints related to the mutual interchanges of populations between pairs of patches or in the control gain parameterizations. The paper is organized as follows. Section 2 describes the proposed SIR epidemic model in a patchy environment of n patches under vaccination control laws which consist of constant and proportional to the susceptible subpopulation actions and which are implemented at each compartment of the patchy structure.

The model has travel matrices which take into account the acquisitions and loses of the subpopulations from the other patches due to populations travelling interchanges between each particular patches. The complete model is described in the presence of a feedback vaccination law which contains, in general, constant and feedback linear information on the susceptible subpopulations. It is assumed, in the most general case, that each community health centre can have either a total, a partial, or none information about the susceptible subpopulations of the remaining patches. Such an information can be suitably used, if desired, to generate the whole vaccination control law. Such a law might take into account at each patch not only the subpopulation information of such a concrete patch but, eventually, a total or a partial information of the remaining patches in the whole disposal. These above cases related to the control synthesis rely on the well-known frameworks of centralized control, partially decentralized control, or (fully) decentralized control which are usually invoked in classical Control Theory research [10], especially when the controlled system is complex or distributed in patches which can be physically distributed [10,18,19]. Section 2 also studies the non-negativity of the solutions with initial conditions in the first orthant of the state space and the allocation and uniqueness of the disease-free equilibrium point. Section 3 characterizes the basic reproduction number of the disease by defining the next generation matrix and using its spectral radius as well as the local and global stability and instability properties of the disease-free equilibrium point according to the value of the disease reproduction number compared to unity. The disease-free equilibrium point is calculated as being explicitly dependent on the disease parameters in the model and the control gains. Special particular results are focused on in the cases when some of the relevant travel matrices are irreducible. The endemic equilibrium points are also studied. It is proved that there is at least one endemic equilibrium which is positive and stable (then attainable, that is, allocated within the first orthant of the state space) if the reproduction number equals or exceeds unity. Such an equilibrium point is confluent with the disease- free one if the reproduction number is unity. It is seen, in particular, that if the infectious travel matrix is irreducible, then either all the infectious subpopulation are zero or none of them is zero. This is a very relevant result since with such a kind of conditions, it can be argued that the infectious subpopulations are non-zero at any patches for any endemic equilibrium point. Parallel results are observed in cases when the susceptible travel matrix is irreducible. The characterization of the whole set of endemic equilibrium points is described via the Moore–Penrose pseudoinverse matrix tools [32] by defining a linear algebraic system which contains a partial information of the potential existing set of endemic equilibrium points by neglecting the influence of the quadratic terms associated with the coefficient transmission rates. A complementary nonlinear equation system which is informative about the quadratic terms taking account from the contacts susceptible-infectious in all the patches is then coupled to the above linear system as an extra constraint. If such an algebraic system is compatible indeterminate then there are infinitely many endemic equilibrium solutions including the attainable and un-attainable ones. Section 4 is devoted to the study of the proposed vaccination controls and their implementation in a fully or partly decentralized control context. In particular, the proportional vaccination to the susceptible subpopulation at each patch can be applied only on the susceptible of that patch by taking into account the susceptible subpopulations of those of the other patches which supply it with such an information. The main objective is to distribute the whole set of available vaccines among all the community health centres by sharing such an information. Another potential strategy can be the implementation of vaccination control strategies at each particular health centre of a concrete patch not only on its assigned recorded susceptible but on the travelling susceptible subpopulations coming into it from other patches. Simulated Examples are given and discussed in Section 5. Finally, conclusions end the paper. The proofs of some of the involved results of Section 3 are given in the Appendices A and B.

Notation

$\overline{n} = \{1, 2, \ldots, n\}$, e_i is the i-th unity Euclidean canonical vector of $\boldsymbol{R}^n$ and I_n is the n-th identity matrix.

$\boldsymbol{R}_+ = \boldsymbol{R}_{0+} \cup \{0\}$; $\boldsymbol{R}_{0+}$ $\{z \in \boldsymbol{R}: z \geq 0\}$ are the sets of positive and non-negative real numbers, respectively.

$\boldsymbol{Z}_+ = \boldsymbol{Z}_{0+} \cup \{0\}$; $\boldsymbol{Z}_{0+}$ $\{z \in \boldsymbol{R}: z \geq 0\}$ are the sets of positive and non-negative integer numbers, respectively.

$A \in \boldsymbol{R}^{n \times n}$ is a Metzler matrix, denoted by $A \in M_E^{n \times n}$, if all its off-diagonal entries are non-negative.

$A \succeq 0$ (in words, A is non-negative) means that the real matrix $A = (a_{ij})$ has non-negative entries; $A \succ 0$ (in words, A is positive) means that $a_{ij} \geq 0$; $\forall i, j \in \overline{n}$ and there is some $(i, j) \in \overline{n} \times \overline{n}$ such that $a_{ij} > 0$; and $A \succ\succ 0$ (in words, A is strictly positive) means that all the entries of the real matrix or real vector A are positive. Similar notations are kept for vectors being non-negative (all the components are non-negative), positive (if non-negative with at least one positive component), and strictly positive (all the components are positive).

$A \succeq B$, respectively $A \succ B$, respectively, $A \succ\succ B$ means that $A - B \succeq 0$, respectively $A - B \succ 0$, respectively, $A - B \succ\succ 0$. On the other hand, $A \prec 0$ is identical to $-A \succ 0$, and $A \prec B$ to $B \succ A$. Similar considerations stand "mutatis–mutandis" for the various notations with the symbols "$\preceq$", "$\prec\prec$".

e_i is the i-th canonical Euclidean vector of the real space $\boldsymbol{R}^r$ whose i-th canonical is unity where the dimension r depends on context.

The superscripts T and $\dagger$ stand for transpose and Moore–Penrose pseudoinverses, respectively. If A is a square real non-singular matrix then the transpose of the inverse, identical to inverse of the transpose is denoted by A^{-T}.

The symbols $\vee$ and $\wedge$ stand for logic disjunction and conjunction, respectively.

If $A = (A_{ij})$ is a real matrix $|A| = (|A_{ij}|)$. If $A = (A_1, A_2, \ldots., A_n)^T$ is a real vector, then $|A| = (|A_1|, |A_2|, \ldots., |A_n|)^T$.

If A is a square matrix then $\rho(A)$ is its spectral radius, $\|A\|_2$ is the ℓ_2 (or spectral) norm and $\lambda_{\max}(A)$, and respectively, $\lambda_{\min}(A)$ is its maximum, and respectively, minimum eigenvalue provided that it is real. $\|A\|_1$ and $\|A\|_\infty$ denote, respectively, the ℓ_1 and ℓ_∞ norms.

The time argument in the time-varying variables of differential equations is suppressed for the sake of simplicity when no confusion is expected.

We point out that patches could also be referred to as "nodes" (villages, suburbs, towns or regions, each one with a health centre) while "compartment" is each individual subpopulation of susceptible infectious or recovered at each node and "subsystem" is each SIR epidemic mathematical model located at each node in the sense that its describes the self-dynamics at any patch of the whole model including the effects of couplings to other compartments or subsystems. Thus, in our model, the whole system has n subsystems, each one located at one of the n patches, and each subsystem has three compartments, one for each subpopulation.

2. SIR Epidemic Model in a Patchy Environment Under Constant and Proportional Vaccination Controls

Consider the following epidemic model in a patchy environment with constant and proportional to the susceptible vaccination controls, which are assumed being monitored in a patchy environment as well:

$$\begin{aligned} \dot{S}_i(t) &= \Lambda_i - \beta_i S_i(t) I_i(t) - d_i^S S_i(t) + \textstyle\sum_{j(\neq i)=1}^n \left(a_{ij} S_j(t) - a_{ji} S_i(t)\right) - V_i \\ (t) \dot{I}_i(t) &= \beta_i S_i(t) I_i(t) - \left(d_i^I + \gamma_i\right) I_i(t) + \textstyle\sum_{j(\neq i)=1}^n \left(b_{ij} I_j(t) - b_{ji} I_i(t)\right) \\ \dot{R}_i(t) &= \gamma_i I_i(t) - d_i^R R_i(t) + \textstyle\sum_{j(\neq i)=1}^n \left(c_{ij} R_j(t) - c_{ji} R_i(t)\right) + V_i(t), \end{aligned} \tag{1}$$

$\forall i \in \overline{n}$, subject to initial conditions $S_{i0} = S_i(0) \geq 0$, $I_{i0} = I_i(0) \geq 0$ and $R_{i0} = R_i(0) \geq 0$. In the above model, $S_i(t)$, $I_i(t)$ and $R_i(t)$ are the susceptible, infectious and recovered (or immune) subpopulations in the i-th patch for $i \in \overline{n}$, respectively, while β_i and γ_i are, respectively, the disease transmission coefficient rate between susceptible and infectious individuals and the recovery rate of the infectious in the i-th patch. The parameter Λ_i is the influx of population into the i-th patch. It can be mentioned that in the real word, the influx may also include infectious and immunized subpopulations. However, the influx to infectious and immunized subpopulations is smaller in general than the one to the susceptible subpopulation. In this way, the model only considers the influx affecting the susceptible. The parameters d_i^S, d_i^I and d_i^R are death rates of the susceptible, infectious and recovered, respectively, in the i-th patch. All the parameters of the epidemic model (1) are assumed non-negative and, furthermore, Λ_i, β_i, d_i^S, d_i^I and d_i^R are assumed to be positive for any $i \in \overline{n}$. The travel matrices $A = (a_{ij}) \succeq 0$, $B = (b_{ij}) \succeq 0$ and $C = (c_{ij}) \succeq 0$ are not necessarily symmetric and this fact does not affect to the problem formulation. Note that the immigration and outmigration amounts are proportional to the subpopulation values at the various patches. However, the stationary populations never reach zero values at any patch if the respective influx term is nonzero. The description of (1) can be made through the susceptible, infectious and recovered vectors $S(t) = (S_1(t), S_2(t), \dots, S_n(t))^T$, $I(t) = (I_1(t), I_2(t), \dots, I_n(t))^T$ and $R(t) = (R_1(t), R_2(t), \dots, R_n(t))^T$, respectively. The vaccination controls are assumed to be monitored via linear feedback information from the susceptible and have the form:

$$V_i(t) = V_{i0} + \sum_{j=1}^{n} K_{ij} S_j(t),\ i = 1, 2, \cdots, n (n \geq 2) \tag{2}$$

for given prefixed control gains K_{ij}. The replacement of (2) into (1) yields:

$$\begin{aligned} \dot{S}_i(t) &= -\beta_i S_i(t) I_i(t) - d_i^S S_i(t) + \textstyle\sum_{j(\neq i)=1}^{n} \left((a_{ij} - K_{ij}) S_j(t) - a_{ji} S_i(t) \right) + \Lambda_i - V_{i0} - K_{ii} S_i(t) \\ \dot{I}_i(t) &= \beta_i S_i(t) I_i(t) - \left(d_i^I + \gamma_i\right) I_i(t) + \textstyle\sum_{j(\neq i)=1}^{n} \left(b_{ij} I_j(t) - b_{ji} I_i(t)\right) \\ \dot{R}_i(t) &= \gamma_i I_i(t) - d_i^R R_i(t) + \textstyle\sum_{j(\neq i)=1}^{n} \left(c_{ij} R_j(t) - c_{ji} R_i(t) + K_{ij} S_j(t)\right) + V_{i0} + K_{ii} S_i(t), \end{aligned} \tag{3}$$

$\forall i \in \overline{n}$. In the sequel, and for the sake of simplicity, the dependence of the variables from time is deleted in the notation when no confusion is expected. The first part of the subsequent result relies on the existence, uniqueness and attainability (or reachability), in the sense that it has no negative component, of the disease-free equilibrium point. The second part of such a result establishes that, for identically zero infection levels through time, the disease-free equilibrium point is globally exponentially stable. The proof is based on the fact that the opposed matrix to an M-matrix is a Metzler matrix and a Metzler matrix is a stability matrix if and only if it is non-singular and its minus inverse is positive:

Theorem 1. *Define two real vectors P and* Λ *and a real square matrix D* as follows:

$$P = \left[S^T, R^T\right]^T \in \mathbf{R}^{2n};\ \Lambda = \left[\Lambda_S^T, \Lambda_R^T\right]^T \in \mathbf{R}^{2n};\ D = \begin{bmatrix} D_{SS} & D_{SR} \\ D_{RS} & D_{RR} \end{bmatrix} \in \mathbf{R}^{2n \times 2n} \tag{4}$$

where:

$$S = [S_1, S_2, \dots, S_n]^T;\ R = [R_1, R_2, \dots, R_n]^T \tag{5}$$

$$\Lambda_S = \Lambda - \Lambda_R;\ \Lambda = [\Lambda_1, \Lambda_2, \cdots, \Lambda_n]^T;\ \Lambda_R = V_0 = [V_{10}, V_{20}, \cdots, V_{n0}]^T \tag{6}$$

$$D_{SS} = \begin{bmatrix} d_1^S + \sum_{j=2}^{n} a_{j1} + K_{11} & K_{12} - a_{12} & \cdots & K_{1n} - a_{1n} \\ K_{21} - a_{21} & d_2^S + \sum_{j(\neq 2)=1}^{n} a_{j2} + K_{22} & \cdots & K_{2n} - a_{2n} \\ \vdots & \vdots & \ddots & \vdots \\ K_{n1} - a_{n1} & K_{n2} - a_{n2} & \cdots & d_n^S + \sum_{j=1}^{n-1} a_{jn} + K_{nn} \end{bmatrix} \tag{7}$$

$$D_{RR}=\begin{bmatrix} d_1^R+\sum_{j=2}^{n} c_{j1} & -c_{12} & \cdots & -c_{1n} \\ -c_{21} & d_2^R+\sum_{j(\neq 2)=1}^{n} c_{j2} & \cdots & -c_{2n} \\ \vdots & \vdots & \ddots & \vdots \\ -c_{n1} & -c_{n2} & \cdots & d_n^R+\sum_{j=1}^{n-1} c_{jn} \end{bmatrix}$$
$$D_{RS}=-K=\begin{bmatrix} -K_{11} & -K_{12} & \cdots & -K_{1n} \\ -K_{21} & -K_{22} & \cdots & -K_{2n} \\ \vdots & \vdots & \ddots & \vdots \\ -K_{n1} & -K_{n2} & \cdots & -K_{nn} \end{bmatrix};\ D_{SR}=0\in R^{n\times n} \quad (8)$$

and assume that the control gains are fixed as follows:

$$V_{i0}\in[0,\Lambda_i];\ K_{ij}\in[0,a_{ij}];\ \forall i,j(\neq i)\in\overline{n}$$
$$K_{ii}>-\left(d_i^S+\sum_{j(\neq i)=1}^{n} a_{ij}\right);\ K_{ii}\geq-\sum_{j(\neq i)=1}^{n} K_{ij};\ \forall i,j(\neq i)\in\overline{n} \quad (9)$$

such that $V_{i0}=\Lambda_i$ for some $i\in\overline{n}$. Then, the following properties hold:

(i) *The disease-free equilibrium point of Equation (1), under the vaccination control Equation (2) exists, it is unique and attainable, and given by*

$$x_{df}^{*}=\left(x_{df}^{*1\,T},\ x_{df}^{*2\,T},\ \ldots,\ x_{df}^{*n\,T}\right)^{T};\ x_{df}^{*i\,T}=\left(S_{idf}^{*},\ 0,\ R_{idf}^{*}\right);\ \forall i\in\overline{n} \quad (10)$$

with $S_{idf}^{*}=e_i^T S_{df}^{*}$, $R_{idf}^{*}=e_i^T R_{df}^{*}$; $\forall i\in\overline{n}$, where:

$$S_{df}^{*}=\left(S_{1df}^{*},\ S_{2df}^{*},\ \cdots,\ S_{ndf}^{*}\right)^{T}=D_{SS}^{-1}\Lambda_S \quad (11)$$

$$R_{df}^{*}=\left(R_{1df}^{*},\ R_{2df}^{*},\ \cdots,\ R_{ndf}^{*}\right)^{T}=D_{RR}^{-1}\left(\Lambda_R+|D_{RS}|\,S_{df}^{*}\right)=D_{RR}^{-1}\left(|D_{RS}|\,D_{SS}^{-1}\Lambda_S+V_0\right)$$

leading to a disease-free equilibrium total population vector:

$$N_{df}^{*}=\left(N_{1df}^{*},\ N_{2df}^{*},\ \cdots,\ N_{ndf}^{*}\right)^{T}=S_{df}^{*}+R_{df}^{*}=\left(I_n+D_{RR}^{-1}|D_{RS}|\right)D_{SS}^{-1}\Lambda_S+D_{RR}^{-1}V_0 \quad (12)$$

and, in the particular case that $d_i=d_i^S=d_i^R$; $\forall i\in\overline{n}$ to the following disease-free equilibrium total population amount:

$$N_{Tdf}^{*}=\sum_{i=1}^{n}N_{idf}^{*}=\sum_{i=1}^{n}\frac{\Lambda_i}{d_i} \quad (13)$$

This limit total population is also reached under any existing endemic equilibrium points. Furthermore, the total population $N(t)$ is bounded for any finite initial conditions and all $t\geq 0$.

(ii) *The solution trajectory of the linearized system around the disease-free equilibrium point of the model Equation (3) within the zero-infective ($I\equiv 0\in \mathbf{R}^n$)2n-dimensional subspace of $\mathbf{R}^{3n}$ is non-negative for any non-negative initial conditions $S_i(0)$, $R_i(0)$; $\forall i\in\overline{n}$ and it is also globally exponentially stable irrespective of the vaccination controls.*

Proof. Note that the epidemic model (1) is subject to the parametrical constraints that Λ_i, β_i, d_i^S, d_i^I and d_i^R are positive for any $i\in\overline{n}$, and $A=(a_{ij})\succeq 0$, $B=(b_{ij})\succeq 0$ and $C=(c_{ij})\succeq 0$ under the vaccination controls (2) subject to (9). Therefore each two terms $a_{ii}S_i$ and each two terms $c_{ii}R_i$, with opposed signs, become cancelled, respectively, in the first and third equation of Equations (3) for all $i\in\overline{n}$. Then, one can fix $a_{ii} = c_{ii} = 0$ for $i \in \overline{n}$ in Equations (7) and (8) with no loss in generality by keeping the summations from one to n. The disease-free equilibrium point satisfies the constraints:

$$-d_i^S S_i + \sum_{j=1}^{n}\left(\left(a_{ij} - K_{ij}\right)S_j - a_{ji}S_j\right) + \Lambda_i - V_{i0} = 0$$

$$-d_i^R R_i + \sum_{j=1}^{n}\left(c_{ij}R_j - c_{ji}R_i + K_{ij}S_j\right) + V_{i0} = 0;$$

$\forall i \in \overline{n}$, by fixing $a_{ii} = c_{ii} = 0$ for $i \in \overline{n}$. Note that D_{RR} has non-positive off-diagonal entries with the sum of all the entries per column being positive. Thus, it is a non-singular M-matrix with $D_{RR}^{-1} \succeq 0$. Also, D_{SS} is has non-positive off-diagonal entries with the sum of all the entries per column being positive from Equation (9). Thus, it is a non-singular M-matrix with $D_{SS}^{-1} \succeq 0$ [1]. Furthermore, $-D_{RS} = |D_{RS}| \succeq 0$. Therefore, the disease-free equilibrium point is unique and defined by Equations (10) and (11) subject to Equations (4)–(9). The total disease-free equilibrium population Equation (12) follows directly from Equation (11) and the disease-free total population vector is $N_{df}^* = S_{df}^* + R_{df}^*$. It is attainable in the sense that it has no negative components and it is also nonzero, since D_{SS} and D_{RR} are non-singular from Equation (11), subject to Equations (4)–(9). Equation (13) follows since the total population satisfies the constraint:

$$\dot{N}_T = \sum_{i=1}^{n} N_i = \sum_{i=1}^{n}\left(S_i + I_i + R_i\right) = \sum_{i=1}^{n}\left[\Lambda_i - \left(d_i^S S_i + d_i^I I_i + d_i^R R_i\right)\right]$$

and, for the disease-free equilibrium point with $d_i = d_i^S = d_i^R; \forall i \in \overline{n}$,

$$\begin{aligned}\dot{N}_{Tdf}^* &= \textstyle\sum_{i=1}^{n}\left[\Lambda_i - d_i N_{idf}^*\right] + \sum_{i=1}^{n}\sum_{j=1}^{n}\left[\left(a_{ij}S_j - a_{ji}S_i\right) + \left(b_{ij}I_j - b_{ji}I_i\right) + \left(c_{ij}R_j - c_{ji}R_i\right)\right] \\ &= \textstyle\sum_{i=1}^{n}\left[\Lambda_i - d_i N_{idf}^*\right] + 0 = \sum_{i=1}^{n}(0) = 0\end{aligned}$$

so that $N_{df}^* = \sum_{i=1}^{n} N_{idf}^* = \sum_{i=1}^{n} \frac{\Lambda_i}{d_i}$. It follows that $N(t)$ is bounded for any finite initial conditions for all $t \geq 0$ and $N(t) \to N_{Tdf}^*$ as $t \to \infty$. Property **(i)** has been proved. To prove Property **(ii)**, first note that the Jacobian matrix of the linearized system (1), subject to Equation (2), or equivalently Equation (3), about x_{df}^* within the manifold $I \equiv 0$ is $J_{df}^* = -D$. Since the conditions Equations (9) hold then D is an M-matrix with $D^{-1} \succ 0$. Thus, $J_{df}^* \in M_E^{n \times n}$ so that the linearized solution trajectory is non-negative for any given set of non-negative initial conditions since a time-invariant linear system has a non-negative solution trajectory irrespective of any given non-negative initial conditions if and only if its matrix of dynamics is a Metzler matrix [11,12]. Furthermore, the Jacobian matrix is invertible satisfying $-J_{df}^{*^{-1}} = D^{-1} \succ 0$. Since a Metzler matrix is a stability matrix if and only if it is non-singular and its minus inverse is positive, one concludes that the linearized system around the disease-free equilibrium point is globally exponentially stable since it is time-invariant so that the asymptotic stability is also exponential. □

If, for generality purposes and coherency with the generality of the model, it is supposed in Theorem 1 **(i)**, Equation (13), that, in general, $d_i \neq d_i^R$, with $d_i = d_i^R = d_i^S + \tilde{d}_i; \forall i \in \overline{n}$ in the sense that if the parameters differ from each other, then the mortality of the recovered who already suffered the disease is slightly higher than that of the susceptible since they suffered from the illness. Thus, one gets:

$$N_{Tdf}^* = \sum_{i=1}^{n} N_{idf}^* = \sum_{i=1}^{n} \frac{\Lambda_i - \tilde{d}_i R_{idf}^*}{d_i}(<) \approx \sum_{i=1}^{n} \frac{\Lambda_i}{d_i}$$

Remark 1. *Note from Equation (1) and Equation (2) that if* $I_i(0) = 0$; *for some* $i \in \overline{n}$ *then* $I_i(t) = 0$; $\forall i \in \overline{n}$, $t \geq 0$. *Under these conditions Theorem 1* **(ii)** *applies.*

Remark 2. *Note from Equations (2), (3), (4) and (9) that, although* $K_{ij} \geq 0$; $\forall i \in \overline{n}$ *in the vaccination law, it is not requested for any particular gain* K_{ii} *to be positive.*

The subsequent result relies on some disease-free equilibrium point results based on the positivity and irreducibility of some relevant travel matrices and constraints on the vaccination control describing population fluxes between patches of the model.

Theorem 2. *The following properties hold:*

(i) *Assume that* $B = (b_{ij})$ *is irreducible. Then,* $I_i(t) = 0$; $\forall t \in [t_1, t_2]$ *for some* $i \in \overline{n}$ *implies that* $I_j(t) = 0$; $\forall t \in [t_1, t_2]$, $\forall j \in \overline{n}$ *irrespectively of the vaccination control law.*

(ii) *Assume that* $V_{i0} = \Lambda_i$; $\forall i \in \overline{n}$ *and assume also that* $A - K = (a_{ij} - K_{ij})$ *is irreducible with* $A \succ K$. *Then,* $S_j(t) = 0$; $\forall t \in [t_1, t_2]$, $\forall j \in \overline{n}$ *if* $S_i(t) = 0$; $\forall t \in [t_1, t_2]$ *for some* $i \in \overline{n}$. *If* $B = (b_{ij})$ *and* $C = (c_{ij})$ *are irreducible,* $K = 0$ *and* $V_{i0} = 0$; $\forall i \in \overline{n}$ *then* $R_j(t) = 0$; $\forall t \in [t_1, t_2]$, $\forall j \in \overline{n}$ *if* $R_i(t) = I_i(t) = 0$; $\forall t \in [t_1, t_2]$ *for some* $i \in \overline{n}$.

(iii) *Assume that the conditions of Property* **(ii)** *hold and that, furthermore,* $K_{ij} \in [0, a_{ij}]$; $\forall i, j(\neq i) \in \overline{n}$, $K_{ii} > -\left(d_i^S + \sum_{j(\neq i)=1}^{n} a_{ij}\right)$ *and* $K_{ii} \geq -\sum_{j(\neq i)=1}^{n} K_{ij}$; $\forall i \in \overline{n}$. *Then,* $N_{df}^* = R_{df}^*$ *and* $N_{Tdf}^* = \sum_{i=1}^{n} R_{idf}^*$, *that is the total population is recovered at the disease-free equilibrium point.*

Proof. Assume that $I_i(t) = 0$; $\forall t \in [t_1, t_2]$, then $\dot{I}_i(t) = 0$; $\forall t \in (t_1, t_2)$ for some $i \in \overline{n}$, $\forall t \in (t_1, t_2)$ and assume also that there are $j(\neq i) \in \overline{n}$ and $t \in [t_1, t_2]$ such that $I_j(t) \neq 0$. One concludes from the second equation of (3), if $I_i(t) = 0$ for $t \in [t_1, t_2]$, so that $\dot{I}_i(t) = 0$ for $t \in (t_1, t_2)$, that $\sum_{j(\neq i)=1}^{n} b_{ij} I_j(t) = \sum_{j=1}^{n} b_{ij} I_j(t) = BI(t) = 0$; $\forall t \in [t_1, t_2]$. Then, $\left(\sum_{j=0}^{n-1} B^j\right) I(t) = 0$; $\forall t \in [t_1, t_2]$. But B is irreducible if and only if $\sum_{j=0}^{n-1} B^j \succ\succ 0$, since $B \succ 0$, and then $\left(\sum_{j=0}^{n-1} B^j\right) I(t) \succ\succ 0$ for any $t \in [t_1, t_2]$ if there is at least one $I_j(t) \neq 0$ for some $j(\neq i) \in \overline{n}$ and some $t \in [t_1, t_2]$, a contradiction to $\sum_{j(\neq i)=1}^{n} b_{ij} I_j(t) = 0$; $\forall t \in [t_1, t_2]$. Then, $I_j(t) = 0$; $\forall t \in [t_1, t_2]$ so that $\dot{I}_j(t) = 0$; $\forall t \in (t_1, t_2)$, $\forall j \in \overline{n}$. Property **(i)** has been proved. On the other hand, one concludes from the first equation of (3) if $S_i(t) = 0$ for $t \in [t_1, t_2]$, so that $\dot{S}_i(t) = 0$ for $t \in (t_1, t_2)$, that

$$\sum_{j(\neq i)=1}^{n} (a_{ij} - K_{ij}) S_j = \sum_{j=1}^{n} (a_{ij} - K_{ij}) S_j = 0; \ \forall t \in [t_1, t_2]$$

provided that $V_{i0} = \Lambda_i$; $\forall i \in \overline{n}$ and, if $R_i(t) = 0$ and $I_i(t) = 0$ for $t \in [t_1, t_2]$, so that $\dot{R}_i(t) = 0$ for $t \in (t_1, t_2)$, one concludes that, if in addition $V_{i0} = 0$; $\forall i \in \overline{n}$ then $\sum_{j(\neq i)=1}^{n} (c_{ij} R_j - c_{ji} R_i) = 0$. The proof of Property **(ii)** is completed under similar reasoning as that used in the proof of Property **(i)**. Finally, Property **(iii)** follows directly from Property **(ii)** and Theorem 1 **(i)** via Equation (9). □

It has to be pointed out that a particular version of Theorem 2 **(i)** for the case of absence of vaccination controls has been proved in another way in [1]. In the total absence of vaccination parameterized by the vector $\Omega = 0$, the vectors and matrices of Equations (4)–(8) are subject to the following replacements $\Lambda_R \to 0$, $D_{SS} \to D_{SS0}$, $D_{RS} \to 0$; and D_{RR} and $D_{SR} = 0$ are kept identical with:

$$D_{SS0} = D_{SS}]_{\Omega=0} = \begin{bmatrix} d_1^S + \sum_{j=2}^n a_{j1} & -a_{12} & \cdots & -a_{1n} \\ -a_{21} & d_2^S + \sum_{j(\neq 2)=1}^n a_{j2} & \cdots & -a_{2n} \\ \vdots & \vdots & \ddots & \vdots \\ -a_{n1} & -a_{n2} & \cdots & d_n^S + \sum_{j=1}^{n-1} a_{jn} \end{bmatrix}$$

3. Basic Reproduction Number: Attainability of the Endemic Equilibrium versus Instability of the Disease-Free One

Define the following matrices:

$$F = Diag\left(\beta_1 S_{1df}^*, \beta_2 S_{2df}^*, \cdots, \beta_n S_{ndf}^*\right) \tag{14}$$

$$U = \begin{bmatrix} d_1^I + \gamma_1 + \sum_{j=2}^n b_{j1} & -b_{12} & \cdots & -b_{1n} \\ -b_{21} & d_2^I + \gamma_2 + \sum_{j(\neq 2)=1}^n b_{j2} & \cdots & -b_{2n} \\ \vdots & \vdots & \ddots & \vdots \\ -b_{n1} & -b_{n2} & \cdots & d_n^I + \gamma_n + \sum_{j=1}^{n-1} b_{jn} \end{bmatrix} \tag{15}$$

The basic reproduction number is $R_0 = \rho(FU^{-1})$, where $(-U)$ is the transition matrix, F is the transmission matrix and FU^{-1} is the next generation matrix. The following positivity and stability result, proven in Appendix A, holds:

Theorem 3. *The following properties hold:*

(i) $(-U) \in M_E^{n\times n}$ *is stability matrix.*

(ii) *If* $\beta_i = 0; \forall i \in \overline{n}$ *then the disease-free equilibrium point is globally exponentially stable and any solution trajectory is non-negative for all time for any given non-negative initial conditions.*

(iii) *If* $R_0 < 1$ *then the disease-free equilibrium point* x_{df}^* *is locally asymptotically stable and, if* $R_0 > 1$*, such an equilibrium point is unstable.*

(iv) *The reproduction number satisfies the subsequent upper-bounding constraint:*

$$R_0 \leq \overline{R}_{01} = \beta \max_{1\leq i\leq n} (\beta_{ir}) \|D_{SS}^{-1} (\Lambda - V_0)\|_2 \rho^{1/2} \left(U^T U\right)^{-1} \tag{16}$$

where $\beta_{ir} = \beta_i/\beta; \forall i \in \overline{n}$*, are relative transmission coefficient rates. Assume, in addition, that* $\|U_{0d}\|_2 < 1/\|U_d^{-1}\|_2$ *and* $\|D_{SS0d}\|_2 < 1/\|D_{SSd}^{-1}\|_2$*, where* U_d *and* U_{0d} *are the diagonal and off-diagonal parts part of* $U = U_d + U_{0d}$ *and* D_{SSd} *and* D_{SSod} *are the diagonal and off-diagonal parts of* D_{SS}*. Then,*

$$R_0 \leq \overline{R}_{02} = \beta \max_{1\leq i\leq n} (\beta_{ir}) \frac{\|D_{SSd}^{-1}\|_2}{1 - \|D_{SSd}^{-1}\|_2 \|D_{SS0d}\|_2} \frac{\|U_d^{-1}\|_2}{1 - \|U_d^{-1}\|_2 \|U_{0d}\|_2} \|\Lambda - V_0\|_2 \tag{17}$$

with $\beta \geq 0$ *being a prefixed reference value of the coefficient transmission rate.*

(v) $\overline{R}_{02}$ *is minimized for any given model parameterization and any given constant vaccination vector* V_0 *if the vaccination control gains for the susceptible are chosen as* $K_{ij} = a_{ij}; \forall i, j \in \overline{n} \backslash \{1\}$*. Such a reproduction number upper-bound is zeroed if each whole influx of population in all patches are vaccinated by constant controls.*

Remark 3. *Note that* β *can be, in practice, one of the coefficient rates (for instance, its maximum or minimum value). Note that the choice* $\beta = 0$ *is feasible if and only if* $\beta_i = 0; \forall i \in \overline{n}$*.*

The non-negativity of the linearized solution proved in Theorem 1 ***(ii)*** *also applies to the whole non-linear system under weak conditions as follows.*

Theorem 4. *Assume that the vaccination control constrains Equations (9) hold and that* $A \succeq K$*. Then, the following properties hold:*

(i) *Any solution trajectory of the whole non-linear system Equation (1) is non-negative and bounded for all time for any given finite non-negative initial conditions*

(ii) *Assume, furthermore, that* $R_0 \geq 1$*. Then, there exists at least one endemic equilibrium point. If, in addition,* $B \succ 0$ *then any endemic equilibrium point has a positive infective population at any patch. If* $A - K \succ 0$ *is irreducible then any endemic equilibrium point has a positive susceptible population at any patch even under a maximum constant vaccination* $V_{i0} = \Lambda_i; \forall i \in \overline{n}$*.*

(iii) *There is no attainable endemic equilibrium point if* $R_0 < 1$ *while, if* $R_0 \leq 1$*, then the unique disease-free equilibrium point is globally asymptotically stable. If* $R_0 = 1$ *then such a disease-free equilibrium point coincides with one of the existing attainable endemic equilibrium points.*

Proof. From Theorem 1 **(i)**, the total population $N(t)$ is bounded for all time. By inspecting Equation (1), one concludes that if any susceptible, infectious or recovered subpopulation at any patch and time instant is zero then its time-derivative cannot be negative since $A \succeq K$, $B \succeq 0$ and $C \succeq 0$ and Equation (9) hold. Therefore,

$$\min_{i \in \overline{n}} \left(S_i(t)\,,\, I_i(t)\,,\, R_i(t)\right) \geq 0 \Rightarrow \min_{i \in \overline{n}} \left(S_i(t)\,,\, I_i(t)\,,\, R_i(t)\right) \geq 0;\ \forall t \geq 0.$$

If, furthermore, $\max_{i \in \overline{n}} \left(S_i(0)\,,\, I_i(0)\,,\, R_i(0)\right) < +\infty$ then, $\sup_{t \in \mathbf{R}_{0+}} \max_{i \in \overline{n}} \left(S_i(t)\,,\, I_i(t)\,,\, R_i(t)\right) < +\infty$ since $N(t) < +\infty; \forall t \geq 0$. Property **(i)** has been proved. Property **(ii)** is proved by contradiction for the case $R_0 > 1$. Assume that $R_0 > 1$ and since no endemic equilibrium point exists. Thus, the disease-free equilibrium point is unstable, any state solution trajectory has bounded non-negative components for any time and any finite non-negative initial conditions, and no endemic equilibrium point exists. Thus, it follows from Poincaré's index that a stable bounded limit cycle should surround the disease-free equilibrium point which is the unique (unstable) equilibrium point which has a unity Poincaré's index. But this feature contradicts that the state solution trajectory is non-negative for all time and any non-negative initial conditions so that no stable limit cycle can surround the unstable disease-free equilibrium point. Therefore, at least one endemic equilibrium point muss exist if $R_0 > 1$. The first part of Property **(ii)** has been proved. Now, if, in addition, B is irreducible then any zero infectious subpopulation at any patch implies that the infectious total population is zero from Theorem 2 **(i)**. By its equivalent contra-positive implication logic proposition, since the endemic equilibrium point has a nonzero total infectious population, any endemic equilibrium infectious subpopulation is nonzero at any patch. Thus, the infectious subpopulation is nonzero at any patch at the endemic equilibrium points. It follows in the same way that, if $(A - K) \succ 0$ is irreducible, then the endemic susceptible subpopulation has to be nonzero at any patch. Property **(ii)** has been proved for $R_0 > 1$. Now, assume that $R_0 = 1$. In this case, the disease-free equilibrium point is critically stable so that it has at least either one centre (i.e., a critical point with two imaginary complex eigenvalues in one of the two-dimensional partial Jacobian matrices) or one spurious patch (i.e., a critical point with one zero eigenvalue and the other one real positive in one of the two-dimensional partial Jacobian matrices) in at least a two-dimensional hyperplane of the phase space. This situation is also incompatible with the non-negativity of the solution trajectory so that the conclusion on the existence of an endemic equilibrium point is similar to the former part of the proof of this property. Proposition **(ii)** has been proved. To prove Property **(iii)**, assume that there is an attainable (i.e., with no negative component) endemic equilibrium point if $R_0 < 1$ and note, from Equations (1), (14) and (15), that

$$I^*_{iend} + e_i^T (F - U)^{-1} \left(-\beta_1 S^*_{1end} I^*_{1end}, -\beta_2 S^*_{2end} I^*_{2end}, \ldots, -\beta_n S^*_{nend} I^*_{nend}\right)^T;\ \forall i \in \overline{n} \tag{18}$$

where $(F-U)^{-1}$ exists and $-(F-U)^{-1} \succ 0$ since $(F-U) \in M_E^{n\times n}$ is a stability matrix since $(-U) \in M_E^{n\times n}$ is a stability matrix, so $U^{-1} \succ 0$, and $R_0 = \rho(FU^{-1}) < 1$. Thus, $(F-U)^{-1}$ has at least one positive entry per column and one positive entry per row. Then, the above equation holds for $\min_{i\in\overline{n}} \beta_i > 0$ with $I_{iend}^* > 0;\ \forall i \in \overline{n}$ if and only if $S_{jend}^* < 0$ for at least a $j \in \overline{n}$. Thus, there is no attainable endemic equilibrium point if $R_0 < 1$ and $\min_{i\in\overline{n}} \beta_i > 0$. Since an endemic equilibrium point exists for $R_0 = 1$ from Property **(ii)**, the fact that Equation (18) also holds for $R_0 = 1$, as a result, and the fact that the subsequent constraint stands for the disease-free equilibrium point if $R_0 < 1$:

$$I_{idf}^* = e_i^T(-U)^{-1}\left(-\beta_1 S_{1df}^* I_{1df}^*, -\beta_2 S_{2df}^* I_{2df}^*, \ldots, -\beta_n S_{ndf}^* I_{ndf}^*\right) = 0;\ \forall i \in \overline{n} \tag{19}$$

it follows from continuity arguments of the equilibrium points with respect to R_0 that one of the endemic equilibrium points necessarily coincide with the disease-free one for $R_0 = 1$. Now since: (a) the disease-free equilibrium point is unique and the unique attainable equilibrium point for $R_0 < 1$ (Theorem 1 **(i)**); and (b) such a point is furthermore locally asymptotically stable, since its linearized version around it is asymptotically stable (Theorem 3 **(iii)**, one concludes that the disease-free equilibrium point is globally asymptotically stable if $R_0 \le 1$. Property **(iii)** has been proved. □

Remark 4. *Theorem 4 (**ii**) establishes that, if the disease-free equilibrium point is unstable or critically stable, then an endemic equilibrium point has to exist. With some extra irreducibility-type conditions on the B-travel matrix and on the $(A-K)$-travel matrix, it is proved that the infectious and susceptible endemic equilibrium amounts are nonzero at any patch. It can be argued that the matrix of proportional vaccination gains K can modify the irreducibility or reducibility properties of the travel matrix A related to the respective properties of $(A-K)$. This fact can imply that, if in the absence of proportional vaccination to the susceptible subpopulation, the endemic equilibrium point has nonzero susceptible (respectively, zero amounts of susceptible at least at one patch) subpopulations at any patch, then, under some kind of proportional vaccination law even for a constant vaccination constraint $V_{i0} = \Lambda_i;\ \forall i \in \overline{n}$, the endemic susceptible could be zeroed at least at one patch but not in all patches. To visualize the above argument, note that the matrix constraint $\sum_{i=0}^{n-1} A^i \succ\succ -\sum_{i=0}^{n-1}\sum_{j=1}^{i}\binom{i}{j} A^{i-j}(-K)^j$ guarantees that $(A-K)$ is irreducible since*

$$\sum_{i=0}^{n-1}(A-K)^i = \sum_{i=0}^{n-1}\sum_{j=0}^{i}\binom{i}{j} A^{i-j}(-K)^j = \sum_{i=0}^{n-1} A^i + \sum_{i=0}^{n-1}\sum_{j=1}^{i}\binom{i}{j} A^{i-j}(-K)^j \succ\succ 0.$$

The characterization of the whole set of endemic equilibrium points is addressed in the following result, which is proved in Appendix B, by using algebraic tools:

Theorem 5. *Assume that $R_0 \ge 1$ and define the following matrices:*

$$A_S = \begin{bmatrix} \overline{a}_{11} & -a_{12} & \cdots & -a_{1n} \\ -a_{21} & \overline{a}_{22} & \cdots & -a_{2n} \\ \cdots & \cdots & \cdots & \cdots \\ -a_{n1} & -a_{n2} & \cdots & \overline{a}_{nn} \end{bmatrix};\ A_I = \begin{bmatrix} \overline{b}_{11} & -b_{12} & \cdots & -b_{1n} \\ -b_{21} & \overline{b}_{22} & \cdots & -b_{2n} \\ \cdots & \cdots & \cdots & \cdots \\ -b_{n1} & -b_{n2} & \cdots & \overline{b}_{nn} \end{bmatrix} \tag{20}$$

$$A_{RI} = Diag\left[-\gamma_1, -\gamma_2, \ldots, -\gamma_n\right];\ A_R = \begin{bmatrix} \overline{c}_{11} & -c_{12} & \cdots & -c_{1n} \\ -c_{21} & \overline{c}_{22} & \cdots & -c_{2n} \\ \cdots & \cdots & \cdots & \cdots \\ -c_{n1} & -c_{n2} & \cdots & \overline{c}_{nn} \end{bmatrix} \tag{21}$$

$$\Lambda_i - V_{i0} = \overline{a}_{ii}S^*_{iend} + \overline{b}_{ii}I^*_{iend} - \sum_{j(\neq i)=1}^{n}\left(\left(a_{ij} - K_{ij}\right)S^*_{jend} + b_{ij}I^*_{jend}\right) \quad (22)$$

$$V_{i0} = \overline{c}_{ii}R^*_{iend} + \gamma_i I^*_{iend} - \sum_{j(\neq i)=1}^{n} c_{ij}R^*_{jend} + K_{ij}S^*_{jend} \quad (23)$$

where

$$\overline{a}_{ii} = d_i^S + K_{ii} + \sum_{j(\neq i)=1}^{n} a_{ji},\ \overline{b}_{ii} = d_i^I + \gamma_i + \sum_{j(\neq i)=1}^{n} b_{ji},\ \overline{c}_{ii} = d_i^R + \sum_{j(\neq i)=1}^{n} c_{ji};\ \forall i \in \overline{n} \quad (24)$$

Then, the following properties hold:

(i) *The following rank condition holds:*

$$rank(b, A) = rank\, A \quad (25)$$

where the limit total population is N^* *irrespective of the equilibrium point as time tends to infinity, and*

$$A = \begin{bmatrix} 1 \ 1 \ \cdots \ 1 \ 1 \\ A_S \quad A_I \quad 0 \\ 0 \quad A_{RI} \quad A_R \end{bmatrix} \in \mathbf{R}^{(2n+1)\times 3n};\ b = \begin{bmatrix} N^* \\ \Lambda_1 - V_{10} \\ \vdots \\ \Lambda_n - V_{n0} \\ V_{10} \\ \vdots \\ V_{n0} \end{bmatrix} \in \mathbf{R}^{2n+1} \quad (26)$$

The whole set of endemic equilibrium solutions, including both the attainable and unattainable ones, is given by

$$x(y) = A^{+}b + (I_{3n} - A^{+}A)y \quad (27)$$

subject to the n constraints:

$$\beta_i = \frac{\left[\left(d_i^I + \gamma_i + \sum_{j(\neq i)=1}^{n} b_{ji}\right)e_{n+i}^T - \sum_{j(\neq i)=1}^{n} b_{ij}e_{n+j}^T\right]\left[A^{+}b + (I_{3n} - A^{+}A)y\right]}{e_i^T[A^{+}b + (I_{3n} - A^{+}A)y][A^{+}b + (I_{3n} - A^{+}A)y]^T e_{n+i}};\ \forall i \in \overline{n} \quad (28)$$

with $x(y) = \left(S^*_{1end}(y), S^*_{2end}(y), \ldots S^*_{nend}(y), I^*_{1end}(y), I^*_{2end}(y), \ldots I^*_{nend}(y), R^*_{1end}(y), R^*_{2end}(y), \ldots R^*_{nend}(y)\right)^T$ *and* e_i *is the Euclidean canonical vector whose its ith component is unity;* $\forall i \in \overline{3n}$, $A^{\mathrm{n}} = D^T\left(DD^T\right)^{-1}\left(C^TC\right)^{-1}C^T \in \mathbf{R}^{3n\times(2n+1)}$ *is the Moore–Penrose pseudoinverse of A, provided that A of rank* $p \leq 2n+1$ *is factorized as* $A = CD$ *with existing matrices* $C \in \mathbf{R}^{2(n+1)\times p}$ *and* $D \in \mathbf{R}^{p\times(2n+1)}$ *both or rank p, and* $y \in \mathbf{R}^{3n}$ *is arbitrary except that it is subject to fulfill Equation (28) for the given coefficient transmission rates* β_i *for* $i \in \overline{n}$, *where* $A^{+T} = A^T$ *[32]. The set of attainable endemic equilibrium points is given by Equation (27) subject to the constraints Equation (28) for any* $y \in Y$ *with* $Y = \{z \in \mathbf{R}^{3n}: x(z)(\in \mathbf{R}^{3n})\ 0\}$.

(ii) *If* $B = \left(b_{ij}\right)$ *is irreducible then the set of attainable endemic equilibrium points is given by (27), subject to the constraints (28), for any* $y \in Y_a$ *with*

$$Y_a = \{z \in \mathbf{R}^{3n}: (x(z)(\in \mathbf{R}^{3n})\ 0) \wedge (x_i(z) > 0; i = n+1, n+2, \ldots, 2n)\} \subset Y$$

(iii) $V_{i0} = \Lambda_i;\ \forall i \in \overline{n}$ *and both* $B = \left(b_{ij}\right)$ *and* $A - K = \left(a_{ij} - K_{ij}\right)$, *with* $A \succ K$, *are irreducible then the set of attainable endemic equilibrium points is given by Equation (27), subject to the constraints Equation (28), for any* $y \in Y_b$ *with*

$$Y_b = \{z \in \mathbf{R}^{3n}: (x(z)(\in \mathbf{R}^{3n})\ 0) \wedge ([x_i(z) > 0; i = 1,2,\ldots,n] \vee [x_i(z) = 0; i = 1,2,\ldots,n]) \wedge (x_i(z) > 0; i = n+1,n+2,\ldots,2n)\} \subset Y_a$$

(iv) *If* $K = 0$, $V_{i0} = \Lambda_i = 0$; $\forall i \in \overline{n}$ *and* $B = (b_{ij})$, $A - K = (a_{ij} - K_{ij})$, *with* $A \succ K$, *and* $C = (c_{ij})$ *are irreducible, then the set of attainable endemic equilibrium points is given by Equation (27) subject to the constraints Equation (28) for any* $y \in Y_c \subset Y_b$ *with* $Y_c = \{ z \in \boldsymbol{R}^{3n} : (\ell_1 \wedge \ell_2 \wedge \ell_3 \wedge \ell_4) \text{ holds} \}$

$$\ell_1 := x(z) \left(\in \boldsymbol{R}^{3n} \right) \succ 0$$

$$\ell_2 := (x_i(z) > 0; i = 1, 2, ..., n) \vee (x_i(z) = 0; i = 1, 2, ..., n)$$

$$\ell_3 := (x_i(z) > 0; i = n+1, n+2, ..., 2n+1)$$

$$\ell_4 := (x_i(z) > 0; i = 2n+1, 2n+2, ..., 3n) \vee (x_i(z) = 0; i = 2n+1, 2n+2, ..., 3n)$$

The conditions for the uniqueness of the existing attainable endemic equilibrium point for $R_0 \geq 1$ are given in the following result which is a direct conclusion of Theorem 5:

Corollary 1. *Assume that* $R_0 \geq 1$. *Then, the attainable equilibrium point is unique if and only there is a* $y \in \boldsymbol{R}^{3n}$ *such that*

(1) $y + A^{\dagger}(b - Ay) \succ 0$,
(2) $E(y + A^{\dagger}(b - Ay)) \succ 0$*(respectively,* $\succ\succ 0$ *if B is irreducible), where* $E = [\boldsymbol{0}_{n\times n} \quad I_{n\times n} \quad \boldsymbol{0}_{n\times n}] \in \boldsymbol{R}^{n\times 3n}$,
(3) *The n constraints (28) hold.*

One such a vector $y \in \boldsymbol{R}^{3n}$ *always exists.*

The following counterpart result to Theorem 5 and Corollary 1 holds for the case when there is only one patch in the epidemic model so that the transportation matrices are zero. The result, proved in Appendix B, gives a nice physical interpretation of the basic reproduction number and its relation to the stability properties and to the attainability of the endemic equilibrium point.

Theorem 6. *Assume that there is only one patch (i.e.,* $n = 1$*) and that* $\Lambda > V$ *with V being a constant vaccination effort. Then, there is a unique stable attainable endemic equilibrium point if the coefficient transmission rate fulfills* $\beta \geq \beta_c = \frac{d^S(d^I+\gamma)}{\Lambda - V}$, *equivalently, if the reproduction number,* $R_0 = \frac{S^*_{df}}{S^*_{end}} \geq 1$, *where* $S^*_{df} = \frac{\Lambda - V}{d^S}$ *is the susceptible subpopulation at the disease-free equilibrium point, the immune one at the disease-free equilibrium being* $R^*_{df} = \frac{V}{d^R}$. *Such an endemic equilibrium point is:*

$$S^*_{end} = \frac{d^I+\gamma}{\beta};\ I^*_{end} = \frac{\beta(\Lambda - V) - d^S(d^I+\gamma)}{\beta(d^I+\gamma)};R^*_{end} = \frac{\beta(d^I+\gamma)V + \gamma[\beta(\Lambda - V) - d^S(d^I+\gamma)]}{\beta d^R(d^I+\gamma)} \tag{29}$$

And the following properties hold:

(i) *If* $\Lambda = V$ *then there is a unique disease-free equilibrium point* $S^*_{df} = I^*_{df} = 0$; $R^*_{df} = \frac{V}{d^R}$ *while the endemic one does not exist.*
(ii) *If* $R_0 = 1$ *then the disease-free and the endemic equilibrium points coincide.*
(iii) *If* $R_0 < 1$ *then the disease-free equilibrium point is globally asymptotically stable and the endemic one is not attainable.*
(iv) *If* $V(t) = V_0 + KS(t)$ *then* $S^*_{df} = \frac{\Lambda - V_0}{d^S + K}$, $R^*_{df} = \frac{K\Lambda + d_S V_0}{d^R(d^S+K)}$, *and*

$$N^*_{df} = \frac{\Lambda}{d^S} + \frac{(d^S - d^R)(d^S V_0 + K\Lambda)}{d^S d^R (d^S + K)} = \frac{\Lambda}{d^S}\left(1 + \frac{K(d^S - d^R)}{d^R(d^S + K)}\right) + \frac{(d^S - d^R)V_0}{d^R(d^S + K)}$$

In the absence of vaccination, $N^*_{df} = S^*_{df} = \frac{\Lambda}{d^S}$ *and* $R^*_{df} = 0$.

The following result, which is proved in Appendix C, relies on the feature that the reproduction number can be reduced by the vaccination controls. This feature implies that the global asymptotic stability towards the disease-free equilibrium point can be guaranteed under smaller values of the coefficient transmission rates via an appropriate monitoring of such controls. Although the proposed model has an identical transmission matrix U for the vaccination-free and vaccinated models, it is assumed for analysis generality purposes that that associated to the vaccination case U_c can be distinct to that associated to the vaccination-free one U_{un}. This is the case, for instance, if an additional treatment control is injected on the infectious subpopulation. See, for instance [14,15].

Theorem 7. *Define* $U_c = U_{un} + \widetilde{U}$ *and* $F_c = F_{un} + \widetilde{F}$, *where* $\widetilde{F}$ *and* $\left(-\widetilde{U}\right)$ *are the disturbed transmission and transition matrix of the controlled epidemic model under a vaccination control law with respect to those of the uncontrolled (i.e., for the case when the vaccination control is null) one. Define* $R_{0un} = \rho\left(F_{un}U_{un}^{-1}\right)$ *and* $R_{0c} = \rho\left(F_cU_c^{-1}\right)$ *as the respective reproduction numbers in the vaccination-free and under vaccination. Assume that the following constraints hold:*

(1) $(-U_{un}) \in M_E^{3n\times 3n}$ *is a stability matrix,*

(2) $F_{un} \succ 0$,

(3) $\|\widetilde{U}\|_2 < 1/2\,\|U_{un}^{-1}\|_2$,

(4) $-F_{un} \prec \widetilde{F} \prec F_{un}\, U_{un}^{-1}\widetilde{U}\left(I_{3n} + U_{un}^{-1}\widetilde{U}\right)\left(I_{3n} - \widetilde{U}\left(I_{3n} + U_{un}^{-1}\widetilde{U}\right)^{-1}U_{un}^{-1}\right)^{-1}U_{un}.$

Then, $U_c \in M_E^{3n\times 3n}$ *is a stability matrix and the following properties hold:*

(i) $R_{0c} \le R_{0un}$.

(ii) *If, the conditions Equations (1)–(3) hold,* $\widetilde{F} = -\left|\widetilde{F}\right| \prec 0$ *and the constraint equation (4) is replaced with following constraints:*

(4′) $-F_{un}\, U_{un}^{-1}\widetilde{U}\left(I_{3n} + U_{un}^{-1}\widetilde{U}\right)\left(I_{3n} - \widetilde{U}\left(I_{3n} + U_{un}^{-1}\widetilde{U}\right)^{-1}U_{un}^{-1}\right)^{-1}U_{un} \prec \left|\widetilde{F}\right| \prec F_{un}.$

Then $R_{0c} \le R_{0un}$. *In addition,* $R_{0c} < R_{0un}$ *if either* $F_{un}U_{un}^{-1}$ *or* $\left|\widetilde{F}\right|U_{un}^{-1}$ *is irreducible. This property result still holds if one but not both) of the two "$\prec$"-symbols of the above equation is replaced with "$\preceq$".*

Remark 5. *Note that the applicability of Theorem 7* ***(ii)*** *is very feasible in practice according to the following considerations. Assume that the pairs* (F_{un}, U_{un}) *and* (F_c, U_c) *are the pairs defining the vaccination-free and vaccination cases linear dynamics around the disease-free equilibrium point which depends on the control gains such that* $U = U_c = U_{un}$ *from (14) and (15) for the model dealt with.(Note that Theorem 7 has been worked for the more general case when* $U_c \neq U_{un}$*). Now,* $\widetilde{F} = -\left|\widetilde{F}\right| \prec 0$ *if* $F_c \prec F_{un}$, *that is, if* $S_{df}^*(F_c) < S_{df}^*(F_{un})$. *This is directly achievable by using appropriate control gains (see Theorem 1). In the simplest case of just one patch in the model (i.e.,* $n = 1$*), note that this is achievable by choosing* $\max\left(V_0, K\right) > 0$ *from Theorem 6* ***(iv)****. The choices of the values of the control gains* V_0 *and* K *monitor the susceptible amounts* $S_{df}^*(F_c)$ *at the disease-free equilibrium. Now, assume that* $R_{0un} = 1$. *This value of the reproduction number corresponds to a certain critical disease transmission rate* β_{cun} *for given remaining modeling parameters in the vaccination-free case. This fact leads to the coincidence of the disease-free equilibrium point with the attainable endemic one and the critical stability of the disease-free equilibrium point. However, under Theorem 7, and since* $\widetilde{F} < 0$, *the vaccination control leads to the asymptotic stability of the modified disease-free equilibrium point and the un-attainability of the endemic one since* $R_{0c} < R_{0un} = 1$. *Therefore, a properly designed vaccination law increases the range of the stability boundary of the disease-free equilibrium point to reach a larger critical disease transmission rate compared to the vaccination-free case.*

4. Use of Available Patch-Crossed Information in Decentralized Vaccination Control Designs

The following situations can occur related to the vaccination controls monitoring actions:

(a) *Centralized Vaccination Control* (*CVC*). Each subsystem has the information available about the susceptible numbers of all the compartments and uses it for feedback vaccination control.
(b) *Decentralized Vaccination Control* (*DVC*) if $K_{ij} = 0; \forall i, j(\neq i) \in \overline{n}$ and $K_{ii} \neq 0; \forall i \in \overline{n}$. Each subsystem uses only self-information for control but there is no use of the susceptible number of other compartments.
(c) *Partially Decentralized Vaccination Control* (*PDVC*) if $K_{ii} \neq 0; \forall i \in \overline{n}, K_{ij} \neq 0; \forall (i,j) \in n_p \times n_q$ and $K_{ij} = 0; \forall (i,j) \in \overline{n} \times \overline{n} \backslash n_p \times n_q$, where n_p and n_q are nonempty proper subsets of $\overline{n}$.
(d) n_w-*Weak Decentralized Vaccination Control* (n_w-*WDVC*) if $K_{ij} = 0; \forall i, j(\neq i) \in \overline{n}, K_{ii} \neq 0; \forall i \in n_p$ and $K_{ii} = 0; \forall i \in \overline{n} \backslash n_p$. That is, at least one compartment of susceptible does not uses susceptible self-information for feedback in the vaccination control law which has a decentralized structure.
(e) n_w-Weak Partially Decentralized Vaccination Control (n_w-WPDVC) if in the definition of n_w-WDVC, $K_{ij} \neq 0$ for some $i, j(\neq i) \in \overline{n}$.

Note that the various concepts of "centralized control" versus "decentralized control" refer to the complete or partial shared information between dynamic subsystems and, in particular, subsystems of the patchy model or just the use of own self- information for control rather than to the physical disposal (generic one or local for each subsystem) of the controller. This is a widely admitted principle in decentralized control of dynamic systems. See, for instance [10]. Two vaccination strategies are now discussed if the vaccination controls are assumed to be monitored via linear feedback information from the susceptible by using available information at each patch from some other patches:

Strategy 1. Only the susceptible subpopulation of each patch, even if travelling population from other patches exists, is a candidate to be vaccinated while some total or partial information from the corresponding subpopulations in other patches is known and monitored for the susceptible vaccination through the crossed control gains associated with the control law (2). Such an information is used to restrict the influence of the immigration from the remaining patches into the own susceptible subpopulation of a patch in accordance with Equation (3). The control law Equation (2) is assumed to be subject to the following constraints:

$$0 \leq K_{ii} \leq M_i + \sum_{j(\neq i)=1}^{n} K_{ji} - \sum_{j(\neq i)=1}^{n} K_{ij},\ 0 \leq K_{ji} \leq a_{ji},\ V_{i0} \leq M_{i0} < \Lambda_i;\ \forall i, j(\neq i) \in \overline{n} \tag{30}$$

where $M_i > 0$ and $M_{i0} > 0$are upper-bounding constant taking into account the vaccines availability at the i-th patch for $i \in \overline{n}$. The first constraint of Equation (30) reflects that a fraction of the travelling susceptible populations coming from the remaining patches is vaccinated while the leaving one to other patches is not vaccinated. The second constraint takes into account that D_{SS} in Equation (7) is an M-matrix so that its inverse exists and is positive, so that the disease-free equilibrium point is a non-negative vector of the state space and locally asymptotically stable since $(-D_{SS}) \in M_E^{n \times n}$.

Strategy 2. Only the susceptible subpopulation proper of each patch is a candidate for vaccination but there is some partial or total information from the susceptible subpopulations from other patches. The available information on the coming in and leaving travelling susceptible subpopulations from the various patches is used to control the distribution of the vaccines to be administrated between the various patches. Such an information is used to restrict the number of administered vaccines at each patch. In this case, the vaccination control law Equation (2) is modified as follows:

$$V_i(t) = V_{i0} + K_i(t) S_i(t);\ \forall i \in \overline{n} \tag{31}$$

and the vaccination control proportional gains are given by:

$$K_i(t) = K_i(S(t)) = K_{ii} + \sum_{j(\neq i)=1}^{n} K_{ij}^0(t);\ \forall i \in \overline{n} \tag{32}$$

where:

$$K_{ij}^{0}(t)=K_{ij}^{0}(S_i(t),\,S_j(t))=\begin{cases}\frac{K_{ij}S_j(t)}{S_i(t)} & if \quad S_i(t)>\varepsilon_i \\ 0 \quad if \quad S_i(t)\le\varepsilon_i & \end{cases} \quad ;\ \forall i,j\in\overline{n} \tag{33}$$

for given prefixed control gains K_{ij} and design constants $\varepsilon_i \in \mathbf{R}_{0+};\ \forall i,j \in \overline{n}$. It turns out from Equations (31)–(33) that coupled information between distinct patch pairs can be available or not in the vaccination controls. As a result, the vaccination control (31)–(33) becomes:

$$V_i(t)=\begin{cases} V_{i0}+\sum_{j=1}^{n}K_{ij}S_j(t) & \text{if} \quad S_i(t)>\varepsilon_i \\ V_{i0}+K_{ii}S_i(t) & \text{if} \quad S_i(t)\le\varepsilon_i\end{cases} \quad ;\ \forall i\in\overline{n} \tag{34}$$

The constraints Equation (30) become modified as follows for each $i \in \overline{n}$ allowing some negative crossed control gains:

$$K_{ij}\le 0;\ 0\le K_{ji}\le a_{ji};\ ;\ \forall j(\neq i)\in\overline{n} \tag{35}$$

$$\frac{K_{ii}}{\sum_{j(\neq i)=1}^{n}\left|K_{ij}\right|}\ge \sup_{t\in\mathbf{R}_{0+}}\max_{1\le j(\neq i)\le n}\left(\frac{S_j(t)}{S_i(t)}\right);\ K_{ii}\le M_i+\sum_{j(\neq i)=1}^{n}\left|K_{ji}\right|\inf_{t\in\mathbf{R}_{0+}}\min_{1\le j(\neq i)\le n}\left(\frac{S_j(t)}{S_i(t)}\right) \tag{36}$$

Note that Equations (35) and (36) may be jointly expressed as follows:

$$\left(\sum_{j(\neq i)=1}^{n}\left|K_{ii}\right|\right)\sup_{t\in\mathbf{R}_{0+}}\max_{1\le j(\neq i)\le n}\left(\frac{S_j(t)}{S_i(t)}\right)\le K_{ii}\le M_i+\sum_{j(\neq i)=1}^{n}\left|K_{ji}\right|\inf_{t\in\mathbf{R}_{0+}}\min_{1\le j(\neq i)\le n}\left(\frac{S_j(t)}{S_i(t)}\right) \tag{37}$$

provided that the following necessary condition holds:

$$\left(\sum_{j(\neq i)=1}^{n}\left|K_{ij}\right|\right)\le\frac{M_i}{\sup_{t\in\mathbf{R}_{0+}}\max_{1\le j(\neq i)\le n}\left(\frac{S_j(t)}{S_i(t)}\right)-\inf_{t\in\mathbf{R}_{0+}}\min_{1\le j(\neq i)\le n}\left(\frac{S_j(t)}{S_i(t)}\right)} \tag{38}$$

Note the following facts:

(1) If $S_i(t)=\varepsilon_i$ and $S_i(t^{+})>\varepsilon_i$ fore some $i\in\overline{n}$ then $V_i(t)$ switches from a constant term to a combined constant plus a linear feedback term except if the control gains $K_{ij}=0;\ \forall j\in\overline{n}$ and such a $i\in\overline{n}$. In this case, the closed-loop linearized dynamic systems around any potential equilibrium points, which are defined by their corresponding Jacobian matrices at such points after absorbing the linear feedback from the susceptible subpopulations, are not time-invariant through time.

(2) If either $\inf_{t\in\mathbf{R}_{0+}} S_i(t)>\varepsilon_i$ or $\sup_{t\in\mathbf{R}_{0+}} S_i(t)\le\varepsilon_i;\ \forall\, i\in\overline{n}$, then the vaccination control law does not switch from a combined constant plus a linear feedback term to a constant term or vice-versa at any patch and at any time instant.

(3) Concerning the Centralized/Decentralized control frameworks, note that a CVC strategy is implementable if the available information allows the use of gains $K_{ij}\neq 0;\ \forall i,j(\neq i)\in\overline{n}$ since all the susceptible subpopulation and its distribution between the various patches is known at each patch. A PDVC, or a DVC strategy is adopted when some or, respectively, all the gains K_{ij} are zeroed; $\forall i,j(\neq i)\in\overline{n}$ because the global information on susceptible is not known, or not used, at each patch. The (n_w-$WDVC$) and (n_w-$WPDVC$) vaccination strategies are implemented if some of the self-proportional gains are not used at some patches (i.e., there is no vaccination action at some health centre on its own susceptible subpopulation) or, if, in addition some of the crossed susceptible information between the various patches is not available or simply not used. It can be convenient to adopt vaccination strategies which allow to guarantee a worst-case minimization, in some sense, of the disease-free equilibrium subpopulations in order to achieve

a corresponding maximization of the recovered subpopulation when the infection is removed. This idea is addressed in the sequel. Note that

$$\|D_{SS}\|_1 = \max_{1\leq i\leq n}\left[K_{ii} + d_I^S + \sum_{j(\neq i)=1}^{n}(2a_{ij} - K_{ij})\right] \tag{39}$$

$$\|D_{SS}\|_\infty = \max_{1\leq i\leq n}\left[K_{ii} + d_I^S + \sum_{j(\neq i)=1}^{n}(a_{ij} + a_{ji} - K_{ji})\right] \tag{40}$$

Then, one has from (11) via Equations (6) and (7) and using the constraints (30) for Strategy 1, by taking into account the bounded relations between the matrix and vector spectral (ℓ_2) and ℓ_1 and ℓ_∞norms, that the following lower-bounds stand for the disease-free equilibrium susceptible vector:

$$\begin{aligned}\left\|S_{df}^*\right\|_\infty &= \max_{1\leq i\leq n} S_{idf}^* \geq \frac{\|\Lambda_S\|_\infty}{\|D_{SS}\|_\infty} = \frac{\max\limits_{1\leq i\leq n}(\Lambda_i - V_{i0})}{\max\limits_{1\leq i\leq n}\left[K_{ii}+d_I^S+\sum_{j(\neq i)=1}^{n}\left(a_{ij}+a_{ji}-K_{ji}\right)\right]} \\ &\geq \frac{\max\limits_{1\leq i\leq n}(\Lambda_i - M_{i0})}{M_i - \sum_{j(\neq i)=1}^{n} K_{ij} + d_I^S + \sum_{j(\neq i)=1}^{n}\left(a_{ij}+a_{ji}\right)}\end{aligned} \tag{41}$$

$$\begin{aligned}\left\|S_{df}^*\right\|_1 &= \sum_{i=1}^{n} S_{idf}^* \geq \frac{\|\Lambda_S\|_1}{\|D_{SS}\|_1} = \frac{\sum_{i=1}^{n}(\Lambda_i - V_{i0})}{\max\limits_{1\leq i\leq n}\left[K_{ii}+d_I^S+\sum_{j(\neq i)=1}^{n}\left(2a_{ij}-K_{ij}\right)\right]} \\ \geq &\frac{\sum_{i=1}^{n}(\Lambda_i - V_{i0})}{\max\limits_{1\leq i\leq n}\left[M_i+\sum_{j(\neq i)=1}^{n} K_{ji}+d_I^S+2\left(\sum_{j(\neq i)=1}^{n}\left(a_{ij}-K_{ij}\right)\right)\right]} \geq \frac{\sum_{i=1}^{n}(\Lambda_i - V_{i0})}{\max\limits_{1\leq i\leq n}\left[M_i+\sum_{j(\neq i)=1}^{n} K_{ji}+d_I^S+2\left(\sum_{j(\neq i)=1}^{n} a_{ij}\right)\right]}\end{aligned} \tag{42}$$

$$\left\|S_{df}^*\right\|_2 = \sqrt{\sum_{i=1}^{n} S_{idf}^{*2}} \geq \frac{\|\Lambda_S\|_2}{\|D_{SS}\|_2} = \frac{\|\Lambda_S\|_2}{\lambda_{\max}^{1/2}\left(D_{SS}^T D_{SS}\right)} \geq \sqrt{\frac{\sum_{i=1}^{n}(\Lambda_i - V_{i0})^2}{n}}\max\left(\frac{1}{\|D_{SS}\|_1}, \frac{1}{\|D_{SS}\|_\infty}\right) \tag{43}$$

Remark 6. *In view of Equations (41)–(43), one concludes that available lower-bounds susceptible subpopulations at the disease-free equilibrium points can be reduced in a suboptimal worst-case design which keeps the maximum available vaccines and jointly minimizes the ℓ_1, ℓ_∞ and ℓ_2 norms by choosing:*

$$V_{i0} = M_{i0};\ K_{ij} = 0;\ K_{ji} = a_{ji};\ \forall i, j \in \overline{n}$$

$$K_{ii} = M_i + \sum_{j(\neq i)=1}^{n} K_{ji} - \sum_{j(\neq i)=1}^{n} K_{ij} = M_i + \sum_{j(\neq i)=1}^{n} a_{ji};\ \forall i \in \overline{n}$$

In the case that some outsider travelers from other patches to a certain patch $i \in \overline{n}$ have to be vaccinated for needs of global fulfillment of objectives, one can use normalizing factors $\ell_{ij} \in [0, 1]$ so that $K_{ij} = \ell_{ij}a_{ij}$ replaces the standard strategy $K_{ij} = 0; \forall j \in \overline{n}$.

In the case that some travelers from a certain patch $i \in \overline{n}$ to other patches should be vaccinated, one can use normalizing factors $\ell_{ji} \in [0, 1]$ so that $K_{ji} = \ell_{ji}a_{ji}$ replaces the standard strategy $K_{ji} = a_{ji}; \forall j \in \overline{n}$.

Note from (31) to (34) that, in the case of Strategy 2, the vaccination control parameterization is time-varying (see, for instance [20]), since there can exist switches if the susceptible subpopulation at any patch is close to zero. The following two technical results are of usefulness for Strategy 2.

Lemma 1. *Let $A \in \mathbf{R}^{n\times n}$ be a stability matrix of stability abscissa $-\rho_a < 0$ and let be $\widetilde{A} : \mathbf{R}_{0+} \to \mathbf{R}^{n\times n}$ a piecewise continuous uniformly bounded matrix function. Then, the matrix function $B : \mathbf{R}_{0+} \to \mathbf{R}^{n\times n}$, being $B(t) = A + \widetilde{A}(t); \forall t \in \mathbf{R}_{0+}$ is stable if $(\rho_a/K_a)\,t > \int_0^t\left\|\widetilde{A}(\tau)\right\|d\tau; \forall t \in \mathbf{R}_+$, guaranteed if $\sup\limits_{t\in\mathbf{R}_{0+}}\|\widetilde{A}(t)\| < \frac{\rho_a}{K_a}$, for some norm-dependent real constant $K_a \geq 1$.*

Proof. Consider the n-th differential system $\dot{z}(t) = B(t)z(t);\ z(0) = z_0$ with $\|z_0\| < \infty$. It turns out that there exists $K_a \geq 1$ such that

$$\| z(t) \| \leq K_a e^{-\rho_a t} \left(\| z_0 \| + \int_0^t e^{\rho_a \tau} \| \widetilde{A}(\tau) \| \, \|z(\tau)\| \, d\tau \right);\ \forall t \in \mathbf{R}_{0+} \tag{44}$$

so that $\|z(t)\| \leq K_a \|z_0\| \, e^{-\int_0^t (\rho_a - K_a \|\widetilde{A}(\tau)\|) d\tau}$ which follows from (44), the constraint $(\rho_a / K_a)\, t > \int_0^t \left\| \widetilde{A}(\tau) \right\| d\tau;\ \forall t \in \mathbf{R}_+$ and Gronwall's Lemma [33] so that $\|z(t)\| \leq K_a \|z_0\|;\ \forall t \in \mathbf{R}_{0+}$ and $z(t) \to 0$ as $t \to \infty$. □

The condition $(\rho_a / K_a)\, t > \int_0^t \left\| \widetilde{A}(\tau) \right\| d\tau$ of Lemma 1 may be weakened to $(\rho_a / K_a)\,(t - t_0) > \int_{t_0}^t \left\| \widetilde{A}(\tau) \right\| d\tau$ for any $t (> t_0) \in \mathbf{R}_+$ and some $t_0 \in \mathbf{R}_{0+}$. Lemma 1 yields to the following result:

Theorem 8. *Consider (14) and (15) with $-U \in M_E^{n \times n}$ a stability matrix and $F (\succ 0) \in \mathbf{R}^{n \times n}$ such that $\rho(FU^{-1}) < 1$ and let $\widetilde{F} : \mathbf{R}_{0+} \to \mathbf{R}^{n \times n}$ be uniformly bounded piecewise continuous and asymptotically convergent to $\widetilde{F}_e \in \mathbf{R}^{n \times n}$. Then, there exists some norm-dependent real constant $K_a \geq 1$ such that $F + \widetilde{F} - U : \mathbf{R}_{0+} \to \mathbf{R}^{n \times n}$ is stable provided that $\sup_{t \in \mathbf{R}_{0+}} \| \widetilde{F}(t) \| < \frac{\rho(F-U)}{K_a}$.*

If, furthermore, $\widetilde{F}(t) \succeq -F;\ \forall t \in \mathbf{R}_{0+}$ then the differential system $\dot{y}(t) = \left(F + \widetilde{F}(t) - U \right) y(t)$ is positive in the sense that it has a solution trajectory within the first open orthant of the state space for any initial condition $y(0) = y_0 \succeq 0$.

Proof. Since $-U \in M_E^{n \times n}$, $F (\succ 0) \in \mathbf{R}^{n \times n}$ and $\rho(FU^{-1}) < 1$ then $(F - U) \in M_E^{n \times n}$ so that it has a maximal real eigenvalue which is stable since $(F - U)$ is stable since $-U$ is stable and $\rho(FU^{-1}) < 1$. Thus, the minus stability abscissa of $(F - U)$ is also its spectral radius, that is, $\rho_a(F - U) = \rho(F - U)$ and $\|e^{(F-U)t}\| \leq K_a e^{-\rho t}$ for any $t \in \mathbf{R}$ and some $K_a \geq 1$. If $\sup_{t \in \mathbf{R}_{0+}} \| \widetilde{F}(t) \| < \frac{\rho(F-U)}{K_a}$ for such an existing norm-dependent real constant K_a, then one has that the time-varying matrix $\left(F + \widetilde{F}(t) - U \right)$ is stable from Lemma 1 and it converges asymptotically to the stability matrix $\left(F + \widetilde{F}_e - U \right)$. On the other hand, the differential system $\dot{y}(t) = \left(F + \widetilde{F}(t) - U \right) y(t)$ has a unique solution for any given $y(0) = y_0 \in \mathbf{R}^n$ given by:

$$y(t) = e^{-Ut} y_0 + \int_0^t e^{-U(t-\tau)} \left(F + \widetilde{F}(\tau) \right) y(\tau) d\tau \tag{45}$$

Since $-U \in M_E^{n \times n}$ then $e^{-Ut} \succ 0$ for any $t \in \mathbf{R}_{0+}$ [12]. Now, note by direct inspection of Equation (45) that $\left([y_0 \succeq 0] \wedge \left[\widetilde{F}(t) \succeq -F;\ \forall t \in \mathbf{R}_{0+} \right] \right) \Rightarrow \left(y(t) \succeq 0;\ \forall t \in \mathbf{R}_{0+} \right)$. □

Remark 7. *A practical implementation of the vaccination control law Equations (31)–(33) is to choose the design constants ε_i for $i \in \overline{n}$ being very close to zero and to make null all the proportional vaccination gains $K_{ij}^0(t)$ at patch i for the crossed susceptible information from other patches $j \neq i$ and any $t \geq t_i$ in the event that $S(t_i) < \varepsilon_i$ at some time instant t_i. In this way, the maximum number of switches is n, the last eventual one occurring in a finite time T_f. Then, the stability conditions of Theorem 8 are simplified to simpler conditions for a time-invariant system on $\left[T_f, +\infty \right)$ by deleting the conditions $\sup_{t \in \mathbf{R}_{0+}} \| \widetilde{F}(t) \| < \frac{\rho(F-U)}{K_a}$ and $\widetilde{F}(t) \to \widetilde{F}_e$ as $t \to \infty$, since $\widetilde{F}(t) = \widetilde{F}_e;\ \forall t \geq T_f$ and the finite time interval $\left[0, T_f \right)$ is irrelevant for stability analysis, and modifying the condition $\rho(FU^{-1}) < 1$ to $\rho((F + F_e)U^{-1}) < 1$.*

5. Simulation Examples

This section contains some numerical simulation examples related to the results presented in the previous sections. The examples are concerned with the existence of equilibrium points along with the effect of the vaccination control strategies proposed in Section 4 on the epidemic spreading. In this case, it will be shown how the vaccination controllers are able to reduce the incidence of an infection within a population.

Example 1. *Consider the SIR patchy system defined by three patches or populations, $n = 3$, with parameters given by:*

$$d = [d_i^X] = \begin{bmatrix} 1/3 & 1/3.1 & 1/3.2 \end{bmatrix} years^{-1}, \beta = [\beta_i] = \begin{bmatrix} 3.24 & 3.08 & 3.16 \end{bmatrix} \times 10^{-2}$$
$$\Lambda = 30d,\ \gamma = [\gamma_i] = \begin{bmatrix} 1.78 & 1.82 & 1.75 \end{bmatrix}$$

in units of week^{-1} except otherwise indicated. The symbol d^X stands for any parameter d^S, d^I, d^R. Notice that it is very typical that different outbreaks of the same epidemic have different reproduction numbers [34,35] since the spreading of the epidemic, and therefore its severity, depends on many factors such as the geographical distribution of the individuals, the probability of an infected individual contact a healthy one, etc. The initial conditions are given by:

$$S_1(0) = 25; \quad I_1(0) = 10; \quad R_1(0) = 0$$
$$S_2(0) = 30; \quad I_2(0) = 10; \quad R_2(0) = 0$$
$$S_3(0) = 20; \quad I_3(0) = 5; \quad R_3(0) = 0$$

while the travel matrices are given by:

$$A = \begin{pmatrix} 0 & 0.2 & 0.3 \\ 0.16 & 0 & 0.3 \\ 0.35 & 0.14 & 0 \end{pmatrix}, \quad B = \begin{pmatrix} 0 & 0.22 & 0.4 \\ 0.15 & 0 & 0.05 \\ 0.15 & 0.15 & 0 \end{pmatrix}, \quad C = \begin{pmatrix} 0 & 0.17 & 0.25 \\ 0.3 & 0 & 0.12 \\ 0.3 & 0.2 & 0 \end{pmatrix}$$

The dynamics of the system without vaccination is depicted in Figures 1–3:

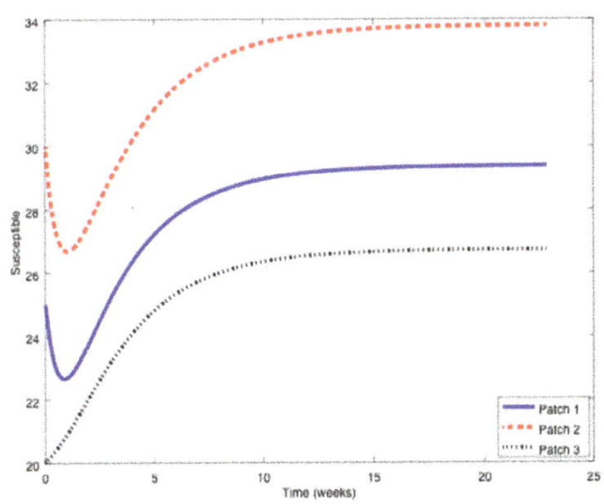

Figure 1. Evolution of the susceptible within each patch without vaccination.

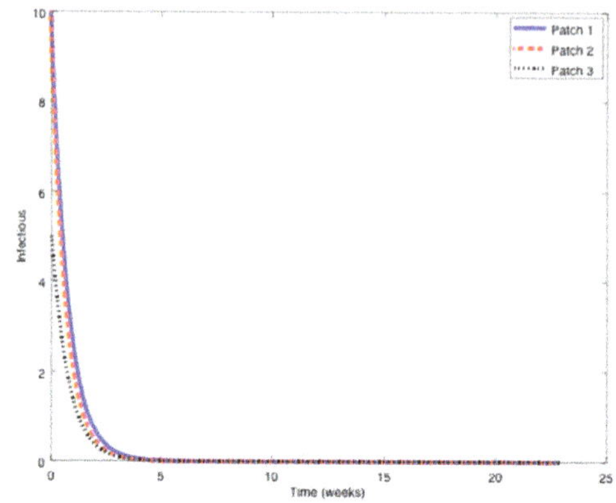

Figure 2. Evolution of the infectious within each patch without vaccination.

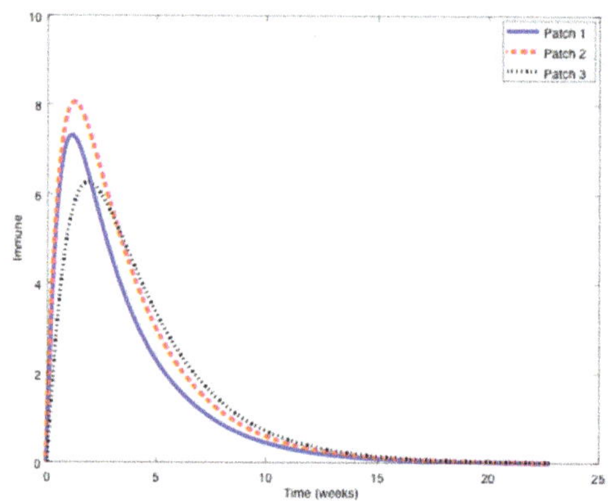

Figure 3. Evolution of the immune within each patch without vaccination.

From Figures 1–3 it can be observed that the above parameters correspond to the case when the reproduction number is less than unity, $R_0 < 1$. Thus, the solution trajectory of the system is non-negative, remains globally bounded and the disease-free equilibrium point is asymptotically stable, as claimed in Theorem 3 (**iii**). Moreover, $I_{dfi} = 0$ and $R_{dfi} = 0$ for $i = 1, 2, 3$ while the values of S_{dfi} are provided in Table 1. In this way, Table 1 displays and compares the value of the equilibrium points obtained from the numerical simulation and theoretically from Equations (10) and (11).

Table 1. Simulated and calculated values for the vaccination-free, disease-free equilibrium point.

	Theoretical Value	Simulated Value
S_{df1}	29.383	29.377
S_{df2}	33.804	33.796
S_{df3}	26.731	26.725

Table 1 shows a good agreement between the theoretical values and the ones obtained by simulation, confirming Theorem 1 results. The total population is given by $N_T = 89.897$. Furthermore, we add now a feedback vaccination term of the form (2) with $V_0 = 0.9\Lambda$, $K = A$. The evolution of the system with this control action is displayed in Figures 4–6.

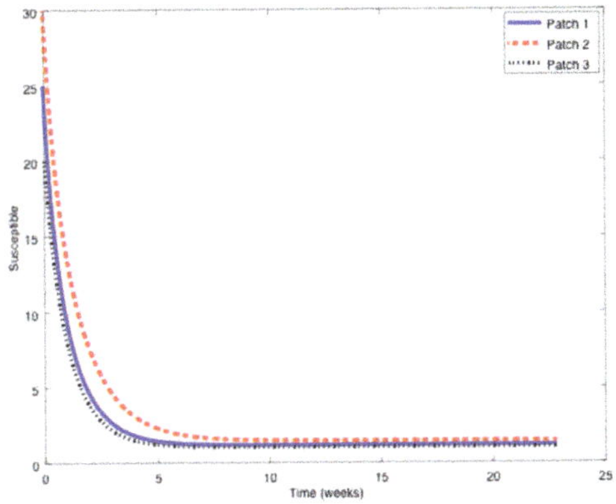

Figure 4. Evolution of the susceptible within each patch with vaccination.

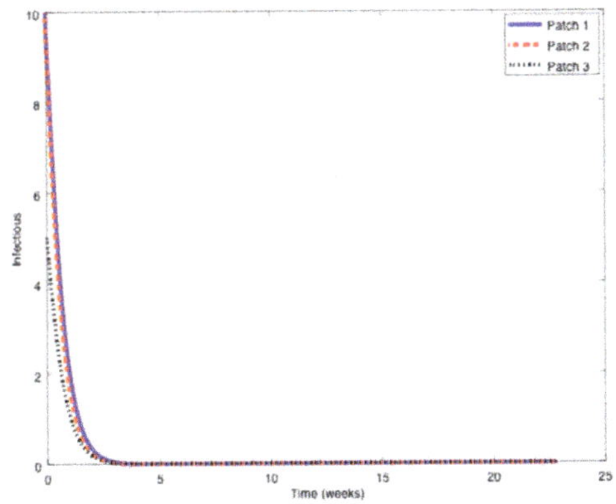

Figure 5. Evolution of the infectious within each patch with vaccination.

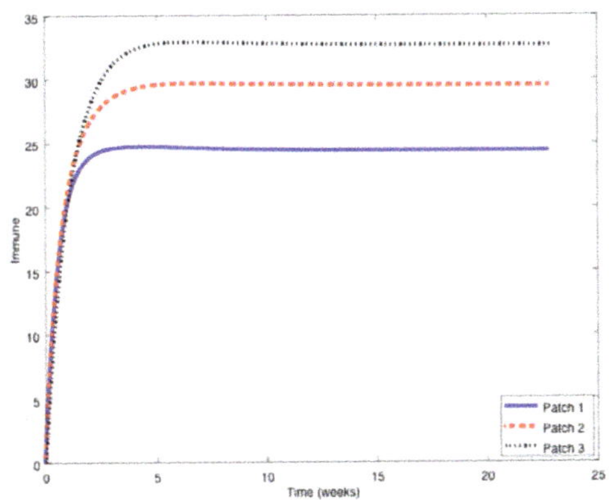

Figure 6. Evolution of the immune within each patch with vaccination.

In this case, the infectious again vanish asymptotically while the disease-free equilibrium point location is contained in Table 2.

Table 2. Simulated and calculated values for disease-free equilibrium point with vaccination.

Disease-free Eqilibrium Point	Theoretical Value	Simulated Value
S_{df1}	1.186	1.186
S_{df2}	1.461	1.461
S_{df3}	1.027	1.027
R_{df1}	24.410	24.412
R_{df2}	29.526	29.528
R_{df3}	32.652	32.655

The total population obtained by numerical simulation is $N_T = 90.268$. As it happened in the previous case, the Table 2 confirms the results provided in Theorem 1 regarding the disease-free equilibrium point location. Moreover, it is verified that the total population at equilibrium does not depend on the particular value of vaccination.

Example 2. *Now, the value of β is increased eight times the value of Example 1 to obtain:*

$$\beta = [\beta_i] = 8\begin{bmatrix} 3.24 & 3.08 & 3.16 \end{bmatrix} \times 10^{-2}$$

so that the reproduction number is now larger than unity, $R_0 > 1$. In this case, the disease-free equilibrium point is unstable and an asymptotically stable endemic equilibrium point appears. The following Figures 7–9 display the evolution of the system in this case when no vaccination is applied.

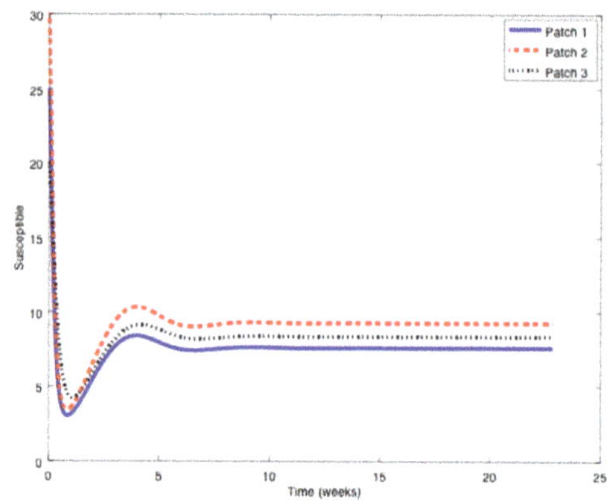

Figure 7. Evolution of the susceptible in all patches when $R_0 > 1$.

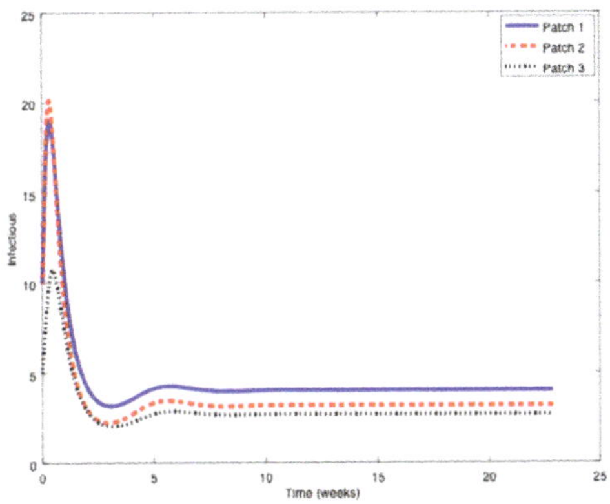

Figure 8. Evolution of the infectious in all patches when $R_0 > 1$.

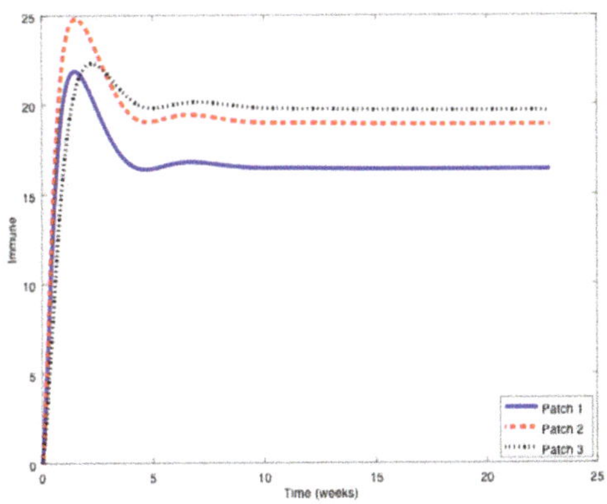

Figure 9. Evolution of the immune in all patches when $R_0 > 1$.

It can be observed that the infectious do not vanish now. The endemic equilibrium point is given by $(S_{end1}, S_{end2}, S_{end3}) = (7.61, 9.26, 8.36)$, $(I_{end1}, I_{end2}, I_{end3}) = (4.03, 3.19, 2.70)$, and $(R_{end1}, R_{end2}, R_{end3}) = (16.40, 18.89, 19.67)$. A series of numerical experiments are conducted now to analyze the effect of parameters and initial conditions in the location of the endemic point. Thus, the initial values of the populations are now changed to:

$$\begin{array}{lll} S_1(0) = 55; & I_1(0) = 15; & R_1(0) = 2 \\ S_2(0) = 40; & I_2(0) = 8; & R_2(0) = 1 \\ S_3(0) = 22; & I_3(0) = 5; & R_3(0) = 2 \end{array}$$

The evolution of the system with different initial conditions is shown in Figures 10–12.

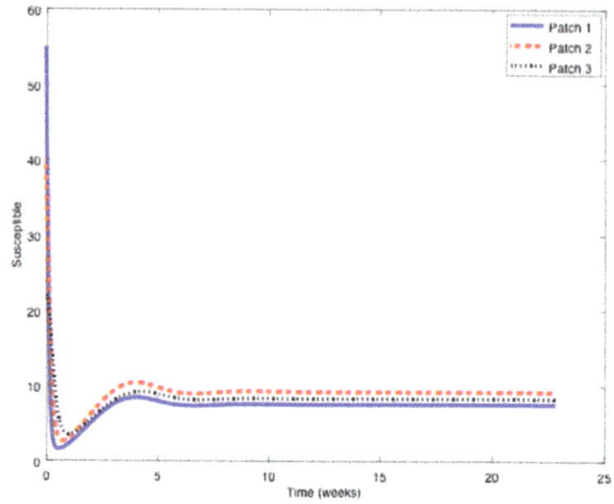

Figure 10. Evolution of the susceptible in all patches when $R_0 > 1$ and different initial conditions.

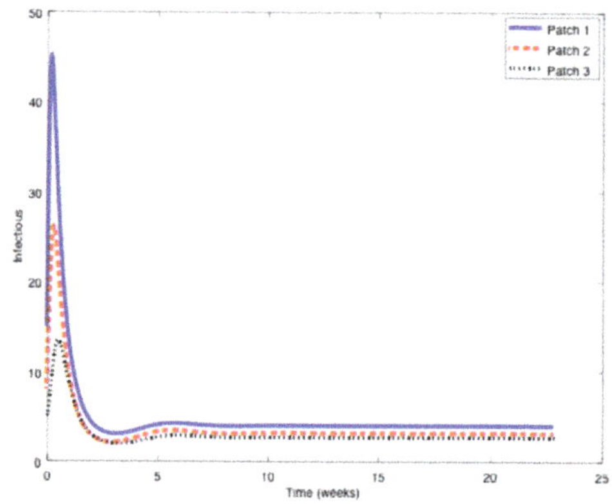

Figure 11. Evolution of the infectious in all patches when $R_0 > 1$ and different initial conditions.

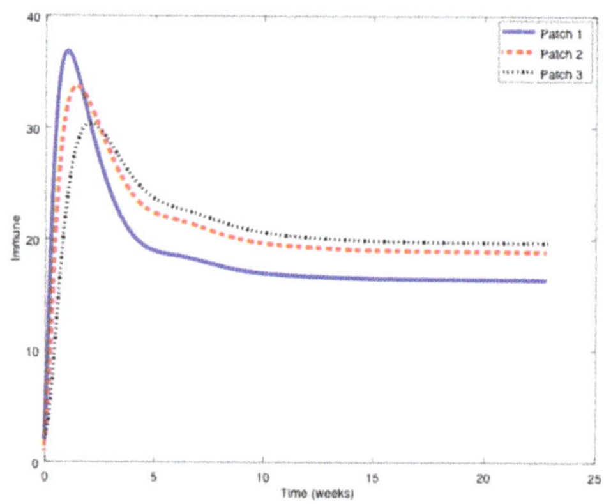

Figure 12. Evolution of the immune in all patches when $R_0 > 1$ and different initial conditions.

The endemic equilibrium point is given by the same values indicated before. Thus, the location of the endemic equilibrium point is not altered by a change in the initial values. Afterwards, the value of β_3 is perturbed (while the others β_1 and β_2 remain unchanged) and the location of the endemic equilibrium point for each case is provided in Table 3.

Table 3. Location of the endemic equilibrium point for different values of β_3.

β_3	$(S_{end1}, S_{end2}, S_{end3})$	$(I_{end1}, I_{end2}, I_{end3})$	$(R_{end1}, R_{end2}, R_{end3})$
28.44×10^{-2}	$(7.51, 9.22, 7.57)$	$(4.00, 3.10, 2.96)$	$(16.53, 18.85, 20.40)$
37.92×10^{-2}	$(7.29, 9.13, 5.86)$	$(3.93, 2.91, 3.53)$	$(16.79, 18.74, 21.95)$
63.20×10^{-2}	$(7.00, 9.00, 3.61)$	$(3.84, 2.67, 4.28)$	$(17.15, 18.61, 23.99)$
94.80×10^{-2}	$(6.84, 8.92, 2.44)$	$(3.80, 2.55, 4.67)$	$(17.34, 18.55, 25.06)$

As it can be deduced from Table 3, the location of the endemic equilibrium point changes according to the change in β_3. To conclude this example, consider now the values of $(\beta_1, \beta_2, \beta_3)$ included in Table 4 and the corresponding endemic points.

Table 4. Location of the endemic equilibrium point for $\beta = 29.92 \times 10^{-2}$.

$(\beta_1, \beta_2, \beta_3)$	$(S_{end1}, S_{end2}, S_{end3})$	$(I_{end1}, I_{end2}, I_{end3})$	$(R_{end1}, R_{end2}, R_{end3})$
(β, β, β)	$(7.56, 8.83, 8.16)$	$(4.00, 3.27, 2.74)$	$(16.46, 19.19, 19.91)$
$(10\beta, \beta, \beta)$	$(0.83, 8.37, 7.31)$	$(6.09, 2.98, 2.20)$	$(20.95, 23.52, 20.87)$
$(\beta, 10\beta, \beta)$	$(7.01, 0.92, 7.65)$	$(3.61, 5.23, 2.51)$	$(17.11, 24.86, 21.24)$
$(\beta, \beta, 10\beta)$	$(6.61, 8.42, 0.90)$	$(3.73, 2.48, 5.16)$	$(17.62, 18.76, 26.50)$

It can be observed in Table 4 how the location of the endemic point changes as the value 10β moves from one position to another one within the vector $[\beta_1, \beta_2, \beta_3]$. Overall, it is concluded that the endemic point does not change with variations of initial conditions, but it generally does with parameter changes.

Example 3. *Finally, consider the Hong Kong influenza epidemic in New York City in 1968–1969. This influenza outbreak is modeled by an SIR epidemic model with the following parameters [36]:*

$$\beta = 3.24 \times 10^{-7}, \ \gamma = 1.78$$

in units of week^{-1}. *The patchy environment is inspired on this real case and it is composed of three cities (or patches),* $n = 3$, *with spreading parameters similar to the above ones and given by:*

$$\Lambda = [\Lambda_i] = \begin{bmatrix} 5 & 4.5 & 5.5 \end{bmatrix} \times 10^3, \beta = [\beta_i] = \begin{bmatrix} 3.24 & 3.18 & 3.08 \end{bmatrix} \times 10^{-7}$$
$$d^X = [d_i^X] = \begin{bmatrix} 1/70 & 1/71 & 1/72 \end{bmatrix} years^{-1}, \gamma = [\gamma_i] = \begin{bmatrix} 1.78 & 1.82 & 1.75 \end{bmatrix}$$

in units of week^{-1} *except otherwise indicated and the symbol* d^X *stands for* d^S, d^I, d^R. *The initial conditions for the populations are given by the 1970 New York City census as:*

$$S_1(0) = 7,960,000; \quad I_1(0) = 15,000; \quad R_1(0) = 0$$

while the initial conditions for the remaining patches are given, similarly, by:

$$S_2(0) = 8,600,000; \quad I_2(0) = 20,000; \quad R_2(0) = 0$$
$$S_3(0) = 7,200,000; \quad I_3(0) = 19,000; \quad R_3(0) = 0$$

The travel matrices are defined by:

$$A = 10^{-2} \times \begin{pmatrix} 0 & 1.2 & 0.3 \\ 1.1 & 0 & 1 \\ 1.2 & 1.4 & 0 \end{pmatrix}, \quad B = 10^{-2} \times \begin{pmatrix} 0 & 1.12 & 0.4 \\ 1.22 & 0 & 0.85 \\ 1 & 1.14 & 0 \end{pmatrix}, \quad C = 10^{-2} \times \begin{pmatrix} 0 & 0.78 & 0.56 \\ 1 & 0 & 0.95 \\ 1.2 & 0.94 & 0 \end{pmatrix}$$

The aim of this example is to show the effect of the vaccination strategies introduced in Section 4. The evolution of the system without vaccination is displayed in Figures 13–15.

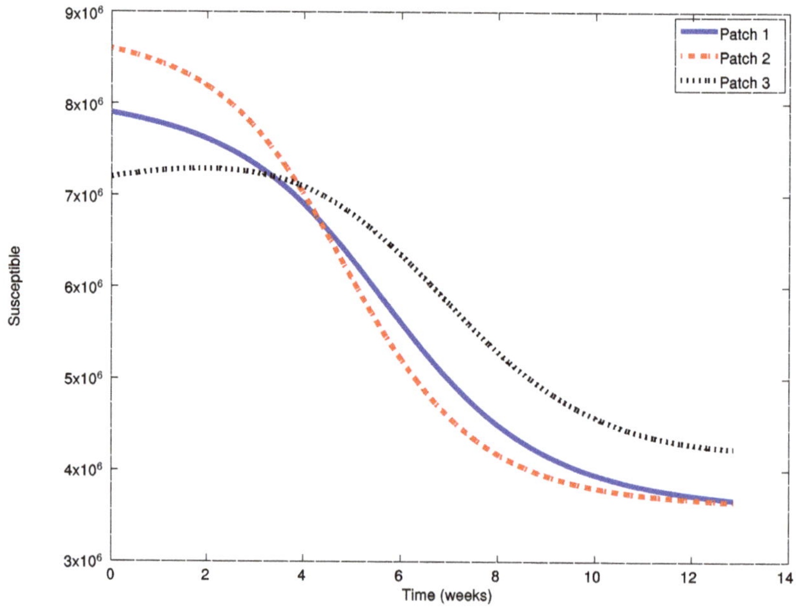

Figure 13. Evolution of the susceptible subpopulation within each patch.

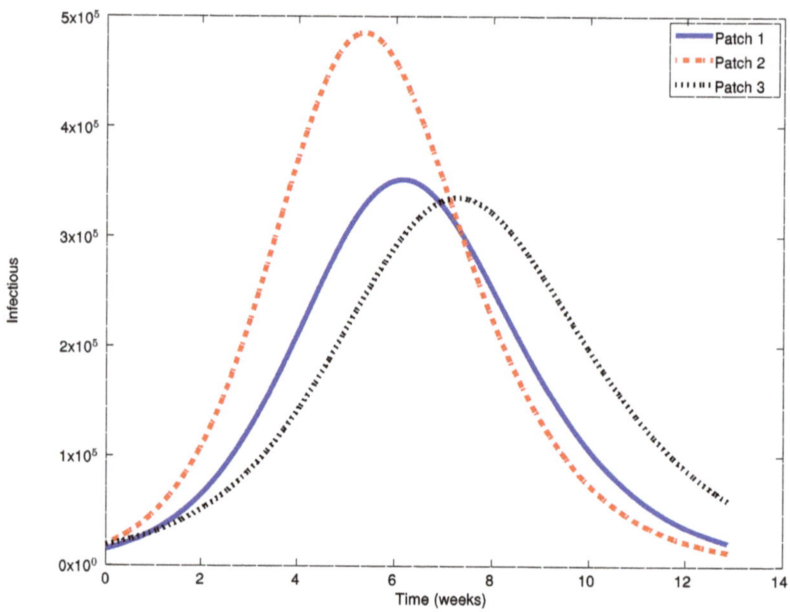

Figure 14. Evolution of the infectious subpopulation within each patch.

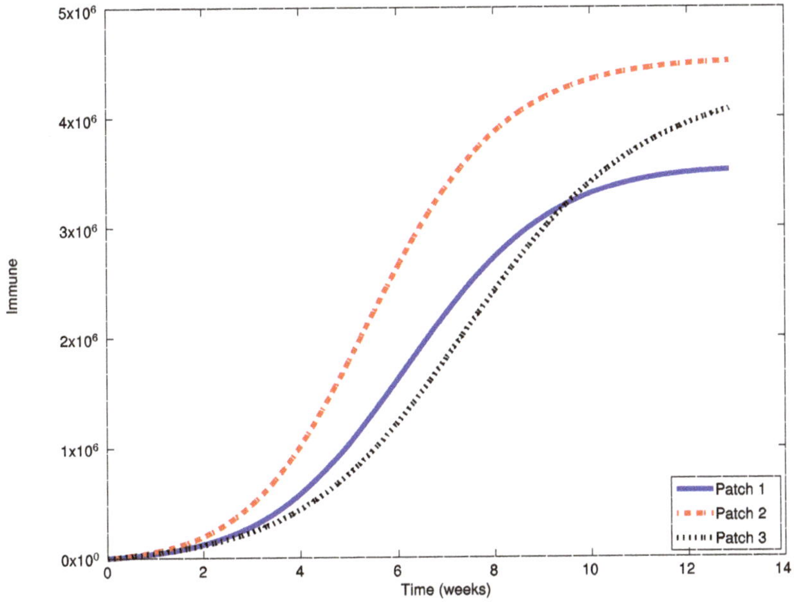

Figure 15. Evolution of the immune subpopulation within each patch.

As it can be observed in Figure 14, the influenza outbreak reaches a peak during the spreading of the infection. In order to reduce the severity of the outbreak, the two vaccination strategies proposed in Section 4 are now applied and compared. To this end, consider the control matrices given by:

$$K = A + Diag\left(\left[10^{-2}, 0.6 \times 10^{-2}, 0.9 \times 10^{-2}\right]\right); \quad M_i = 5 \times 10^5; \quad M_0 = 0.9\Lambda; \quad V_0 = M_0$$

It can be readily seen that the above selection satisfies the constraints imposed by (30). Moreover, the thresholds to be used in Strategy 2 are given by $\varepsilon_1 = 4.3 \times 10^6$; $\varepsilon_2 = 5.1 \times 10^6$; $\varepsilon_3 = 4.7 \times 10^6$. The Figures 16–21 display the evolution of various infectious subpopulations in agreement with the implemented vaccination controls. The Figure 16, Figure 18, and Figure 20 show the evolution of the infectious subpopulation at each patch without vaccination and when both vaccination strategies introduced in Section 4 are employed. Furthermore, the Figure 16, Figure 18, and Figure 20 show the vaccination commands generated by both strategies at each patch. It can be seen that the solution trajectory of the infectious is non-negative and globally bounded as it is proved in Theorem 4. From Figure 16, Figure 18, and Figure 20 it can also be concluded that the application of a judicious vaccination campaign significantly reduces the peak caused by the outbreak. In addition, Figure 17, Figure 19, and Figure 21 show that Strategies 1 and 2 generate very similar infectious subpopulation profiles, where the plots for both cases are almost superimposed. However, the vaccination law profile through time is different for Strategies 1 and 2, fact that can be observed in Figure 17, Figure 19, and Figure 21. During the first weeks, both control laws are the same but when the susceptible reach the corresponding prescribed threshold, the susceptible feedback term of Strategy 2's vaccination law is switched off and only a constant vaccination is applied. The shutting down of the feedback term causes a noticeable decrease of the control command while the evolution of the infectious subpopulations is similar. Consequently, the vaccination Strategy 2 is able to reduce the outbreak peak, saving vaccination effort. Notice that, in this experiment, each patch disposes of full information of the remaining ones since the values of the susceptible subpopulation at the others patches are used to calculate the amount of vaccination according to Equations (31)–(33).

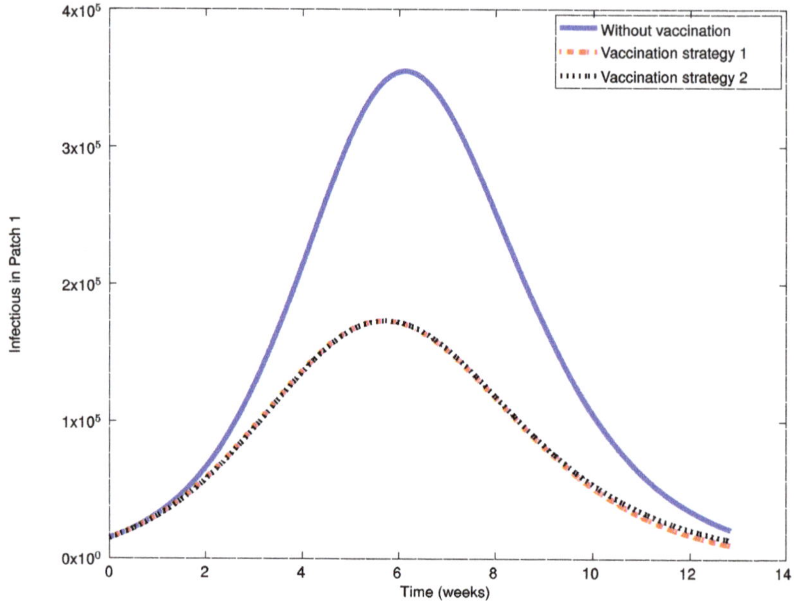

Figure 16. Evolution of the infectious subpopulation within patch 1 under different vaccination strategies.

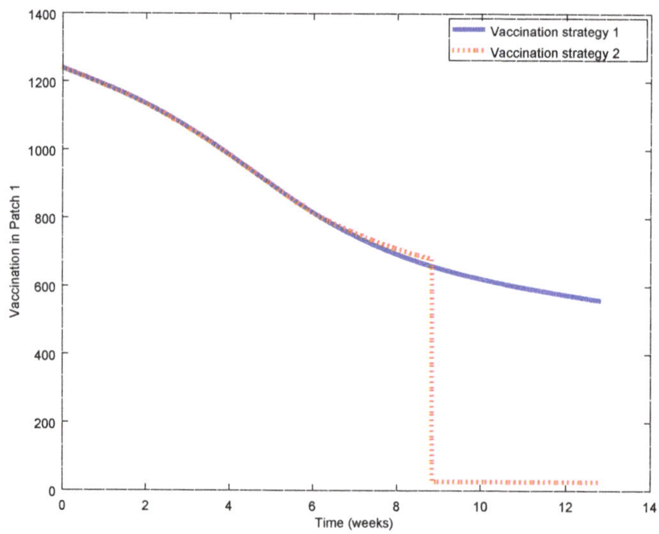

Figure 17. Vaccination law in patch 1 for Strategies 1 and 2.

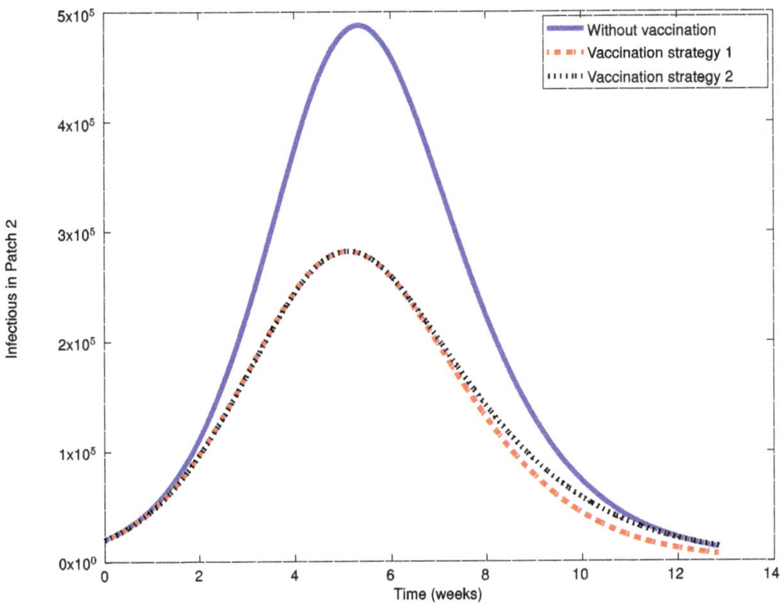

Figure 18. Evolution of the infectious subpopulation within patch 2 under different vaccination strategies.

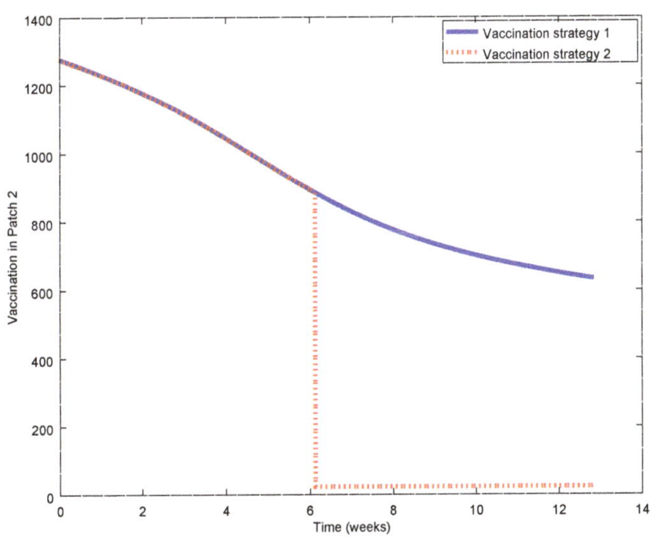

Figure 19. Vaccination law in patch 2 for Strategies 1 and 2.

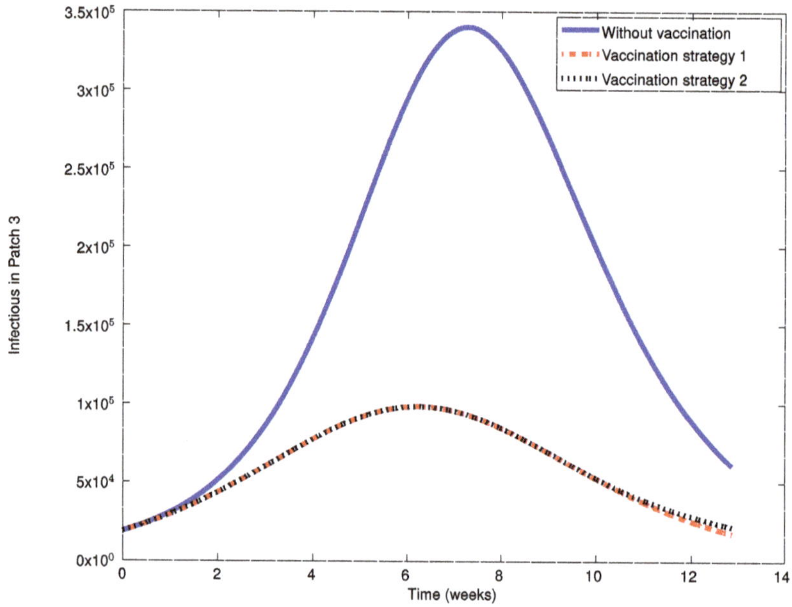

Figure 20. Evolution of the infectious subpopulation within patch 3 under different vaccination strategies.

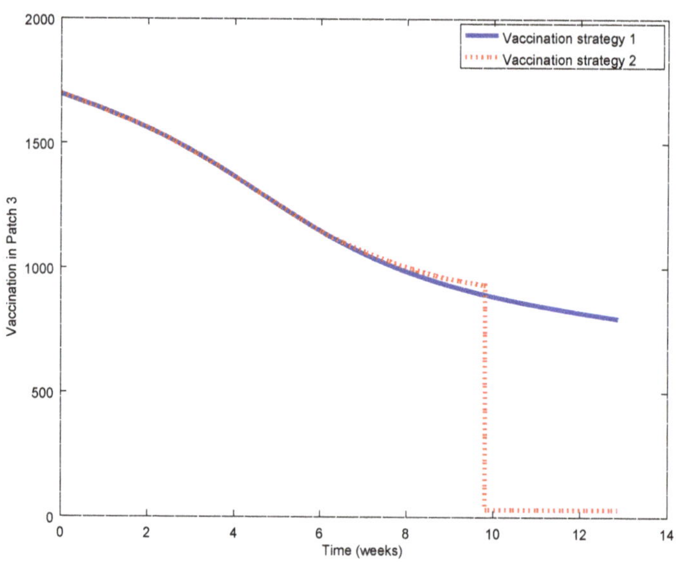

Figure 21. Vaccination law in patch 3 for Strategies 1 and 2.

Now, we will change the matrix K so that it takes the following upper-triangular form:

$$K = \begin{pmatrix} 10^{-2} & 0.1A_{12} & 0.1A_{13} \\ 0 & 0.6 \times 10^{-2} & 0.1A_{23} \\ 0 & 0 & 0.9 \times 10^{-2} \end{pmatrix}$$

In this case, the first patch has available information of the second and third patches, the second patch has only information of the third patch which has only self-information. This structure implies for the first patch, for instance, that the vaccination law considers an amount of 10% of individuals coming into the patch from the second and third ones in order to calculate the total administered vaccination. It is important to notice the difference with respect to the previous example, where all the amount of travelling individuals (coming in and going out of the patch) is considered to calculate the vaccination. The illness evolution is displayed in the various Figures 22–27. In particular, the evolution of the infectious under these circumstances is depicted for each patch in Figure 22, Figure 24, and Figure 26. On the other hand, the vaccination generated by each one of the strategies is displayed for each patch in Figure 23, Figure 25, and Figure 27. The main conclusions drawn before regarding the effect of applying an appropriate vaccination to individuals as well as those related to the comparison of Strategies 1 and 2 hold here too. However, in this case the peak in the infectious in reduced less by applying vaccination than in the previous example. The main reason for this issue is that with the new control matrix, K, the number of administered vaccines is much lower now than in the previous case. This fact can be observed by comparing the Figures 17 and 23, Figures 19 and 25, and Figures 21 and 27. This result shows the importance of vaccination campaigns in order to control an epidemic outbreak in a patchy environment.

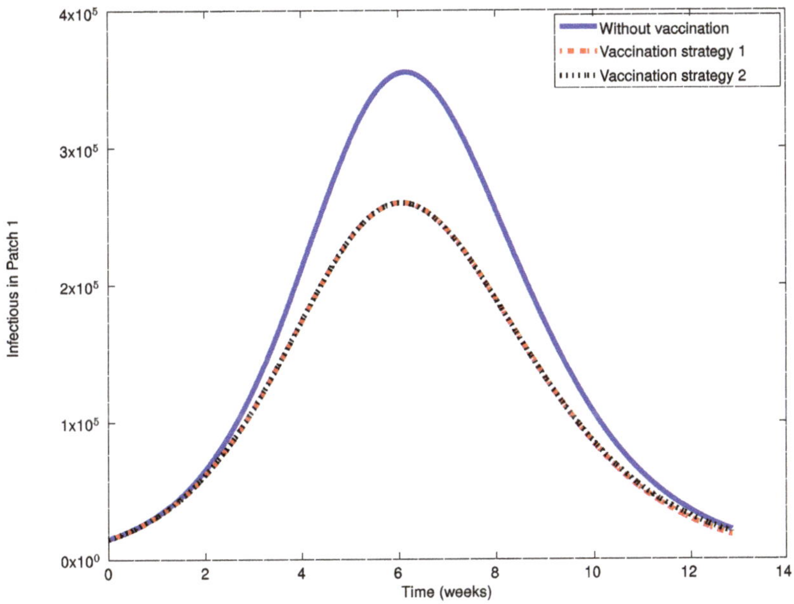

Figure 22. Evolution of the infectious subpopulation within patch 1 under different vaccination strategies and upper-triangular matrix K.

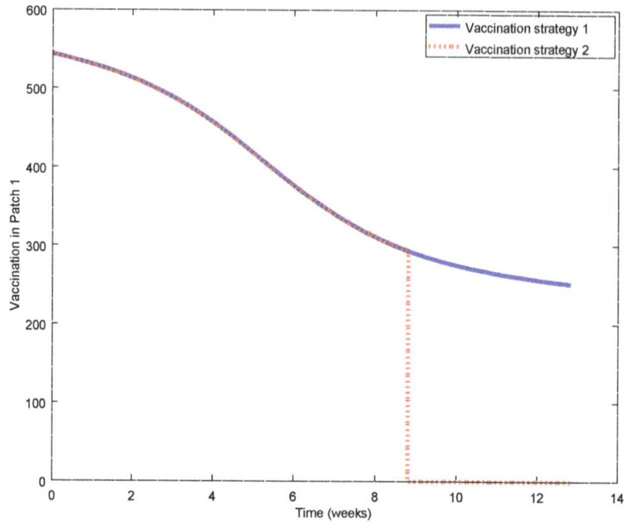

Figure 23. Vaccination law in patch 1 for Strategies 1 and 2 with upper-triangular matrix *K*.

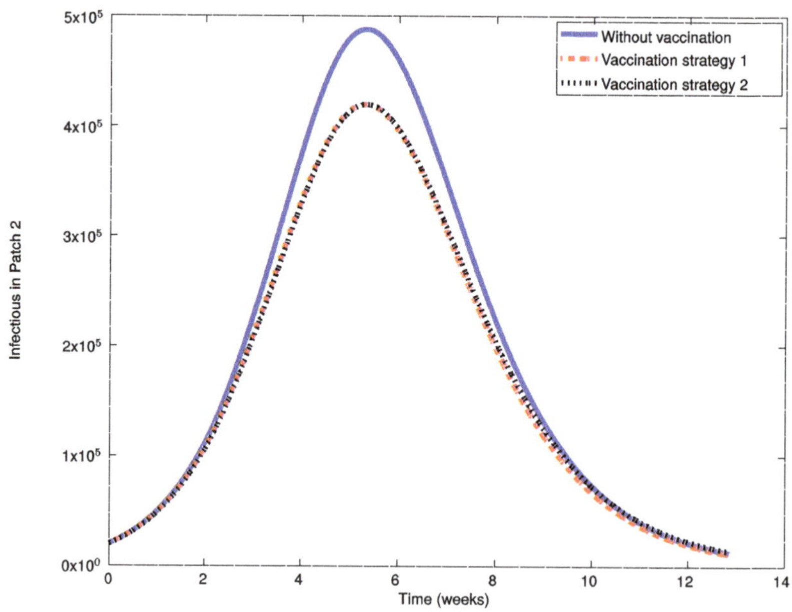

Figure 24. Evolution of the infectious subpopulation within patch 2 under different vaccination strategies and upper-triangular matrix *K*.

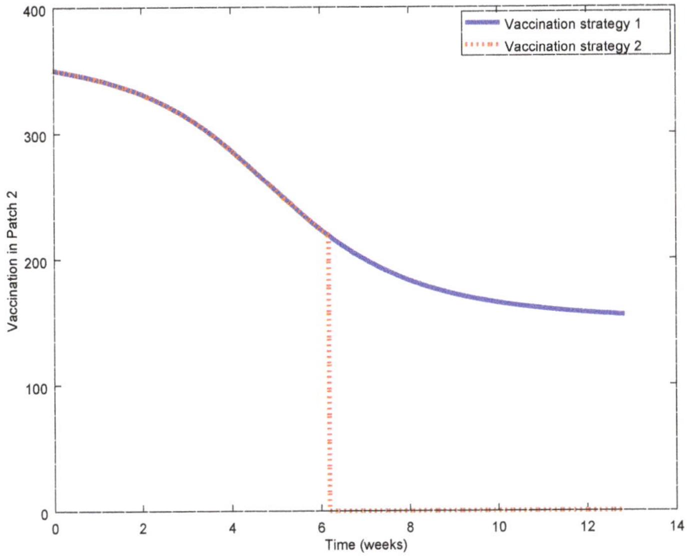

Figure 25. Vaccination law in patch 2 for Strategies 1 and 2 with upper-triangular matrix *K*.

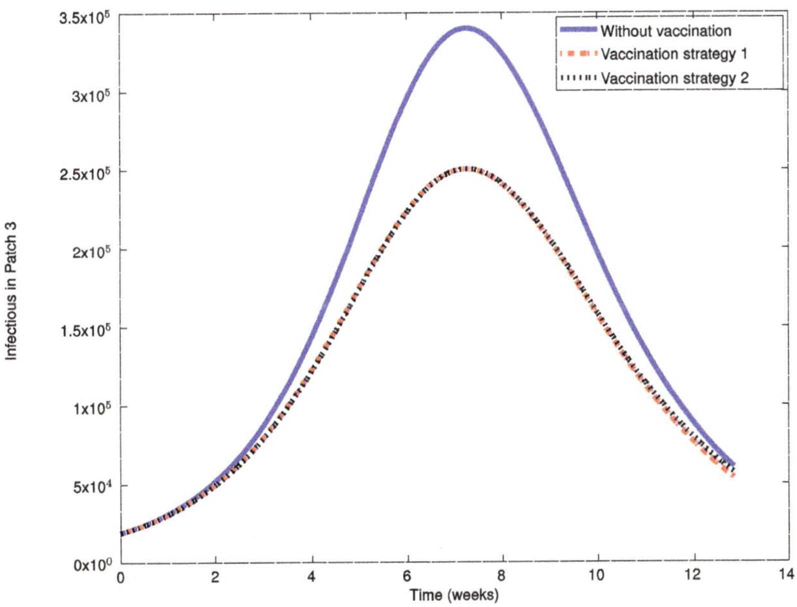

Figure 26. Evolution of the infectious subpopulation within patch 3 under different vaccination strategies and upper-triangular matrix *K*.

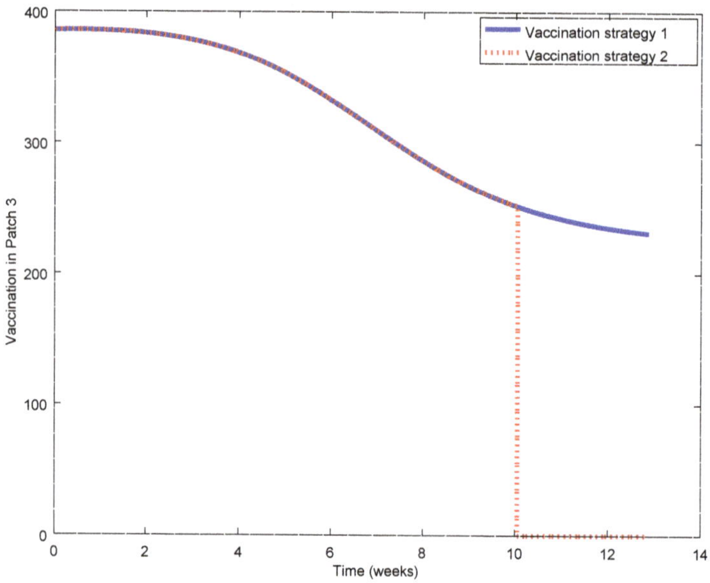

Figure 27. Vaccination law in patch 3 for Strategies 1 and 2 with upper-triangular matrix K.

6. Conclusions

This paper has considered a SIR epidemic model in a patchy environment, each patch being assumed to have its own health or medical centre. It has been assumed that there are potential travellers coming into and leaving each patch which are interchanged with the remaining patches. It has been assumed that the vaccination controls are exerted at each community health centre while either the total information or a partial information of the total subpopulations, including the interchanging ones, is shared by all the set of health centres of the whole environment under study. In this way, vaccination control laws involving constant terms and feedback information on the susceptible subpopulations have been proposed and discussed to be administrated at each health centre. In the cases that not all the information of the subpopulations distributions at other patches is known by the health centre of each particular patch, the feedback vaccination rule is considered to have a decentralized nature. Since there the control laws involved crossed gains to take into account or not (if such gains are zeroed) the couplings between patches, the vaccination action can be of either a centralized or of a (totally or partially) decentralized nature. The paper has also investigated the existence, allocation (depending on the vaccination control gains) and uniqueness of the disease-free equilibrium point as well as the existence of at least an attainable and stable endemic equilibrium point. A formal analytic characterization of the potential whole set of endemic equilibrium points has also being given based on algebraic mathematical tools for the solvability of algebraic systems of equations.

Author Contributions: M.D.S., S.A.-Q. and R.N. conceived the whole model; M.D.S. performed the theoretical analysis snd the mathematical proofs as well as the main paper elaboration and writing ; A.I. conceived the experiments and performed the simulations and paper corrections ; S.A.-Q. corrected the whole paper equations and collaborated in the conceptual framework; R.N. also contributed to the practical discussions related to the given examples in accordance with the theoretical framework.

Funding: This rsearch was funded by the Spanish Government through Grants DPI2015-64766-R and DPI2016-77271-R (MINECO/FEDER, UE), and by UPV/EHU through Grant PGC 17/33.

Acknowledgments: The authors are grateful to the Spanish Government for Grants DPI2015-64766-R and DPI2016-77271-R (MINECO/FEDER, UE), and to UPV/EHU for Grant PGC 17/33.

Conflicts of Interest: The authors declare that they do not have any competing interests.

Appendix A

Proof of Theorem 3. Note that U is nonsingular since it has non-positive off-diagonal entries with the sum of all the entries per column being positive. Thus, $(-U) \in M_E^{n\times n}$ is non-singular matrix with $U^{-1} \succ 0$ so that $(-U)$ is a stability matrix. Property **(i)** has been proved. On the other hand, note that the Jacobian matrix of the linearized system solution trajectory of the infectious subpopulations around the disease-free equilibrium point is $(F-U)$ where $(-U)$ is a Metzler stability matrix from Property (i). Therefore, such a linearized system is globally exponentially stable if $F=0$, that is if $\beta_i = 0$ (fully absence of illness) ; $\forall i \in \overline{n}$. Since the constraints $\beta_i = 0$; $\forall i \in \overline{n}$ remove the quadratic terms from the model dynamics, it follows also that the stability is asymptotically global for the whole model. Property (ii) has been proved. Now, note that $F-U = \left(I_n - FU^{-1}\right)(-U)$ since $(-U)$ is a Metzler stability matrix, then non-singular, from Property **(i)**. If $F=0$, $F-U=-U$ is a stability matrix and it continuous to be a stability matrix from the continuity of its eigenvalues as functions of its entries for any $F \succeq 0$ such that $R_0 = \rho\left(FU^{-1}\right) < 1$. Therefore, the disease-free equilibrium point is locally asymptotically stable if $R_0 < 1$. It has a critically stable eigenvalue for $R_0 = 1$ and it is unstable if $R_0 > 1$. Property (iii) has been proved. On the other hand, decompose $U = U_d + U_{od}$, where U_d is the diagonal part of U and U_{od} is its off-diagonal part. Since U_d and U are non-singular, one gets:

$$U = U_d + U_{od} = U_d\left(I_n + U_d^{-1}U_{od}\right) \tag{A1}$$

Note also that the matrix D_{SS} is non-singular from Theorem 1 and it can be decomposed as the sum of its diagonal D_{SSd}, which is also non-singular, and non-diagonal D_{SSod}, parts to yield:

$$D_{SS} = D_{SSd} + D_{SSod} = D_{SSd}\left(I_n + D_{SSd}^{-1}D_{SSod}\right) \tag{A2}$$

so that

$$U^{-1} = \left(I_n + U_d^{-1}U_{od}\right)^{-1}U_d^{-1};\ D_{SS}^{-1} = \left(I_n + D_{SSd}^{-1}D_{SSod}\right)^{-1}D_{SSd}^{-1} \tag{A3}$$

Assume that $\|U_{od}\|_2 < 1/\|U_d^{-1}\|_2$ and $\|D_{SSod}\|_2 < 1/\|D_{SSd}^{-1}\|_2$. Then, one gets from Banach´s Perturbation Lemma [13]:

$$\|U^{-1}\|_2 \le \frac{\|U_d^{-1}\|_2}{1-\|U_d^{-1}\|_2\,\|U_{od}\|_2};\ \|D_{SS}^{-1}\|_2 \le \frac{\|D_{SSd}^{-1}\|_2}{1-\|D_{SSd}^{-1}\|_2\,\|D_{SSod}\|_2} \tag{A4}$$

Then, by using Equations (3), (4)–(8) and Equation (11), since F is diagonal and $U^{-T}U^{-1}$ is symmetric, the reproduction number satisfies that:

$$R_0 = \rho\left(FU^{-1}\right) \le \|FU^{-1}\|_2 \le \|F\|_2\|U^{-1}\|_2 = \rho(F)\sqrt{\lambda_{\max}\left(U^{-T}U^{-1}\right)} = \rho(F)\rho^{1/2}\left(U^{-T}U^{-1}\right) \tag{A5}$$

which leads to Equation (16). One gets also from Equation (16) and Equation (A4) that:

$$R_0 = \rho\left(FU^{-1}\right) \le \beta \max_{1\le i\le n}(\beta_{ir})\frac{\|D_{SSd}^{-1}\|_2}{1-\|D_{SSd}^{-1}\|_2\,\|D_{SSod}\|_2}\|\Lambda - V_0\|_2\|\left(U^{-1}U^{-T}\right)^2\|_2^{1/4} \tag{A6}$$

which leads to Equation (17). Property (iv) has been proved. Finally, note from Equation (17) that $\overline{R}_{02}$ is minimized if $K_{ij} = a_{ij}$; $\forall i,j \in \overline{n}\backslash\{1\}$, implying that $D_{SSod} = 0$ for any given model parameters and constant vaccination vector V_0. On the other hand, it follows from Equation (17) that $\overline{R}_{02} = 0$, then $R_0 = 0$, if $\Lambda_i = V_{i0}$; $\forall i \in \overline{n}$ is the influx of population into the i-th patch. Property (v) is proved. □

Appendix B

Proof of Theorem 5. One firstly sums up the two first equations of Equation (1), so as to primarily delete the influence of the disease transmission rates towards a linearization study. Secondly, one expands the obtained result jointly with the third equation in a single compacted algebraic system while taking into account Equation (2). Then, one gets that Equations (22)–(23), subject to Equations (20), (21) and (24), hold. From Theorem 1 **(i)**, the limit total population N^* is unique for the disease-free equilibrium point and any endemic attainable existing equilibrium point and this amount is allocated as first element in the linear system Equation (27). Then, one has to solve the auxiliary linear system $Ax = b$ in $x = x(y)$ with A and b defined in (26) which gives the endemic equilibrium points. It is known that there is (at least) one attainable endemic equilibrium point from Theorem 4 (ii) since $R_0 \geq 1$. Therefore, the above algebraic system has, at least, an attainable endemic solution and, from the Rouché–Froebenius theorem from Linear Algebra, Equation (25) holds. The whole set of endemic equilibrium solutions, including the attainable and unattainable ones, has to satisfy Equation (27). But note that the above algebraic system has only a partial information on the epidemic model Equations (1)–(2) since it does not include the information on the influence of the disease coefficient rates because of summing up action on the two first equations of Equation (1) leading to cancel the nonlinear common term. Therefore, the constraints Equation (28) are got by incorporating to Equation (27) the second equation of Equation (1) including the nonlinear term excluded from Equations (22) and (23). So, the particular vector y of the general solution Equation (27) is constrained to fulfill Equation (28). Property **(i)** has been proved.

On the other hand, if B is irreducible, one deduces from Theorem 2 **(i)** that at any attainable endemic equilibrium point, the limit endemic infectious subpopulations at any patches are nonzero since if they are zero then there is no endemic infection. So, Property **(ii)** follows from the proof of Property **(i)** with y being restricted to belong to the set Y_a. In the same way, Property **(iii)** follows with y restricted to belong to Y_b since B is irreducible, $A - K$ is irreducible and positive and $V_{i0} = \Lambda_i; \forall i \in \overline{n}$ so that the endemic equilibrium infectious population is positive at any patch and the susceptible ones at all patches are either all of them zero and or all of them nonzero from the first part of Theorem 2 (ii). Finally, Property (iv) follows under similar arguments from the second part of Theorem 2 (ii) involving the joint irreducibility of the positive matrices B, $A - K$ and C. □

Proof of Corollary 1. If $R_0 \geq 1$ then $\overline{y} \in \boldsymbol{R}^{3n}$ always exists such that $x = A^{+}b + (I_{3n} - A^{+}A)\overline{y} \succ 0$ is an endemic attainable equilibrium point from Theorem 4 (ii). Then, $x = A^{+}b + (I_{3n} - A^{+}A)y$, where $y = \overline{y} + y'$ for any $y' \in Ker(I_{3n} - A^{+}A)$. Since the whole endemic infectious subpopulation being the sum of all the infection subpopulations in all the patches is non-zero, it holds that $E(y + A^{+}(b - Ay)) \succ 0$, that is, the endemic infectious subpopulation in at least one patch has to be positive. If $B$is irreducible, then the infectious subpopulations at the endemic steady-state are nonzero in all patches since, otherwise, the infections total endemic equilibrium subpopulation would be identically zero (Theorem 2.1 **(i)**). For uniqueness, of such an equilibrium point, the constraints Equation (28) should also hold (Theorem 5 **(i)**). The necessity of the constraints 1 to 3 for the uniqueness of any existing stable attainable endemic equilibrium point have been proved and it is also known that such a point always exists since $R_0 \geq 1$ (Theorem 4 **(ii)** and **(iii)**). Now, group the constraints Equation (28), as components of a vector $\beta = (\beta_1, \beta_2, \cdots, \beta_n)^T$, resulting the following vector equation:

$$\beta = \gamma_1\gamma_2(y)\gamma_3(y) = \gamma_1\gamma_2(\overline{y})\gamma_3(\overline{y}) \tag{A7}$$

for any $y = \overline{y} + y'$ with $y' \in Ker(I_{3n} - A^{+}A)$, where

$$\gamma_1 = \begin{bmatrix} \gamma_{11}^T \\ \gamma_{12}^T \\ \vdots \\ \gamma_{1n}^T \end{bmatrix}; \ \gamma_3(\overline{y}) = \begin{bmatrix} 1/\gamma_{31}(\overline{y}) \\ 1/\gamma_{32}(\overline{y}) \\ \vdots \\ 1/\gamma_{3n}(\overline{y}) \end{bmatrix} \tag{A8}$$

$$\gamma_2(\overline{y}) = Diag[\gamma_{21}(\overline{y}), \gamma_{22}(\overline{y}), \ldots, \gamma_{2n}(\overline{y})] \tag{A9}$$

$$\gamma_{1i}^T = \left(d_i^I + \gamma_i + \sum_{j(\neq i)=1}^{n} b_{ji} \right) e_{n+i}^T - \sum_{j(\neq i)=1}^{n} b_{ij} e_{n+j}^T; \ i \in \overline{n} \tag{A10}$$

$$\gamma_{2i}(\overline{y}) = e_i^T A^+ b + \left(1 - e_i^T A^+ A\right) \overline{y}; \ i \in \overline{n} \tag{A11}$$

$$\gamma_{3i}(\overline{y}) = e_i^T \left[A^+ b + \left(I_{3n} - A^+ A\right) \overline{y}\right] \left[A^+ b + \left(I_{3n} - A^+ A\right) \overline{y}\right]^T e_{n+i}; \ i \in \overline{n} \tag{A12}$$

If the endemic equilibrium solution x is unique for $y = \overline{y} + y'$ then $y' \in Ker(I_{3n} - A^+ A)$ and the given constant vector $\boldsymbol{\beta}$ of coefficient transmission rates satisfies Equation (28) for any $y' \in Ker(I_{3n} - A^+ A)$. If the constraint 3 is fulfilled for some $y' \notin Ker(I_{3n} - A^+ A)$ then x is not unique and Equation (28) is violated for $y = \overline{y} + y'$. Therefore, the endemic equilibrium solution is unique under the constraints 1 to 3 if and only if $\Delta\boldsymbol{\beta} = \left(\nabla_{\overline{y}^T} \boldsymbol{\beta}\right) \Delta y \neq 0$ for the gradient matrix:

$$\nabla_{\overline{y}^T} \beta = \begin{bmatrix} \frac{\partial \beta_1}{\partial \overline{y}_1} & \frac{\partial \beta_1}{\partial \overline{y}_2} & \cdots & \frac{\partial \beta_1}{\partial \overline{y}_{3n}} \\ \frac{\partial \beta_2}{\partial \overline{y}_1} & \frac{\partial \beta_2}{\partial \overline{y}_2} & \cdots & \frac{\partial \beta_2}{\partial \overline{y}_{3n}} \\ \cdots & \cdots & \cdots & \cdots \\ \frac{\partial \beta_n}{\partial \overline{y}_1} & \frac{\partial \beta_n}{\partial \overline{y}_2} & \cdots & \frac{\partial \beta_n}{\partial \overline{y}_{3n}} \end{bmatrix} \tag{A13}$$

for any $\Delta y \notin Ker(I_{3n} - A^+ A)$. In other words, and from the equivalence of a logic proposition with its contra-positive one, if and only if, $Ker\left(\nabla_{\overline{y}^T} \boldsymbol{\beta}\right) \subseteq Ker(I_{3n} - A^+ A)$. Note that

$$\gamma_2(\overline{y})\gamma_3(\overline{y}) = \begin{bmatrix} \gamma_{21}(\overline{y})/\gamma_{31}(\overline{y}) \\ \gamma_{22}(\overline{y})/\gamma_{32}(\overline{y}) \\ \vdots \\ \gamma_{2n}(\overline{y})/\gamma_{3n}(\overline{y}) \end{bmatrix} \tag{A14}$$

Thus, in order to operate with the needed gradients in a closed form, define also the vector $\hat{\gamma}_2(\overline{y})$ associated with the matrix $\gamma_2(\overline{y})$ and the matrix $\hat{\gamma}_3(\overline{y})$ associated with the vector $\gamma_3(\overline{y})$ as follows:

$$\hat{\gamma}_2(\overline{y}) = \begin{bmatrix} \gamma_{21}(\overline{y}) \\ \gamma(\overline{y})_{22} \\ \vdots \\ \gamma_{2n}(\overline{y}) \end{bmatrix}; \ \hat{\gamma}_3(\overline{y}) = Diag[1/\gamma_{31}(\overline{y}), 1/\gamma_{32}(\overline{y}), \ldots, 1/\gamma_{3n}(\overline{y})] \tag{A15}$$

Since the transposition and Moore–Penrose inversion can be permuted for any matrix, Equation (A12) can be expressed equivalently as follows:

$$\gamma_{3i}(\overline{y}) = \left[b^T A^{+T} + \overline{y}^T \left(I_{3n} - A^T A^{+T}\right)\right] e_i e_{n+i}^T \left[A^+ b + \left(I_{3n} - A^+ A\right) \overline{y}\right]; \ i \in \overline{n} \tag{A16}$$

Note from Equations (A8)–(A12) via Equation (A14) subject to Equation (A15) and Equation (A16) that

$$\begin{aligned}
&\nabla_{\overline{y}^T}\gamma_1 = 0;\ \nabla_{\overline{y}^T}\hat{\gamma}_2(\overline{y}) = Diag\left(1-(A^+A)_{11}, 1-(A^+A)_{22}, \cdots, 1-(A^+A)_{3n\times 3n}\right),\\
&\nabla_{\overline{y}^T}\gamma_{3i}(\overline{y}) = 2\left[b^T A^{+^T} e_i e_{n+i}^T\left(I_{3n}-A^+A\right) - \overline{y}^T\left(I_{3n}-A^T A^{+^T}\right)e_i e_{n+i}^T A^+ A\right],\\
&\nabla_{\overline{y}^T}\gamma_3(\overline{y}) = 2\begin{bmatrix} b^T A^{+^T} e_1 e_{n+1}^T\left(I_{3n}-A^+A\right) - \overline{y}^T\left(I_{3n}-A^T A^{+^T}\right)e_1 e_{n+1}^T A^+ A \\ b^T A^{+^T} e_2 e_{n+2}^T\left(I_{3n}-A^+A\right) - \overline{y}^T\left(I_{3n}-A^T A^{+^T}\right)e_2 e_{n+2}^T A^+ A \\ \vdots \\ b^T A^{+^T} e_n e_{2n}^T\left(I_{3n}-A^+A\right) - \overline{y}^T\left(I_{3n}-A^T A^{+^T}\right)e_n e_{2n}^T A^+ A \end{bmatrix}
\end{aligned} \tag{A17}$$

and direct gradient calculations yield:

$$\begin{aligned}
\left(\nabla_{\overline{y}^T}\beta\right)\Delta y &= \left(\nabla_{\overline{y}^T}[\gamma_1\gamma_2(\overline{y})\gamma_3(\overline{y})]\right)\Delta y = \gamma_1.\nabla_{\overline{y}^T}[\gamma_2(\overline{y})\gamma_3(\overline{y})]\Delta y\\
&= \gamma_1\left(\gamma_2(\overline{y}).\nabla_{\overline{y}^T}\gamma_3(\overline{y}) + \nabla_{\overline{y}^T}\left[\hat{\gamma}_2(\overline{y})\right]\hat{\gamma}_3(\overline{y})\right)\Delta y
\end{aligned} \tag{A18}$$

Then, the endemic equilibrium point is unique if and only if

$$Ker\left(\gamma_1\left(\gamma_2(\overline{y}).\nabla_{\overline{y}^T}\gamma_3(\overline{y}) + \nabla_{\overline{y}^T}\left[\hat{\gamma}_2(\overline{y})\right]\hat{\gamma}_3(\overline{y})\right)\right) \subseteq Ker\left(I_{3n+1} - A^+A\right) \tag{A19}$$

provided that the constraints 1–3 hold. □

Proof of Theorem 6. For the endemic equilibrium point to exist and be attainable, there exists a non-negative real number ν such that $S_{end}^* = \nu I_{end}^*$. If $n = 1$ the travel matrices in Equation (1) are zeroed and one has at the endemic equilibrium point that:

$$\Lambda - \beta\nu\, I_{end}^{*2} - d^S\nu I_{end}^* - V = 0 \tag{A20}$$

$$\beta\nu\, I_{end}^{*2} - \left(d^I+\gamma\right)I_{end}^* = 0 \tag{A21}$$

$$\gamma\, I_{end}^* - d^R R_{end}^* + V = 0 \tag{A22}$$

One gets from Equation (A21) for $I_{end}^* \neq 0$, since $\nu = \frac{S_{end}^*}{I_{end}^*}$ that $I_{end}^* = \frac{d^I+\gamma}{\beta\nu} = \frac{\left(d^I+\gamma\right)I_{end}^*}{\beta S_{end}^*}$ leading to $S_{end}^* = \frac{d^I+\gamma}{\beta}$. Replacing this value in Equation (A20) leads to $I_{end}^* = \frac{\beta(\Lambda-V)-d^S\left(d^I+\gamma\right)}{\beta\left(d^I+\gamma\right)}$. Note that $S_{end}^* > 0$ and also that if $I_{end}^* \geq 0$, then $v > 0$ and $I_{end}^* \geq 0$ (respectively, $I_{end}^* > 0$) if $\beta \geq \beta_c$ (respectively, $\beta > \beta_c$). It is direct to see that the disease-free equilibrium point is $S_{df}^* = \frac{\Lambda-V}{d^S}$, $I_{df}^* = 0$ and $R_{df}^* = \frac{V}{d^R}$, and that $\beta \geq \beta_c$ is fully equivalent to $R_0 = \frac{S_{df}^*}{S_{end}^*} \geq 1$ implying the attainability of the endemic equilibrium point. Note from Equation (A22) that $R_{end}^* = \frac{V+\gamma I_{end}^*}{d^R}$ which leads to $R_{end}^* = \frac{\beta\left(d^I+\gamma\right)V+\gamma\left[\beta(\Lambda-V)-d^S\left(d^I+\gamma\right)\right]}{\beta d^R\left(d^I+\gamma\right)}$.

After replacing the calculated endemic infectious amount. Note also that:

(1) If $R_0 = 1$ then the endemic equilibrium point is confluent with the disease-free one which is locally asymptotically stable.
(2) If $R_0 < 1$ then the endemic equilibrium point is not attainable since it has negative component.
(3) If $R_0 < 1$ then the disease- free equilibrium point is locally asymptotically stable since the state-solution trajectory of the Jacobian matrix at such a point is a stability matrix. It is also globally asymptotically stable since: (a) it is the unique attainable equilibrium point which is,

furthermore, locally asymptotically stable; (b) the total population is bounded; and (c) all the subpopulations are non-negative for all time implying that all of them are bounded for all time as result; (d) if it would be potentially surrounded by some limit cycle, such a cycle should be unstable since the critical point is asymptotically stable.

On the other hand, if $V(t) = V_0 + KS(t)$ then $S^*_{df} = \frac{\Lambda - V^*_{df}}{d^S} = \frac{\Lambda - V_0 - KS^*_{df}}{d^S}$ leading to $S^*_{df} = \frac{\Lambda - V_0}{d^S + K}$, and $R^*_{df} = \frac{V^*_{df}}{d^R} = \frac{V_0 + KS^*_{df}}{d^R} = \frac{V_0 + K(\Lambda - V_0)/(d^S + K)}{d^R}$ leading to $R^*_{df} = \frac{K\Lambda + d_S V_0}{d^R(d^S + K)}$. If $V_0 = K = 0$ then $S^*_{df} = \frac{\Lambda}{d^S}$ and $R^*_{df} = 0$ The result has been fully proved after calculating the total equilibrium population by summing up the susceptible and immune equilibrium subpopulations. □

Appendix C

Proof of Theorem 7. Note that there exists $U_{un}^{-1} \succ 0$, what is obvious since $(-U_{un})$ is a Metzler stability matrix. If $\rho\left(U_{un}^{-1}\widetilde{U}\right) < 1$then there exists $\left(U_{un} + \widetilde{U}\right)^{-1} = \left(U_{un}\left(I_{3n} + U_{un}^{-1}\widetilde{U}\right)\right)^{-1} = \left(I_{3n} + U_{un}^{-1}\widetilde{U}\right)^{-1} U_{un}^{-1}$. Thus,

$$R_{0c} = \rho\left(F_c U_c^{-1}\right) = \rho\left[\left(F_{un} + \widetilde{F}\right)\left(U_{un}\left(I_{3n} + U_{un}^{-1}\widetilde{U}\right)\right)^{-1}\right] = \rho\left[\left(F_{un} + \widetilde{F}\right) M U_{un}^{-1}\right] \tag{A23}$$

where $M = \left(I_{3n} + U_{un}^{-1}\widetilde{U}\right)^{-1} = U_c^{-1}U_{un}$. Since $M\left(I_{3n} + U_{un}^{-1}\widetilde{U}\right) = \left(I_{3n} + U_{un}^{-1}\widetilde{U}\right)M = I_{3n}$ then

$$M = I_{3n} - U_{un}^{-1}\widetilde{U}M = I_{3n} - U_{un}^{-1}\widetilde{U}\left(I_{3n} + U_{un}^{-1}\widetilde{U}\right)^{-1} = I_{3n} - U_{un}^{-1}\widetilde{U}U_{un} \tag{A24}$$

Thus, the following matrix equalities hold from:

$$F_c U_c^{-1} = F_{un} M U_{un}^{-1} + \widetilde{F} M U_{un}^{-1} = \left(F_{un} + \widetilde{F}\right) U_{un}^{-1}\left(I_{3n} - \widetilde{U}U_c^{-1}\right) \tag{A25}$$

Now, one has

$$\left(I_{3n} + U_{un}^{-1}\widetilde{U}\right)^{-1} = \left(U_{un}^{-1}U_{un} + U_{un}^{-1}\widetilde{U}\right)^{-1} = \left(U_{un} + \widetilde{U}\right)^{-1} U_{un} = U_c^{-1}U_{un} \tag{A26}$$

$$-\left(F_{un} + \widetilde{F}\right) U_{un}^{-1}\widetilde{U}\left(I_{3n} + U_{un}^{-1}\widetilde{U}\right)^{-1} = -\left(F_{un} + \widetilde{F}\right) U_{un}^{-1}\widetilde{U}U_c^{-1}U_{un} \tag{A27}$$

and one has from Equation (A27) that:
$R_{0c} \le R_{0un}$ if

$$0 \prec F_{un} U_{un}^{-1}\left(I_{3n} - \widetilde{U}U_c^{-1}\right) + \widetilde{F}U_{un}^{-1}\left(I_{3n} - \widetilde{U}U_c^{-1}\right) \prec F_{un} U_{un}^{-1} \tag{A28}$$

Note that Equation (A28) is equivalent to:

$$-F_{un} U_{un}^{-1} \prec -F_{un} U_{un}^{-1}\widetilde{U}U_c^{-1} + \widetilde{F}U_{un}^{-1}\left(I_{3n} - \widetilde{U}U_c^{-1}\right) \prec 0 \tag{A29}$$

then to

$$-F_{un} U_{un}^{-1}\left(I_{3n} - \widetilde{U}U_c^{-1}\right) \prec \widetilde{F}U_{un}^{-1}\left(I_{3n} - \widetilde{U}U_c^{-1}\right) \prec F_{un} U_{un}^{-1}\widetilde{U}U_c^{-1} \tag{A30}$$

The constraints Equation (A30) can be written in equality form as follows:

$$-F_{un} U_{un}^{-1}\left(I_{3n} - \widetilde{U}U_c^{-1}\right) + M_1 = \widetilde{F}U_{un}^{-1}\left(I_{3n} - \widetilde{U}\right) = F_{un} U_{un}^{-1}\widetilde{U}U_c^{-1} - |M_2| \tag{A31}$$

for some given real $3n$ matrices $M_1 \succ 0$ and $M_2 \succ 0$. Since $\left(I_{3n} + U_{un}^{-1}\widetilde{U}\right)^{-1} = U_c^{-1}U_{un}$, note that

$$\widetilde{U}\, U_c^{-1}U_{un} = \widetilde{U}\left[U_{un}^{-1}\left(U_{un} + \widetilde{U}\right)\right]^{-1} = \widetilde{U}\left(I_{3n} + U_{un}^{-1}\widetilde{U}\right)^{-1} \tag{A32}$$

so that, if $\rho\left(U_{un}^{-1}\widetilde{U}\right) < 1$, one has that $\left(I_{3n} + U_{un}^{-1}\widetilde{U}\right)^{-1}$ exists and

$$I_{3n} - \widetilde{U}U_c^{-1}U_{un} = I_{3n} - \widetilde{U}\left(U_{un} + \widetilde{U}\right)^{-1}U_{un} = I_{3n} - \widetilde{U}\left(I_{3n} + U_{un}^{-1}\widetilde{U}\right)^{-1} \tag{A33}$$

is also nonsingular if $\rho\left(\widetilde{U}U_c^{-1}\right) = \rho\left[\widetilde{U}\left(I_{3n} + U_{un}^{-1}\widetilde{U}\right)^{-1}\right] < 1$. From Banach Perturbation Lemma [13],

$$\|\widetilde{U}\left(U_{un} + \widetilde{U}\right)^{-1}U_{un}\|_2 = \|\widetilde{U}\left(I_{3n} + U_{un}^{-1}\widetilde{U}\right)^{-1}\|_2 \le \|\widetilde{U}\|_2 \frac{1}{1 - \|\widetilde{U}\|_2\|U_{un}^{-1}\|_2} < 1 \tag{A34}$$

that is, if $\|\widetilde{U}\|_2 < 1/2\, \|U_{un}^{-1}\|_2$ which ensures both the previous condition $\rho\left(U_{un}^{-1}\widetilde{U}\right) < 1$ guaranteeing that $\left(I_{3n} + U_{un}^{-1}\widetilde{U}\right)^{-1}$ exists and that $\left(I_{3n} - \widetilde{U}U_c^{-1}U_{un}\right)^{-1} = \left(I_{3n} - \widetilde{U}\left(I_{3n} + U_{un}^{-1}\widetilde{U}\right)^{-1}U_{un}^{-1}\right)^{-1}$ exist. Thus, since , Equation (A31) is equivalent to

$$\begin{aligned} -F_{un}\, U_{un}^{-1} + M_1\left(I_{3n} - \widetilde{U}\left(I_{3n} + U_{un}^{-1}\widetilde{U}\right)^{-1}U_{un}^{-1}\right)^{-1} = \widetilde{F}U_{un}^{-1} \\ = F_{un}\, U_{un}^{-1}\widetilde{U}U_c^{-1}\left(I_{3n} - \widetilde{U}\left(I_{3n} + U_{un}^{-1}\widetilde{U}\right)^{-1}U_{un}^{-1}\right)^{-1} - |\, M_2|\left(I_{3n} - \widetilde{U}\left(I_{3n} + U_{un}^{-1}\widetilde{U}\right)^{-1}U_{un}^{-1}\right)^{-1} \end{aligned} \tag{A35}$$

Recovering again the matrix inequality form for Equation (A35) and $\widetilde{U}\, U_c^{-1} = \widetilde{U}\left(I_{3n} + U_{un}^{-1}\widetilde{U}\right)$, since M_1 and M_2 are arbitrary, yields that the condition 4 is equivalent to Equation (A35), which is also equivalent to Equation (A28), since $U_{un}^{-1} \succ 0$ if $\|\widetilde{U}\|_2 < 1/2\, \|U_{un}^{-1}\|_2$. Property **(i)** has been proved. Now, assume that$\widetilde{F} = -\left|\widetilde{F}\right| \prec 0$. Then, Equation (A35) holds, and then Equation (A28) also holds, if, for the given pair $(F_{un}\,,\, U_{un})$, the pair $\left(\left|\widetilde{F}\right| = -F\,,\, \widetilde{U}\right)$ fulfils the matrix constraints:

$$-\,F_{un}\, U_{un}^{-1}\widetilde{U}\left(I_{3n} + U_{un}^{-1}\widetilde{U}\right)\left(I_{3n} - \widetilde{U}\left(I_{3n} + U_{un}^{-1}\widetilde{U}\right)^{-1}U_{un}^{-1}\right)^{-1}U_{un} \prec \left|\widetilde{F}\right| \prec F_{un} \tag{A36}$$

Since $F_{un}U_{un}^{-1} \succ 0$ and $|F|\, U_{un}^{-1} \succ 0$ then the matrix inequalities Equation (A36) imply that Property **(ii)** holds since $R_{0c} \le R_{0un}$ and, furthermore, $R_{0c} < R_{0un}$ if either $F_{un}U_{un}^{-1}$ or $\left|\widetilde{F}\right| U_{un}^{-1}$ is irreducible. In the last case, one (but not both) of the symbols $'' \prec ''$ might be replaced with $''\preceq''$. This result is a direct application of Corollary 1.2 in [12] since if A and B are real matrices of the same order with $A \succeq B (\neq A) \succ 0$, equivalently, $A \succ B \succ 0$ then the maximal eigenvalue of A is larger than that of B if A is irreducible but they can be identical if A is reducible. □

References

1. Li, M.Y.; Shuai, Z. Global stability of an epidemic model in a patchy environment. *Can. Appl. Math. Q.* **2009**, *17*, 175–187.
2. Wang, W.; Zhao, X.Q. An epidemic model in a patchy environment. *Math. Biosci.* **2004**, *190*, 97–112. [CrossRef] [PubMed]
3. Muroya, Y.; Enatsu, Y.; Kuniya, Y. Global stability of extended multi-group SIR epidemic models with patches through migration and cross patch infection. *Acta Math. Sci.* **2013**, *33*, 341–3612. [CrossRef]

4. Iggidr, A.; Sallet, G.; Tsanou, B. Global stability analysis of a metapopulation SIS epidemic model. *Math. Popul. Stud.* **2012**, *19*, 115–129. [CrossRef]
5. Jin, Y.; Wang, W. The effect of population dispersal on the spread of a disease. *J. Math. Anal. Appl.* **2005**, *308*, 343–364. [CrossRef]
6. Sattenspiel, L.; Dietz, K. A structured epidemic model incorporating geographic mobility among regions. *Math. Biosci.* **1995**, *128*, 71–91.
7. Takaguchi, T.; Lambiotte, R. Sufficient conditions of endemic threshold on metapopulation networks. *arXiv* **2015**, arXiv:1410.5116v2. [CrossRef]
8. Chalub, F.A.C.C.; Costa, T.J.; Patricio, P. Migrations, vaccinations and epidemic control. *arXiv* **2017**, arXiv:1712.07918v1.
9. Khaleghian, P. Decentralization and public services: The case of immunization. *Soc. Sci. Med.* **2004**, *59*, 163–183. [CrossRef]
10. Singh, M.G. *Decentralised Control*; North-Holland Systems and Control Series; North Holland Publishing Company: New York, NY, USA, 1981; Volume 1.
11. Berman, A.; Plemmons, R.J. *Nonnegative Matrices in the Mathematical Sciences*; Academic Press: New York, NY, USA, 1979.
12. Kaczorek, T. *Positive 1D and 2D Systems*; Communications and Control Engineering Series; Springer: London, UK, 2002.
13. Ortega, J.M. *Numerical Analysis*; Academic Press: New York, NY, USA, 1972.
14. de la Sen, M.; Agarwal, R.P.; Nistal, R.; Alonso-Quesada, S.; Ibeas, A. A switched multicontroller for an SEIADR epidemic model with monitored equilibrium points and supervised transients and vaccination costs. *Adv. Differ. Equ.* **2018**, *2018*, 390. [CrossRef]
15. Nistal, R.; de la Sen, M.; Alonso-Quesada, S.; Ibeas, A. On a new discrete SEIADR model with mixed controls: Study of its properties. *Mathematics* **2019**, *7*, 18. [CrossRef]
16. Alonso-Quesada, S.; de la Sen, M.; Nistal, R. On vaccination strategies for a SISV epidemic model guaranteeing the nonexistence of endemic solutions. *Discr. Dyn. Nat. Soc.* **2018**, *2018*, 9484121. [CrossRef]
17. Xia, W.; Kundu, S.; Maitra, S. Dynamics of a delayed SEIQ epidemic model. *Adv. Differ. Equ.* **2018**, *2018*, 36. [CrossRef]
18. Barambones, O.; Garrido, A.J.; Garrido, I. Robust speed estimation and control of an induction motor drive based on artificial neural networks. *Int. J. Adapt. Control Signal Process.* **2008**, *22*, 440–464. [CrossRef]
19. Bakule, L.; de la Sen, M. Decentralized stabilization of networked complex composite systems with nonlinear perturbations. In Proceedings of the 2009 International Conference on Control and Automation, Christchurch, New Zealand, 9–11 December 2009; Volumes 1–3, pp. 2272–2277.
20. Ibeas, A.; de la Sen, M. Robustly stable adaptive control of a tandem of master-slave robotic manipulators with force reflection by using a multiestimation scheme. *IEEE Trans. Cybern. Part B-Cybern.* **2006**, *36*, 1162–1179. [CrossRef]
21. Kiouach, D.; Sabbar, Y. Stability and threshold of a stochastic SIRS epidemic model with vertical transmission and transfer from infectious to susceptible individuals. *Discr. Dyn. Nat. Soc.* **2018**, *2018*. [CrossRef]
22. Lee, C.; Garbett, A.; Wilkinson, D.J. A network epidemic model for online commissioning data. *Stat. Comput.* **2018**, *28*, 891–904. [CrossRef]
23. Sabbar, Y.; Kiouach, D. Long-time behavior of stochastic SIQD epidemic model with intervention strategies. In Proceedings of the International Conference on Fixed Point Theory and Applications ICFPTA´18, Mohammedia, Morocco, 8 May 2018; pp. 133–136.
24. Shamsi, N.G.; Torabi, S.A.; Shakouri, H.G. An option contract for vaccine procurement using the SIR epidemic model. *Eur. J. Oper. Res.* **2018**, *267*, 1122–1140. [CrossRef]
25. Jia, N.; Ding, L.; Liu, Y.J.; Hu, P. Global stability and optimal control of epidemic spreading on multiplex networks with nonlinear mutual interaction. *Phys. A Stat. Mech. Its Appl.* **2018**, *502*, 93–105. [CrossRef]
26. Kiouach, D.; Boulaasair, L. Stationary distribution and dynamic behaviour of a stochastic SIVR epidemic model with imperfect vaccine. *J. Appl. Math.* **2018**, *2018*. [CrossRef]
27. Das, A.; Pal, M. A mathematical study of an imprecise SIR epidemic model treatment control. *J. Appl. Math. Comput.* **2018**, *56*, 477–500. [CrossRef]

28. Alonso-Quesada, S.; de la Sen, M.; Nistal, R. A state feedback vaccination strategy applied to a SISV model for avoiding endemic equilibrium points. roceedings of the 2018 15th International Conference on Control, Automation, Robotics and Vision (ICARCV), Singapore, 18–21 November 2018; pp. 466–473.
29. Brockmann, D.; Helbing, D. The hidden geometry of a complex, network-driven contagion phenomena. *Science* **2013**, *342*, 1337–1342. [CrossRef] [PubMed]
30. Pei, S.; Kandula, S.; Yang, W.; Shaman, J. Forecasting the spatial transmission of influenza in United States. *Proc. Natl. Acad. Sci. USA* **2018**, *115*, 2753–2757. [CrossRef]
31. Okongo, M.O. The local and global stability of the disease free equilibrium in a co infection model of HIV/AIDS, tubercolosis and malaria. *IOSR J. Math.* **2015**, *11*, 33–43.
32. Barnett, S. *Matrices in Control Theory with Applications to Linear Programming*; Van Nostrand Reinhold Company: London, UK, 1971.
33. Bellman, R. The stability of solutions of linear differential equations. *Duke Math. J.* **1943**, *10*, 643–647. [CrossRef]
34. van den Driessche, P. Reproduction numbers of infectious disease models. *Infect. Dis. Model.* **2017**, *2*, 288–303. [CrossRef] [PubMed]
35. Biggerstaff, M.; Cauchemez, S.; Reed, C.; Gambhir, M.; Finelli, L. Estimates of the reproduction number for seasonal, pandemic, and zoonotic influenza: A systematic review of the literature. *BMC Infect. Dis.* **2014**, *14*, 480. [CrossRef] [PubMed]
36. Magal, P.; Webb, G. The parameter identification problem for SIR epidemic models: Identifying unreported cases. *J. Math. Biol.* **2018**, *77*, 1629–1648. [CrossRef]

symmetry

MDPI

Article

Evolution of Conformity Dynamics in Complex Social Networks

Yuhui Gong [1,2] and Qian Yu [3,*]

1 School of Economics and Management, China University of Geoscience, Wuhan 430074, China; gyh@whut.edu.cn
2 School of Art and Design, Wuhan University of Technology, Wuhan 430070, China
3 School of Economics, Wuhan University of Technology, Wuhan 430070, China
* Correspondence: yuqian@whut.edu.cn

Received: 10 January 2019; Accepted: 21 February 2019; Published: 28 February 2019

Abstract: Conformity is a common phenomenon among people in social networks. In this paper, we focus on customers' conformity behaviors in a symmetry market where customers are located in a social network. We establish a conformity model and analyze it in ring network, random network, small-world network, and scale-free network. Our simulations shown that topology structure, network size, and initial market share have significant effects on the evolution of customers' conformity behaviors. The market will likely converge to a monopoly state in small-world networks but will form a duopoly market in scale networks. As the size of the network increases, there is a greater possibility of forming a dominant group of preferences in small-world network, and the market will converge to the monopoly of the product which has the initial selector in the market. Also, network density will become gradually significant in small-world networks.

Keywords: conformity; evolutionary dynamics; ring network; small-world network; scale-free network

1. Introduction

Humans are highly susceptible to social influence. When a person's judgment conflicts with a group's, that person often conforms their judgment to that of the group [1]. This diffuse occurrence that individuals change their behaviors and attitudes to match the majority's behavior is known as conformity. Conformity is everywhere in our lives. For example, consumption caused by conformity will be influenced by the evaluation of others and driven by other people's behavior. People's pursuit of fashion often leads to the pursuit of a certain type of goods and forms a trend of popularity. Merchants often take advantage of consumers' conformity and the fact that they will imitate idols and follow fashion trends to promote their products.

The earliest study on conformity influence dates back to the 1930s and was carried out by social psychologists Jenness [2] and Sherif [3]. Since then, extensive studies have shown that conformity can affect individual choice behavior. The well-known experiment by Asch [4] in 1955 showed that over 75% of people tend to conform to others in varying degrees. Bernheim [5] analyzed conformity by a theoretical model in which individuals care about both consumption and social status. Existing works carried out by Kelman et al. [6] and Akert [7] have repeatedly verified the significant effect of conformity influence in our social life. Krüger [8] determined the conditions needed to reach a consensus in a double-clique network with conformity and anticonformity as types of social influence.

After then, many researches have measured the degree of conformity. Mehrabian and Ksionzky [9] used a two-dimensional scheme to represent affinitive characteristics to build a function of conformity. Luce and Fishburn [10] built a conformity function which was proved to be the only possible function, which is concave for gains and convex for loses, while satisfying the requirements of segregation and binary prospect theory. Also, researchers have made efforts to establish models for exploring the factors and effects of conformity, mostly focusing on the relationship between conformity and personal characters. Crutchfield [11] put forward the idea that conformity is involved in individual differences such as age and gender. Reitan and Shaw [12] found conformity was also related to the group size. Egebark [13] did a research to show that appreciation, expressed as a single "like" on social networks, from a single stranger increased the size of the influencing group and doubled the probability that subjects expressed positive support. Zollman [14] found that conformity effects induce reliability in some contexts and, surprisingly, this happens even when it is counterproductive. Often, the methods for reducing its effects are not helpful. He attempted to determine the epistemic effects of conformity by analyzing a mathematical model of this behavior. Lascu and Zinkhan [15] proposed a classical conformity model and classified the influencing factors of conformity. The influencing factors of conformity were classified in: individual factors, group factors, product factors, and work factors. Among them, individual factors mainly refer to personality characteristics, knowledge and experience, cultural differences, personality characteristics, social status, and so on; group factors mainly refer to group size, group cohesion, individual position, consistency of group opinion, group authority, and so on. Product factors are its visibility, the specific features of the product, and the functions of the product [16]; work factors mainly refer to the fuzziness of information, the openness of conformity, and the influence of authority [17]. In addition, Deutsch [18] divided the influencing factors of conformity into normative effects and informational effects.

In the study of evolutionary games, an interesting question is why spatial topology can provide the beneficial environment for the evolution of cooperation [19,20]. Various works have demonstrated the effect of spatial reciprocity on the promotion of cooperation dynamics [21]. Conformity has been attracting much attention also recently. Szolnoki and Perc designated a fraction of population as being driven by conformity rather than payoff maximization [22]. Yang [23] and Niu [24] found that conformity-driven reproductive ability, especially rational conformity behavior, can greatly enhance cooperation compared with homogeneous reproductive ability in the evolutionary spatial prisoner's dilemma game. Kabir et al. [25] found that the shape of the cooperative cluster and the ability to expand a single perfect C-cluster are the two factors that bolster the effect of network reciprocity.

In economics, customers are affected not only by the quality and price of the product, but also by the behaviors of other people in the market. For example, Bernheim [5] proposed a theory of social conformity and presented a model to describe the conformity process that has been replicated a large number of times across age groups and cultures by researchers. Corazzini [26] studied the role of social preferences and conformity in herding behavior in anonymous risky environments. Park and Kwanghee [27] identified three features of clothing conformity, i.e., normative, informative, and identifiable conformity. They found the interest in appearance was relatively high, and normative conformity was the most common among the three factors of clothing conformity. By analyzing quarterly data for the 1975–94 period, Wermers [28] investigated whether mutual funds tend to "herd" when trading stocks and examined the impact of herding on stock prices and whether it is stabilizing or destabilizing.

As it is well known, the social network is a theoretical construct useful in social sciences to study relationships between individuals, groups, or organizations. Most researchers use random networks, small-world networks, or scale-free networks to represent the social networks. The random networks, which was studied by Erdös and Rényi [29], is generated by a disconnected set of nodes that are then paired randomly. Other two well-known and much studied classes of complex networks are small-world networks and scale-free networks. Watts and Strogatz [30] found the small-world networks, which are characterized by specific structural features—short path lengths and high clustering—are more suitable than the random networks for describing social networks. They can be generated from a k-regular ring graph, and each edge of the network will rewire randomly according to a certain probability. Scale-free networks, proposed by Barabasi and Albert [31], are more appropriate for representing real networks of hyperlinks between websites and online social networks, when the preferential attachment mechanism works, and the power-law degree distributions of degrees are obeyed.

Along with the developing of the complex network theory, many research issues of social problems were proposed. Zanette [32] reported numerical evidence that an epidemic-like model, which can be interpreted as the propagation of a rumor, exhibits critical behavior at a finite randomness of the underlying small-world network. Moreno [33] derived mean-field equations characterizing the dynamics of a rumor process that takes place on top of complex heterogeneous networks. Nekovee [34] showed that scale-free social networks are prone to the spreading of rumors, just as they are to the spreading of infections. They are relevant to the spreading dynamics of chain emails, viral advertising, and large-scale information dissemination algorithms on the Internet. Zhang [35] researched how the conformity tendency of a person changes with their role, as defined by their structural properties in a social network. Li [36] reviewed the models for characterizing the information diffusion in online social networks. Martinčić-Ipšić et al. [37] established two weighted similarity measures to analyze link prediction among co-occurrence language networks based on hashtags and all the words in tweets. Carrera [38] developed a probabilistic approach to discovering information diffusion among network communities based on an extended hidden Markov model (HMM).

Although a large number of theoretical, empirical, and experimental studies, particularly in psychology and sociology, have investigated the presence of conformity observations with regard to human society, there are few works investigating conformity in social network. It is in this light that we examined the effect of consumers' conformity in the different structures of social networks. On the basis of the studies of conformity and complex networks, we explored the effect of social networks' structure on conformity. We modeled the social networks as random networks, small-world networks, and scale-free networks and then checked the evolution of conformity behavior on these networks.

This paper is organized as follows. Section 1 describes the concepts and the assumptions used in the study. The conformity in ring networks is analyzed in Section 2, and the conformity in complex networks is analyzed in Section 3. Finally, a summary of our study is presented in Section 4.

2. Conformity in Ring Networks

In order to investigate the effects of conformity in social networks, we used the model of conformity effects which was constructed by Chalip and Green [39]. While this model is impoverished in many ways relative to the actual effect of conformity, its simplicity provides a convenient avenue for investigating the effects of the phenomena.

Suppose there are two products, product 1 and product 2. Each customer in the market will be informed about them with an assigned probability. Customers are arranged on a social network, and time is divided into a series of discrete time periods. In each period, every customer simultaneously surveys his neighbors' behavior. If the majority of people make the opposite choice to theirs, the customers change their mind. In the opposite case, the customers maintain their choice. It is said that the customers in the market have a high level of conformity and would be largely influenced by other people in their neighborhood. For convenience, the following several assumptions are made:

Assumption 1. At the first round, the event of a customer being informed by either firm 1 or firm 2 is exclusive, and a customer will choose to buy from the informer.

Assumption 2. Suppose conformity is the only factor influencing a customer's decision. Whether a customer in the following rounds will change its mind merely depends on other customers in the network, regardless of the product price and quality.

Assumption 3. Not all customers are equally social, some may be in contact only with a few neighbors, while others may be connected to relatively larger groups.

Assumption 4. Every customer is affiliated to a group, the size of any network should be $n \geq 2$, and any network is closed.

Assumption 5. The decision-making process of customers can continue for infinite rounds.

At the beginning, every customer will be informed of the products from either firm 1 or firm 2 and then will make their choice, which might be changed in the following rounds. Assume that any individual is informed by firm 1 with probability $(1-\varepsilon)$ and by firm 2 with probability ε.

For customer i, its choice at time t is S_i^t. We define $N_i^t(1) = \left|\{a \in N : S_a^t = 1\}\right|$ as the number in i's neighborhood who would choose to buy products from firm 1; $N_i^t(2)$ is defined accordingly. We can define the choice function S_i^t of a customer based on conformity:

$$S_i^t = \begin{cases} 1 \text{ if } N_i^{t-1}(1) > N_i^{t-1}(2) \\ 1 \text{ if } N_i^{t-1}(1) = N_i^{t-1}(2) \text{ and } S_i^{t-1} = 1 \\ 2 \text{ otherwise} \end{cases}, \tag{1}$$

In order to simplify our research, we began the study on a ring network model which has some similarities to the model of Zollman [14]. Actually, in a regular grid topology, each node in the network is connected with two neighbors along one or more dimensions. If the network is one-dimensional and the chain of nodes is connected to form a circular loop, the resulting topology is known as a ring. Imagine customers who lead lonely lives and only know about two customers among their neighbors. The social network of these customers would be a ring, as follow (Figure 1):

Figure 1. A sample market on a ring network.

We set white icons to be firm 1 choosers, and black icons to be firm 2 choosers. According to our choice function S_i^t, a customer originally informed by firm 1 will change its decision if both of its neighbors choose to buy from firm 2. We can thus conclude that a ring will converge to choose firm 2 as shown in Figure 2, if every firm 1 chooser is surrounded by firm 2 choosers. If every ring follows this rule, then firm 2 will take up the market eventually.

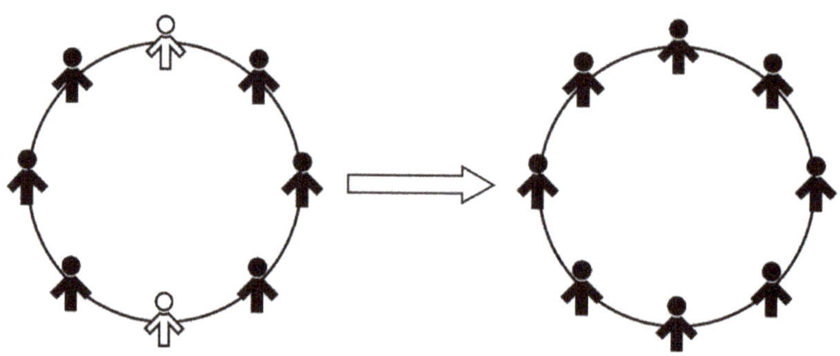

Figure 2. Conformity effect on customer's choices in the ring network.

In addition, an interesting phenomenon will occur if every firm 1 chooser and firm 2 chooser appear alternatively: in this case, the ring would never converge, every node would change from 1 to 2 and then back to 1 (Figure 3), which means any individual will change its mind after every round.

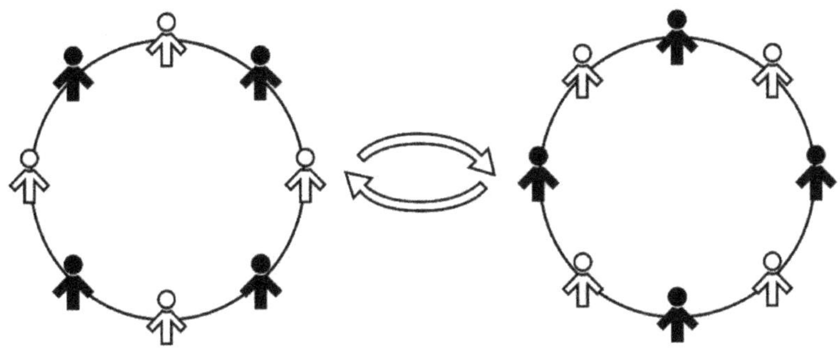

Figure 3. Switch loop of customer's choices affected by conformity.

The following propositions can be obtained for analysis.

Proposition 1. *The chooser in a ring network will converge to firm 1 if, and only if, every customer has two firm 1 choosers as neighbors and does not have two firm 2 choosers as neighbors.*

Proof. This proposition seems to be reasonable as we can see from the rings above: every customer with firm 1 choosers as neighbors will become a firm 1 chooser in the next round. As the process goes on, it is possible that all customers would become firm 1 choosers. With this in hand, our task now is to explore if a ring will converge to either firm 1 or firm 2 in the end.

To make it simple, we broke the ring and extended it to a linear model. The probability of not having two consecutive firm 2 choosers as neighbors is set to be Q. The customer itself has a probability $(1-\varepsilon)$ to become a firm 1 chooser and a probability ε to become a firm 2 chooser. When there is only one customer in the market, $Q = 1$; when there are two customers, $Q(2,\varepsilon) = (1-\varepsilon) + \varepsilon(1-\varepsilon)$. As the number of customers increases, we divide the situation into two parts. If the first customer chooses firm 1, then the rest $(n-1)$ customers should not have two firm 2 choosers as neighbors in a row, for which the probability $Q(n-1,\varepsilon)$ is required; if the first one is a firm 2 chooser, then the second person should choose firm 1, and for the rest $(n-2)$ customers, the probability must satisfy $Q(n-2,\varepsilon)$. Based on this, we have:

$$Q(n,\varepsilon) = (1-\varepsilon)Q(n-1,\varepsilon) + \varepsilon(1-\varepsilon)Q(n-2,\varepsilon), \tag{2}$$

□

Proposition 2. *For* $0 < \varepsilon < 1$, $Q(n,\varepsilon)$ *is strictly decreasing.*

Proof. In the above linear model, the probability of not having two consecutive firm 2 choosers decreases as the number of customers increases according to proposition 2. As in a ring, the only difference between this and the linear model is one connection between the first customer and the last one. In this case, if the first person is a firm 1 chooser, we require $Q(n-1,\varepsilon)$; if the first customer chooses firm 2, then the second and the last customers should all choose firm 1 to meet the requirement; for the rest $(n-3)$ customers, the probability should be $Q(n-3,\varepsilon)$.

So, the adjusted probability $P(n,\varepsilon)$ based on $Q(n,\varepsilon)$ can be given as:

$$P(n,\varepsilon) = (1-\varepsilon)Q(n-1,\varepsilon) + \varepsilon\left(1-\varepsilon^2\right)Q(n-3,\varepsilon), \tag{3}$$

It has a similar character as $Q(n,\varepsilon)$ according to another proposition. □

Proposition 3. *For* $0 < \varepsilon < 1, P(n, \varepsilon)$ *is strictly decreasing.*

Proof. As the size of the ring network increases, the probability of one customer having two firm 2 choosers as neighbors will increase, and according to Proposition1, there would finally be some customers choosing firm 2, and the ring will never converge to firm 1. Two firms would be able to maintain a duopoly market as they share the customers together. In the simplest network (the ring), conformity may not be able to affect all customers because of lack of communication between people. It is worth mentioning that a small ring has higher chances to converge to either firm than a large one, according to the probability function $P(n, \varepsilon)$ which is strictly decreasing. However, in a larger ring, the majority would have higher chances to cause convergence to firm 1 or firm 2 and thus to lead to a monopoly market. It is also reasonable and provable that the firm with higher ε would be able to gain more customers. In order to increase their value, firms may adopt other methods such as informative advertising to increase the value of ε [14]. □

In order to illustrate the above propositions, we calculated the probability of not having two choosers of firm 2 in a row as neighbors for different values of n and ε. The size of the network was set to be in the range from 2 to 20; we obtained four curves with different ε. The values of both probability functions are ploted in Figures 4 and 5:

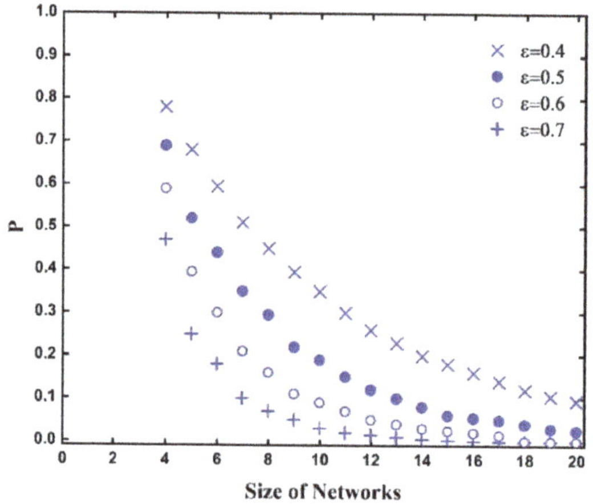

Figure 4. $Q(n, \varepsilon)$ function decreases as n increases. $Q(n, \varepsilon)$ with higher ε converges more quickly to 0.

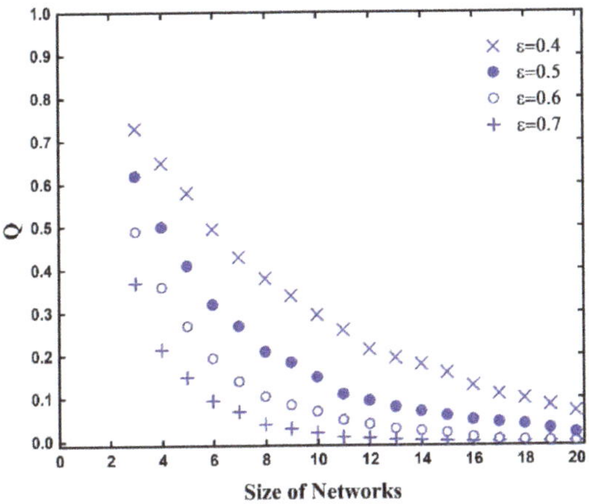

Figure 5. $P(n, \varepsilon)$ function decreases as n increases. $P(n, \varepsilon)$ with higher ε converges more quickly to 0.

As mentioned above, for any ε, $Q(1, \varepsilon) = P(1, \varepsilon) = 1$. In this sense, we neglected the icons when $n = 1$ and chose to start with $n = 2$ in $Q(n, \varepsilon)$, $n = 3$ in $P(n, \varepsilon)$. It is evident from Figures 4 and 5 that both probability functions $P(n, \varepsilon)$ and $Q(n, \varepsilon)$ decrease as n increases, which indicates the larger a social group is, the more difficult it will be to converge to one firm. It is also worth mentioning that firms with higher probability to inform customers (i.e., with higher ε) are less likely to take up the whole market in a ring network. This seems a little counterintuitive, but we should notice that firms of this type have higher chances to defeat their opponents and own more shares in the market.

3. Analysis of Conformity in Complex Networks

Most real-world networks, especially social networks, are complex. Complex networks are networks with non-trivial topological features that are neither purely regular nor purely random. For example, small-world networks, according to Watts and Strogatz, are distinguished from other networks by two specific properties: high clustering among nodes and short path lengths. A scale-free network, such as collaboration networks and inter-bank payment networks, is a network whose degree of distribution follows a power law. Hence, in order to further study the influence of customer conformity in the social networks, we conducted a comparative study in random networks, small-world networks, and scale-free networks.

We supposed there are N nodes in the network $G = (N, V)$, in which $N = \{1, 2, 3, \cdots, n\}$ is a finite set of nodes, and $V = \{(i, j)|i, j \in N\}$ is a set of connection lines between all nodes, and in which $(i, j) \in V$ represents an associated relationship between customer i and j. Also, each customer is set at a 0 or 1 status, $s_i \in \{0, 1\}$, where s_1 indicates that production of firm1 is selected, and s_0 indicates that production of firm 2 is selected. Then, the state space of a social network with n customers can be described as $\Theta^n = \{s_0, s_1\}^n$, at any time t.

When the network size n is large enough and the nodes in the network obey the homogeneous mixed distribution, the mean-field equation can be applied to analyze the evolution dynamics of the customer's choice behavior system. Suppose $\rho_k(t)$ represents the proportion of nodes in which the customer with degree k in the network selects "firm 1" at time t, and $\frac{kP(k)}{\langle k \rangle}$ represents the probability of the node connecting with the customer node with degree k in the network, where $\langle k \rangle = \sum_{k \geq 1} kP(k)$ is the network average degree. Then, at time t, the probability of any customer selecting firm 1 is: $\varphi(t) = \sum_{k \geq 1}(kP(k)\rho_k(t)) / \langle k \rangle$. Thus, the probability of a node that chooses firm 1 among the nodes connected with customers with degree k is: $\begin{pmatrix} k \\ a \end{pmatrix} \varphi(t)^a (1 - \varphi(t))^{k-a}$, defined as $P(1|a, k, \varphi(t))$. It can be seen that a mean field parameter of $\varphi(t)$ is applicable to any network node and is not affected by network connectivity.

Let $\delta > 0$, δ is a conformity coefficient, indicating the probability of customers to adjust their choice behavior. The probability that the customer choice state changes from firm 1 to firm 0 is:

$$r(1|k,\varphi(t)) = \sum_{a=0}^{k} \delta f(k,a) P(1|a,k,\varphi(t)), \tag{4}$$

where $f(k,a) = \frac{a}{k}$ represents the proportion of the specific behavior of the associated node.

The probability of state changing from firm 0 to firm 1 is:

$$r(0|k,\varphi(t)) = \sum_{a=0}^{k} \delta f(k,a) P(1|a,k,1-\varphi(t)), \tag{5}$$

Then, $r(0|k,\varphi(t)) + r(1|k,\varphi(t)) = \delta$.

Proposition 4. *The equilibrium state in the market is* $\rho_k = \delta^{-1} r(1|k,\varphi(t))$.

Proof. The change rate of the customer's choice behavior to firm1 in the network can be expressed by the mean-field equation:

$$\frac{d\rho_k(t)}{dt} = -\rho_k(t) r(1|k,\varphi(t)) + (1-\rho_k(t)) r(0|k,\varphi(t)), \tag{6}$$

□

The above formula shows that the change rate of the behavior of the customer selecting firm 1 mainly depends on the change rate of the selection behavior and the current state of the customer's behavior, regardless of the specific time, and can be regarded as the Markov process of the continuous time system. For $k \geq 1$ and $\frac{d\rho_k(t)}{dt} = 0$, in equilibrium state, the proportion of customers whose selection behavior is 1 can be obtained by $\rho_k = \frac{r(1|k,\varphi(t))}{r(0|k,\varphi(t)) + r(1|k,\varphi(t))} = \delta^{-1} r\left(1 \middle| k,\varphi(t)\right)$.

Following the assumptions about the market and consumers in the previous section, we carried out a series of simulation experiments to examine the influence of network structure, network size, and the initial market share on the evolution of customer conformity behaviors. In each simulation, the social network was generated with a different structure, obtaining random networks, small-world networks, and scale-free networks. Then, each experiment was iterated for 20 rounds and performed 100 times.

The results of the series of experiments are as follows:

(1) Evolution of herding behavior under different network structure types and network size.

The initial market shares ε in the social networks were set randomly with the expected value of 0.5, and the size were set as 100, 300, and 500. Then, simulations were carried out in the generated random network, small-world network, and scale-free network, respectively. The experiment results are reported in Figure 6.

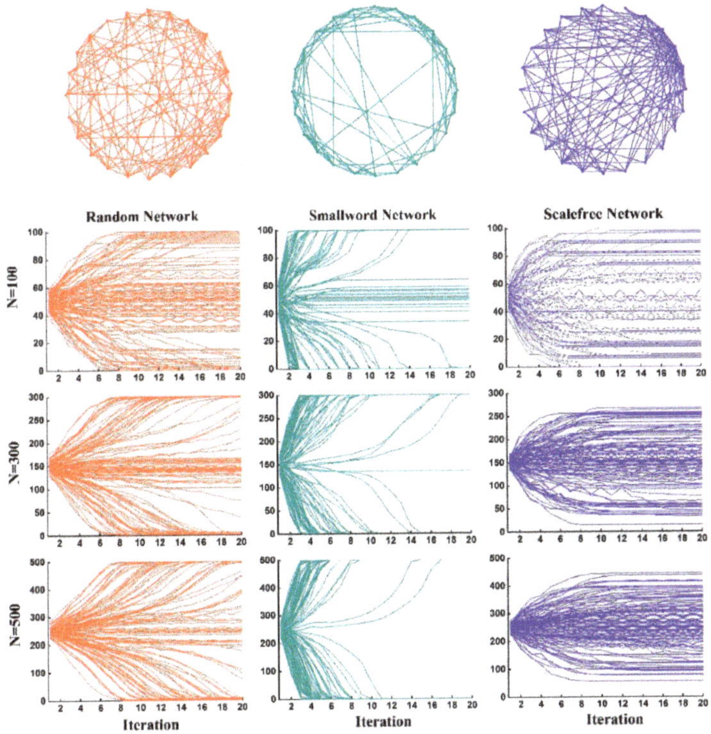

Figure 6. Evolution of market share with effect of conformity on networks having different structures and sizes.

The Figure 6 shows that the topology structure of the networks has a significant impact on the evolution of the consumers' conformity behavior. In the random network, the market share, which represents the distribution value of the consumers' choice, may converge to one of the products, that is, the market share may reach the highest value of 1 or fall to 0, which appears as the line of iterations rises to the top or falls to the bottom; it is also possible that the line of iterations will oscillate around the middle. In the small-world network, the market share will soon converge to a stable state, showing that all consumers tend to have a common choice. In the scale-free network, the distribution value of the consumers' choice will oscillate to varying degrees with a high probability rather than converge to a monopoly state. In addition, the scale of the network has a significant impact on the evolution of customers' conformity behaviors. The larger the scale is, the stronger the effect of the network structure will be.

The simulation results showed that in the small-scale random network, due to the randomness of social connection, it is difficult to play a decisive role. The evolution of consumers' behavior has a strong randomness, and the market has a certain probability to converge to the monopoly of one product. However, as the scale of the network increases, there is a great possibility of forming a dominant group of preferences in social relations, and the market share will converge to a certain monopoly state. In the small-world network, because of the shortest path and high agglomeration, consumers are likely to form a power of public opinion, which causes a high probability of converging to a monopoly of certain products in the market. In a small-scale network, there is a certain probability in a chaotic state, but as the scale increases, the convergence increases, and the market will quickly converge to a monopoly of a certain product. In a scale-free network, a non-hub node may be affected by different selected hub nodes and shift to different options, so it is difficult to converge to a monopoly state in the overall network.

(2) Evolution of conformity behavior in small-world networks with different network densities and network scales.

The initial market shares ε in the small-world networks were set randomly with the expected value of 0.5, and the size were set to 100, 300, and 500. Then, simulations were carried out in the generated networks with average degree of 10, 30, and 50, respectively. The experiment results are shown in Figure 7 as follows:

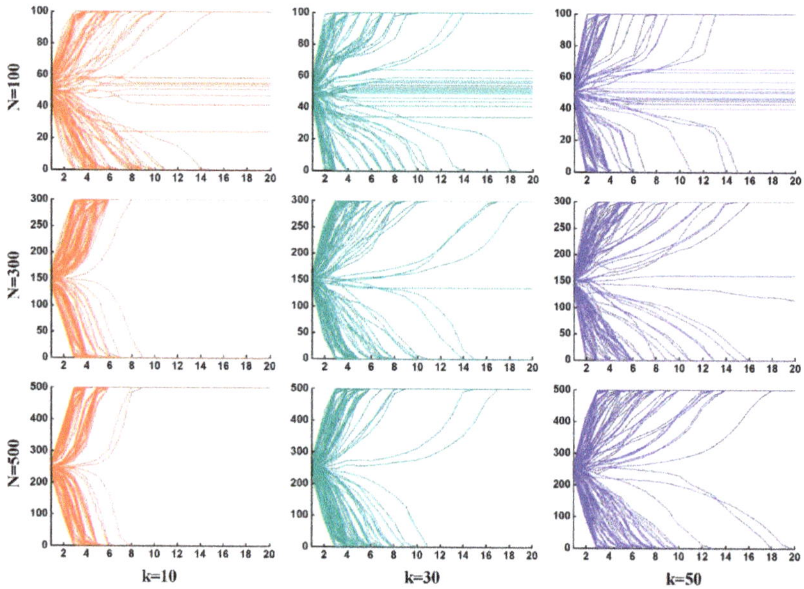

Figure 7. Evolution of market share with effect of conformity on small-world networks having different average degrees and sizes.

The simulation results showed that the network density (which is closely related to the average number of neighbors) has a little effect on the evolution of consumers' conformity behavior in small-world networks when the scale is small, while it will become gradually significant when the scale of the networks gets bigger. With the increase of the network density, the convergence of the market share becomes slower and weaker and may even fail to converge to the monopoly of a certain product. The possible reason is that, when the network density of small-world networks increases, the interference effect among different clustering groups is stronger, which will result in an unstable sway of some nodes in conformity selection.

(3) The influence of different initial distributions on the evolution of herd behavior.

To check the effect of the initial market share on customer conformity, we set the social networks as random network, small-world network, and scale-free network, with size ranging from 100 to 500. Then, the experiments were carried out with initial distribution of 0.4, 0.45, 0.5, 0.55, and 0.6, respectively. The experimental results are shown in Figure 8 as follows:

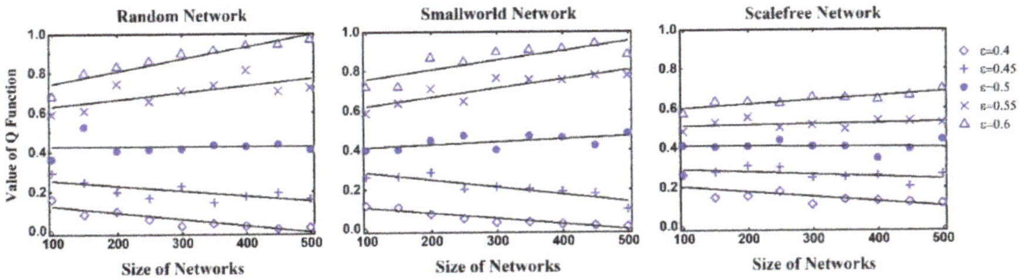

Figure 8. Monotonicity of Q' in different networks and with different ε.

In order to compare the results with those of the ring networks, we similarly defined $Q' = \underset{t}{E}(\varphi(t)) = \frac{\sum_t \sum_{k\geq 1} kP(k)\rho_k(t)}{\langle k\rangle T}$, where Q' represents the mean probability of choosing firm 1 during evolution.

When Q' approaches 1, the distribution of consumer choices in the market tends to be monopolized by product 1. When Q' approaches 0, the distribution of consumer choices in the market tends to be monopolized by product 2. The experimental results show that in all three kinds of social network, Q' value will increase as ε increases; when the given ε is greater than 0.5, Q' will increase with the size of the network. When the given ε is less than 0.5, Q' will decrease with the increase of network size. When ε is equal to 0.5, Q' will stabilize between 0.4 and 0.5. The above trends are particularly significant in random networks and are less obvious in scale-free networks.

The above results show that the dominant consumer groups in the initial selection distribution tend to influence the herd behavior. In turn, the market will converge to the monopoly of the product. In addition, when the initial market share is quite balanced, the distribution of consumer choice may appear as chaotic in a social network with a small scale. In a large-scale social network, it is more likely that the market will converge to the monopoly of dominant products. The above results are obvious in random networks and small-world networks, but in scale-free networks, the chaotic state is more likely to occur.

(4) The evolution of Conformity behavior on Facebook social network.

To verify the simulation results, we used a real-world social network dataset. The following Figure 9 shows the real-world social network on Facebook, whose dataset was collected from survey participants using the Facebook app and freely provided by the SNAP library of Stanford university. This dataset consists of 'circles' (or 'friends lists') from Facebook and includes node features (profiles), circles, and ego networks. The network consists of 4039 nodes and 88234 edges, the average clustering coefficient is 0.617, the average path length is 3.693, and the average degree is 43.691, which meet the characteristics of small-world network.

Figure 9. Image of the social network on Facebook.

The experiments were carried out with the initial distribution ε as 0.4, 0.45, 0.5, 0.55, and 0.6, and each experiment was run 30 times. The experimental results (Figure 10) showed that the market will converge to the monopoly of a certain product which is dominant in the initial market. When the initial distribution ε is larger than 0.55 or less than 0.45, the market will converge to a monopoly state more quickly. When the initial distribution ε is around 0.5, the market will not to converge to a monopoly state in the overall network and will exhibit a chaotic state. These results are consistent with experiments carried out on simulation networks.

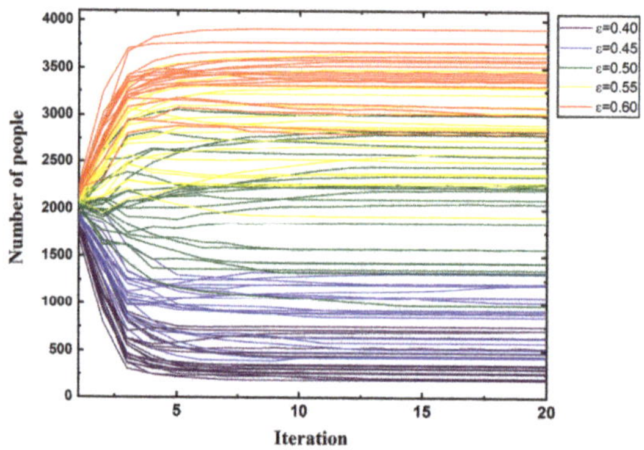

Figure 10. Evolution of market dynamics on Facebook social networks with different ε.

4. Conclusions

In this paper, we established the conformity model in networks where the customers in the market have strong conformity and are largely influenced by their neighborhood in social networks. We studied the influence of customer conformity in different structures of social networks, including ring network, random network, small-world network, and scale-free network.

When the network is a ring network, the market share will more likely converge to a monopoly state, and the firm with a higher market share will be able to gain more customers. However, conformity does not always lead to convergency as the network size increases. The reason may be that a customer might not be able to assert its influence on other buyers because of limited communication in the social network.

The simulations on random networks, small-world networks, and scale-free networks showed that the topology structure, network size, and initial market share will have significant effects on the evolution of customers' conformity behaviors. Firstly, when the social network is a small network, the market will more likely converge into a monopoly market due to its shortest path and high agglomeration, and public opinion will be more powerful in the social network. However, in the scale-free network and random network, it is difficult to converge to a monopoly state in the overall network.

Secondly, the size of networks also plays an important role in determining conformity in social networks. As the scale of the network increases, there is a greater possibility of forming a dominant group of preferences in social networks. However, in scale-free networks, this is difficult to happen, and the network will exhibit a chaotic state. Furthermore, the initial market share will dramatically determine the final results. The market will converge to the monopoly of the product which has the initial selector in market.

Finally, the results also showed that the network density will has an effect on consumers' conformity in social networks, especially in small-world networks, and the impact will be increasingly significant when the scale of the networks gradually becomes larger.

Author Contributions: Methodology, Q.Y.; Visualization, Y.G.; Writing—original draft preparation, Y.G.; Writing—review and editing, Q.Y.

Funding: This work was supported by the National Natural Science Foundation of China (No. 71774128).

Conflicts of Interest: The authors declare no conflict of interest.

References

1. Berns, G.S. Price, placebo, and the brain. *J. Mark. Res.* **2005**, *42*, 399–400. [CrossRef]
2. Jenness, A. Social influences in the change of opinion. *J. Abnorm. Soc. Psychol.* **1932**, *27*, 29–34. [CrossRef]
3. Sherif, M. A study of some social factors in perception. *Arch. Psychol.* **1935**, *187*, 5–61.
4. Asch, S.E. Studies of independence and conformity: I. A minority of one against a unanimous majority. *Psychol. Monogr.* **1956**, *70*, 1–70. [CrossRef]
5. Bernheim, B.D. A Theory of conformity. *J. Political Econ.* **1994**, *102*, 841–877. [CrossRef]
6. Kelman, H. Communing and relating. *Am. J. Psychoanal.* **1958**, *18*, 77–98. [CrossRef]
7. Aronson, E.; Wilson, T.D.; Akert, R.M. *Social Psychology: International Edition*; Pearson Schweiz Ag: Zug, Switzerland, 2011.
8. Krüger, T.; Szwabiński, J.; Weron, T. Conformity, anticonformity and polarization of opinions: Insights from a mathematical model of opinion dynamics. *Entropy* **2017**, *19*, 371. [CrossRef]
9. Mehrabian, A.; Ksionzky, S. Models for affiliative and conformity behavior. *Psychol. Bull.* **1970**, *74*, 110–126. [CrossRef]
10. Luce, R.D.; Fishburn, P.C. A note on deriving rank-dependent utility using additive joint receipts. *J. Risk Uncertain.* **1995**, *11*, 5–16. [CrossRef]
11. Crutchfield, R.S. Conformity and character. *Am. Psychol.* **1955**, *10*, 191–198. [CrossRef]
12. Reitan, H.T.; Shaw, M.E. Group membership, sex-composition of the group, and conformity behavior. *J. Soc. Psychol.* **1964**, *64*, 45–51. [CrossRef] [PubMed]
13. Egebark, J.; Ekstrm, M. Liking what others "Like": Using Facebook to identify determinants of conformity. *Exp. Econ.* **2018**, *21*, 793–814. [CrossRef]
14. Zollman, K.J.S. Social structure and the effects of conformity. *Synthese* **2010**, *172*, 317–340. [CrossRef]
15. Lascu, D.N.; Zinkhan, G. Consumer conformity: Review and applications for marketing theory and practice. *J. Mark. Theory Pract.* **1999**, *7*, 1–12. [CrossRef]
16. Kuenzel, J.; Musters, P. Social interaction and low involvement products. *J. Bus. Res.* **2007**, *60*, 876–883. [CrossRef]
17. Sharma, S.; Bikhchandani, S. Herd behavior in financial markets; A Review. *IMF Work. Pap.* **2000**, *47*, 279–310. [CrossRef]
18. Deutsch, M.; Gerard, H.B. A study of normative and informational social influences upon individual judgment. *J. Abnorm. Psychol.* **1955**, *51*, 629–636. [CrossRef] [PubMed]
19. Tanimoto, J. *Fundamentals of Evolutionary Game Theory and Its Applications*; Springer: Tokyo, Japan, 2015.
20. Tanimoto, J. *Evolutionary Games with Sociophysics: Analysis of Traffic Flow and Epidemics*; Springer: Singapore, 2018.
21. Wang, Z.; Kokubo, S.; Tanimoto, J.; Fukuda, E.; Shigaki, K. Insight into the so-called spatial reciprocity. *Phys. Rev. E* **2013**, *88*, 042145. [CrossRef] [PubMed]
22. Szolnoki, A.; Perc, M. Conformity enhances network reciprocity in evolutionary social dilemmas. *J. R. Soc. Interface* **2014**, *12*, 20141299. [CrossRef] [PubMed]
23. Yang, H.X.; Tian, L. Enhancement of cooperation through conformity-driven reproductive ability. *Chaos Solitons Fractals* **2017**, *103*, 159–162. [CrossRef]
24. Niu, Z.; Xu, J.; Dai, D.; Liang, T.; Mao, D.; Zhao, D. Rational conformity behavior can promote cooperation in the prisoner's dilemma game. *Chaos Solitons Fractals* **2018**, *112*, 92–96. [CrossRef]
25. Kabir, K.A.; Tanimoto, J.; Wang, Z. Influence of bolstering network reciprocity in the evolutionary spatial prisoner's dilemma game: A perspective. *Eur. Phys. J. B* **2018**, *91*, 312. [CrossRef]
26. Corazzini, L.; Greiner, B. Herding, social preferences and (non-)conformity. *Econ. Lett.* **2007**, *97*, 76–80. [CrossRef]
27. Park, K.H.; Yoo, H.S. Effects of appearance interest and demographic characteristics on clothing conformity. *Fash. Text. Res. J.* **2013**, *15*, 210–218. [CrossRef]

28. Wermers, R. Mutual fund herding and the impact on stock prices. *J. Financ.* **1999**, *54*, 581–622. [CrossRef]
29. Erdös, P.; Rényi, A. Some problems and results on consecutive primes. *Simon Stevin* **1950**, *27*, 115–125.
30. Watts, D.J.; Strogatz, S.H. Collective dynamics of 'small-world' networks. *Nature* **1998**, *393*, 440–442. [CrossRef] [PubMed]
31. Barabasi, A.L.; Albert, R. Emergence of scaling in random networks. *Scinece* **1999**, *286*, 509–512.
32. Zanette, D.H. Critical behavior of propagation on small-world networks. *Phys. Rev. E Stat. Nonlinear Soft Matter Phys.* **2001**, *64*, 050901. [CrossRef] [PubMed]
33. Moreno, Y.; Nekovee, M.; Pacheco, A.F. Dynamics of rumor spreading in complex networks. *Phys. Rev. E Stat. Nonlinear Soft Matter Phys.* **2004**, *69*, 066130. [CrossRef] [PubMed]
34. Nekovee, M.; Moreno, Y.; Bianconi, G.; Marsili, M. Theory of rumour spreading in complex social networks. *Phys. A Stat. Mech. Its Appl.* **2008**, *374*, 457–470. [CrossRef]
35. Zhang, J.; Tang, J.; Zhuang, H.; Leung, C.W.; Li, J. Role-aware conformity influence modeling and analysis in social networks. In Proceedings of the Twenty-Eighth AAAI Conference on Artificial Intelligence, Québec City, QC, Canada, 27–31 July 2014; AAAI Press: Palo Alto, CA, USA, 2014.
36. Li, M.; Wang, X.; Gao, K.; Zhang, S. A survey on information diffusion in online social networks: Models and methods. *Information* **2017**, *8*, 118.
37. Martinčić-Ipšić, S.; Močibob, E.; Perc, M. Link prediction on Twitter. *PLoS ONE* **2017**, *12*, e0181079. [CrossRef] [PubMed]
38. Carrera, B.; Jung, J.-Y. SentiFlow: An information diffusion process discovery based on topic and sentiment from online social networks. *Sustainability* **2018**, *10*, 2731. [CrossRef]
39. Chalip, L.; Green, B.C. Establishing and maintaining a modified youth sport program: Lessons from Hotelling's location game. *Sociol. Sport J.* **1998**, *15*, 326–342. [CrossRef]

symmetry

MDPI

Article

A Novel Computational Technique for Impulsive Fractional Differential Equations

Changyou Ma *

College of Mathematics and Information Science, Neijiang Normal University, Neijiang 641100, China; 30000771@njtc.edu.cn

Received: 24 December 2018; Accepted: 25 January 2019; Published: 13 February 2019

Abstract: A computational technique for impulsive fractional differential equations is proposed in this paper. Adomian decomposition method plays an efficient role for approximate analytical solutions for ordinary or fractional calculus. Semi-analytical method is proposed by use of the Adomian polynomials. The method successively updates the initial values and gives the numerical solutions on different impulsive intervals. As one of the numerical examples, an impulsive fractional logistic differential equation is given to illustrate the method.

Keywords: fractional derivative; Adomian method; computational technique

1. Introduction

Fractional calculus appears frequently in various applied topics [1–7] and pure mathematics [8–12]. They are employed to depict the long-interaction of different statues of the systems. The fractional order controls the amount of dependence on past information and shows the quantity of the memory. On the other hand, as a result, it holds both quantitative and qualitative aspects. It shows some features that are not present in other tools. This is the main reason for the popularity of the fractional calculus as a modeling tool for the memory process.

Impulse theory is often used in control methods of differential equations. The impulsive point changes dynamics of continuous time systems locally. Then, the solution has a jump and becomes a piecewise continuous function on the whole interval, and impulsive points are the endpoints of each short interval. A differential equation containing impulses is also called a system with a jump. Hence, the impulsive differential equation is not a continuous time system but the one combining both continuous and discrete point information. It depicts the impact of external conditions which may be negative or positive. Impulsive fractional differential equations have received much attention recently [13–19]. It can illustrate totally distinct dynamics in comparison to standard fractional systems, and this property has often been adopted in fractional impulsive control. Many analytical methods have been efficiently developed for differential equations [20–27]. However, the less numerical method and analytical method were developed for impulsive fractional differential equations. In this study, our main purpose is to extend methods from the integer order to the fractional order.

The Adomian decomposition method (ADM) has been applied in various nonlinear problems, and the Adomian polynomials play a crucial role in the treatment of the nonlinear terms in fractional differential equations. Recently, Duan et al. proposed a new way to calculate the polynomials which can derive the same results but greatly improve computational speed and save time in comparison with the classical one. Hence, various novel algorithms based on the new Adomian polynomials can be considered now. It was successfully used in fractional differential equations [28] where a semi-analytical method was developed.

In this paper, a novel computational technique is proposed for the following equation by use of new Adomian polynomials [21–23]:

$$\begin{cases} {}_{t_0}^{C}D_t^{\alpha}x(t) = F(t,x), t \in J\prime := J\backslash[t_{N_1}, \cdots, t_{N_M}], J := [t_0, T], \\ \Delta x_{N_k} = I_k\left(x_{N_k}^-\right) = x\left(t_{N_k}^+\right) - x\left(t_{N_k}^-\right) = y_k, 1 \le k \le M, \\ x_k = x(t_k), x\left(t_{N_k}^+\right) = \lim_{\hbar \to \infty} x\left(t_{N_k} + \hbar\right), x\left(t_k^-\right) = \lim_{\hbar \to 0} x\left(t_{N_k} - \hbar\right), \\ \hbar > 0, x(t_0) = x_0. \end{cases} \tag{1}$$

The Adomian polynomials are used in fractional differential equations. However, to the best of our knowledge, we did not find any work on semi-analytical solutions for impulsive fractional differential equations. This paper combines both analytical and numerical solutions' features to develop a semi-analytical method.

2. Preliminaries

2.1. Definitions and Properties of Fractional Calculus

The fractional calculus is defined as the following:

Definition 1 [1]. *The Riemann–Liouville (R-L) integral of* α *order is defined by*

$$ {}_{t_0}I_t^{\alpha} f(t) = \frac{1}{\Gamma(\alpha)} \int_{t_0}^{t} (t-\tau)^{\alpha-1} f(\tau)\, d\tau, \ 0 < t, 0 < \alpha. \tag{2}$$

Definition 2 [1]. *The R–L derivative is defined as*

$$\begin{aligned} &{}_{t_0}D_t^{\alpha} f = \tfrac{1}{\Gamma(m-\alpha)} \tfrac{d^m}{dt^m} \textstyle\int_{t_0}^{t} \tfrac{1}{(t-\tau)^{\alpha-m+1}} f(\tau)\, d\tau, \\ &t_0 < t, 0 < \alpha,\ m = [\alpha] + 1, \end{aligned} \tag{3}$$

where Γ is the Gamma function.

Definition 3 [1]. *The Caputo derivative is defined as*

$$\begin{aligned} &{}_{t_0}^{C}D_t^{\alpha} f = {}_{t_0}D_t^{\alpha}\big(f(t) - \textstyle\sum_{k=0}^{m-1} \tfrac{(t-t_0)^k}{k!} f^{(k)}(t_0)\big), \\ &0 < t, 0 < \alpha,\ m = [\alpha] + 1. \end{aligned} \tag{4}$$

Remarks: For *Definition 3, the Caputo derivative of a constant is zero;*

If $f(t) \in C^m([t_0, \infty), R)$, then the Caputo derivative can be rewritten as

$$ {}_{t_0}^{C}D_t^{\alpha} f = \frac{1}{\Gamma(m-\alpha)} \int_{t_0}^{t} \frac{1}{(t-\tau)^{\alpha-m+1}} \frac{d^m}{d\tau^m} f(\tau)\, d\tau. \tag{5}$$

In Definition 3, the function $f(t)$ can be discrete if it is integrable such that the fractional impulsive equation makes sense at the impulsive point.

In the sequel, we all use the definition of the Caputo derivative. We need the integral transform so that the fractional differential equation can be reduced to an integral one and the integral methods can be applied straightforward.

Property 1. *The Leibniz integral law holds*

$$\begin{aligned}&{}_{t_0}I_t^{\alpha C}{}_{t_0}D_t^{\alpha} f(t) = f(t) - f(t_0),\\ &t_0 \le t, 0 < \alpha \le 1.\end{aligned} \tag{6}$$

Lemma 1 [14]. *The impulsive fractional differential Equation (1) is equivalent to the following integral equation of fractional order*

$$\begin{cases} x(t) = x_0 + {}_{t_0}I_t^{\alpha} F(t,x),\ t \in [t_0, t_{N_1}],\\ x(t) = x_0 + y_1 + {}_{t_0}I_t^{\alpha} F(t,x),\ t \in (t_{N_1}, t_{N_2}],\\ x(t) = x_0 + y_1 + y_2 + {}_{t_0}I_t^{\alpha} F(t,x),\ t \in (t_{N_2}, t_{N_3}],\\ \quad\vdots\\ x(t) = x_0 + \sum\limits_{j=1}^{k} y_j + {}_{t_0}I_t^{\alpha} F(t,x),\ \ t \in (t_{N_k}, t_{N_{k+1}}],\\ \quad\vdots\\ x(t) = x_0 + \sum\limits_{j=1}^{M} y_j + {}_{t_0}I_t^{\alpha} F(t,x),\ t \in (t_{N_M}, T],\ 1 < N_M. \end{cases} \tag{7}$$

2.2. Adomian Polynomials

Considering a nonlinear equation

$$x(t) = G(x(t)) \tag{8}$$

for the nonlinear term $G(x(t))$, the Adomian polynomial named after G. Adomian [29] can be obtained by

$$A_n = \frac{1}{n!}\frac{\partial^n}{\partial \lambda^n}(G[\sum_{k=0}^{\infty} x_k \lambda^k])|_{\lambda=0}. \tag{9}$$

With the known values of $x_0, \cdots, x_n$, we can successively obtain A_n.

Duan [21–23] newly proposed a fast Adomian polynomial as the following

$$A_n = \frac{1}{n}\sum_{k=0}^{n-1}(k+1)x_{k+1}\frac{\partial A_{n-1-k}}{\partial x_0}, A_0 = G(x_0) \tag{10}$$

Generally, the one of the z-variable is calculated by

$$A_n = \frac{1}{n}\sum_{i=1}^{z}\sum_{k=0}^{n-1}(k+1)x_{i,k+1}\frac{\partial A_{n-1-k}}{\partial x_{i,0}}, i = 1, \ldots, z. \tag{11}$$

Although both lead to the same analytical solution, the new one is given in a more concise form and saves computational time. This is very important for solutions of the fractional calculus since the fractional derivative has the memory effects and can possess a large storage space.

3. Semi-Analytical Method Based on Adomian Polynomials

Consider the following fractional system with impulse (1). Using the idea by Duan [21], we give steps of a novel algorithm for impulsive fractional differential equations.

- Assume the solution in a series form as

$$x(t) = \sum_{i=0}^{\infty} c_i (t - t_0)^{i\alpha} \tag{12}$$

and x_n is assumed as $\sum_{i=0}^{n} c_i (t - t_0)^{i\alpha}$ accordingly.

- Substituting (12) into (8), with Adomian polynomials, the coefficients of c_i are obtained as

$$\begin{cases} c_{n+1} = \frac{\Gamma(1+n\alpha)}{\Gamma(1+(n+1)\alpha)} A_n[c_0, c_1, \ldots, c_n], 0 \le n, \\ c_0 = x_0 + \sum_{j=1}^{k} y_j. \end{cases} \tag{13}$$

- x_n can be obtained as

$$x_n = \psi(c_0, t_0, \sum_{j=1}^{k} y_j, t), t \in (t_{N_k}, t_{N_{k+1}}]. \tag{14}$$

- Set $t \in [t_0, T]$, $t = ih$, $H = \frac{T}{N}$, $h = \frac{H}{K}$, $i = 0, 1, \ldots, NK$ and let $x_i^* = \psi(x_{i-1}^*, t_{i-1}, \sum_{j=1}^{k} y_j, t_i)$, where $x_0^* = c_0$. We can obtain the numerical solutions $x_0^*, \ldots, x_i^*$.

4. Numerical Solutions based on Adomian Polynomials

In this section, we consider an application of the method to Caputo fractional differential equations with impulses

$$\begin{cases} {}_{t_0}^{C}D_t^{\alpha} x(t) = \mu x(t)(1 - x(t)), \\ t \in J\prime := J \backslash [t_{N_1}, \ldots, t_{N_M}], J := [t_0, T] \\ \Delta x_k = y_k, x_k = x(t_k), y_k = 0.1 \\ x(t_0) = x_0 = 0.2, t_0 = 0. \end{cases} \tag{15}$$

By use of Lemma 1, we have an integral equation as

$$\begin{cases} x(t) = x_0 + \mu_{t_0} I_t^{\alpha} x(t)(1 - x(t)), \ t \in [t_0, t_{N_1}], \\ x(t) = x_0 + y_1 + \mu_{t_0} I_t^{\alpha} x(t)(1 - x(t)), \ t \in (t_{N_1}, t_{N_2}], \\ x(t) = x_0 + y_1 + y_2 + \mu_{t_0} I_t^{\alpha} x(t)(1 - x(t)), \ t \in (t_{N_2}, t_{N_3}], \\ \quad \vdots \\ x(t) = x_0 + \sum_{j=1}^{k} y_j + \mu_{t_0} I_t^{\alpha} x(t)(1 - x(t)), t \in (t_{N_k}, t_{N_{k+1}}], \\ \quad \vdots \\ x(t) = x_0 + \sum_{j=1}^{M} y_j + \mu_{t_0} I_t^{\alpha} x(t)(1 - x(t)), \ t \in (t_{N_M}, T], \ 1 < N_M. \end{cases} \tag{16}$$

Adopt the semi-analytical method in Section 3. We have the recurrence relationship of the coefficients as

$$\begin{cases} c_{n+1} = \frac{\Gamma(1+n\alpha)}{\Gamma(1+(n+1)\alpha)} A_n[c_0, c_1, \ldots, c_n], 0 \le n, \\ c_0 = x_0 + \sum_{j=1}^{k} y_k, 1 \le k \le M. \end{cases} \tag{17}$$

We give the first few coefficients here

$$
\begin{cases}
c_1 = \dfrac{\Gamma(1)}{\Gamma(1+\alpha)}(\mu c_0 - \mu c_0^2), \\
c_2 = \dfrac{\Gamma(1+\alpha)}{\Gamma(1+2\alpha)}(\mu^2 c_0 - 3\mu^2 c_0^2 + 2\mu^2 c_0^3), \\
c_3 = \dfrac{\Gamma(1+2\alpha)}{\Gamma(1+3\alpha)}(-6\mu^3 c_0^2 + 10\mu^3 c_0^3 - 5\mu^3 c_0^4 + \mu^3 c_0), \\
\qquad \vdots
\end{cases}
\tag{18}
$$

such that we determine the approximate analytical expression of series solutions.

We vary the parameters α and M to observe the behavior. In Figure 1, the fractional order $\alpha = 0.9$ and the number of impulsive points is set to 5. We can see that, with the increase in M (See Figures 2 and 3), the solutions' values also increase if all of the impulse is positive. Figure 4 illustrates the stable solution without an impulse in the same fractional case. From all of the figures, we can observe that our semi-analytical solutions are plotted on the interval [0,10] which holds a longer time domain than the standard one.

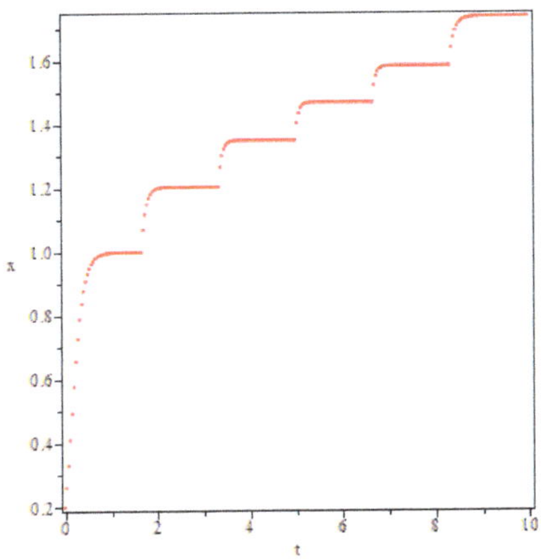

Figure 1. Numerical simulation:$\alpha = 0.9, M = 5$.

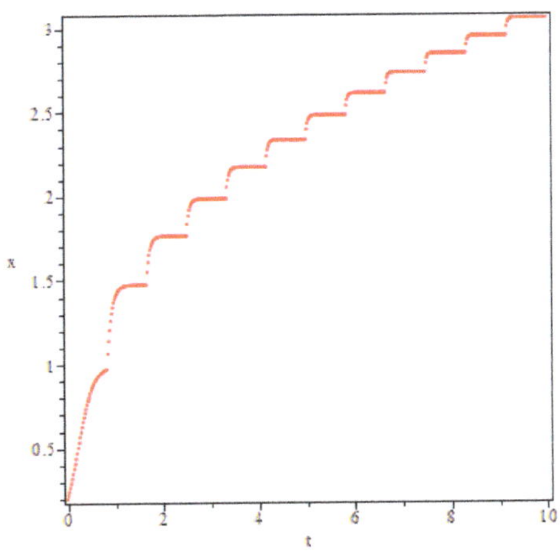

Figure 2. Numerical simulation:$\alpha = 0.9, M = 11$.

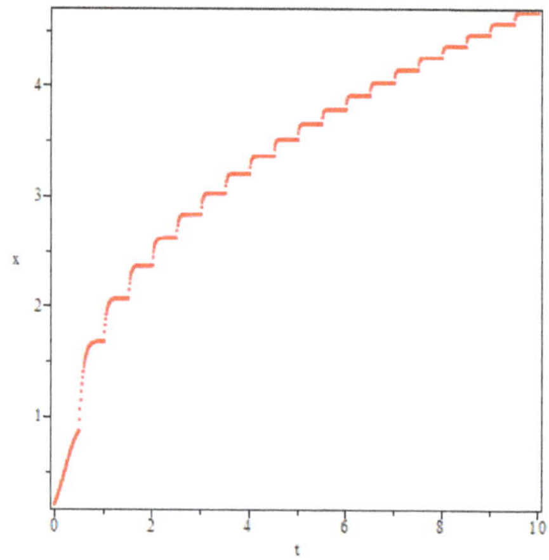

Figure 3. Numerical simulation:$\alpha = 0.9, M = 19$.

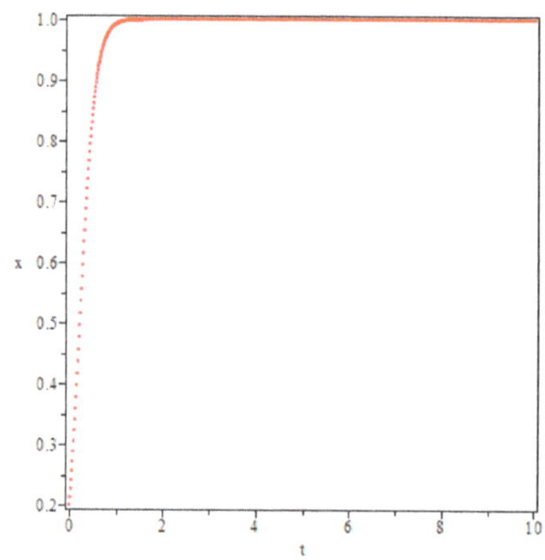

Figure 4. Numerical simulation: $\alpha = 0.9$, without impulse.

5. Conclusions

Impulsive fractional differential equation has recently become an important topic, but less work has focused on numerical or analytical methods. In this paper, we develop an efficient method for nonlinear equations. New Adomian polynomials are adopted to treat the nonlinear terms, and a semi-analytical method is developed. Firstly, the impulsive fractional differential equation is given equivalently in an integral equation. Fractional Taylor series is implemented to derive a recurrence relationship. Since there is no differential or integral calculus, it becomes very quick and saves computational time to derive the analytical or numerical solutions in comparison with classical ADM. The semi-analytical solution shows that the method is very efficient. However, there are still some difficulties that we need to overcome in future. The following topics are also disadvantages that we will try to address:

1. It is still challenging work to do error analysis. For many nonlinear cases, the exact solution is unknown and numerical errors cannot be obtained. We will pay attention to this topic in the near future;
2. In this method, we generally adopt a fractional series expansion which is a fractional analogy of the Taylor series. What about other expansions which satisfy the features of the new polynomials?

For example, how can series solutions be found for boundary value problems? Hence, it is very important to develop new ideas for this topic.

Author Contributions: The author C.M. takes full responsibility for this study and finished this paper by himself.

Funding: There is no financial program supporting this study.

Acknowledgments: The author feels grateful to the editor and the referees' kind help to improve this study.

Conflicts of Interest: The author declares no conflict of interest.

References

1. Kilbas, A.A.; Srivastava, H.M.; Trujillo, J.J. *Theory and Applications of the Fractional Differential Equations*; North-Holland Mathematics Studies; Elsevier Science Limited: New York, NY, USA, 2006.
2. Diethelm, K. *The Analysis of Fractional Differential Equations*; Springer: New York, NY, USA, 2010.
3. Machado, J.T.; Kiryakova, V.; Mainardi, F. Recent history of fractional calculus. *Commun. Nonlinear Sci. Numer. Simul.* **2011**, *16*, 1140–1153. [CrossRef]
4. Agarwal, R.; Hristova, S.; O'Regan, D. A survey of Lyapunov functions, stability and impulsive Caputo fractional differential equations. *Fract. Calc. Appl. Anal.* **2016**, *19*, 290–318. [CrossRef]
5. Baleanu, D.; Diethelm, K.; Scalas, E.; Trujillo, J.J. *Fractional Calculus: Models and Numerical Methods*; World Scientific: Singapore, 2012.
6. Mohammadi, F.; Cattani, C. A generalized fractional-order Legendre wavelet Tau method for solving fractional differential equations. *J. Comput. Appl. Math.* **2018**, *339*, 306–316. [CrossRef]
7. Cattani, C. Sinc-Fractional Operator on Shannon Wavelet Space. *Front. Phys.* **2018**, *6*, 118. [CrossRef]
8. Li, C.; Dao, X.; Guo, P. Fractional derivatives in complex planes. *Nonlinear Anal.* **2009**, *71*, 1857–1869. [CrossRef]
9. Guariglia, E. Fractional Derivative of the Riemann Zeta Function. In *Fractional Dynamics*; Cattani, C., Srivastava, H.M., Yang, X.J., Eds.; De GruyterOpen: Warsaw/Berlin, Germany, 2015; pp. 357–368.
10. Atangana, A.; Baleanu, D. New Fractional Derivatives with Nonlocal and Non-Singular Kernel: Theory and Application to Heat Transfer Model. *Thermal Sci.* **2016**, *20*, 1–7. [CrossRef]
11. Guariglia, E.; Silvestrov, S. A functional equation for the Riemann zeta fractional derivative. *AIP Conf. Proc.* **2017**, *1798*, 020063.
12. Ortigueira, M.D.; Rodríguez-Germá, L.; Trujillo, J.J. Complex Grünwald-Letnikov, Liouville, Riemann-Liouville, and Caputo derivatives for analytic functions. *Commun. Nonlinear Sci. Numer. Simul.* **2011**, *16*, 4174–4182. [CrossRef]
13. Mophou, G.M. Existence and uniqueness of mild solutions to impulsive fractional differential equations. *Nonlinear Anal. TMA* **2010**, *72*, 1604–1615. [CrossRef]
14. Feckan, M.; Zhou, Y.; Wang, J. On the concept and existence of solution for impulsive fractional differential equations. *Commun. Nonlinear Sci. Numer. Simul.* **2012**, *17*, 3050–3060. [CrossRef]
15. Stamova, I.; Stamov, G. Stability analysis of impulsive functional systems of fractional order. *Commun. Nonlinear Sci. Numer. Simul.* **2014**, *19*, 702–709. [CrossRef]
16. Wang, J.R.; Feckan, M.; Zhou, Y. On the new concept of solutions and existence results for impulsive fractional evolution equations. *Dyn. Part. Differ. Equ.* **2011**, *8*, 345–361.
17. Wang, J.R.; Zhou, Y.; Feckan, M. Nonlinear impulsive problems for fractional differential equations and Ulam stability. *Comput. Math. Appl.* **2012**, *64*, 3389–3405. [CrossRef]
18. Zhang, X.M. On the concept of general solution for impulsive differential equations of fractional-order q is an element of (1,2). *Appl. Math. Comput.* **2015**, *268*, 103–120.
19. Zhang, X.M. On impulsive partial differential equations with Caputo-Hadamard fractional derivatives. *Adv. Differ. Equ.* **2016**, *2016*, 281. [CrossRef]
20. He, J.H. Approximate analytical solution for seepage flow with fractional derivatives in porous media, Comput. Method. *Appl. Mech. Eng.* **1998**, *167*, 57–68. [CrossRef]
21. Duan, J.S. An efficient algorithm for the multivariable Adomian polynomials. *Appl. Math. Comput.* **2010**, *217*, 2456–2467. [CrossRef]
22. Duan, J.S. Recurrence triangle for Adomian polynomials. *Appl. Math. Comput.* **2010**, *216*, 1235–1241. [CrossRef]

23. Duan, J.S. Convenient analytic recurrence algorithms for the Adomian polynomials. *Appl. Math. Comput.* **2011**, *217*, 6337–6348. [CrossRef]
24. He, W. Adomian decomposition method for fractional differential equations of Caputo-Hadamard type. *J. Comput. Complex. Appl.* **2016**, *2*, 160–162.
25. Wu, G.C.; Baleanu, D.; Deng, Z.G. Variational iteration method as a kernel constructive technique. *Appl. Math. Model* **2015**, *39*, 4378–4384. [CrossRef]
26. Zeng, Y. Approximate solutions of three integral equations by the new Adomian decomposition method. *J. Comput. Complex. Appl.* **2016**, *2*, 38–43.
27. Daftardar-Gejji, V.; Jafari, H. Adomian decomposition: A tool for solving a system of fractional differential equations. *J. Math. Anal. Appl.* **2005**, *301*, 508–518. [CrossRef]
28. Duan, J.S. The Adomian decomposition method with convergence acceleration techniques for nonlinear fractional differential equations. *Comput. Math. Appl.* **2013**, *66*, 728–736. [CrossRef]
29. Adomian, G. *Solving Frontier Problems of Physics: The Decomposition Method*; Kluwer Academic Publishers: Dordrecht, The Netherlands, 1994.

symmetry

MDPI

Article

An Intelligent Approach for Handling Complexity by Migrating from Conventional Databases to Big Data

Shabana Ramzan [1], Imran Sarwar Bajwa [1,*] and Rafaqut Kazmi [2]

1 Department of Computer Science & IT, Islamia University of Bahawalpur, Bahawalpur 63100, Pakistan; shabana@gscwu.edu.pk

2 School of Computing, University of Technology Malaysia, Johor 81310, Malaysia; rafaqutkazmi@gmail.com

* Correspondence: imran.sarwar@iub.edu.pk

Received: 26 October 2018; Accepted: 14 November 2018; Published: 3 December 2018

Abstract: Handling complexity in the data of information systems has emerged into a serious challenge in recent times. The typical relational databases have limited ability to manage the discrete and heterogenous nature of modern data. Additionally, the complexity of data in relational databases is so high that the efficient retrieval of information has become a bottleneck in traditional information systems. On the side, Big Data has emerged into a decent solution for heterogenous and complex data (structured, semi-structured and unstructured data) by providing architectural support to handle complex data and by providing a tool-kit for efficient analysis of complex data. For the organizations that are sticking to relational databases and are facing the challenge of handling complex data, they need to migrate their data to a Big Data solution to get benefits such as horizontal scalability, real-time interaction, handling high volume data, etc. However, such migration from relational databases to Big Data is in itself a challenge due to the complexity of data. In this paper, we introduce a novel approach that handles complexity of automatic transformation of existing relational database (MySQL) into a Big data solution (Oracle NoSQL). The used approach supports a bi-fold transformation (schema-to-schema and data-to-data) to minimize the complexity of data and to allow improved analysis of data. A software prototype for this transformation is also developed as a proof of concept. The results of the experiments show the correctness of our transformations that outperform the other similar approaches.

Keywords: big data; complexity; NoSQL databases; Oracle NoSQL; data migration

1. Introduction

The modern information systems have to deal with high-dimension data in terms of gigantic size, and the heterogenous and complex nature of the data. Similarly, the cloud applications and social media applications also have to store, manage and process a massive amount of data. However, the Relational Databases (RDBs) have fixed schema and allow storage and handling of only structured data in the form of tuples or relations [1]. Additionally, the RDBs only provide vertical scalability (vertical scalability allows only vertical growth of a data-structure by adding only new records at run-time.) at higher hardware cost but no horizontal scalability (horizontal scalability allows horizontal growth of a data-structure by also allowing the addition of fields at run-time.) is provided by the RDBs. Since horizontal scalability is needed by today's software applications to handle high-speed heterogenous data; currently, the relational databases have to face various challenges at the application development level and operational level. At the application development level, the system developer needs high coding velocity to handle large number of users; however, such capability is not available in relational databases. Additionally, modern complex and heterogenous data needs horizontal scaling but that feature is also not provided by the relational databases and consequently, they fail to cope with the needs of modern data-intensive software applications.

Once of the key challenges in recent times has been to handle high-speed data, as there is a rapid increase in digital information, exponentially growing (see Figure 1) to Petabytes (PB) PB = 1000 TB) from Terabytes (TB) 1TB = 1000 GB, and even to Exabytes (EB) 1EB = 1000 TB as shown in Figure 1. John Gantz and David Reinse also predicted this phenomenon [2]. Typical relational database systems have shown their limits for such exponential growth of data. The shortcomings of typical relational databases are addressed by Big data solutions such as NoSQL databases [3–5]. Here, NoSQL stands for "Not Only SQL". Such databases are currently the main focus of research due to the fast and persistent growth of data. The NoSQL was introduced in 1998 by Carlo, and the name given to his relational database solution that was due to not using Structured Query Language (SQL) [6]. The idea of NoSQL was redefined in 2009 and became the competitor of RDBs. Now they have become the backbone of large-sized enterprises such as Google, Twitter, Facebook, Amazon, etc. due to its peculiar features such as high availability (when a data is automatically distributed evenly across a cluster with no single master.), efficient performance, horizontal scalability, and the support of a variety of data models and queries. Moreover, the rapid growth of cloud computing has highlighted the problems that are endured in handling large volumes of data. However, NoSQL databases can handle "Big Data" problems efficiently rather than RDBs. These databases are becoming popular because they are providing a high level of scalability. Additionally, they are very efficient in handling the unstructured data to facilitate universal data communication [7] in modern information systems. Relational databases follow the ACID (Atomicity, Consistency, Isolation, Durability) and BASE (Basically available, Soft-state, Eventual-consistency) properties. Whereas, NoSQL databases exist in a spectrum between ACID and BASE alliance.

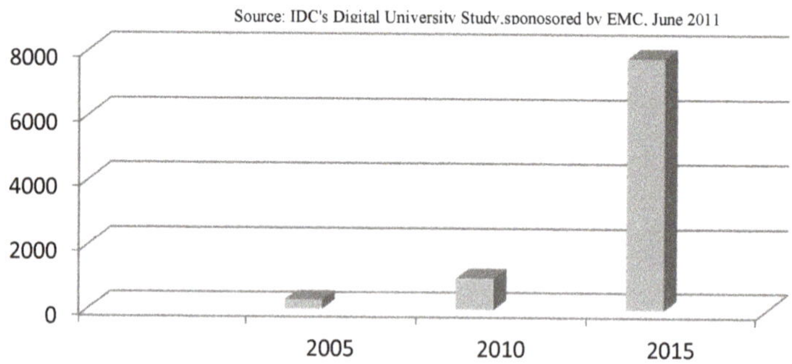

Figure 1. Exponential growth of digital information, going to Exabyte.

The NoSQL databases have typically four different models: (1) key-value store, (2) column store, (3) graph store, and (4) document store [8]. Each NoSQL database model has its own distinct schema of storing data [9]. The most simple and flexible model is a key value store that is used in our study and an overview of key-value stores is given below:

1.1. Key-Value Stores

A Key-value store is a database that stores data in the form of associative arrays known as a hash or a dictionary. Each dictionary has a collection of records that have different fields of data. They store data as a key-value (record) pair as shown in Figure 2. Each value is stored and retrieved through a unique key. A value is a data of an arbitrary type, size and structure. Here, value can be anything such as a number, text, image, programming code (such as PHP), markup code (such as HTML), etc. They do not have any query language, only use get, put and delete operations [10]. A key can be simple (filename, hash or URL) or a composite key (such as in Oracle NoSQL) [11].

A set of operations are used to interact with key-value stores such as Get operation is used to retrieve a value that is stored against a unique key and put operation is used to insert the key-value pair. However, manipulation of multiple values in a single operation is not allowed by these single-key operations. These operations facilitate the users that do not have proper knowledge of query language to easily retrieve data. A key-value store handles the process data retrieval manually at the application level. Here, lookup structures are used that are based on keys such as Log-Structured Merge-trees (LSM-trees) and Distributed Hash Tables (DHTs) [12], and are highly suitable for applications that can access data through a single key, such as web session information, user profile/configuration and online shopping cart. Key-value stores are the only databases that provide efficient data retrieval and storage mechanisms to cloud-based applications [13]. Key-value stores provide features like easy partitioning and high scalability. Figure 2 shows a storage model of a typical key-value store.

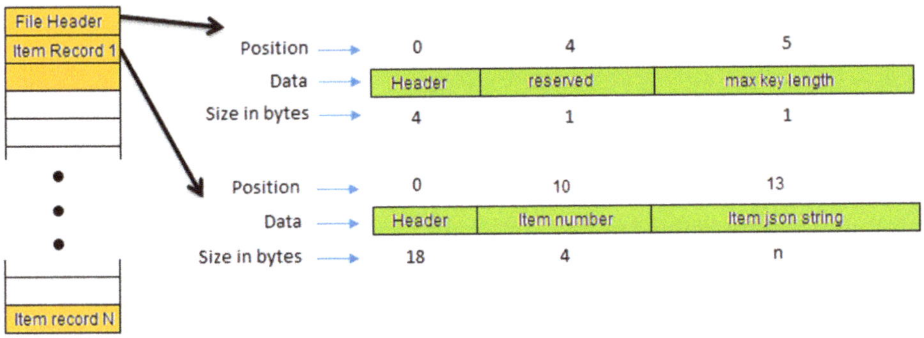

Figure 2. Key-Value Store data model.

On the base of storage models, key-value stores can be divided into three types of stores including permanent, temporary and hybrid. In permanent stores, all the data is stored on the hard disk but I/O operations are slow. The temporary key-value store ensures fast data access as all data is stored in memory; however, if the system is down, the data will be lost. Whereas, a hybrid store is a combination of the positive features of both permanent and temporary stores as it supports data storage in memory and when specified conditions are met, the date is written to the hard disk. The most popular key-value stores are hyperdex, Redis, Riak, Oracle NoSQL, BerkelyDB, Yahoo Pnuts and Project Voldemort. The following section provides an overview of Oracle NoSQL.

1.2. Oracle NoSQL

Oracle NoSQL is a distributed type of key-value store. It provides important features like horizontal scalability, monitoring, transactional semantics for improved data manipulation, and simple administration of data. The Oracle NoSQL has a very simple data model. Each row is a key-value pair; value is associated with a unique key. Value is of arbitrary length. It has tables, rows and fields which are equivalent to tables, rows and columns of relational databases but has a different concept. The following are the key features of Oracle NoSQL stores:

- Oracle NoSQL table is schema free but relational databases' tables have predefined schema.
- Each column has a separate schema but in relational databases each table has a schema.
- Each row in Oracle NoSQL database can have unrelated fields but in relational databases each row is a collection of related items.

Figure 3 shows the relational and Oracle NoSQL key-value store databases. Oracle Berkeley DB Java Edition high-availability storage engine is the basis of Oracle NoSQL. It provides sharding, replication, transparent load balancing, high availability and fault tolerance. It is a free schema and supports various programming languages such as C++, C, C#, Ruby, Scala, Java, Javascript(Node.js), and Python. Oracle NoSQL supports simple data types (java string float, integer, long, boolean, double) as well as complex data types (array, enum, fixed binary, map, records).

No.	EmpID	Ename
1	1.	Mark
2	2	Jhon
3	3	Micheal
4	4	Tony
5	5	David

(a)

Key	Value
user 1: employee	{65,865,9634}
user2: employee	{34,85,76,94}
user3: employee	{desg:manager, branchcode: 345}
user4: employee	{ name: mark, empid:346}
user5: employee	{258,36,45}

(b)

Figure 3. Relational database (**a**) and Oracle NoSQL Key-value store database (**b**).

Considering the challenges of big data, the modern organizations are rapidly shifting to NoSQL databases from conventional RDBs. Relational to relational database like MySQL to Oracle conversion is possible because they are based on mathematical theory [14]. On the other hand, NoSQL databases are non-relational and their scheme design is completely different. So RDB-trained staff has difficulty in converting existing RDB system to NoSQL databases [15,16]. However, the lack of methodological and tool support for automated migration from RDB to NoSQL has been a real challenge in recent times. In this paper, a methodology is presented to transform the existing relational database into a NoSQL database. It automatically transforms both the data and schema. The proposed approach transforms the MySQL database into an Oracle NoSQL database by handling the complexity of data.

The rest of the paper is organized as follows. Section 2 introduces related work in the fields of RDBs, NoSQL and migration between these two generations of databases. Section 3 describes the used approach and Section 4 discusses the implementation details of our approach. The results of experiments and discussion is given in Section 5 along the evaluation details. Finally, Section 6 concludes the results with possible future work.

2. Related Work

A data transformation from a relational database to an NOSQL database depends on different factors such as mapping styles, query structures, storage structures, etc. Additionally, querying data from a relational database and a NoSQL database at the same time is difficult but now it is applicable by using the data adapter technique [17]. Here, a method DB converter is described that transforms relational data into NoSQL data for querying results. However, this conversion is temporary, only for query execution [17], since NoSQL databases are efficient in data storage and provide high levels of scalability and availability. There are several different studies on NoSQL databases [4], such as BigTable [13], Cassandra [18], HBase [15], MongoDB [19] and for big data [20]. It is studied that the schema conversion from relational databases to NoSQL is difficult because relational databases use JOINs but the NoSQL databases do not support it. In NoSQL databases, nesting tables are used as an alternative to JOINs. This method is designed to improve the performance of cross table query. In nesting table technique, the parent-child layer is designed with references as relationships between the tables. Here, the referred table is defined as a child and the other one is defined as a parent [21]. The cross table query is important in SQL databases, but in NoSQL, a question arises on how to use JOIN type or alternative queries to retrieve data from NoSQL databases. Column-oriented databases provide a solution for these types of queries because column-oriented databases have a design principle of DDI (Denormalization, Duplication and Intelligent keys). This method works as: initially, denormalization of the database and its transformation into a big table; then identification of unique keys in a big table; and finally, the selection of the most suitable key as primary key. In this method, MySQL database is transformed into a column-oriented database [22]. Most of the web-based applications and Content Management System (CMS) solutions are using relational databases for data management, but users of internet and clouds are growing rapidly, so it is difficult for relational databases to handle the huge data traffic. The designed approach transforms the real CMS SQL database to a NoSQL database [23]. This approach has two steps, first to denormalize the SQL database and then to choose a unique identifier key as a primary key for a big table. In this approach, MySQL database is migrated to a column-oriented Hbase database.

Another method is designed to transform data from a relational database (MySQL) to NoSQL (MongoDB). Migration from relational to NoSQL has a few steps; initially, MySQL database connection is created, after connectivity, the details of the database are accessed through prototype software. In next step, mapping is performed between the relational database MySQL to NoSQL MongoDB [24,25].

For transformation from RDBs to NoSQL, another application is developed which deals with the transformation of relational database schema to NoSQL schema. This application is able to handle both the DDL and DML commands of relational schema and transform these commands into equaling commands of NoSQL [26]. To access the NoSQL database, a subset of SQL commands is used. CQL is the query language for Cassandra, where CQL and SQL are quite similar. Cassandra and MongoDB are integrated because MongoDB is capable of performing complex queries. Therefore, authors designed a system for translation of SQL commands to NoSQL. This system is implemented by middleware in C# [27]. Table 1 shows the comparison of existing approaches. The majority of the researchers tend to use HBase and MonogoDB as a target database but no one used Oracle NoSQL. The facts tabulated in the following table clearly show the research gap that currently no approach or tool supports automated transformation of MySQL to Oracle NoSQL for both data and schema transformation.

Table 1. Comparison of transformation approaches.

Source Database	Target Database	Schema Conversion	Data Conversion	Conversion Time	Data Set	Technique	Study Reference
MySQL	MongoDB	Yes	No	No	72 Tables	Transform algorithm	Zhao et al.
MySQL	HBase	Yes	No	No	Hush database 1 thousand transactio-ns)	Automatic transformation Mechanism based on NoSQL DDI Design Principle.	Lee et al.
MySQL	MongoDB	Yes	Yes	No	Two datasets 1. Twitter App. 2.W3Scho-ols App	framework (1) migration module (2) mapping module	Rocha et al.
MySQL	MongoDB	No details about schema conversion	Yes	No	—	Migration Methodology (1) Extracting logical structure (2) Mapping between databases.	Hanine et al.
SQL	any key-oriented NoSQL DB	Yes	No	No	European Air quality database	Transformation layer	Schreiner et al.
SQL	HBase	Yes	No	No	15 GB Dell DVD Store relational database	Heuristic based approach	Serrano et al.
RDB	HBase	Yes	No	No	RDB schema with 7 tables.	Extracting conversion rules and applied conversion rules.	Ouanouki et al.
RDB	HBase	Yes	No	No	—	heuristic-based approach	Li et al.
RDB	document-oriented NoSQL,	Yes	Yes	No	Different databases of different sizes.	Column-level Denormalization and Atomic Aggregates	Yoo et al.
MySQL	**Oracle NoSQL**	**Yes**	**Yes**	**Yes**	**Five different databases**	**Automatic Transformation**	**Proposed Scheme**

There is another methodology for the conversion of a relational database to HBase in four steps [28]. First, create a single merge table in HBase and convert all one-to-one and one-to-many relationships into that table. Second, merge neighboring tables through a recursive method. Third, a row key design, and fourth, create access patterns views. The set of rules for schema conversion between an existing relational database to Hbase are defined [29]. First, experimentally justify the need of conversion rules by observing the conversions without conversion rules. This first experiment is used as a baseline of the second experiment. Second, the experiment is performed convert the existing relational database application to Hbase using a first list of conversion rules. This conversion proves that the conversion rules reduce the difficulty of the whole conversion process. Another approach is presented for RDBs to NoSQL migration that has two phases [30], the first phase transforms relational database schema to HBase schema and also provides guidelines to develop HBase application. In the second phase, schema mappings are used to create a set of programs to automatically transform the data of the source database to the target database. Similarly, this proposed a solution for migrating RDBMS schema to document oriented NoSQL database schema [16]. This method provides atomicity using atomic aggregate and avoids join operations. It uses the column-level denormalization in order to minimize the disadvantages of table-level denormalization.

Data extraction from Big data has been one of the major research challenges [31,32] in recent times. Since, Big data has discrete and heterogenous types of data, this challenge becomes more difficult. However, various contributions [32–34] are made to address this challenge. Suciu [33] discussed the extraction of knowledge from Big data and [34] discussed how conceptual modeling can help in addressing this challenge. The NoSQLayer tool presented is proposed for migrating from a relational database to a NoSQL database; this approach has two modules. The data migration module migrates the SQL database to the NoSQL database. Here, the metadata of the MySQL database is accessed during Java Metadata API. The data mapping module transforms data from a relational database to a NoSQL database seamlessly [35].

To the best of our knowledge, there is no approach or tool available to handle the complexity of automatic conversion of a RDB (such as MySQL) to Big data solutions (such as Oracle NoSQL database) and the major contribution of this paper is to present a novel approach that is intelligent enough to handle the complexity of data and automatically transform MySQL database to Oracle NoSQL for both data and schema conversion.

3. Used Approach for Handling Complexity of RDB to Big Data Conversion

The used approach is based on a rule-based system that has two modules: The first module handles the conversion of a relational database (such as MySQL) schema to a NoSQL database schema which is very flexible in nature (Schema Conversion). Whereas, the second module handles the conversion of the data from the relational database (such as MySQL) to NoSQL database (such as Oracle NoSQL). The working of the first module in the proposed approach is shown in Figure 4, that performs the schema transformation.

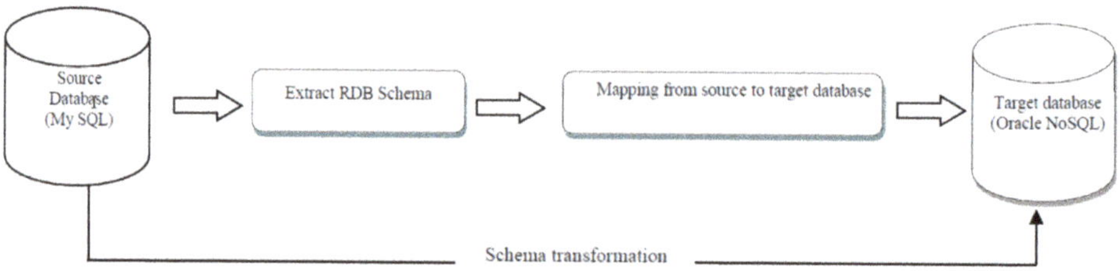

Figure 4. Schema transformation module.

The following text explains the components used in both modules (schema transformation and data transformation) and their working.

3.1. Schema Transformation

This module handles the transformation of MySQL schema to Oracle NoSQL schema. For this purpose, a relational database in MySQL is taken as an input and parses in Java to extract metadata of MySQL databases such as table-names, their attribute-names, attribute-data-types, relationship-names, indexes from the database, etc. Here, the JDBC driver and the Java metadata base class library is used to access the schema of tables from MySQL database. For relationship metadata extraction, the primary and foreign key constraints of each table were used as the relationship information can be helpful for schema conversion. Here, Java metadata class library provides different methods to extract schema information in different aspects, e.g., if we want to get information about a primary key then we use metadata's primary key function. The methods used for this approach are listed in Table 2. When all required metadata of the tables' schemas are extracted, the mapping given in Table 3 is used to transform the MySQL metadata to Oracle NoSQL key-value store.

Table 2. List of methods used in schema transformation.

Methods	Description
getTables()	This method returns all the tables of the database. The list returned by this method is traversed to get information of each table.
getColumns()	The names of all attributes that are defined by the parameters and their characteristics are retrieved through this method.
getMetaData()	This method returns all other information about relational databases, such as data type and constraints.
getIndexInfo()	All indexes that are created on the relational database return through this method.
getImportedKeys()	This method retrieves a description of the primary key columns that are referenced by a table's foreign key columns.
Getprimary keys()	This method retrieves a description of the primary key columns of the given table.

This mapping is implemented in Java to accomplish the schema level transformation of MySQL to Oracle NoSQL database. Once the schema transformation is accomplished, the data level transformation is carried out; this is described in the following section.

Table 3. Migration mapping RDBs to Oracle NoSQL.

MySQL	Oracle NoSQL
Table	Record or Table
Column Name	Field Name
Column Data Type	Field Type
Column	Field
Users	Users
Permisssions	Priviliges
Index	Index
JOIN	Parent-Child link
Foreign Key	Reference (parent-child link)

In our approach, we have created an online test preparation of a student database in MySQL. A subset of this MySQL database is shown in Figure 5 that is used to explain the schema conversion methodology of our approach.

Relational database has an important feature of JOIN. But Oracle NoSQL database uses parent-child relationship instead of JOIN. MySQL database Student table (parent table) is linked with Marks table (child table) and Marks table (parent table) is linked with Course table (child table) and Course table (parent table) is linked with lecturer table (child table) and lecturer table (parent table) is linked with Classes (child table). In Figure 6, we see that Student is a parent table, Marks is a child table as well as the Course is a sub-child table and Course table is a parent table and Lecturer is a child table and classes are sub-child table.

In Oracle NoSQL key-value store, these tables are stores in the form of Avro schema. Avro schema supports both APIs of Oracle NoSQL, and that is why the entire key-value store is based on this schema. Figure 7 shows the Avro Schema of tables stored in Oracle NoSQL store and Figure 8 shows Avro Schema of Marks table. Table API of the Oracle NoSQL key-value store is just a front-end layer to provide a user friendly environment.

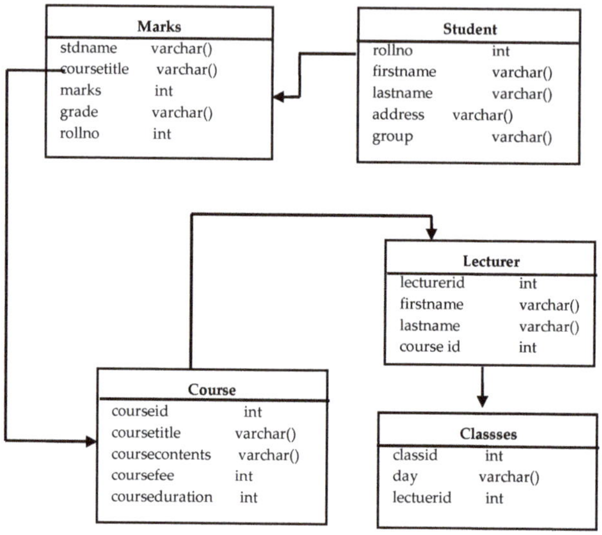

Figure 5. Database model used as test database.

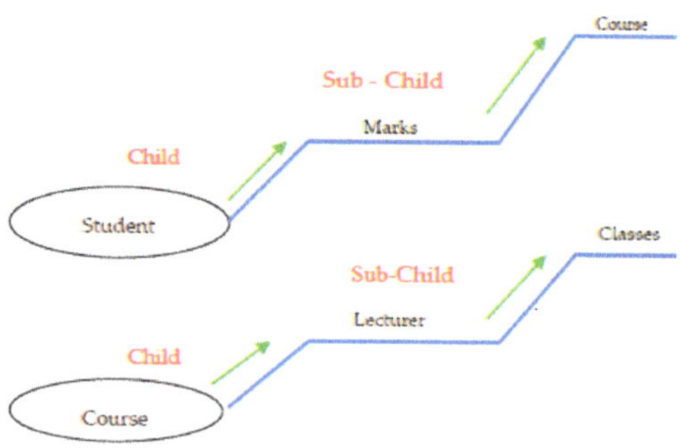

Figure 6. Relationships of Oracle NoSQL database tables.

Tables:

CourseID

SYS$IndexStatsLease

SYS$JoinStatsLease

SYS$TableStatsIndex

SYS$TableStatsJoin

Lecturer.firstname

Lecturer.lastname

Lecturer.courseid

Lecturers

Figure 7. Detail of tables stored in Oracle NoSQL store.

```
{
 "Type"  : "Table",
"Name" :  "Marks",
"Owner": "null",
"shardkey": ["Coursetitle"],
"primarykey": ["Coursetitle"],
"fields": [{
        "name": "StdName",
        "type": "Integer",
        "nullable": "true",
        "default": null
         },{
        "name": "Coursetitle",
        "type": "String",
        "nullable": "true",
        "default": null

        "name": "Mark",
        "type": "Integer",
       "nullable": "true",
         "default": null
         },{
        "name": "Grade",
        "type": "STRING",
         "nullable": "true",
        "default": null
        }, {

        "name": "RollNo",
        "type": "INTEGER",
        "nullable": "true",
        "default": null
       }   ]
}
```

Figure 8. Avro Schema of Marks Table.

3.2. *Data Transformation*

For data transformation, a table in source MySQL database is selected from which the data is to be extracted and transformed into JSON format for final storage in an oracle NoSQL database. For this purpose, the ETL (Extract, Transform and Load) methodology [31] is used. The data transformation module is shown in Figure 9.

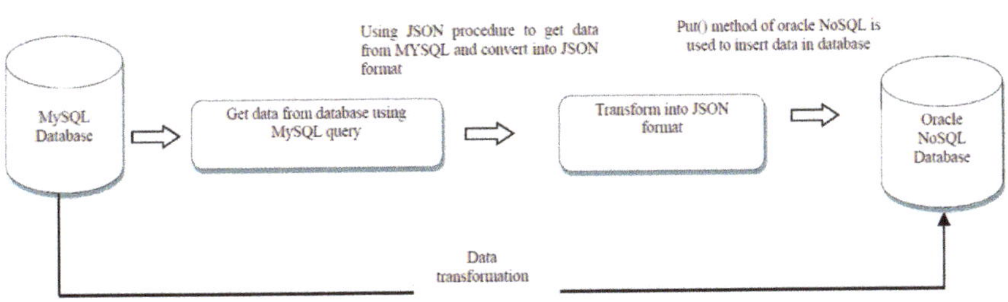

Figure 9. Data transformation module.

JSON format is a light weight, data interchange format, and is easy to understand. JSON is used in Oracle NoSQL database as a data type and also as a schema; the schema created through JSON is called Avro Schema. Before inserting data into a key value store, it is necessary to create an Avro schema for the record which defines the structure of the data.

In the first step, the required table is selected, from which data will be extracted. The next step is to convert data of the table row by row from MySQL database and store it in a text file; when all the data of the table is transformed, then this text file is called by another module that stores it into the Oracle NoSQL database. For data transformation, we need to know about some data types used in MySQL database to compare it with data types of the Oracle NoSQL database as shown in Table 4.

Table 4. Data type comparison.

MySQL	Oracle NoSQL
int, bigint	Integer
Long	Long
Array	Array
Bolean	Bolean
Float	Float
Double	Double
String	String, Java String
BLOB	Binary

In the data transformation process, every column data type of the MySQL database table is compared with the Oracle NoSQL data types. Such a comparison helps in finding the exact match of MySQL data type. Figure 10 shows the mapping of category table from RDBs to Oracle NoSQL.

ID	S_Name	Class	Subject	Marks
1	James	1st year	OOP	84
2	Jane	2nd year	Java	70

{"id":1, “S-name”: “james”, "Class": "ist year", "Subject": “OOP”, "Marks": 84}
{"id":2, “S-name”: “jane”, "Class": "2nd year", "Subject": “Java”, "marks": 70}

Figure 10. Mapping of relational table to Oracle NoSQL table.

The designed software prototype transforms all data of the required table into JSON format and temporarily stores this data into a text file. After converting data into JSON format, the next step is to insert data into Oracle NoSQL destination schema table. Finally, put() function is called to read JSON data from the file and store it into an Oracle NoSQL schema table.

4. Implementation Details

The approach discussed in the previous section is implemented in Java as Eclipse plugin. The implemented system starts working with the connectivity with a source relational database. After connectivity, two options will be displayed for conversion/transformation process. One is Schema Conversion and the other is Data Conversion. Which option is used depends on the user or administrator. If the user selects schema conversion, then the first step is to select the required database to transform into the Oracle NoSQL database. The next step is hidden from the user, which actually maps the databases and makes a conversion. In the last step, the transformed database schema will be displayed on the form. After creating schema from the relational to NoSQL database, a user can also transform data from MySQL to NoSQL. The second option is Data Conversion, if the user goes for this option, firstly, he will select the database and the required table from the database afterwards. The next step is to transform the data from the relational database to the Oracle NoSQL database. Consequently, a stored procedure is designed which transforms relational database data into Avro schema base JSON data. The entire working of the proposed system is shown in Figure 11.

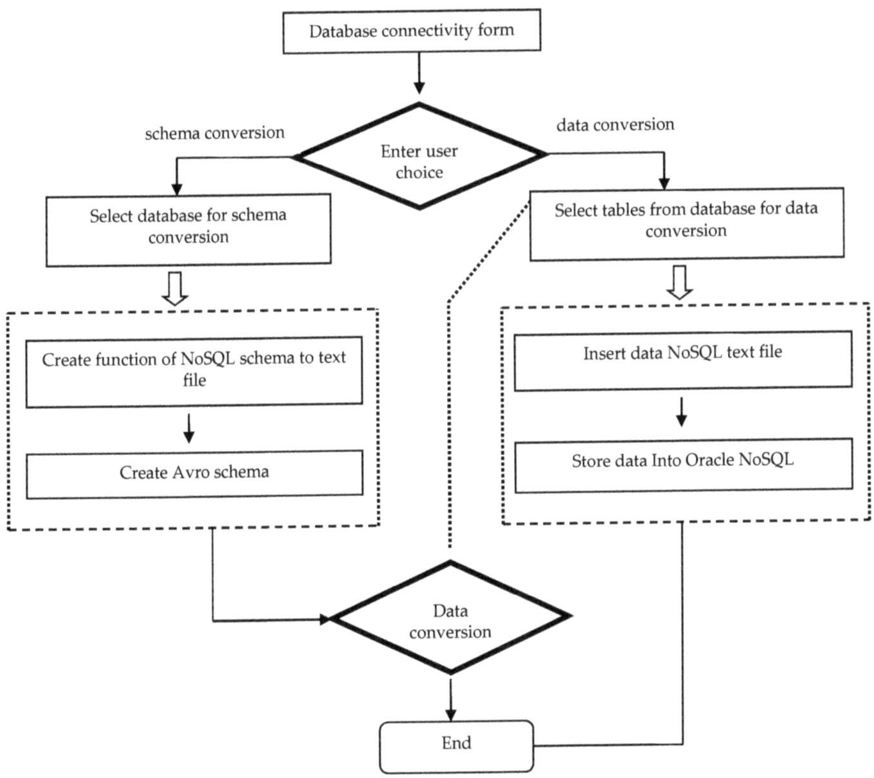

Figure 11. Framework of the proposed approach.

4.1. Module A. Creating Schema of Mysql Table and Store It in the File

The function is used, which transforms the MySQL database tables into Avro Schema as shown in Figure 7. This function has the following steps.

1. Get metadata of MySQL database tables:

In this step, after database connectivity, MySQL tables are selected one by one and a function of Java metadatabase class library is called, which selects the table and its attributes.

2. Create a file:

For this step, a function of file Java class library is used, and to write data to this file PrintWriterfunction or PrintWriter Java class library are used.

3. Get keys from metadata:

In this part, the primary and foreign keys of are table are used for primary and foreign key mapping with NoSQL Avro Schema.

4. Map the data types of both databases and create fields in avro style:

In this step, the mapping process is defined, col.next() built-in function is used to select the column names of the tables one by one and compare the data type of MySQL table with Oracle NoSQL data types.

5. Create JOIN like parent-child relationship:

The fk.next() is a result set which hold the foreign key details and its corresponding tables. The table name of the foreign key is selected as a parent table name for the under-process table.

According to Oracle NoSQL, a parent-child relationship is like:

Parenttable.childtable (attributes with data types and primary keys of both tables).

4.2. Module B. Schema from File to NoSQL

In this module, the file is called, which has temporarily stores the schema of tables; after that, an object of Oracle NoSQL key-value store is created to access the NoSQL database. In the next step, an object of Table API is created for new table creation. Now the function runs while looped and gets data from files and sends it to KVstoreexecuteSync() function for table creation in the Oracle NoSQL store.

4.3. Module C. Transform Mysql Data into Oracle NoSQL Data

This module performs two tasks:

1. Create procedure: A procedure is created in a generalize format to get data from the database and create its JSON schema.
2. Call procedure to Transform data into JSON: The above function is called and executed, and then it gets data from the database and creates its JSON values row by row. Completing this task, the data that comes in the procedure is stored in a file and the data of this file will be sent to NoSQL database to store data in the database.

5. Results and Discussion

The implemented system was tested with a number of examples to verify the working of the tool and the accuracy of the transformation output. Here, MySQL is used as a source database and Oracle NoSQL is used as a target database. The experiment is performed on different datasets of databases. Our developed system has two parts, one is schema conversion, and the other is data transformation. In the Schema Conversion part, the software will work according to these steps:

1. In this step, Oracle NoSQL is started by running a set of commands in the CLI interface.
2. When designed software starts running, the user connects to a MySQL database through the software.
3. After MySQL connectivity, the next step is to select the required database from the connected databases of MySQL for conversion into the NoSQL database.
4. When a database is selected, then the software will give two options, one schema conversion and the other is data conversion; when the user clicks on the Schema Conversion button, then software will automatically create the schema of database tables. If there is a relationship between tables, then all tables linked with one another will be selected. The software will automatically convert this relationship into a parent-child relationship of Oracle NoSQL database. This parent-child relationship is an alternative for JOINs.
5. For schema conversion, a function is called, which get the table schema from a MySQL database and then stores it in a text file in the application. After completing this function, another function is executed; it gets data from the file and starts mapping the MySQL table schema, its attributes and data type with oracle NoSQL attribute style and data type. Later on, this table schema converted into Avro Schema and is stored into the Oracle NoSQL database. Avro schema is used in the Oracle NoSQL database for creating a record or table schema in which the data will be stored. Details of converted tables will be displayed in the form as shown in Figure 12, if any error is found in the conversion process, it will also be displayed in the form.

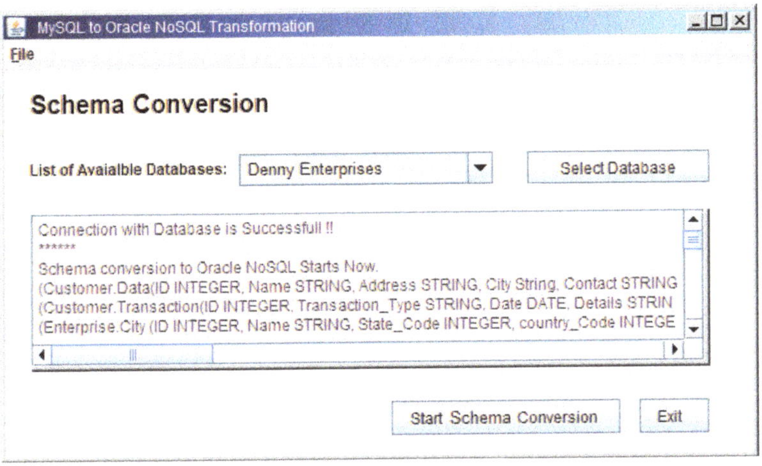

Figure 12. Schema conversion details.

In the data transformation part, all steps are the same except for Step.5. The Step.5 of data transmission is started when a table is selected from the given tables. After selecting the table, and clicking on the Data transformation button, a function is called to get the data of the table from MySQL database and store it in a text file. After that, another function is called which gets data from a text file and stores it in specific schema table of the Oracle NoSQL database. The form will display the details of the converted table. The conversion detail of the city table is shown in Figure 13.

The proposed system is tested on a Ci5 2.4 Ghz processor with 4GB RAM with Ubuntu 14 OS on VM; five different databases are tested in the proposed system to check the effectiveness of the proposed system. The databases used for the evaluation of our proposed methodology are given below:

- World: This database has three tables (Country, City, Language) [size: 1.2 GB]
- OnlineQuiz: This database has five tables (Category, SubCategory, Quiz, Users, TestDetails.. [size: 2.3 GB]
- Accounts and products: This database has six tables (User, Accounts, Transactions, redemption, ebaycard, products). [size: 3.2 GB]
- Employees: This database contains five tables (Emp, Dept, Products, Sale, Accounts). [size: 1.4 GB]
- Classicmodels: This database has eight tables (customers, offices, emp, orderdet, order, payments, productline, products). [size: 1.1 GB]
- Denny Enterprises: This database has seven tables (Customer, Transactions, City, Products, Payment, Stock, Order). [size: 1.78 GB]

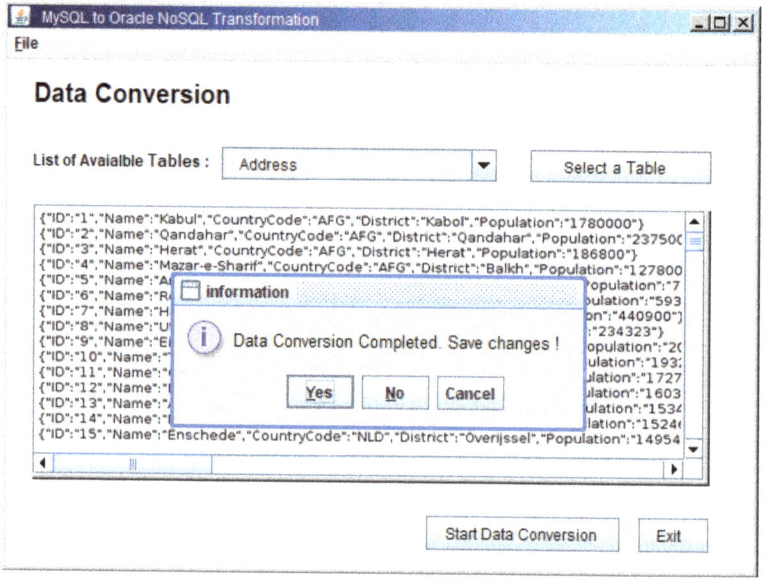

Figure 13. Data conversion details.

Figure 13 shows the screenshot of the data conversion module that allows the user to select a table name of the source database and it converts it into the Oracle NoSQL database table. Here, for the conversion, the approach discussed in Section 3.2 is applied and the results of the conversion are also shown in Figure 13.

5.1. Evaluation Methodology

The working, and the results, of the presented approach have been discussed in the previous section. To evaluate the performance of our approach, an evaluation methodology was designed to find how accurately the relational database schema and data is transformed into Key-value NoSQL format. An evaluation methodology, for the performance evaluation of intelligent tools, is used, and was originally proposed by Hirschman, L., Thompson in 1995 [36]. The following section describes the evaluation methodology used to evaluate the performance of our approach.

5.1.1. Criterion for Evaluation

A criterion was defined for the quantitative evaluation of the designed approach to find how accurately it transforms the source database to the target database. Accuracy of the transformation is measured by finding how close the output is of our approach to the opinion of a human expert (named total results). In this study, the opinion of a human expert for the target input was taken and used as a total result for the sake of evaluation.

5.1.2. Method of Evaluation

For the quantitative evaluation of the results of the used approach, each correct transformation (tables, fields, views, keys, etc.) was matched with the expert's opinion ($N_{total\_transformations}$). The results of all transformations were matched as all the transformations that matched the expert's opinion were declared correct ($N_{correct\_transformations}$), and otherwise, were considered incorrect ($N_{incorrect\_transforamtions}$).

5.1.3. Measures of Evaluation

A set of evaluation measures used in our evaluation methodology are: recall, precision, and F-Measure. The details of these three evaluation measures are given below:

Recall. The recall can be attributed as the completeness of the results produced by system. In our methodology, Recall (R) is calculated by finding the number of correct transformation from the total number of transformations. In Equation (1), $N_{correct\_transformations}$ is the number of correct transformations generated by the approach and $N_{total\_transformations}$ is the number of total correct transformations.

$$R = \frac{N_{correct\_transformations}}{N_{total\_transformations}} \tag{1}$$

Precision. The precision can be attributed to as the accuracy of the designed system. Precision is measured by comparing the designed system's number of correct results by all (incorrect and correct) results produced by the system, calculated as: In Equation (2), $N_{correct\_transformations}$ is the number of correct transformations generated by the approach and $N_{incorrect\_transformations}$ is the number of total incorrect transformations.

$$P = \frac{N_{correct\_transformations}}{N_{correct\_transformations} + N_{incorrect\_transformations}} \tag{2}$$

F-measure: The F-measure can be attributed as a harmonic mean of Precision and Recall. F-measure is the harmonic mean or the "standard" average of total, correct, and incorrect results. By using harmonic mean, Sasaki (2007) [24] calculated F-measure using the following formula:

$$F = \frac{2(P)(R)}{P+R} \tag{3}$$

5.2. *Quantitate Evaluation*

A set of five cases were selected to test the accuracy of the transformation. The selected cases have a set of MySQL databases with different numbers of respective tables in each database. All these five cases were processed with our tool for schema transformation and then data transformation. Table 5 shows the results of schema transformation whereas, each metadata element was considered on the transformation element.

Table 5. Calculate the values of P, R and F-measure.

Case	Total Transforma-tions	Correct Transforma-tions	Incorrect Transforma-tions	Missed Transforma-tions	Precision (P) %	Recall (R) %	F-Measure (F) %
1	24	21	2	1	87.50	91.30	89.35
2	37	33	3	1	89.18	91.66	90.40
3	20	18	1	1	90.00	94.73	92.30
4	26	23	1	2	88.46	95.83	91.99
5	19	16	2	1	84.21	88.89	86.48

Table 5 shows that in all five experiments of RDB to NoSQL migration, the rate of success of transformation in terms of Recall was as high as 88 to 96%. There were rare incorrect transformations as well. In our approach, the conversion rules are supporting the maximum type of conversions from RDB to NoSQL.

Figure 14 shows the results of recall, precision and F-measure of all five different case studies. The results of these measurements show the accuracy and performance of the system under the different database loads. The results shown in Figure 14 depict that the schema to schema transformation and data to data transformation are carried out successfully.

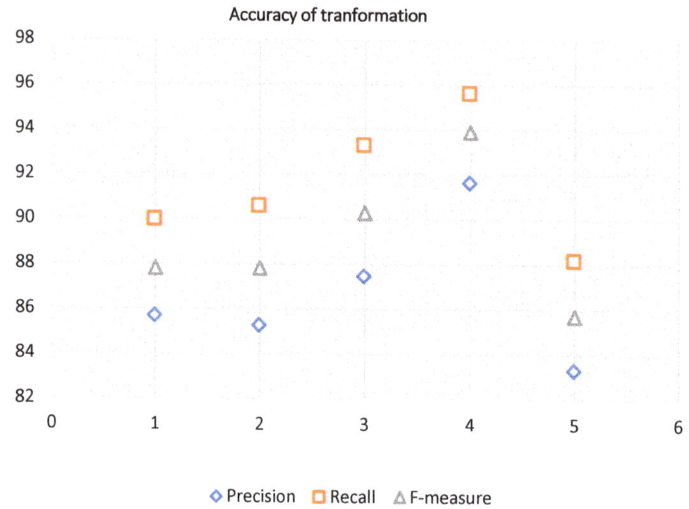

Figure 14. Evaluation results from Relational to NoSQL key-value store.

5.3. Qualitative Evaluation

These databases first tested on schema conversion and the details of this conversion are shown as a graph in Figure 15. In this graph, the world database has three tables, onlineQuiz has five tables and accounts has six tables; however, the conversion time taken for OnlineQuiz accounts for more than double the world class because the OnlineQuiz and accounts database have multi parent child relationship, and therefore, it takes the max time.

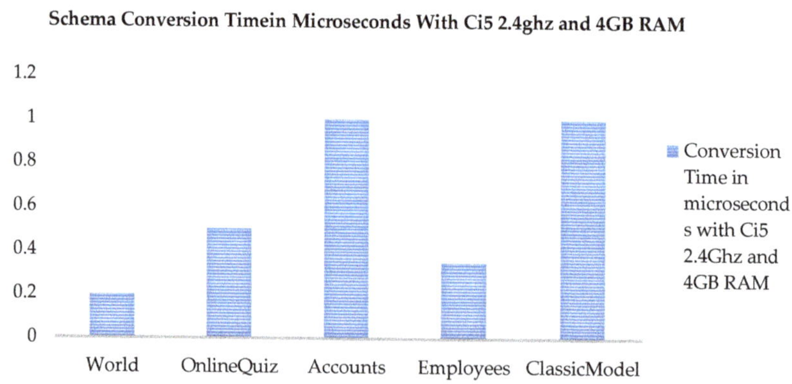

Figure 15. Schema Conversion time in microseconds.

The data conversion time of databases is shown in Figure 16. The single table data conversion time is the same, in the parent-child relationship there is a minor difference in time, but in the multi parent-child relationship, the time is double the parent-child relationship. This is because in the proposed software prototype, the process of mapping parent-child relationship is time consuming. In this process phase, the system finds all foreign keys which are linked with the child table then it creates NoSQL key value store (JSON) storage schema of data and sends it to databases.

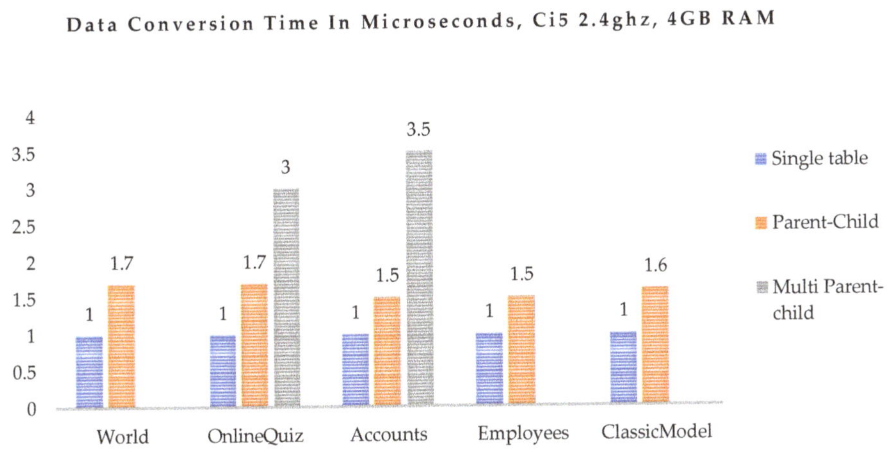

Figure 16. Data conversion time in microseconds.

We could not compare the results of our prototype tool to other tools as no other tool is available that can generate Oracle NoSQL database from the relational database. However, we have compared the results of our prototype tool to a few tools that migrate the relational database to different types of NoSQL databases. Table 6 shows a comparison of performance with the previous approaches:

Table 6. Comparison of our approach with previous approaches.

	Source Database	Target Database	Time	Dataset Size
NoSQLayer [35]	MySQL	MongoDB	1.66 min	50 K records
DigiBrowser [37]	MySQL	NoSQL	10 min	1.5 million records [4.2 GB]
ODBAPI [38]	MySQL	CouchDB	-	-
Kuderu, et al. [39]	RDB	NoSQL	13 min	5000 Transactions
Our Approach	MySQL	Oracle NoSQL	3.5 min	3.2 GB

In this paper, the used approach is novel and automatically transforms the existing database in MySQL to Oracle NoSQL database and provides a highly accurate transformation. The used approach uses a rule-based system to perform transformation at the schema level as well as at the data level. A software prototype for this transformation is also developed as a proof of concept. The results of the experiments show the correctness of our transformations, and outperforms the other similar approaches.

6. Conclusions and Future Work

This study has presented a system to automatically transform relational database into a NoSQL key-value store. The developed system does conversion at the schema level as well as at the data level. The user chooses the type of conversion one wants to perform. In the schema conversion part, the structure of the whole database tables with relationships will be converted to the Oracle NoSQL schema. In the data conversion part, the data of the required tables are converted to Oracle NoSQL supported data types. JSON schema is used for this conversion methodology. The software prototype is developed in Java language. The system has been implemented and evaluated on different sample databases. The results show that the transformation process is very efficient and accurate.

As a future direction, our approach will be able to enhance advance technologies to support all other relational databases and NoSQL databases. The transformation time can be further reduced by using direct entry method (from MySQL to Oracle NoSQL without using middle storage medium).

Here, a model transformation to map RDB elements to NoSQL elements can also improve the accuracy and efficiency of the said migration.

Author Contributions: S.R. contributed in design, implementation, and experimentation of this research and writing this manuscript. I.S.B. supervised this work and also edited this manuscript. R.K. contributed in experiments and evaluation of this research.

Funding: This research received no external funding.

Conflicts of Interest: The authors declare no conflict of interest.

References

1. Codd, E.F. Relational database: A practical foundation for productivity. In *Readings in artificial Intelligence and Databases*; Elsevier: Heidelberg, Germany, 1988; pp. 60–68.
2. Gantz, J.; Reinsel, D. Extracting value from chaos. *IDC Iview* **2011**, *1142*, 1–12.
3. Hecht, R.; Jablonski, S. Nosql evaluation: A use case oriented survey. In Proceedings of the 2011 International Conference on Cloud and Service Computing (CSC), Hong Kong, China, 12–14 December 2011; pp. 336–341.
4. Han, J.; Haihong, E.; Le, G.; Du, J. Survey on NoSQL database. In Proceedings of the 2011 6th International Conference on Pervasive Computing and Applications (ICPCA), Port Elizabeth, South Africa, 26–28 October 2011; pp. 363–366.
5. Sadalage, P.J.; Fowler, M. *Nosql Distilled: A Brief Guide to the Emerging World of Polyglot Persistence*; Pearson Education: Upper Saddle River, NJ, USA, 2012.
6. Strozzi, C. Nosql-a relational database management system. *Lainattu* **1998**, *5*, 2014.
7. Iwazume, M.; Iwase, T.; Tanaka, K.; Fujii, H.; Hijiya, M.; Haraguchi, H. Big data in memory: Benchimarking in memory database using the distributed key-value store for machine to machine communication. In Proceedings of the 2014 15th IEEE/ACIS International Conference on Software Engineering, Artificial Intelligence, Networking and Parallel/Distributed Computing (SNPD), Las Vegas, NV, USA, 30 June–2 July 2014; pp. 1–7.
8. Mohamed, M.A.; Altrafi, O.G.; Ismail, M.O. Relational vs. NoSQL databases: A survey. *Int. J. Comput. Inf. Technol.* **2014**, *3*, 598–601.
9. Scherzinger, S.; Klettke, M.; Störl, U. Managing schema evolution in NoSQL data stores. *arXiv* **2013**, arXiv:1308.0514.
10. DeCandia, G.; Hastorun, D.; Jampani, M.; Kakulapati, G.; Lakshman, A.; Pilchin, A.; Sivasubramanian, S.; Vosshall, P.; Vogels, W. Dynamo: Amazon's highly available key-value store. *ACM SIGOPS Oper. Syst. Rev.* **2007**, *41*, 205–220. [CrossRef]
11. Davoudian, A.; Chen, L.; Liu, M. A survey on NoSQL stores. *ACM Comput. Surv.* **2018**, *51*, 40. [CrossRef]
12. O'Neil, P.; Cheng, E.; Gawlick, D.; O'Neil, E. The log-structured merge-tree (LSM-tree). *Acta Inform.* **1996**, *33*, 351–385. [CrossRef]
13. Chang, F.; Dean, J.; Ghemawat, S.; Hsieh, W.C.; Wallach, D.A.; Burrows, M.; Chandra, T.; Fikes, A.; Gruber, R.E. Bigtable: A distributed storage system for structured data. *ACM Trans. Comput. Syst.* **2008**, *26*, 4. [CrossRef]
14. Li, N.; Xu, B.; Zhao, X.; Deng, Z. Database conversion based on relationship schema mapping. In Proceedings of the 2011 International Conference on Internet Technology and Applications (iTAP), Wuhan, China, 16–18 August 2011; pp. 1–5.
15. Vora, M.N. Hadoop-HBASE for large-scale data. In Proceedings of the 2011 International Conference on Computer Science and Network Technology (ICCSNT), Harbin, China, 24–26 December 2011; pp. 601–605.
16. Yoo, J.; Lee, K.-H.; Jeon, Y.-H. Migration from RDBMS to NoSQL using column-level denormalization and atomic aggregates. *J. Inf. Sci. Eng.* **2018**, *34*, 1–17.
17. Liao, Y.-T.; Zhou, J.; Lu, C.-H.; Chen, S.-C.; Hsu, C.-H.; Chen, W.; Jiang, M.-F.; Chung, Y.-C. Data adapter for querying and transformation between SQL and NoSQL database. *Future Gen. Comput. Syst.* **2016**, *65*, 111–121. [CrossRef]
18. Lakshman, A.; Malik, P. Cassandra: A decentralized structured storage system. *ACM SIGOPS Oper. Syst. Rev.* **2010**, *44*, 35–40. [CrossRef]

19. Chodorow, K. *Mongodb: The Definitive Guide: Powerful and Scalable Data Storage*; O'Reilly Media, Inc.: Sebastopol, CA, USA, 2013.
20. Manyika, J.; Chui, M.; Brown, B.; Bughin, J.; Dobbs, R.; Roxburgh, C.; Byers, A.H. *Big Data: The Next Frontier for Innovation, Competition, and Productivity*; McKinsey Global Institute: San Francisco, CA, USA, 2011.
21. Zhao, G.; Lin, Q.; Li, L.; Li, Z. Schema conversion model of SQL database to NoSQL. In Proceedings of the 2014 Ninth International Conference on P2P, Parallel, Grid, Cloud and Internet Computing (3PGCIC), Krakow, Poland, 4–6 November 2014; pp. 355–362.
22. Lee, C.-H.; Zheng, Y.-L. Automatic SQL-to-NoSQL schema transformation over the MYSQL and HBASE databases. In Proceedings of the 2015 IEEE International Conference on Consumer Electronics-Taiwan (ICCE-TW), Taipei City, Taiwan, 6–8 June 2015; pp. 426–427.
23. Lee, C.-H.; Zheng, Y.-L. SQL-to-NoSQL schema denormalization and migration: A study on content management systems. In Proceedings of the 2015 IEEE International Conference on Systems, Man, and Cybernetics (SMC), Hong Kong, China, 9–12 October 2015; pp. 2022–2026.
24. Sasaki, Y. The Truth of the F-Measure. University of Manchester, Technical Report, Version. Available online: http://www.flowdx.com/F-measure-YS-26Oct07.pdf (accessed on 20 January 2018).
25. Hanine, M.; Bendarag, A.; Boutkhoum, O. Data migration methodology from relational to NoSQL databases. *Int. J. Comput. Electr. Autom. Control Inf. Eng.* **2016**, *9*, 2566–2570.
26. Schreiner, G.A.; Duarte, D.; dos Santos Mello, R. Sqltokeynosql: A layer for relational to key-based NoSQL database mapping. In Proceedings of the 17th International Conference on Information Integration and Web-Based Applications & Services, Brussels, Belgium, 11–13 December 2015; p. 74.
27. Rith, J.; Lehmayr, P.S.; Meyer-Wegener, K. Speaking in tongues: SQL access to NoSQL systems. In Proceedings of the 29th Annual ACM Symposium on Applied Computing, Gyeongju, Korea, 24–28 March 2014; pp. 855–857.
28. Serrano, D.; Han, D.; Stroulia, E. From relations to multi-dimensional maps: Towards an SQL-to-hbase transformation methodology. In Proceedings of the 2015 IEEE 8th International Conference on Cloud Computing (CLOUD), New York, NY, USA, 27 June 27–2 July 2015; pp. 81–89.
29. Ouanouki, R.; April, A.; Abran, A.; Gomez, A.; Desharnais, J. Toward building rdb to hbase conversion rules. *J. Big Data* **2017**, *4*, 10. [CrossRef]
30. Li, C. Transforming relational database into HBASE: A case study. In Proceedings of the 2010 IEEE International Conference on Software Engineering and Service Sciences (ICSESS), Beijing, China, 16–18 July 2010; pp. 683–687.
31. Radonić, M.; Mekterović, I. Etlator-a scripting ETL framework. In Proceedings of the 2017 40th International Convention on Information and Communication Technology, Electronics and Microelectronics (MIPRO), Opatija, Croatia, 22–26 May 2017; pp. 1349–1354.
32. Yang, C.; Huang, Q.; Li, Z.; Liu, K.; Hu, F. Big Data and cloud computing: Innovation opportunities and challenges. *Int. J. Dig. Earth* **2017**, *10*, 13–53. [CrossRef]
33. Suciu, G.; Dobre, C.; Suciu, V.; Todoran, G.; Vulpe, A.; Apostu, A. Cloud computing for extracting price knowledge from big data. In Proceedings of the 2015 Ninth International Conference on Complex, Intelligent, and Software Intensive Systems (CISIS), Santa Catarina, Brazil, 8–10 July 2015; pp. 314–317.
34. Storey, V.C.; Song, I.-Y. Big data technologies and management: What conceptual modeling can do. *Data Knowl. Eng.* **2017**, *108*, 50–67. [CrossRef]
35. Rocha, L.; Vale, F.; Cirilo, E.; Barbosa, D.; Mourão, F. A framework for migrating relational datasets to NoSQL1. *Procedia Comput. Sci.* **2015**, *51*, 2593–2602. [CrossRef]
36. Hirschman, L.; Thompson, H.S. Chapter 13 evaluation: Overview of evaluation in speech and natural language processing. In *Survey of the State of the Art in Human Language Technology*; Cambridge University Press: New York, NY, USA, 1995.
37. Karnitis, G.; Arnicans, G. Migration of relational database to document-oriented database: Structure denormalization and data transformation. In Proceedings of the 2015 7th International Conference on Computational Intelligence, Communication Systems and Networks (CICSyN), Riga, Latvia, 3–5 June 2015; pp. 113–118.

38. Sellami, R.; Bhiri, S.; Defude, B. ODBAPI: A unified REST API for relational and NoSQL data stores. In Proceedings of the 2014 IEEE International Congress on Big Data (BigData Congress), Anchorage, AK, USA, 27 June–2 July 2014; pp. 653–660.
39. Kuderu, N.; Kumari, V. Relational Database to NoSQL Conversion by Schema Migration and Mapping. *Int. J. Comput. Eng. Res. Trends* **2016**, *3*, 506–513. [CrossRef]

MDPI
St. Alban-Anlage 66
4052 Basel
Switzerland
Tel. +41 61 683 77 34
Fax +41 61 302 89 18
www.mdpi.com

Symmetry Editorial Office
E-mail: symmetry@mdpi.com
www.mdpi.com/journal/symmetry

www.ingramcontent.com/pod-product-compliance
Lightning Source LLC
Chambersburg PA
CBHW061804210926
51850CB00006B/611
9783039368440